Planning, Development and Management of Sustainable Cities

Planning, Development and Management of Sustainable Cities

Special Issue Editors

Tan Yigitcanlar
Md. Kamruzzaman

MDPI • Basel • Beijing • Wuhan • Barcelona • Belgrade

Special Issue Editors
Tan Yigitcanlar
Queensland University of Technology
Australia

Md. Kamruzzaman
Monash University
Australia

Editorial Office
MDPI
St. Alban-Anlage 66
4052 Basel, Switzerland

This is a reprint of articles from the Special Issue published online in the open access journal *Sustainability* (ISSN 2071-1050) from 2014 to 2015 (available at: https://www.mdpi.com/journal/sustainability/special_issues/sustainable_cities).

For citation purposes, cite each article independently as indicated on the article page online and as indicated below:

LastName, A.A.; LastName, B.B.; LastName, C.C. Article Title. *Journal Name* **Year**, *Article Number*, Page Range.

ISBN 978-3-03897-906-7 (Pbk)
ISBN 978-3-03897-907-4 (PDF)

Contents

About the Special Issue Editors

Tan Yigitcanlar is an Associate Professor at the School of Civil Engineering and Built Environment, Queensland University of Technology, Brisbane (Australia). Along with this position, he also carries out an Executive Director role at the World Capital Institute, Monterrey (Mexico); an Honorary Professor position at the Federal University of Santa Catarina, Florianopolis (Brazil); and an Adjunct Professor role at the Open Innovation Academy, Daegu (Korea). He has been responsible for research, teaching, training, and capacity-building programs in the fields of urban planning and development in esteemed Australian, Brazilian, Korean, Finnish, Japanese, and Turkish universities. He is an eminent Australian scholar with international recognition, reputation, and impact on policy, practice, and society. His research aims to address contemporary urban planning and development challenges that are economic, societally, spatially, governance, or technology-related in nature. He is the Editor-in-Chief of 'Elsevier's Smart Cities Book Series', Honorary Editor-in-Chief of the 'International Journal of Knowledge-Based Development', Section Editor-in-Chief of 'Sustainability', Regional Editor of the 'Journal of Knowledge Management', and Associate Editor of the 'International Journal of Environmental Science and Technology', 'Global Journal of Environmental Science and Management', 'Journal of Open Innovation: Technology, Market, and Complexity', and the 'Asia Pacific Journal of Innovation and Entrepreneurship'. He undertakes the Chairman role of the annual 'Knowledge Cities World Summit' series. Under this conference brand, he has organized 12 conferences in many locations of the world. He has also delivered over 50 keynote and invited talks at prestigious international academic conferences and industry events. He has published his research findings extensively. These publications include over 200 articles and chapters published in high-impact journals and books, alongside the 12 key reference books published by esteemed international publishing houses.

Md. Kamruzzaman is an Associate Professor of Urban Planning and Design at Monash University. He is also an Honorary Associate Professor of Global, Urban, and Social Studies at RMIT University. Liton has a PhD in Transport Planning, an MSc (with Distinction) in Geoinformation Science and Earth Observation, a Bachelor's degree in Urban and Regional Planning, and a Graduate Certificate in Academic Practice. His research interest spans over three key areas of urban/transport planning: (a) Effectiveness of strategic urban policies, e.g., transit-oriented development (TOD), innovation precincts, urban form, and structure; (b) Socioeconomic and travel behavior impacts of transport infrastructure, e.g., bicycle-sharing schemes, light rail, and airports; (c) Envisioning the future of cities, e.g., smart cities, autonomous vehicles, and climate vulnerability such as the urban heat island effect. Liton has vast experience in teaching in transport and land use planning, GIS, and remote sensing. Prior to joining Monash University, he taught across three universities: Queensland University of Technology, Australia; University of Ulster, UK; and Jahangirnagar University, Bangladesh. Liton is an Editorial Board Member of the 'Journal of Transport and Land Use' and a Section Editor of 'Sustainability'. He maintains close collaborations with numerous professional and research bodies, including the World Conference on Transport Research Society (WCTRS), World Society for Transport and Land Use Research (WSTLUR), Planning Institute of Australia (PIA), Bangladesh Institute of Planners, and The Chartered Institute of Logistics and Transport (UK).

sustainability

MDPI

Editorial

Planning, Development and Management of Sustainable Cities: A Commentary from the Guest Editors

Tan Yigitcanlar * and Md. Kamruzzaman

School of Civil Engineering and Built Environment, Science and Engineering Faculty, Queensland University of Technology (QUT), 2 George Street, Brisbane QLD 4001, Australia; md.kamruzzaman@qut.edu.au

* Author to whom correspondence should be addressed; tan.yigitcanlar@qut.edu.au; Tel.: +61-7-3138-2418; Fax: +61-7-3138-1170.

Academic Editor: Marc A. Rosen

Received: 27 October 2015; Accepted: 27 October 2015; Published: 3 November 2015

Abstract: Cities are the most dramatic manifestations of human activities on the surface of the earth. These human-dominated organisms—*i.e.*, cities—degrade natural habitats, simplify species composition, disrupt hydrological systems, and modify energy flow and nutrient cycling. Today, these consequential impacts of human activities, originated from population increase, rapid urbanization, high private motor vehicle dependency, deregulated industrialization and mass livestock production, are increasing exponentially and causing great deal of environmental, social, and economic challenges both at global and local scales. In such a situation, establishment of sustainable cities, through sustainable urban development practices, is seen as a potential panacea to combat these challenges responsibly, effectively, and efficiently. This paper offers a critical review of the key literature on the issues relating to planning, development and management of sustainable cities, introduces the contributions from the Special Issue, and speculates on the prospective research directions to place necessary mechanisms to secure a sustainable urban future for all.

Keywords: sustainability; sustainable development; sustainable urban development; sustainable city; sustainable urban ecosystems; sustainability assessment; smart cities

1. Background and Literature Review

Environmental externalities mainly generated from population increase, rapid urbanization, high private motor vehicle dependency, deregulated industrialization, and mass livestock production have placed serious concerns for the future of our wellbeing, and even our existence in the long run [1–4]. Realization of the fact that urgent measures must be taken to combat environmental externalities responsibly, effectively, and efficiently have resulted in the rediscovery of the need for more eco-friendly practices [5–7]. Subsequently, during the last few decades, "sustainability" and "sustainable development" have become popular topics not only for scholars, particularly in the fields of environmental economics, technology and science, urban planning, development, and management, but also for urban policy makers and professional practitioners [8–10]. The emergence of these new concepts starting from early 1970s is an outcome of the response to the growing concerns about the impacts of development practices on the state of the environment [11–13].

As a result of both conceptual and empirical research conducted over the last five decades, presently the notion of "sustainability" has been clearly understood; however, a precise definition with a broad consensus is yet to be devised [14,15]. In generic terms, sustainability is a normative concept that indicates the way humans should act towards nature, and the way they should be responsible towards one another and future generations [16,17]. Berkes *et al.* [18] see sustainability as a process to determine the relationship among ecological, social and economic dimensions. This relationship is ascertained based on resilience to both natural and anthropogenic disturbances, desirability to human societies, and temporal and spatial scale boundaries [19]. The idea of meeting the reasonable needs of the current generation, while enhancing the lives and ecosystems of future generations has been discussed extensively in academic debates [20]. This understanding is considered as a fundamental characteristic of sustainability [21]. Nevertheless, not all practices are completely sustainable or unsustainable; there are many shades of grey in between these two opposite poles. Allen [22] coins the concept of strong-*versus*-weak sustainability; where weak sustainability is the substitutability between human and natural capitals in an acceptable level, while strong sustainability is the maintenance and enhancement of the natural capital. This has turned the evaluation of levels or degrees of sustainability [23,24], in other words sustainability assessment, into a major area of research and advocacy for scholars [25–28].

The renowned Brundtland Report [29] gave a significant boost to the "sustainable development" idea and the subsequent efforts. Successively, the concept is adopted in many disciplinary scopes, and also in many countries, regions, cities, and firms worldwide [30,31]. Driven from the sustainability concept, the essence of sustainable development is seen as meeting the fundamental human needs, while preserving the critical life-support systems of our planet [32,33]. It is broadly defined as a development type that can be continued either indefinitely or for the implicit time period of concern [34]. In the debates on sustainable development, the role of societies is critically analysed for managing economic, social, and natural capitals, which are considered non-substitutable and their consumption is irreversible [35–37]. One of the most popular definitions comes from Allen [22]; sustainable development is the development that satisfies the human needs and improves the quality of life in such a way that ecosystems should keep renewing themselves.

A thorough review of the literature reveals that there is no harmony in the operational contents of sustainable development [38]. It covers dynamic environmental, social, and economic concerns at different spatio-temporal scales [39]. This brings about its many interpretations [40]. Scholars and practitioners do not have a solid consensus about what is to be sustained, what is to be developed, how to link environment and development, and for what extent of time [41]. However, the most focused understanding lies with the "the triple bottom line" sustainable development approach, which considers environmental quality, economic prosperity and social justice [42]. As a matter of fact, sustainability represents a nested hierarchy consisting of environment, society and economy as living environment, which enables human society to build an economic system that is not a threat to the environment [43]. Furthermore, in recent years "the quadruple bottom line" sustainable development perspective, also involving the governance domain, has gained a wider acceptance.

Today more than half of the world's population lives in urban areas, and the future of humanity is absolutely urban [44]. Urban areas are the centre of economic development where sustainability is a critical concern [45]. It is not only due to the concentration of the human and financial resources, but also due to the phenomenal growth of urban centres and increasing share of urban population globally [46,47]. In cities, depletion of natural resources and destruction of natural areas are the usual costs of an improved quality of life that disrupts urban ecosystems [48]. There is no other option remaining to achieve the prosperity in urban development without considering the sustainability in planning and development—or "sustainable urban development" [49]. In sustainable urban development, the primary goal is to make cities and their ecosystems healthy and sustainable over time in terms of their environmental, economic and social dimensions [50]. This perspective gives birth to the concepts of "sustainable cities" and "sustainable urban ecosystems". The sustainable cities concept has brought the need for rethinking of sustainable urban development practices considering the size of cities [51].

Böhringer and Jochem [52] present a quite convincing view on the development of sustainable cities; an issue that cannot be clearly measured, and is difficult to improve. Bell and Morse [53] indicate that quantification has limitations, and clearly it is not possible to measure all of the human experience; and clearly there is a trade-off between necessary simplifications and at the same time having sustainability indicators that are meaningful. Today, more and more cities are prioritizing assessment of sustainability of their cities or urban development; and many local governments have made efforts to develop thorough appraisal schemes with consideration for environment, society and economy dimensions—in line with the triple bottom approach [54–56]. However, current research and practice expose that sustainability assessment process itself raises weakness and threats, which need to be improved, especially in comparative analysis [57–60].

Stipulated by Yigitcanlar *et al.* [61], prosperity and environmental sustainability of cities are inextricably linked; therefore, cities can only maintain their prosperity when environmental and social objectives are fully integrated with economic goals. In the light of this view during the last two decades, many cities of the developed world have introduced new frameworks and guidelines to incorporate sustainability in their urban planning and development processes. In Europe, for instance, the report "Sustainable Urban Development in the European Union: A Framework for Action" [62] and the Leipzig Charter [63] are the primary policy documents, which have set up the strategies for urban sustainability to be practiced by national and local governments. Additionally, many national scale documents have placed sustainable urban development as the prime objective of planning and development, such as the Fifth National Policy Document on Spatial Planning in the Netherlands [64], and Planning Policy Statement 1 in the UK [65]. There are also numerous local efforts to promote sustainable city formation via the leverage of urban planning and design. An earlier practice analysis by Berke and Conroy [66] in 30 American cities reveal that no significant differences in how extensively sustainability principles are supported between the plans that state an intention to integrate sustainable urban development and those that do not. However, in recent years, planning efforts of many local governments from Europe, North America and Australia have figured out new and innovative ways to better integrate sustainability principles, technologies and frameworks in their planning schemes [67–70]. Furthermore, some efforts—with limited success so far—are also put in place in the context of developing countries including China, Korea, Malaysia, Turkey, and Vietnam [71–74]. However, a lot more needs to be done to improve the conditions especially in the rapidly emerging economies of the developing world [75,76].

Besides national and international guidelines and frameworks, academia and industry have also contributed to develop sustainability assessment methods and tools [77]. These tools and methods have provided support for making better decisions in the sustainable development processes [78–81]. Furthermore, growing awareness also has an important impact on some of the industry practices today, as we witness the appearance of new sustainable businesses and green technology firms [82,83]. On this very point, Utting [84] identifies the role and responsibilities of corporate businesses for adopting policies and practices to support and promote sustainable (urban) development. The key promoters for implementation of sustainability management in industries include the "World Business Council for Sustainable Development" [85], "Global Reporting Initiative" [86], "Organisation for Economic Co-operation and Development's Sustainable Development Standards" [87], and "United Nation's Transforming Our World: the 2030 Agenda for Sustainable Development" [88]. The United Nation's 2030 Agenda declares that global challenges—e.g., extreme poverty, environmental degradation, and climate change—can only be solved if all parties contribute to implement the "Sustainable Development Goals" [89,90]. This view is highly critical, as the realization of sustainable cities cannot be achieved without all businesses and governments going sustainable along with the communities as a whole in all countries [91]. Furthermore, the rapid advancements in the domain of urban technologies are also seen as a significant contributor to the efforts in dealing with unsustainable outcomes produced from our cities [92]. This technology perspective eventually led to the development of a new city brand—so-called "smart cities" that symbolises a new kind of technology-led sustainable urban utopia [93,94].

Review of the key literature finds that majority of academic research focuses on the planning and development aspects of sustainable cities, while issues related to their governance seems to be in neglect [95]. As much as planning and development of sustainable cities, these cities' management or governance is also a highly crucial and also challenging task—as urban governance and sustainability are rapidly becoming key issues in cities all around the world—thus need further attention from scholars [96]. Yigitcanlar and Teriman [97] suggest a continuous link between urban planning and development processes in order to form an integrated mechanism for achieving sustainable outcomes. While such integration is essential, we also suggest that the management process should be an incorporated mechanism in the planning and development processes.

2. The Special Issue

We are all aware that cities around the globe are being redesigned to become more sustainable. Despite significant research progress in sustainability and cities individually, relatively little investigation has been made by integrating the two themes together. At least three types of environment co-exist in a city—*i.e.*, natural, artefact, and social—and each of these generate both positive and negative externalities for a city [98]. As a result, diverse views prevail in relation to the sustainability of cities. Some scholars argue that the term of sustainable cities is an oxymoron since cities cannot be sustainable at all [99]. Others say that the idea is utopian [100]. Others assert that cities will, must, and are becoming more and more sustainable [101].

Against this background, it is possible to state that there has been growing, but still rather limited, research that systematically investigate sustainable cities, and the specific roles planning, development and management plays in their formation, stimulation and sustained success. Given that there is no formula that can unilaterally be applied in all of the urban environments to achieve sustainability, this Special Issue aims to gather diverse views and report progress towards sustainable cities. A fundamental objective of this Special Issue is to compile and present the cutting edge work of researchers who focus on a joined-up thinking of both themes—*i.e.*, sustainability and city. By doing so, we believe this Special Issue on "Planning, Development and Management of Sustainable Cities" contributes to the knowledge pool in this area, particularly with new evidence driven from empirical research.

Following this guest editorial commentary, the Special Issue includes the following case study, review and research papers:

(1) Article: "Typology of Cities based on City Biodiversity Index: Exploring Biodiversity Potentials and Possible Collaborations among Japanese Cities" by Yuta Uchiyama, Kengo Hayashi and Ryo Kohsaka
(2) Article: "The Influence of Low-Frequency Noise Pollution on the Quality of Life and Place in Sustainable Cities: A Case Study from Northern Portugal" by Juliana Araújo Alves, Lígia Torres Silva and Paula Cristina C. Remoaldo
(3) Article: "Sustainable Water Infrastructure Asset Management: A Gap Analysis of Customer and Service Provider Perspectives" by Sangjong Han, Hwankook Hwang, Seonghoon Kim, Gyu Seok Baek and Joonhong Park
(4) Article: "Moving towards Sustainability: Road Grades and On-Road Emissions of Heavy-Duty Vehicles: A Case Study" by Wendan Zhang, Jian Lu, Ping Xu and Yi Zhang
(5) Article: "Visualization of a City Sustainability Index (CSI): Towards Transdisciplinary Approaches Involving Multiple Stakeholders" by Koichiro Mori, Toyonobu Fujii, Tsuguta Yamashita, Yutaka Mimura, Yuta Uchiyama and Kengo Hayashi
(6) Case Report: "Assessing Sustainability of Mixed Use Neighbourhoods through Residents' Travel Behaviour and Perception: The Case of Nagpur, India" by Sarika Bahadure and Rajashree Kotharkar

(7) Review: "Ecologizing Our Cities: A Particular, Process-Function View of Southern California, from within Complexity" by Ashwani Vasishth
(8) Article: "A Framework for Sustainable Urban Water Management through Demand and Supply Forecasting: The Case of Istanbul" by Murat Yalçıntaş, Melih Bulu, Murat Küçükvar and Hamidreza Samadi
(9) Article: "Application of Environmental Change Efficiency to the Sustainability of Urban Development at the Neighborhood Level" by Hsing-Fu Kuo and Ko-Wan Tsou
(10) Article: "Spatio-Temporal Features of China's Urban Fires: An Investigation with Reference to Gross Domestic Product and Humidity" by Zhenbo Wang, Xiaorui Zhang and Bo Xu
(11) Article: "Critical Connections: The Role of the Built Environment Sector in Delivering Green Cities and a Green Economy" by Peter Newton and Peter Newman
(12) Article: "Framing Processes in the Envisioning of Low-Carbon, Resilient Cities: Results from Two Visioning Exercises" by Stephen McGrail, A. Idil Gaziulusoy and Paul Twomey
(13) Article: "Aligning Public Participation to Stakeholders' Sustainability Literacy: A Case Study on Sustainable Urban Development in Phoenix, Arizona" by Matthew Cohen, Arnim Wiek, Braden Kay and John Harlow
(14) Article: "Managing Knowledge to Promote Sustainability in Australian Transport Infrastructure Projects" by Jay Yang, Mei Yuan, Tan Yigitcanlar, Peter Newman and Frank Schultmann
(15) Article: "Sustainable Urban Transport in the Developing World: Beyond Megacities" by Dorina Pojani and Dominic Stead
(16) Review: "Trees in Canadian Cities: Indispensable Life Form for Urban Sustainability" by Peter N. Duinker, Camilo Ordóñez, James W. N. Steenberg, Kyle H. Miller, Sydney A. Toni and Sophie A. Nitoslawski
(17) Article: "The Bumpy Road toward Low-Energy Urban Mobility: Case Studies from Two UK Cities" by Tim Schwanen
(18) Article: "Scaling-up Strategy as an Appropriate Approach for Sustainable New Town Development? Lessons from Wujin, Changzhou, China" by Hao Chen, Qiyan Wu, Jianquan Cheng, Zhifei Ma and Weixuan Song
(19) Article: "Neighborhood Sustainability Assessment: Evaluating Residential Development Sustainability in a Developing Country Context" by Tan Yigitcanlar, Md. Kamruzzaman and Suharto Teriman
(20) Article: "Regional Open Innovation Roadmapping: A New Framework for Innovation-Based Regional Development" by Wim Schwerdtner, Rosemarie Siebert, Maria Busse and Ulf B. Freisinger
(21) Article: "Incorporating Bio-Physical Sciences into a Decision Support Tool for Sustainable Urban Planning" by Zina Mitraka, Emmanouil Diamantakis, Nektarios Chrysoulakis, Eduardo Anselmo Castro, Roberto San Jose, Ainhoa Gonzalez and Ivan Blecic

3. Concluding Remarks and Research Directions

The Special Issue generates new insights by investigating the sustainable cities from various disciplinary angles (*i.e.*, urban studies, urban planning, urban management, urban design, architecture, civil engineering, construction management, regional science, environmental science, bio-physical sciences, environmental planning, and knowledge management) and country contexts (*i.e.*, Australia, Canada, China, Germany, India, Japan, Malaysia, Portugal, Turkey, the UK, and the USA) as well as international comparisons.

In the light of the sustainability related matters discussed by the contributors of the Special Issue, we compile the following sets of generic research questions focusing on the "planning", "development", and "management" domains of sustainable cities. We believe investigating these issues further in prospective research projects by scholars of this highly interdisciplinary field will shed light on better conceptualization and practice of sustainable urban development and sustainable cities.

- What is a sustainable city supposed to be, and how can benchmarks be determined and set considering sustainability is a vague term?
- What is the current status of cities and the inhibitors and threats on the way towards sustainable urban development?
- What are the commonalities amongst cities that are moving towards sustainability, and what are the factors of success and failure?
- How can sustainability frameworks be developed and applied, recognizing that every city is unique, to the planning of cities?
- How can institutional and social capacities be developed and further enhanced for the formation of sustainable cities?
- How can sustainable cities be governed to make sure that existing high sustainability levels are maintained and improved over time?

Lastly, we wish to thank the authors of the Special Issue papers for accepting our invitation and submitting and revising their manuscripts within a short time frame, and thank the referees for their thorough and timely reviews. We also thank the journal's Assistant Editor, Ms. Yaqiong Guo, for inviting us to serve as the Guest Editors of this Special Issue.

Author Contributions: Tan Yigitcanlar and Md. Kamruzzaman jointly prepared this Special Issue editorial that is written in the format of a literature review paper. Both authors read and approved the final manuscript.

Conflicts of Interest: The authors declare no conflict of interest.

References

1. Mahbub, P.; Goonetilleke, A.; Ayoko, G.; Egodawatta, P.; Yigitcanlar, T. Analysis of build-up of heavy metals and volatile organics on urban roads in Gold Coast, Australia. *Water Sci. Technol.* **2011**, *63*, 2077–2085. [CrossRef] [PubMed]
2. McMichael, A.J. Globalization, climate change, and human health. *N. Engl. J. Med.* **2013**, *368*, 1335–1343. [CrossRef] [PubMed]
3. Kamruzzaman, M.; Hine, J.; Yigitcanlar, T. Investigating the link between carbon dioxide emissions and transport related social exclusion in rural Northern Ireland. *Int. J. Environ. Sci. Technol.* **2015**, *12*, 3463–3478. [CrossRef]
4. Chikaraishi, M.; Fujiwara, A.; Kaneko, S.; Poumanyvong, P.; Komatsu, S.; Kalugin, A. The moderating effects of urbanization on carbon dioxide emissions: A latent class modeling approach. *Technol. Forecast. Soc. Chang.* **2015**, *90*, 302–317. [CrossRef]
5. Cecere, G.; Corrocher, N.; Gossart, C.; Ozman, M. Lock-in and path dependence: An evolutionary approach to eco-innovations. *J. Evolut. Econ.* **2014**, *24*, 1037–1065. [CrossRef]
6. Yigitcanlar, T.; Kamruzzaman, M. Investigating the interplay between transport, land use and the environment: A review of the literature. *Int. J. Environ. Sci. Technol.* **2014**, *11*, 2121–2132. [CrossRef]
7. Kamruzzaman, M.; Yigitcanlar, T.; Washington, S.; Currie, G. Australian baby boomers switched to more environmentally friendly modes of transport during the global financial crisis. *Int. J. Environ. Sci. Technol.* **2014**, *11*, 2133–2144. [CrossRef]
8. Arndt, H.W. Economic development: A semantic history. *Econ. Dev. Cult. Chang.* **1981**, *29*, 457–466. [CrossRef]
9. Bartelmus, P. *Environment and Development*; Routledge: London, UK, 1986.
10. Sachs, J.D. *The Age of Sustainable Development*; Columbia University Press: New York, NY, USA, 2015.
11. Finco, A.; Nijkamp, P. Pathway to urban sustainability. *J. Environ. Policy Plan.* **2001**, *3*, 289–309. [CrossRef]
12. Goonetilleke, A.; Yigitcanlar, T.; Ayoko, G.; Egodawatta, P. *Sustainable Urban Water Environment: Climate, Pollution and Adaptation*; Edward Elgar Publishing: Cheltenham, UK, 2014.

13. Yigitcanlar, T.; Lee, S.H. Korean ubiquitous-eco-city: A smart-sustainable urban form or a branding hoax? *Technol. Forecast. Soc. Chang.* **2014**, *89*, 100–114. [CrossRef]
14. Masnavi, M.R. Measuring urban sustainability: Developing a conceptual framework for bridging the gap between theoretical levels and the operational levels. *Int. J. Environ. Res.* **2007**, *1*, 188–197.
15. Wei, Y.; Huang, C.; Lam, P.T.I.; Yuan, Z. Sustainable urban development: A review on urban carrying capacity assessment. *Habitat Int.* **2015**, *46*, 64–71. [CrossRef]
16. Baumgärtner, S.; Quaas, M.F. What is sustainability economics? *Ecol. Econ.* **2010**, *69*, 445–450. [CrossRef]
17. Yigitcanlar, T.; Dizdaroglu, D. Ecological approaches in planning for sustainable cities: A review of the literature. *Glob. J. Environ. Sci. Manag.* **2015**, *1*, 71–94.
18. Berkes, F.; Folke, C.; Colding, J. *Linking Social and Ecological Systems: Management Practices and Social Mechanisms for Building Resilience*; Cambridge University Press: Cambridge, UK, 1998.
19. Mayer, A.L. Strengths and weaknesses of common sustainability indices for multidimensional systems. *Environ. Int.* **2008**, *34*, 277–291. [CrossRef] [PubMed]
20. Pollalis, S.; Georgoulias, A.; Ramos, S.; Schodek, D. *Infrastructure Sustainability and Design*; Routledge: London, UK, 2012.
21. Hult, A. The circulation of Swedish urban sustainability practices: To China and back. *Environ. Plan. A* **2015**, *47*, 537–553. [CrossRef]
22. Allen, R. *How to Save the World*; Barnes & Noble: Manhattan, NY, USA, 1980.
23. Mori, K.; Christodoulou, A. Review of sustainability indices and indicators: Towards a new city sustainability index (CSI). *Environ. Impact Assess. Rev.* **2012**, *32*, 94–106. [CrossRef]
24. Dizdaroglu, D.; Yigitcanlar, T. Integrating urban ecosystem sustainability assessment into policy-making: Insights from the Gold Coast City. *J. Environ. Plan. Manag.* **2015**. [CrossRef]
25. Ness, B.; Urbel-Piirsalu, E.; Anderberg, S.; Olsson, L. Categorising tools for sustainability assessment. *Ecol. Econ.* **2007**, *60*, 498–508. [CrossRef]
26. Levrel, H.; Kerbiriou, C.; Couvet, D.; Weber, J. OECD pressure-state-response indicators for managing biodiversity: A realistic perspective for a French biosphere reserve. *Biodiver. Conserv.* **2008**, *18*, 1719–1732. [CrossRef]
27. Singh, R.K.; Murty, H.R.; Gupta, S.K.; Dikshit, A.K. An overview of sustainability assessment methodologies. *Ecol. Indic.* **2012**, *15*, 281–299. [CrossRef]
28. Gil, J.; Duarte, J.P. Tools for evaluating the sustainability of urban design: A review. *Proc. ICE Urban Des. Plan.* **2013**, *166*, 311–325. [CrossRef]
29. World Commission on Environment and Development (WCED). *Our Common Future*; WCED: Oxford, UK, 1987.
30. Ali, M. *Sustainability Assessment: Context of Resource and Environmental Policy*; Academic Press: New York, NY, USA, 2013.
31. Li, X.; Yeh, A.G.O. Modelling sustainable urban development by the integration of constrained cellular automata and GIS. *Int. J. Geogr. Inf. Sci.* **2000**, *14*, 131–152. [CrossRef]
32. Kates, R.W.; Clark, W.C.; Corell, R.; Hall, J.M.; Al, E. Sustainability science. *Science* **2001**, *292*, 641–642. [CrossRef] [PubMed]
33. Mörtberg, U.; Haas, J.; Zetterberg, A.; Franklin, J.P.; Jonsson, D.; Deal, B. Urban ecosystems and sustainable urban development: Analysing and assessing interacting systems in the Stockholm region. *Urban Ecosyst.* **2013**, *16*, 763–782. [CrossRef]
34. Lélé, S.M. Sustainable development: A critical review. *World Dev.* **1991**, *19*, 607–621. [CrossRef]
35. Dyllick, T.; Hockerts, K. Beyond the business case for corporate sustainability. *Bus. Strategy Environ.* **2002**, *11*, 130–141. [CrossRef]
36. Yigitcanlar, T. (Ed.) *Sustainable Urban and Regional Infrastructure Development: Technologies, Applications and Management*; IGI Global: Hersey, PA, USA, 2010.
37. Wiek, A.; Kay, B. Learning while transforming: Solution-oriented learning for urban sustainability in Phoenix, Arizona. *Curr. Opin. Environ. Sustain.* **2015**, *16*, 29–36. [CrossRef]
38. Bithas, K.P.; Christofakis, M. Environmentally sustainable cities: Critical review and operational conditions. *Sustain. Dev.* **2006**, *189*, 177–189. [CrossRef]
39. Alberti, M. Measuring urban sustainability. *Environ. Impact Assess. Rev.* **1996**, *16*, 381–424. [CrossRef]

40. Tanguay, G.A.; Rajaonson, J.; Lanoie, P. Measuring the sustainability of cities: An analysis of the use of local indicators. *Ecol. Indic.* **2010**, *10*, 407–418. [CrossRef]
41. Parris, T.M.; Kates, R.W. Characterizing and measuring sustainable development. *Annu. Rev. Environ. Resour.* **2003**, *28*, 559–586. [CrossRef]
42. Elkington, J. *Cannibals with Forks: The Triple Bottom Line of the 21st Century Business*; Capstone Publishing: Oxford, UK, 1997.
43. Winkler, R.; Deller, S.; Marcouiller, D. Recreational housing and community development: A triple bottom line approach. *Growth Chang.* **2015**, *46*, 481–500. [CrossRef]
44. Mclaren, D.; Agyeman, J. *Sharing Cities: A Case for Truly Smart and Sustainable Cities*; MIT Press: Boston, MA, USA, 2015.
45. McCormick, K.; Kiss, B. Learning through renovations for urban sustainability: The case of the Malmö Innovation Platform. *Curr. Opin. Environ. Sustain.* **2015**, *16*, 44–50. [CrossRef]
46. Saveriades, A. Establishing the social tourism carrying capacity for the tourist resorts of the east coast of the Republic of Cyprus. *Tour. Manag.* **2000**, *21*, 147–156. [CrossRef]
47. Abernethy, V.D. Carrying capacity: The tradition and policy implications of limits. *Ethics Sci. Environ. Politics* **2001**, *23*, 9–18.
48. Turner, V.K.; Galletti, C.S. Do sustainable urban designs generate more ecosystem services? A case study of Civano in Tucson, Arizona. *Prof. Geogr.* **2015**, *67*, 204–217. [CrossRef]
49. UN-Habitat. *State of the World's Cities 2012/2013: Prosperity of Cities*; Routledge: New York, NY, USA, 2013.
50. Smith, R.M. Planning for urban sustainability: The geography of LEED®–Neighborhood Development™ (LEED®–ND™) projects in the United States. *Int. J. Urban Sustain. Dev.* **2015**, *7*, 15–32. [CrossRef]
51. Yigitcanlar, T. (Ed.) *Rethinking Sustainable Development: Urban Management, Engineering and Design*; IGI Global: Hersey, PA, USA, 2010.
52. Böhringer, C.; Jochem, P.E.P. Measuring the immeasurable: A survey of sustainability indices. *Ecol. Econ.* **2007**, *63*, 1–8. [CrossRef]
53. Bell, S.; Morse, S. *Sustainability Indicators: Measuring the Immeasurable*; Earthscan: London, UK, 2008.
54. Lee, Y.J.; Huang, C.M. Sustainability index for Taipei. *Environ. Impact Assess. Rev.* **2007**, *27*, 505–521. [CrossRef]
55. Dizdaroglu, D.; Yigitcanlar, T. A parcel-scale assessment tool to measure sustainability through urban ecosystem components: The MUSIX model. *Ecol. Indic.* **2014**, *41*, 115–130. [CrossRef]
56. Yang, J.; Yuan, M.; Yigitcanlar, T.; Newman, P.; Schultmann, F. Managing knowledge to promote sustainability in Australian transport infrastructure projects. *Sustainability* **2015**, *7*, 8132–8150. [CrossRef]
57. Zellner, M.L.; Theis, T.L.; Karunanithi, A.T.; Garmestani, A.S.; Cabezas, H. A new framework for urban sustainability assessments: Linking complexity, information and policy. *Comput. Environ. Urban Syst.* **2008**, *32*, 474–488. [CrossRef]
58. Yigitcanlar, T.; Dur, F. Developing a sustainability assessment model: The sustainable infrastructure, land-use, environment and transport model. *Sustainability* **2010**, *2*, 321–340. [CrossRef]
59. Dur, F.; Yigitcanlar, T.; Bunker, J. A spatial indexing model for measuring neighbourhood level land-use and transport integration. *Environ. Plan. B* **2014**, *41*, 792–812. [CrossRef]
60. Sharifi, A.; Murayama, A. Viability of using global standards for neighbourhood sustainability assessment: Insights from a comparative case study. *J. Environ. Plan. Manag.* **2015**, *58*, 1–23. [CrossRef]
61. Yigitcanlar, T.; Dur, F.; Dizdaroglu, D. Towards prosperous sustainable cities: A multiscalar urban sustainability assessment approach. *Habitat Int.* **2015**, *45*, 36–46. [CrossRef]
62. European Union (EU). *Sustainable Urban Development in the European Union: A Framework for Action*; European Commission: Brussels, Belgium, 1998.
63. European Union (EU). *Leipzig Charter*; European Council: Brussels, Belgium, 2007.
64. Ministerie van Volkshuisvesting, Ruimtelijke Ordening en Milieubeheer (VROM). *Making Space, Sharing Space: The Fifth National Policy Document on Spatial Planning 2000/2020*; VROM: The Hague, The Netherlands, 2001.
65. Department for Communities and Local Government (DCLG). *Planning Policy Statement 1: Delivering Sustainable Development*; DCLG: London, UK, 2005.
66. Berke, P.R.; Conroy, M.M. Are we planning for sustainable development? An evaluation of 30 comprehensive plans. *J. Am. Plan. Assoc.* **2000**, *66*, 21–33. [CrossRef]

67. Bulkeley, H. Urban sustainability: Learning from best practice? *Environ. Plan. A* **2006**, *38*, 1029–1044. [CrossRef]
68. Nijkamp, P.; Perrels, A. *Sustainable Cities in Europe*; Routledge: New York, NY, USA, 2014.
69. Pearson, L.; Newton, P.; Roberts, P. (Eds.) *Resilient Sustainable Cities: A Future*; Routledge: New York, NY, USA, 2014.
70. Raco, M. Sustainable city-building and the new politics of the possible: Reflections on the governance of the London Olympics 2012. *Area* **2015**, *47*, 124–131. [CrossRef]
71. Djoundourian, S. Environmental performance of developing countries: A comparative study. *Top. Middle East. Afr. Econ.* **2012**, *14*, 265–277.
72. Pugh, C. *Sustainable Cities in Developing Countries*; Routledge: New York, NY, USA, 2013.
73. Yigitcanlar, T.; Kamruzzaman, M.; Teriman, S. Neighborhood sustainability assessment: Evaluating residential development sustainability in a developing country context. *Sustainability* **2015**, *7*, 2570–2602. [CrossRef]
74. Yigitcanlar, T.; Bulu, M. Dubaization of Istanbul: Insights from the knowledge-based urban development journey of an emerging local economy. *Environ. Plan. A* **2015**, *47*, 89–107. [CrossRef]
75. Jones, S.; Tefe, M.; Appiah-Opoku, S. Incorporating stakeholder input into transport project selection: A step towards urban prosperity in developing countries? *Habitat Int.* **2015**, *45*, 20–28. [CrossRef]
76. Musakwa, W.; van Niekerk, A. Earth observation for sustainable urban planning in developing countries needs, trends, and future directions. *J. Plan. Lit.* **2015**, *30*, 149–160. [CrossRef]
77. Dizdaroglu, D.; Yigitcanlar, T.; Dawes, L. A micro-level indexing model for assessing urban ecosystem sustainability. *Smart Sustain. Built Environ.* **2012**, *1*, 291–315. [CrossRef]
78. Schmidheiny, S.; Chase, R.; de Simone, L. Business progress toward sustainable development. *Yale F&ES Bull.* **1997**, *101*, 143–156.
79. Bentivegna, V.; Curwell, S.; Deakin, M.; Lombardi, P.; Mitchell, G.; Nijkamp, P. A vision and methodology for integrated sustainable urban development: BEQUEST. *Build. Res. Inf.* **2002**, *30*, 83–94. [CrossRef]
80. Brandon, P.S.; Lombardi, P. *Evaluating Sustainable Development in the Built Environment*; John Wiley & Sons: London, UK, 2010.
81. Dur, F.; Yigitcanlar, T. Assessing land-use and transport integration via a spatial composite indexing model. *Int. J. Environ. Sci. Technol.* **2015**, *12*, 803–816. [CrossRef]
82. Alvarez, J.B.; Mackalski, R.; Loeb, A.; Mazzanti, L. Leading sustainable change: An organizational perspective. *Strateg. Manag. J.* **2015**, *34*, 1209–1231.
83. Gibbs, D.; O'Neill, K. Building a green economy? Sustainability transitions in the UK building sector. *Geoforum* **2015**, *59*, 133–141. [CrossRef]
84. Utting, P. *Business Responsibility for Sustainable Development: Geneva 2000*; United Nations: Geneva, Switzerland, 2000.
85. World Business Council for Sustainable Development (WBCSD). *Signals of Change: Business Progress toward Sustainable Development*; WBCSD: Geneva, Switzerland, 1997.
86. Global Reporting Initiative (GRI). *The Global Reporting Initiative: An Overview*; GRI: Boston, MA, USA, 2002.
87. Organisation for Economic Co-operation and Development (OECD). *An Update of the OECD Composite Leading Indicators*; OECD: Paris, France, 2002.
88. United Nations (UN). Transforming Our World: The 2030 Agenda for Sustainable Development. Available online: https://sustainabledevelopment.un.org (accessed on 8 September 2015).
89. Sachs, J.D. From millennium development goals to sustainable development goals. *Lancet* **2012**, *379*, 2206–2211. [CrossRef]
90. Griggs, D.; Stafford-Smith, M.; Gaffney, O.; Rockström, J.; Öhman, M.C.; Shyamsundar, P.; Noble, I. Policy: Sustainable development goals for people and planet. *Nature* **2013**, *495*, 305–307. [CrossRef] [PubMed]
91. Rauscher, R.; Momtaz, S. *Sustainable Neighbourhoods in Australia: City of Sydney Urban Planning*; Springer: New York, NY, USA, 2015.
92. Marsal-Llacuna, M.L.; Colomer-Llinàs, J.; Meléndez-Frigola, J. Lessons in urban monitoring taken from sustainable and livable cities to better address the Smart Cities initiative. *Technol. Forecast. Soc. Chang.* **2015**, *90*, 611–622. [CrossRef]
93. Caragliu, A.; del Bo, C.; Nijkamp, P. Smart cities in Europe. *J. Urban Technol.* **2011**, *18*, 65–82. [CrossRef]

94. Yigitcanlar, T. Smart cities: An effective urban development and management model? *Aust. Plan.* **2015**, *52*, 27–34. [CrossRef]
95. Zhu, J.; Simarmata, H.A. Formal land rights *versus* informal land rights: Governance for sustainable urbanization in the Jakarta metropolitan region, Indonesia. *Land Use Policy* **2015**, *43*, 63–73. [CrossRef]
96. Evans, B.; Joas, M.; Sundback, S.; Theobald, K. *Governing Sustainable Cities*; Routledge: New York, NY, USA, 2013.
97. Yigitcanlar, T.; Teriman, S. Rethinking sustainable urban development: Towards an integrated planning and development process. *Int. J. Environ. Sci. Technol.* **2015**, *12*, 341–352. [CrossRef]
98. Lang, U. Cultivating the sustainable city: Urban agriculture policies and gardening projects in Minneapolis, Minnesota. *Urban Geogr.* **2014**, *35*, 477–485. [CrossRef]
99. Rees, W.E. Is "sustainable city" an oxymoron? *Local Environ.* **1997**, *2*, 303–310. [CrossRef]
100. Blassingame, L. Sustainable cities: Oxymoron, utopia, or inevitability? *Soc. Sci. J.* **1998**, *35*, 1–13. [CrossRef]
101. Camagni, R.; Capello, R.; Nijkamp, P. Towards sustainable city policy: An economy-environment technology nexus. *Ecol. Econ.* **1998**, *24*, 103–118. [CrossRef]

sustainability

Article

Critical Connections: The Role of the Built Environment Sector in Delivering Green Cities and a Green Economy

Peter Newton [1,*] and Peter Newman [2]

1 Institute for Social Research, Swinburne University of Technology, Melbourne 3122, Australia
2 The CUSP Institute, Curtin University, Perth 6102, Australia; p.newman@curtin.edu.au

* Author to whom correspondence should be addressed; pnewton@swin.edu.au; Tel.: +61-3-9214-4769

Academic Editors: Tan Yigitcanlar and Md. Kamruzzaman

Received: 3 March 2015; Accepted: 10 July 2015; Published: 17 July 2015

Abstract: The green agenda for cities and the economy in general is a major focus of global institutions and is increasingly a major national and urban priority. Core issues and best practice for built environment businesses were collated from published studies and used in a survey of Australian firms to see how committed they were to the green economy. The results show high awareness of the challenges and opportunities with 85% of firms having sustainability as an established agenda with senior management and over 20% of built environment firms deriving more than 50% of their sales from green products and services. This is much higher in design firms and is globally high. Whilst recognizing the scope for more engagement by industry in transitioning to a low carbon green economy, there is doubt within the built environment sector about how to create a business case for innovative green ventures and a lack of certainty or encouragement from government about how to proceed.

Keywords: green economy; built environment sector; eco-cities; sustainable urban development; green innovation; low carbon economy; socio-technical transition

1. Introduction

The 21st century's economy will be urban and green. The urban transition is advancing globally, albeit unsustainably [1,2]. The green economy transition is in its infancy. Yet, like other revolutionary socio-technical transitions before it, there are irrepressible sets of push and pull factors massing that can trigger transformative change. The push factors are those capable of delivering innovation. New technologies are among these as well as associated business strategies and practices and government policies and programs that all need to shift towards facilitating this new economy. The pull factors are also clear and relate, among others, to the challenge of creating sustainable and resilient built environments capable of functioning within the ecosystem limits of a single planet subject to climate change and forecast to be home to 9 billion people by 2050 [3,4].

This study represents a first attempt within Australia to explore the critical connections that exist between cities and the built environment industries that plan and manage them. Particular focus is on the extent to which these industries are operating in a manner that can deliver much needed sustainable regenerative urban development in the 21st century and contribute to the emergence of a green economy more broadly. It is based on a 2013 survey of 173 senior managers in both private and public sector built environment organizations with membership in national industry associations with acknowledged sustainability objectives.

Cities will be critical to sustainable urban development in the 21st century. A transition to sustainable urban development will, however, require a socio-technical transformation of a scale and complexity similar to that of the industrial revolution. The first created a factor of 20 fold increase in productivity. A green revolution will require a further increase by at least a factor of 4 or 5 which means that levels of wealth could double while cutting resource use by 50% or even 80% as suggested and illustrated by Von Weisacker *et al.* [5]. The second, post-industrial revolution is centered on sustainable development as initially articulated by the United Nations [6,7] and by much scholarship and public discussion over subsequent decades (see Figure 1). The graph depicts the volume of articles using the term "sustainability" that were published annually since the Brundtland Report using the search engine Factiva, an on-line tool that aggregates searchable content from over 10,000 licensed sources such as newspapers, journals, magazines, TV and radio transcripts and newswires worldwide, covering most languages.

Sustainability focuses on the need to ensure that human activities and the systems within which they operate (e.g., our human settlements and their populations) can continue into the future—within the earth's planetary boundaries [8].

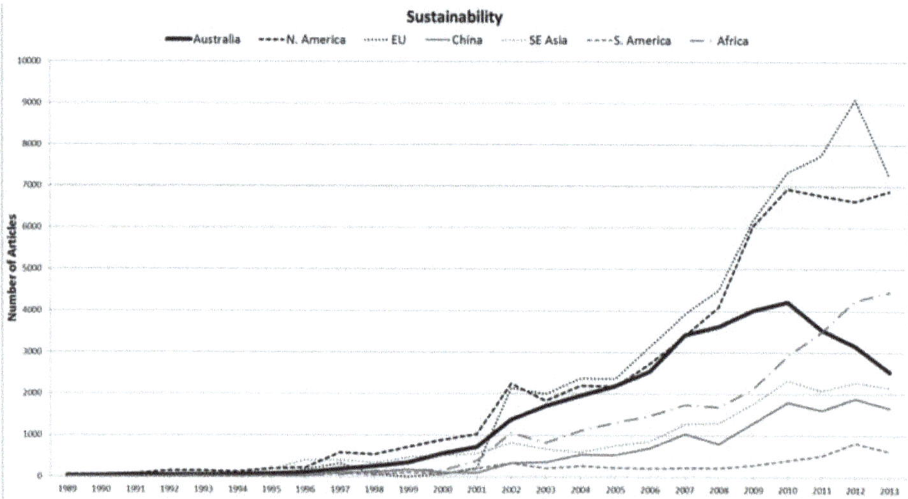

Figure 1. Trajectories of counts of the term "sustainability" in print. (Source: derived from [9] using key word "sustainability".)

The progressively growing interest in the subject of sustainability, especially in those developed societies that have the human, social and financial capital to implement transformative change has somewhat stalled though they remain the main source of green products and services. The Global Financial Crisis in North America in 2007–2008 has had knock-on effects globally, with European governments in particular having additional sovereign risk challenges that diminished fiscal support for investment in innovative (longer term) sustainable production and consumption programs in favor of those deemed "shovel-ready" (and job-saving). During this period, many government departments with responsibility for metropolitan planning and development replaced "sustainability" with "livability" or "productivity" as their principal objective, given their clear economic connection to the attraction of international investment and skilled labor . . . both of which are highly mobile in the 21st century and focused on cities. Cities in North America, Australia and New Zealand and, to a lesser extent, Europe achieve their livability and productivity rankings as a result of their high levels of resource consumption and greenhouse gas emissions [1] though decoupling of wealth from fossil fuels is now progressing [10]. Cities in developing countries such as China are also aspiring to create more livable built environments as they rapidly urbanize with substantial ecological footprints in fast growing cities such as Beijing and Shanghai; however, they too are beginning to decouple like those in developed countries [10].

The principal global challenges are threefold. First, we live in a carbon constrained world which is witnessing increasing concentrations of greenhouse gases in the earth's atmosphere capable of triggering climate change of a scale which could take centuries to reverse [11,12]. How we generate and consume energy is central to this issue. Second, we live in a resource constrained world where peak oil, water shortages, decline in agricultural land and loss of biodiversity are indications that our harvesting of the earth's natural resources is now occurring at a rate which is exceeding replacement rate [13,14]. Our patterns of consumption of housing, travel, water and manufactured products are central to this issue [15]. Population growth—forecast to reach nine billion by 2050—when coupled with per capita consumption defines the magnitude of the sustainability challenge. The task of transition from unsustainable levels of consumption is a challenge for the citizens of developed countries in North America, Western Europe and Australia that have ecological footprints three to four times the global average. Concerns about the environmental impacts of consumption have been registered in the OECD [16] with forecasts that consumption pressure is expected to intensify significantly by 2030. Forecasts have been advanced of major economic and social disruption or collapse associated with continued business as usual operations by industries, governments and communities [17] unless there is radical change. Third, we live in a world of increasingly concentrated populations, with the world's 9 billion-population forecast to be 70% urban by 2050, though this remains a question as to whether it helps or hinders the reduction in footprint [10].

The 21st century sustainability challenge will focus on cities, their future mode of development and redevelopment and their resilience to a mix of exogenous and endogenous forces now in play. In developed societies these have been recently catalogued [18]. The exogenous factors mirror those listed above: resource constraints (land, water, raw materials, oil); climate change (and its link to sea level rise and its impact on urban infrastructure; increased temperatures and changes to rainfall frequency and intensity—flooding in some locales, drought and heat extremes and megafires in others); bio-security (including pandemics) and financial uncertainty. The endogenous factors reflect the contemporary context and dynamics of each individual city: the quality of its existing buildings and infrastructure; its human and social capital and levels of social and spatial disadvantage; its governance structures; its urban environmental quality and the nature and trajectory of its economic base. In this context, a wide range of international, national and metropolitan studies of city performance continue to catalogue deficiencies across the spectrum of human settlement indicators associated with environmental quality and metabolism, economic productivity and competitiveness, liveability and social inclusion—all of which are inter-related dimensions that characterise the current state of sustainability of urban development [19].

A transition to sustainable urban development will require the emergence of a new form of (green) capitalism based on sustainability principles that embraces social and environmental as well as economic objectives capable of redressing this growing list of problems. The challenge confronting this transition is immense and is the topic of much contemporary debate [20]. To be successful, the urban sustainability transition will need to be closely integrated with and be a key driver of the emergence of a green economy. They are critically connected (see Figure 2) and represent the transition arena within which significant urban innovation is required. As current urban sustainability transition theory indicates; however, innovation capable of transformative change currently faces formidable challenges from well-established regimes that represent entrenched industrial, governance and consumer practices [1,21,22].

A range of (green) physical infrastructures are required to support urban living: transport, energy, water, waste, communications and buildings. The consensus is that the sustainability performance of each is currently poor, given that they all emerged in an era where there were few resource constraints and climate constraints. Next generation infrastructures and urban designs will need to demonstrate significantly greater eco-efficiency and resilience in their operation than those that they need to replace [23]. The demand for new urban infrastructures and green services represents the trigger for a raft of innovative infrastructure technologies to move more widely into the urban marketplace [22]. In the energy sector, this relates primarily to renewable energy and the speed with which it can penetrate a currently dominant fossil fuel based regime. The resistance being faced in countries with significant fossil fuel endowments or dependencies is shaped by the threat to business and investors of holding the wrong assets. Recent divestment of fossil fuel stocks is a signal that green capitalism will also be based on creative destruction [24]. However, in this era increasingly centred on new technologies that out-perform existing technologies on sustainability criteria, there will need to be widespread acceptance of rapid change across all industry sectors.

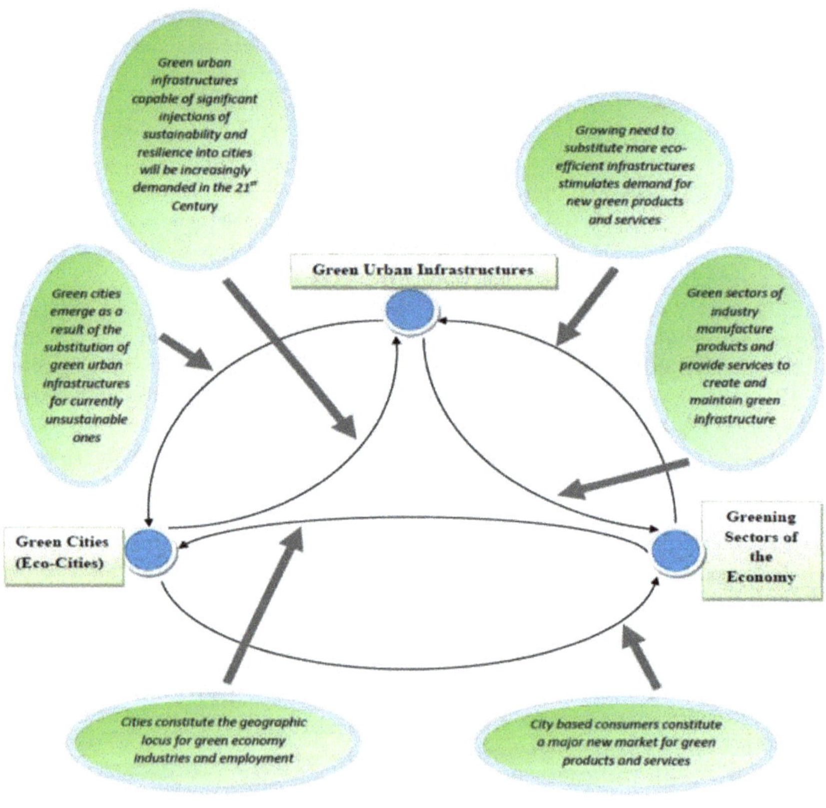

Figure 2. Critical connections: green economy, green urban infrastructure and green (eco) cities. Source: [22].

Figure 2 also indicates that cities will constitute a geographic locus for green industry location, given the innovative capacity agglomeration economies deliver to firms generally as well as providing the extensive customer base for new green products and services. It is this spatial convergence of supply and demand forces that will underpin the emergence of a green urbanism and the *eco-city* [25]. A common feature for the sustainability transition, irrespective of sector, will be the critical normative goals addressed: using resources more efficiently and reducing non-renewable resource consumption, reducing emissions and utilising wastes as resources, restoring environmental quality, enhancing human wellbeing, and developing better functioning cities. In all these arenas, governments have a major role to play in setting performance benchmarks and targets for industry and community to achieve, given the challenges that are to be overcome that are beyond any previous era. Consumers are also central to this transition in terms of the signals they send to both industry and government that they are demanding more sustainable living environments (housing, transport/mobility, energy) and domestic products. At present, these demand signals are not strong [15,26] and as such represent a challenge (risk) for organisational strategies attempting to introduce more sustainable products and services to the market.

In the next section, we attempt a sketch of some of the dimensions of a green economy as it begins to emerge. We then proceed to explore the directions that Australia's built environment industries are taking as reflected in senior management's thinking about their company's sustainability practices and green growth strategies. They are critical to a green economy transition. The manner in which they assess the strengths, weaknesses, opportunities and threats in their operating environment (see Figure 3)—with particular reference to the current and likely future practices of governments and consumers—will influence the rate at which they are likely to innovate and explore new areas for products and services. The results from a survey of built environment firms undertaken by the authors constitute a marker in assessing the progress towards a green economy and green cities in Australia and internationally.

2. Critical Connections

2.1. Green Economy

At a broad conceptual level, a green economy can be defined as one that is sensitive to and not destructive of the environment. It involves a transition from the current model of development that continues to give primacy to economic decisions in isolation and assumes that environmental problems and externalities can be solved or accommodated if the economy is sound. It is based on an integration of ecological thinking and innovation into all social and economic planning by government and industry from the beginning, not after the issues have been framed. It involves recognition that the macro-economy is part of a larger natural ecosystem and resource base, which has capacity constraints that will be severely tested in this century [2,27]. It involves what some authors (e.g., [28]) have termed eco-positive development, where focus is on making positive contributions to the public realm and ecological base in contrast to the present approach of merely attempting to minimise the negatives associated with economic development and city building. Others have termed it restorative or regenerative development [29,30]. We are currently at some distance from embracing this paradigm, though it remains the kind of long-term goal needed to transform our cities.

To date, the most influential advocates for a green economy have been from international agencies such as the United Nations and OECD that are seeking to establish a new global platform for sustainable development. These global policy agendas (e.g., [31–36]) are presented as a backdrop against which more specific national policy agendas are emerging. The Green Growth agenda has been pursued by many emerging nations e.g., Vietnam, Indonesia, China *etc.* following the lead of Korea [37]. A number of initiatives, such as those out of Asia (e.g., [34,35,37]) are focusing on projects capable of delivering blueprints for low carbon green growth at a national scale. Since the era of sustainability strategies developed out of the Rio process, the focus of most Western countries (e.g., USA), has centred more on green energy and green jobs, driven by national interests related to energy security and creating new industries capable of generating new jobs in an era of slow growth. Europe has led the world in green economy strategies through a range of initiatives developed through the EU (again, see Figure 1) [38].

The academic literature in this area has attracted a range of authors from a variety of disciplines, each of whom view the green economy through different sets of lenses [39–44]. Titles on the green economy are being written from a diverse range of perspectives, e.g., economic [45], from an industry location and employment perspective [46], from an education and training perspective [47] and from a broad societal governance perspective. Milani [48] suggests that the green economy will emerge from the bottom up, through public opinion and new social movements. The role of business is explored in various ways: Brown [49] puts the responsibility for bringing about the transition to an eco-economy almost exclusively into the hands of the state (via the three principle levers available to governments: regulation, subsidies/incentives and pricing/taxing instruments); while Hatfield-Dodds *et.al.* [50] argue that both government and industry have a role, especially considering that the business sector is beginning to emerge as a central player in strategic thinking and planning about sustainability. MIT [51] shows who are the leaders in this transition. These are issues that will be explored via an industry survey that follows in the *Greening Business* section.

The emergence of green business rankings [52] (which quantify the ecological footprints, policies and reputations of the US top 500 and Global top 100 businesses) and Clean Tech indexes [53] among others are following in the wake of the earlier and broader-based corporate sustainability (triple bottom line-based) rankings of companies e.g., Dow Jones Sustainability Index; Australian Sustainable Asset Management Index (see [54,55]). The former group (green rankings) are indicative of the changes beginning to enter all industry sectors that will shine a spotlight on corporate environmental performance. In addition, how businesses in the built environment sector in Australia are attempting to measure the performance of their organisations beyond profit and shareholder value dimensions is examined in the following section.

In its narrowest sense, the green economy has been seen to revolve primarily around energy and the transition from fossil fuels to renewables [56,57]. This is sometimes referred to as the "de-coupling" agenda where GDP and GHG are decoupled—wealth grows and footprint declines [10,56]. In its broader sense, it can be seen to embrace all major sectors of the economy creating opportunities for innovations that enable achievement of green goals relevant to industries within those sectors [47] as well as being found in green industrial clusters [58]. In fact the emerging concepts of the green economy cross over the different sectors and are not easily separated under these older classification systems. Some examples are provided in Table 1.

It is clear that there have been major opportunities in all industry sectors for some time to create greener products and services—that are demonstrably superior in performance to those currently in the marketplace. Transition is a socio-technical process, however [1] where multiple barriers exist to the diffusion of a more eco-efficient product. In the auto industry, for example, strategies for transitioning to hybrid and electric vehicle manufacturing would have been on corporate radar screens for decades, but given the power of the entrenched ICE (internal combustion engine) and fossil fuel industries, and an absence of drivers from governments and customers, triggers for change have lagged. Table 1 outlines the key concepts for business that need to be the focus of their agenda if they are to contribute to competitive, productive and liveable cities in the next major economic wave of innovation related to sustainable development. The green goals used in this table were provided to businesses in Australia as the basis of a survey to ascertain their commitment to the green economy, thus helping to define the meaning of "green" to these businesses.

Table 1. Greening the economy: key sectors and green goals.

Industry Sectors	Green Goals Associated with Sector
Manufacturing	Cradle-to-cradle manufacture; closed loop production; industrial ecology clusters; circular economy; green materials manufacture
Energy Utilities	Renewables; distributed (local) generation; energy efficiency; smart grid
Water Utilities	Integrated (stormwater, wastewater) systems; water sensitive urban design; city as catchment
Waste	Reduction, reuse, recycling, and materials reforming; eco-industrial development; closed loop planning
Property and Construction	Smart, green building design; virtual design and construction; constructed as designed; life cycle analysis entrenched together with Building Information Modelling ; green supply chain; design for deconstruction
Trade (Retail/Wholesale)	Zero waste (packaging, food *etc.*); carbon management; carbon neutral products
Transport	Hybrid, electric, hydrogen vehicles; land use-transport integration; active transport; extensive public transport
Finance and Property	Green accounting; urban retrofitting; building accreditation; green loans; green bonds
Services	Zero waste; reduced consumption, carbon management; e-services
Government	Green procurement; de-coupling policies; sector-specific decarbonising schemes; eco-efficient regulation; green performance targets for own operations as well as industry and population; product labelling (declarations), standards, rating

2.2. *Surveying the Greening of Business*

Nations and firms are increasingly aware of the need to be ahead of or at least in touch with the next new wave of innovation [59] and are seeking to identify what will give rise to the new areas of innovation. For a major societal transformation to occur—such as to a green economy—there needs to be an associated critical mass of new and emerging enabling technologies capable of being linked to a clearly recognised and pressing need in the marketplace—in this instance, the "wicked" challenges of 21st century sustainable development (economic, social and environmental). Earlier socio-technical and economic transitions can be seen to have conferred significant first mover advantages to those firms, regions and nations that have been adept at nurturing the development and implementation of innovative technologies, e.g., Detroit and the auto industry; Silicon Valley's role in the information technology revolution; and the more diverse but locally significant spatial clusters in the biotechnology industry [60]. The emergence of green industrial clusters has been identified by Davies [58]. Advantages also accrue to those organisations capable of "sudden catch-up" [59], that is, adopting and adapting the new technology products or processes to current business operations. Slow catch-up (laggards) is to be avoided—by firms as well as governments. Indeed, the window of opportunity for transitioning to a green economy and a more sustainable society—without major social and economic upheaval—is seen to be narrowing [5,14,50,61].

At a macro-economic level, new classes of employment and local economic development are associated with the development of new "engines" of an economy. The attraction of green industries and green jobs is high on the radar screens of governments in many developed and major developing countries—seeking new employment opportunities for economies that have experienced a global recession and have growing populations seeking work [62]. Global trade and national balance of payments is also a significant driver. Even authoritarian governments like the Gulf States and China are implementing green economy strategies. Friedman [63,64] (p. 11), reports on a three sentence summary of China's modern economic history provided to him by C.H. Tung, the first Chief Executive of Hong Kong: "China was asleep during the industrial revolution. She was just waking during the information technology revolution. She intends to participate fully in the green revolution". China has led all nations in the level of investment made in green industries during and after the 2008–2009 global financial crisis [65,66]. In Australia, the federal government from 2007 to 2013 was committed to a range of green economy initiatives including Infrastructure Australia that ensured all infrastructure decisions included commitments to decarbonising the economy and the Clean Energy Package of policies that included a carbon tax and facilitation of renewable energy. The election of a conservative neo-liberal federal government in September 2013 led to a major retreat in public policy and expenditure on most fronts related to climate change (e.g., renewable energy targets were reduced) and the environment. Continued lack of bipartisan agreement in these areas in Australia creates uncertainty and risk for business planning and investment in green industry. This stands in marked contrast to Britain's three main political parties who made a joint commitment at the beginning of 2015 to cut greenhouse gas emissions by 80% (compared to 1990 levels) by 2050. Nevertheless at a state and local level there are many initiatives being taken to assist the development of the green economy in Australia, e.g., a low carbon procurement model is being developed in the state of New South Wales [67].

At a micro-economic level, determining sustainability's strategic worth to companies continues to be an area of increasing focus among leaders of business and is of continued interest to the research community [51,68–70]. From a survey of literature in this area, an attempt was made by the authors to create a "map" of the strategic and operational issues and indicators that could be seen to define the state of progress in an organisation's transition to adoption of sustainability principles and practices to their operations. The key elements of this "map" are found in Figure 3.

These Elements relate to:

(1) An organisation's current and expected future *operating environment* (what transition theorists term "*the landscape*"). Included here are the (external) threats and opportunities as well as the (internal) strengths and weaknesses of the firm as reflected in the current company "profile" … the traditional SWOT factors.

(2) Whether the corporate strategies of the private sector organisations are beginning to catalogue and target new market opportunities connected with emerging environmental challenges; whether sustainability issues are a permanent fixture on the agenda of senior managers; whether the firm's business model has changed as a result of issues surrounding sustainability; establishing the level of difficulty management has in evaluating the business case for sustainability; whether the firm's green services or products can make a net positive contribution to the prosperity of the firm; establishing where "niche innovation" is possible and how can it be financed?

(3) Whether organisations demonstrate sustainable performance of their own operations; e.g., is the ecological footprint of their operations shrinking? And is employee productivity, health and well-being being enhanced? Is there regular monitoring and reporting?

(4) Areas where government can assist business in creating an environment conducive to low carbon green growth. Conditions conducive to longer term planning and investment need to be created by governments where firms can have a measure of confidence that by more aggressively pursuing green strategies they will not unduly raise the risk of being exposed to a sudden change of public policy. Post-2013, Australia's federal political parties have shown a diminished vision and a lack of bipartisanship in relation to green economy thinking and planning, and have instead provided only short-term programs linked to electoral cycles that seem to demand policies that are friendly to voters and established (fossil fuel based) industries. This is not conducive to creating an environment for transitioning to green growth.

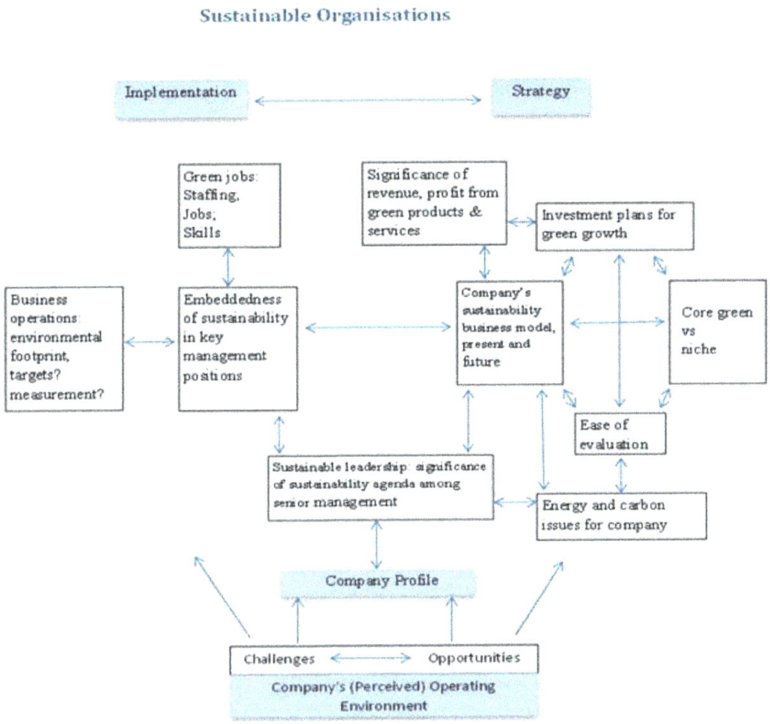

Figure 3. Facets of a sustainable organization.

2.2.1. Survey Population

In order to explore these issues, an on-line survey was undertaken of senior executives and managers belonging to organisations in the built environment sector in Australia who were members of either the Green Building Council of Australia (GBCA), Infrastructure Partnerships Australia (IPA), Infrastructure Sustainability Council of Australia ISCA) or Engineers Australia (EA; more specifically, members of the Environmental College and the Society of Sustainability and Environmental Engineers)—all with knowledge of corporate strategy and performance of their respective organisations. The four industry associations who collaborated in this survey by facilitating contact with the relevant executives and managers also have established reputations for their progressive attitudes to corporate sustainability reporting. A total of 173 completed questionnaires were obtained from the national survey, undertaken in mid 2013: 85% from private sector firms and 15% from public sector agencies. Of the total respondents, 30% operated in a single state, 26% in multiple states, 12% in all seven states, 11% in Australia and the Pacific and 21% in Australia and globally. Ninety percent were headquartered in the capital cities (60% Sydney and Melbourne). Responses were received from organisations across the size spectrum, with 40% from SMEs (fewer than 50 employees), 17%, 50–200, 19%, 200–1000, 17%, 1000–10,000 and 6%, over 10,000. Questionnaires developed for this survey are presented in Newton and Newman (2013). As the built environment sector is represented by a wide spectrum of private sector firms, it was decided to aggregate responses for analysis and reporting purposes into three groups: design (38% of total), building product manufacturers and construction (37%); and property services (25%). Depending upon the topic under analysis, responses will be presented as an analysis of the total sample, private *versus* public sector or comparison of responses by the three private sector groupings. The findings are presented in relation to the set of topics for investigation outlined in the previous section.

2.2.2. Business Perceptions of Operating Environment and Key Challenges

A formidable set of challenges with implications for built environment sustainability and resilience are now confronting all societies and their economies as outlined above. Private sector firms as well as public sector organisations have been confronted with the task of establishing management strategies and making investments that can advance their operations and at the same time contribute net positive benefits to the regions in which they operate. The manner in which the organisations responding to this survey have rated the challenges are instructive (see Table 2).

Table 2. Rating of challenge to the company in the next five years (private firms only).

Area of Challenge	Level of Challenge (%)					
	1 (Low)	2	3 (Moderate)	4	5 (Major)	Not Applicable
Technological (n = 156)						
Acquiring and/or adapting new low carbon technology	18.6	12.2	40.4	13.5	5.8	9.6
Limited Australian R&D capacity	18.6	16.0	21.8	16.0	13.5	14.1
Identifying emerging new technology and assessing relevance to business	10.3	14.7	41.0	22.4	10.3	1.3
Market Related (n = 156)						
Responding to changing market conditions	7.7	11.5	28.8	28.8	19.9	3.2
Increasing competition in Australian markets	8.3	12.2	16.7	30.8	23.7	8.3
Increasing competition in global market	14.1	9.0	22.4	24.4	12.8	17.3
International trade barriers to export of products and services	28.8	11.5	17.9	4.5	3.2	34.0
Whether customers are willing to pay a premium for green products, services	6.4	10.3	26.9	23.7	31.4	1.3
Competitors increasing commitment to sustainability, green products/services faster than expected	18.6	17.3	31.4	17.9	5.1	9.6
Government Related (n = 156)						
Addressing impact of government regulations, standards *etc.*/maintaining "licence to operate" on our business	9.6	16.7	30.8	20.5	15.4	7.1
Lack of certainty regarding government policies/legislation	5.8	10.3	28.8	20.5	30.8	3.8
Financial (n = 156)						
Taxation policy (e.g., carbon tax, corporate tax *etc.*)	14.7	21.2	33.3	18.6	7.7	4.5
Global financial uncertainties	5.8	12.8	32.1	28.8	16.0	4.5
Cost of capital, borrowing	14.7	23.7	30.8	15.4	7.7	7.7
Access to capital	14.7	22.4	29.5	11.5	13.5	8.3
Growing revenue	5.1	10.9	28.2	35.9	15.4	4.5
Staff Related (n = 153)						
Acquiring necessary skills/competing for new talent	6.5	13.7	37.3	24.2	16.3	2.0
Shortage of skilled labour	15.7	19.0	35.3	17.6	10.5	2.0
Attracting talented people	9.8	21.6	27.5	26.8	13.1	1.3
Retaining and motivating existing employees	7.2	23.5	35.3	24.2	7.8	2.0
Business Operations (n = 153)						
Reducing costs and increasing efficiencies	3.3	11.1	31.4	33.3	20.9	0.0
Innovating to achieve competitive differentiation	5.9	8.5	26.8	38.6	15.7	4.6
Responding effectively to threats and opportunities of sustainability	5.9	15.0	36.6	29.4	11.1	2.0
Increasing operating speed and adaptability	2.6	17.6	32.0	30.7	13.1	3.9
Responding effectively to disruption of our business model	5.9	24.2	33.3	24.2	7.2	5.2
Stricter requirements from partners along the value chain	11.8	25.0	29.6	16.4	3.9	13.2
Externalities (n = 153)						
Climate change adaptation (increasing temperature, sea level, rainfall, variability *etc.*) challenges for your company	26.8	22.9	21.6	13.7	5.2	9.8
Shortage (high cost) of material inputs to company operations	21.6	22.2	22.2	13.1	3.9	17.0

In line with previously published studies in this area [71], we found that while 88% of companies indicated that sustainability is permanently on their agenda (see following section; Table 3), when this issue is placed in the context of the large number of challenges that business needs to manage, it is rated down the scale of importance. Responding effectively to threats and opportunities of sustainability to their business are seen by 40% as a significant challenge (category 4 and 5 responses), but green competition is only identified by 23% as such. Green technology is seen as something that can be acquired relatively easily. Resource constraints and carbon constraints are rated as significant by less than 20% of companies. The big challenge is understanding "whether customers are willing to pay a premium for green products and services". Sixty percent of respondents indicate that uncertainty in this area is the key issue. Uncertainty rates high as a key challenge in other areas as well: global financial markets (45%) and consistency of government (over 50%). Uncertainty in these three areas ranks with issues of increased competition in the Australian market (54%), the challenges of growing revenue (51%) and being innovative (54%) as the key targets on the radar screens of senior management in the built environment industry.

2.2.3. Management and Sustainability

The survey findings suggest that issues relating to sustainability and carbon reduction are now a permanent and core fixture on the agenda of built environment's top management for over 40% of surveyed organisations, with a higher representation among private sector companies (see Table 3).

Table 3. Sustainability as an agenda item with senior management. (Source: derived from Author survey; [51].)

	Private *n* = 128	**Public** *n* = 21	**Total** *n* = 149	**MIT** 2010	**MIT** 2011	**Design** *n* = 48	**Mfg&Con** *n* = 47	**Serv** *n* = 33
Already a permanent fixture and core strategic consideration	43.0	38.1	42.3	24.0	28.0	60.4	31.9	33.3
On the agenda permanently, but not core	42.2	47.6	43.0	38.0	42.0	31.3	48.9	48.5
Temporarily on the agenda, but not core	12.5	14.3	12.8	22.0	19.0	6.3	17.0	15.2
Excluded from agenda, because viewed as a passing fad	0.0	0.0	0.0	4.0	2.0	0.0	0.0	0.0
Never considered for the agenda	2.3	0.0	2.0	8.0	8.0	2.1	2.1	3.0
Total	100.0	100.0	100.0	100.0	100.0	100.0	100.0	100.0

Although the sample size of public organisations represented in this survey is small, there are several topics where important differences appear to be emerging that the authors considered should be reported (albeit with the above qualification). Most tables, however, are restricted to reporting private sector responses. When these responses are segmented according to the major categories of built environment industry it is clear that the design firms are significantly further advanced in recognising the strategic importance of incorporating sustainability objectives in their business models for competing for urban development projects. This is an area where new knowledge is emerging capable of being added to other longer established dimensions of design innovation that can now make a major contribution to achieving more eco-efficient outcomes on construction projects. International comparisons suggest that Australian companies have elevated sustainability to a higher level of corporate concern than their overseas counterparts [51] (a possible reflection of their membership of those industry associations collaborating in the survey that have strong environmental objectives).

For 52% of private sector firms and 38% of public organisations, sustainability is being embedded within *all* facets of their operations, and there remains only a very small percentage where sustainability has not been addressed at all (Table 4).

Table 4. Extent of sustainability penetration throughout organization.

	Private n = 124	Public n = 21	Total n = 145	Design	Mfg&Con	Services
Sustainability is being embedded within all aspects of my organization's operations	51.6	38.1	49.7	62.2	48.9	40.6
My organization addresses sustainability as a separate, delineated activity	41.9	57.1	44.1	33.3	48.9	43.8
Sustainability has not been addressed in my organization	6.5	4.8	6.2	4.4	2.1	15.6
Total	100.0	100.0	100.0	100.0	100.0	100.0

Not unexpectedly, perhaps, the design firms are further advanced in ensuring wide penetration of sustainability thinking and processes compared to manufacturing and construction firms and property services firms: 62%, 49% and 41% respectively. As design represents that phase of a project where innovation potential is highest, with capacity to eventually impact all parts of a building or construction project, it can be expected that sustainability outcomes will become more and more mainstreamed into built environment products and processes in future.

It is evident that a significant proportion of built environment organisation's business models have changed as a result of issues surrounding sustainability and low carbon development (Table 5).

Table 5. Whether organizations' business models have changed as a result of issues surrounding sustainability and future low carbon development. (Source: derived from Authors survey; [71].)

	Private n = 129	Public n = 21	Total n = 150	Design	Mfg&Const	Services	MIT 2011
Yes	45.7	42.9	45.3	47.9	55.3	29.4	46.0
No	50.4	42.9	49.3	47.9	40.4	67.6	40.0
I do not know	3.9	14.3	5.3	4.2	4.3	3.9	14.0
Total	100.0	100.0	100.0	100.0	100.0	100.0	100.0

A higher percentage of private sector organisations have effected some change, compared to public sector organisations. A comparison of the three built environment industry segments reveals that manufacturing and construction firms appear to have made more change to their business models and property services least, reflecting the higher exposure to global competition by the former (and larger) firms. Design firms may be (inadvisably) resting on their leadership in sustainability thinking that has characterised this segment of industry to date. In addition, they were found to be not as environmentally progressive when it came to monitoring their own day to day activities (e.g., water and energy use, waste generation). Taken overall, the percentages associated with changed business models are closely aligned to published studies of international organisations.

Part of the reason for lack of a higher level of progress in embedding sustainability more deeply and effectively within private sector organisations is the level of difficulty they expressed in evaluating the business case (see Table 6).

Table 6. Level of difficulty expressed in evaluating the business case for sustainability related strategies (private sector).

Issue	Level of Difficulty (%)					Average Score
	Not	2	Moderately	4	Very	
Establishing financial incentives for considering sustainability	7.3	15.3	23.4	34.7	19.4	3.44
Predicting customer response to sustainability strategies	2.4	19.4	29.8	37.1	11.3	3.35
Quantifying intangible effects of sustainability strategies (e.g., brand reputation, employee hiring, retention and productivity)	7.3	21.8	21.8	31.5	17.7	3.31
Capturing comprehensive metrics about sustainability impact of operations	6.5	18.5	35.5	26.6	12.9	3.21
Lack of model/framework for incorporating sustainability in business cases	10.5	18.5	27.4	31.5	12.1	3.16
Uncertainty about future carbon pricing	13.7	14.5	35.5	18.5	17.7	3.12
Opposition from executives or influential individuals	12.1	22.6	25.8	23.4	16.1	3.09
Evaluating competing priorities	4.8	27.4	37.1	26.6	4.0	2.98
Quantifying sustainability-related risks	8.1	24.2	39.5	19.4	8.9	2.97

Note: n = 124.

All issues listed have a level of difficulty associated with their execution. Those heading the list involve: the challenge of resolving uncertainties surrounding customer response and incentives; predicting customer response to sustainability strategies; and quantifying intangible effects of sustainability strategies ("reputation" being the principal attribute often referred to here; McKinsey, 2014 [70]. An analysis across the three built environment business segments revealed that the property services firms had above average (>3.0 score) difficulty with all nine aspects of the business case, compared to only three for design and seven for manufacturing and construction.

2.2.4. In-House Sustainability

Results of the survey suggest that the private sector is more alert to green business strategies and opportunities external to the firm than the green credentials of its own internal operations. It is in relation to in-house sustainability management and measurement that the public sector currently has the lead (see Table 7).

Across the three built environment industry segments, some important differences emerged. Design firms were ahead in relation to sustainability strategies involving committees, dedicated managers, defined indicators and procurement, but the manufacturing and construction firms were ahead in relation to the routine measurement of the environmental performance of the organisation (a number of which would relate to their licence to operate). A major gap (in the order of 20%–30%) common to all three segments related to the variation in measurement of energy (cost-related) compared to CO_2 emissions (environment-related), a reflection of the absence of targets or sanctions that currently impact business operations in any tangible way.

Table 7. In-house sustainability indicators (percent of responses indicating "present").

Indicator	Private *n* = 124	Public *n* = 21	Total *n* = 145	Design	Mfg&Con	Services
A formal sustainability policy/strategy	75.0	85.7	76.6	80.0	72.3	71.9
A senior sustainability manager role	47.6	52.4	48.3	55.6	38.3	50.0
Sustainability Board or committee	33.1	52.4	35.9	35.6	27.7	37.5
Key sustainability performance indicators (such as feature in the Annual Report or its equivalent)	38.7	71.4	43.4	44.4	34.0	37.5
A sustainability-oriented procurement strategy	46.8	57.1	48.3	53.3	46.8	37.5
Routine measurement of the following:						
Energy used	74.2	95.2	77.2	71.1	78.7	71.9
CO_2 emissions generated	49.2	81.0	53.8	53.3	48.9	43.8
Water used	54.0	95.2	60.0	48.9	63.8	46.9
Other emissions to air (SO_2, NO_2 *etc.*)	16.9	9.5	15.9	4.4	31.9	12.5
Liquid waste	29.8	19.0	28.3	11.1	42.6	37.5
Solid waste	45.2	76.2	49.7	31.1	59.6	43.8

2.3. *Measuring the Green Economy*

A UNEP [32] (p. 454) report on the green economy has suggested that green industries are dominated by service industries and tend to be concentrated in the largest consumer markets. There is, however, an absence of research that attempts to measure how businesses are performing in this respect. In this study we attempt to measure the significance of green business to an industry sector by the proportion of an organisation's incomes that are derived from providing green products or services to meet current needs in the marketplace (Table 8).

Table 8. Proportion of organizations' income derived from providing green products or services to meet sustainability needs in the market place.

Proportion	Private *n* = 145	Public *n* = 26	Total *n* = 171	Design	Mfg&Con	Services
100%	11.0	0.0	9.4	19.6	7.1	5.3
50%–99%	11.0	3.8	9.9	7.8	16.1	7.9
10%–49%	24.1	3.8	21.1	27.5	21.4	23.7
Less than 10%	40.7	42.3	40.9	39.2	41.1	42.1
None	13.1	50.0	18.7	5.9	14.3	21.1
Total	100.0	100.0	100.0	100.0	100.0	100.0

For built environment organisations in the public sector, over 90% derive either less than 10% or no income from this source. In contrast, 22% of private sector firms are currently securing more than half of their income from green products and services; but there appears to be considerable scope for growth with over half currently having little or no involvement. Of those organisations deriving some income from green business, the distribution between products and services is in line with UNEP [32] findings (see Table 9).

Table 9. Approximate proportion of "green income" due to green services as opposed to green products (%).

Proportion	Private n = 123	Public n = 13	Total n = 136	Design	Mfg&Con	Services
More than 90% services/10% products	52.0	61.5	52.9	85.4	13.3	56.7
70% services/30% products	2.4	0.0	2.2	0.0	6.7	0.0
50% services/50% products	6.5	7.7	6.6	2.1	11.1	6.7
30% services/70% products	10.6	7.7	10.3	4.2	20.0	6.7
Less than 10% services/90% products	28.5	23.1	27.9	8.3	48.9	30.0
Total	100.0	100.0	100.0	100.0	100.0	100.0

According to these findings, 52% of private sector firms receive more than 90% of their income from services (*vs.* 62% in the public sector). Across the three built environment industry segments, there is greater contrast: design firms derive over 85% of their green income from services, while for manufacturing and construction firms approximately half is from green products, reflecting a greater *diversification* of green business for this group (*i.e.*, adding a service function to traditional product activity). This is also the case for property services.

When we cross-classify these two measures of green business activity for private sector firms (Table 10), we find that they are at varying stages of developing lines of business in these areas.

Table 10. Measuring green business activity in the built environment sector (private sector only).

Proportion of Total Income Derived from Green Lines of Business	Proportion of Firm Income from Services *Versus* Products			
	Mostly Products (>50%)	Roughly 50:50% Products and Services	Mostly Services (>50%)	Total
100%	3.3	0.0	9.8	13.0
50%–99%	4.9	3.3	4.9	13.0
10%–49%	10.6	0.8	16.3	27.6
<10%	20.3	2.4	23.6	46.3
None	0.0	0.0	0.0	0.0
Total	39.0	6.5	54.5	100.0

Note: n = 123.

The study found that only 13% of companies derive all their revenue from either green products or services, with a further 13% receiving at least half from this area of business. Green services have a larger share of firm income but not by a large margin. The challenge is for the 46% of built environment businesses that currently have less than 10% of their sales in green products and services to be actively looking for such opportunities.

2.4. Locating Green Opportunities by the Private Sector

When built environment organisations were probed as to where they saw major opportunities (Table 11) for delivering green products or services in the built environment sector, most of the responses to this open-ended question did not identify new or cutting edge areas.

Table 11. Targets for green business as identified by built environment industries.

General
Providing green services, products, technologies in response to increased awareness and demand—to meet changing customer needs; wherever possible at a cost point that increases market interest (**)
New Products and Technologies to Drive Business Opportunities
Alternative (renewable, low carbon) energy sources, distributed generation (***) Natural/hybrid ventilation systems; adiabatic cooling, solar shading (*) Low energy lighting Improved energy metering, monitoring, energy management systems, energy efficiency (EE) Low VOC products, enhanced IEQ Recycling of C&D waste/less waste (*) Paperless operations—requires less storage/floorspace
New Services
Sustainability consulting Resource minimization consulting Carbon and/or sustainability rating and profiling of buildings, infrastructures, precincts Financial packages/options for clean energy/EE LCI/LCA assessment of products Climate change planning and design Centers of expertise aligned to needs of low carbon economy
Design
Design leadership; companies should challenge themselves to find new eco-efficient solutions; have a belief that they *can* derive green designs at competitive cost; "find ways to produce cost effective sustainable dwellings that produce a competitive edge for the company" (**) Utilize GBCA products in design (***) Sustainable precinct design (*) Design to prescribed as well as increased levels of sustainability
Construction
Energy efficient retrofits (***) **Facility Management** Optimal building operating performance Building with lower life cycle costs
Leadership
Be seen as a market leader in promoting sustainable built environment technologies (**) Demonstrate credentials/points of difference with competitors
Educating Clients
Educating clients about longer term benefits of green design, green technology as a means of growing business (**)

(*) represents similar comment by more than one respondent; (**) by at least three; (***) by five or more.

They represent incremental rather than the more innovative step change advances which are considered necessary for any transformative urban sustainability transition (see [1] for a wider discussion on the three horizons of urban technology innovation). They were mostly in areas where the level of risk/uncertainty had diminished as a result of a sufficient number of successful implementations over the past 10 to 15 years, which established the viability, and attractiveness of the particular green product or service. Responses to another open-ended question "What needs to change to create opportunities for low carbon green growth" revealed that there are a clearly defined set of persistent and significant barriers that are a challenge to innovation in this sector (see Table 12) and to the emergence of a green economy and eco-efficient cities.

Table 12. What needs to change to create opportunities for business in relation to low carbon living and low carbon green growth.

Government
Clarity, continuity, and commitment of governments (all levels) on direction for low carbon green growth in order for business to adopt sustainability business model
Increase government regulations and audits
Energy/carbon legislative change that is appropriate and effective
Green standards and certification for products and services; life cycle product performance declarations need to be mandatory
Application to life cycle assessment to all aspects of government—drive a whole of life approach
Green procurement
Industry
Locate and increase pools of investment capital to match company's sustainability goals
Once large corporations change their business models, they represent a major opportunity/driver for smaller consulting companies
Demonstration of connection between sustainable built environment and financial sustainability of organisation
Increasing global demand for new green products and services
Attitudes of company executives
Public
A cultural shift towards a greener way of life opens up a multitude of opportunities for new products and services
Increased education of consumers regarding carbon footprints of their behaviour/lifestyle

There are several areas where governments need to begin talking to their built environment industries and businesses and general public more generally about how they can create critical connections for a more sustainable and productive future.

3. The Carbon Challenge

As discussed earlier, in its narrowest sense, the green economy has been seen to revolve primarily around energy and the transition from fossil fuels to renewables. The Australian Labor government (2007–2013) introduced its Clean Energy Strategy in July 2011 containing a number of measures to reduce the nation's greenhouse gas emissions, including a carbon tax. In September 2013, there was a change of (federal) government to a conservative Liberal/National coalition that has subsequently repealed the carbon tax, abolished many of the climate-related agencies and programs, reduced the national renewable energy target and has introduced a "direct action" program in an attempt to mitigate growth in CO_2 emissions (many commentators see this as being less effective compared to a price on carbon, with an inadequate budget and untested processes for project assessment and delivery). This watering down of a national commitment to carbon reduction continues to cause local uncertainty (a major concern in the business sector revealed in this and other studies [72]) as well as international friction in the lead up to the UN Climate Change meeting in Paris in December 2015.

The Economist Intelligence Unit (EIU) undertook a survey [73] of 130 Australia-based senior executives prior to the introduction of the carbon tax in Australia. The present survey was undertaken several months after its introduction, which should have provided organisations with an opportunity to begin to gauge its impact on their operations. The two surveys reveal similar percentages in relation to whether organisations overall have a strategy in place for reducing their carbon footprint (see Table 13). Overall, more than 30% are still lacking such a strategy, with a higher proportion in the private sector.

Table 13. Organization has a strategy in place for reducing its carbon. (Source: derived from author survey and Economist Intelligence Unit (EIU) (2011) [73].)

	Private $n = 90$	Public $n = 17$	Total $n = 107$	EIU $n = 131$	Design	Mfg&Con	Services
No	35.8	14.3	32.6	30.0	34.1	28.3	50.0
Yes	64.2	85.7	67.4	70.0	65.9	71.7	50.0
Total	100.0	100.0	100.0	100.0	100.0	100.0	100.0

Governments currently have a lead role in establishing policies, regulations, pricing and incentives for the green economy. Recent announcements by international scientific groups (IPCC, 2014 [11] as well as financial institutions World Bank, 2012 [74]) all point to a likely "4° world" by the end of this century unless significant greenhouse gas mitigation occurs (e.g., ~80% reductions on 1990 CO_2 levels by 2050). There are multiple pathways that have been advanced to decarbonise the built environment [10,72,75] as well as the economy and society more generally. A list of these, based on EIU, 2011 [73], was incorporated in this survey. The results from private sector respondents are listed in Table 14.

Table 14. Preferred government action on carbon reduction (%).

	Yes	No
Provide subsidies for clean technology investments by companies	87.4	12.6
Establish incentives for corporate behaviour that leads to low carbon business operations	87.4	12.6
Provision of information on sustainable practices for companies	87.4	12.6
Introduce a performance standard/label for all energy generation technologies	85.7	14.3
Provision of education on green practices for consumers	85.7	14.3
Establishment of national carbon emission reduction goals	84.0	16.0
Subsidies for clean technology usage by consumers	82.4	17.6
Establishment of environmental reporting standards for business	72.3	27.7
Link to an international carbon pricing scheme	67.2	32.8
Introduce carbon labelling for all manufactured products	62.2	37.8
Carbon cap and trade scheme	55.5	44.5
Current federal government carbon pricing scheme	52.1	47.9
Establishment of penalties for lack of carbon efficiency compliance by companies	50.4	49.6
Corporate tax on carbon footprint of business operations	43.7	56.3
Consumer/sales tax on carbon footprint of goods/services consumed	39.5	60.5
Establishment of penalties for lack of carbon efficiency compliance by consumers	34.5	65.5
None of the above—government can help most by doing nothing and letting the market come up with solutions	9.2	90.8

Government was clearly endorsed as having a role to play (over 90% of industry respondents in favour), but, from this point, on the directions for intervention varied. The highest proportions of "no" votes by companies were reserved for any imposition of carbon taxes on goods and services consumed as well as consumer-centred compliance, suggesting that more decisive action was required higher up the supply chain. Here, corporate taxes or company carbon compliance charges were among the least favoured actions for governments to take. Incentives and subsidies were clearly favoured over taxes for both consumers and producers. There was strong support for introduction of a performance label for all energy generation technologies as well as incentives for the introduction of clean technologies, both targeting the front end of the energy supply chain—and indicative of "direct action". National carbon emission reduction goals were endorsed by 84% of business, part of the search for certainty on the part of business and a clear signal to government that more consistency is required.

4. Conclusions

The emergent new low carbon green economy is shaping as the next competitive advantage for industry. This study and the survey on which it is based has demonstrated that in the built environment area, Australian business and government are gearing up for this challenge. There are signs that Australian business is highly aware of the issues and has structured itself to prepare, perhaps even more so than their international counterparts, based on the results of global surveys [51,69]. However, the transition is not yet as market-oriented as it could be as many businesses are still waiting to see if consumers will pay a premium for going green; meanwhile, they are doing normal business.

The research revealed that private sector firms are at varying stages of developing green lines of business. The study found that only 13% of companies derive all their revenue from green products or services, with a further 13% receiving at least half from this area of business. Green services have a larger share, but not by a large margin. The challenge is for the 46% of built environment businesses that currently have less than 10% of their sales in green products and services to be actively looking for such opportunities. The significant scope for increasing green revenue is tempered by the high level of uncertainty surrounding a firm's understanding of what customers are willing to pay for green products and services (55% of respondents—"uncertain"). The challenges of competition, growing revenue and being innovative—among several other traditional business metrics—clearly outranked sustainability issues around the management table when put in the full context of contemporary business operations.

These challenges notwithstanding, sustainability is a permanent agenda item with senior management in 85% of both private sector and public sector organisations responding to the survey (43% of private sector firms have sustainability as a permanent and core agenda item compared to 38% in public sector). The private sector appears more alert to sustainability/low carbon agenda issues than the public sector, possibly because they are more exposed externally to the "front line" of the economy. They are embedding sustainability within all aspects of their organisation's operations to a greater extent than the public sector (51.6% to 38.1%).

For in-house sustainability and measurement practices, public sector organisations are currently in the lead in terms of having a formal sustainability policy, sustainability manager, sustainability board/committee, reportable sustainability indicators and a sustainability-oriented procurement strategy. Public sector organisations also had higher levels of routine measurement of energy, water and CO_2 emissions; the exceptions were with noxious emissions, where there has been mandated reporting for such waste discharges for some time by state environmental protection authorities.

Approximately half of the organisations are yet to change their business model in response to sustainability/low carbon development issues. A primary reason for this is the level of difficulty reported to be associated with evaluating "green" business cases. A comparable percentage indicates that sustainability is yet to be embodied within all facets of operations.

The opportunities identified for green business development within the built environment sector were weak and not representative of the range that currently exist in the marketplace, let alone those that are emerging opportunities. Most of the opportunities identified could be classed as "mature", reflecting the lack of leading edge "green" innovation currently represented in most urban development projects in Australia.

Two-thirds of organisations surveyed had a strategy in place for reducing their carbon footprint (similar to results from the EIU survey a year earlier). Ninety percent of respondents indicated that government has a lead role to play in encouraging carbon reduction; although there was significant variability in response as to where government intervention should occur. Most favoured areas were subsidies, incentives, information and education. Least favoured were taxes, either on business or consumers.

This survey has demonstrated that, in the built environment area, Australian business and government are gearing up for the green economy challenge. There are signs that Australian business is highly aware of the issues and has structured itself to prepare, perhaps even more so than businesses in other countries, based on the results of global surveys. But the transition is not yet as market-oriented as it could be since many businesses are still waiting to see if consumers will go green and whether governments can offer more consistent policy direction in this area. Meanwhile, they are conducting business as usual. Publicly funded organisations appear to be less market-aware than business but are demonstrating green economy approaches and outcomes at a high level of commitment. Tipping points in all the critical transition arenas are yet to be reached to enable a more rapid shift to a green economy. If Australia is to be a strong global competitor in the green economy and cities are to be core to this transition, then awareness and commitment will need to increase across the board in private and public sector built environment industries—and among consumers. National governments, in particular, need to take a leadership position and then stay the course.

Acknowledgments: Funding support for this study has been provided by the Australian Research Council Discovery Project DP110100543: Green Shoots? Exploring the Genesis and Development of a Green Economy in Australia. The authors also gratefully acknowledge the support of the Green Building Council of Australia (GBCA), Infrastructure Partnerships Australia (IPA), Infrastructure Sustainability Council of Australia ISCA) and Engineers Australia (EA; more specifically, members of the Environmental College and the Society of Sustainability and Environmental Engineers) in identifying senior managers for interview. All are organisations whose objectives are directed towards evolving a more sustainable built environment industry in Australia for their members.

Author Contributions: Both authors contributed to the design of the research, the questionnaire and the write-up. The on-line survey, data cleaning and tabulation was undertaken by the CATI Unit at Swinburne University of Technology. Both authors have read and approved the final manuscript.

Conflicts of Interest: The authors declare no conflict of interest.

References

1. Newton, P. Liveable and sustainable? Socio-technical challenges for 21st century cities. *Urban Technol.* **2012**, *19*, 82–101.
2. Newman, P.; Beatley, T.; Boyer, H. *Resilient Cities: Responding to Peak Oil and Climate Change*; Island Press: Washington, DC, USA, 2009.
3. Rockström, J.; Steffen, W.; Noone, K.; Parsson, Å.; Chapin, F.S., III; Lambin, E.; Lenton, T.M.; Scheffer, M.; Folke, C.; Schellnhuber, H.J.; *et al.* Planetary boundaries: Exploring the safe operating space for humanity. *Ecol. Soc.* **2009**, *14*, 1–33.
4. Pearson, L.; Newton, P.; Roberts, P. *Resilient Sustainable Cities: A Future*; Routledge: London, UK, 2014.
5. Von Weizsäcker, E.U.; Hargroves, K.C.; Smith, M.H.; Desha, C.; Stasinopoulos, P. *Factor Five: Transforming the Global Economy through 80% Improvements in Resource Productivity*; Earthscan: London, UK, 2009.
6. Ward, B.; Dubos, R. *Only One Earth: The Care and Maintenance of a Small Planet*; Norton and Company: New York, NY, USA, 1972.
7. World Commission on Environment and Development (WCED). *Our Common Future*; Oxford University Press: Oxford, UK, 1987.
8. Rockstrom, J.; Klum, M. *The Human Quest: Prospering within Planetary Boundaries*; Princeton University Press: New Jersey, NJ, USA, 2012.
9. Factavia website. Available online: https://global.factiva.com/factivalogin/login.asp?productname=global (accessed on 11 February 2015).
10. Newman, P.; Kenworthy, J. *The End of Automobile Dependence*; Island Press: Washington, DC, USA, 2015.
11. Intergovernmental Panel on Climate Change (IPCC). *Ipcc Working Group 3—Mitigation of Climate Change: Assessment Report 5*; IPCC: Geneva, Swizerland, 2014.
12. Parry, M.L.; Canziani, O.F.; Palutikof, J.P.; van der Linden, P.J.H.C.E. *Climate Change 2007: Impacts, Adaptation and Vaulnerability*; IPCC Working Group 2: Geneva, Swizerland, 2014.
13. Rees, W.E. Ecological footprints and bio-capacity: Essential elements in sustainability assessment. In *Renewable-Based Technology: Sustainability Assessment*; Dewulf, J., van Langenhove, H., Eds.; Wiley: Chichester, UK, 2006.
14. World Wildlife Fund (WWF). *Living Planet Report 2014: Species and Spaces, People and Places*; WWF: Gland, Swizerland, 2014.

15. Newton, P.; Meyer, D. Exploring the attitudes-action gap in household resource consumption. *Sustainability* **2013**, *12*, 1211–1233. [CrossRef]
16. Organisation for Economic Co-Operation and Development (OECD). *Household Behaviour and the Environment: Reviewing the Evidence*; OECD: Paris, France, 2008.
17. Turner, G. *Is Global Collapse Imminent? An Updated Comparison of the Limits to Growth with Historical Data (Research Paper No. 4)*; Melbourne Sustainable Society Institute: Melbourne, Austrailia, 2014.
18. Newton, P.; Doherty, P. Chapter 2: The challenge to urban sustainability and resilience: What cities need to prepare for. In *Resilient Sustainable Cities: A Future*; Pearson, L., Newton, P., Roberts, P., Eds.; Routledge: London, UK, 2014; pp. 7–18.
19. Worldwatch Institute. *State of the World: Our Urban Future*; Norton and Company: New York, NY, USA, 2007.
20. Worldwatch Institute. *State of the World 2013: Is Sustainability still Possible?* Island Press: Washington, DC, USA, 2013.
21. Geels, F. The multi-level perspective on sustainability transitions: Response to seven criticisms. *Environ. Innov. Soc. Transit.* **2011**, *1*, 24–40. [CrossRef]
22. Newton, P. Chapter 9: City transitions, infrastructure innovation, green economy and the eco-city. In *Resilient Sustainable Cities: A Future*; Pearson, L., Newton, P., Roberts, P., Eds.; Routledge: London, UK, 2014; pp. 91–104.
23. Engineers Australia. *Infrastructure Report Card*; Engineers Australia: Canberra, Australia, 2010.
24. Mcglade, C.; Elkins, P. The geographical distribution of fossil fuels unused when limiting global warming to 2 degrees C. *Nature* **2015**, *517*, 187–190. [CrossRef] [PubMed]
25. Newman, P.; Matan, A. *Green Urbanism in Asia: The emerging Green Tigers*; World Scientific Publishing: Singapore, Singapore, 2013.
26. Pitt and Sherry; Swinburne University. *National Energy Efficient Building Project: Final Report Prepared for the Department of State Development, Adelaide on Behalf of the Department of Industry, Canberra*; Pitt and Sherry: Canberra, Australia, 2014.
27. Daly, H.E. *Beyond Growth: The Economics of Sustainable Development*; Beacon Press: Boston, MA, USA, 1996.
28. Birkeland, J. Resilient and sustainable buildings. In *Resilient Sustainable Cities: A Future*; Pearson, L., Newton, P., Roberts, P., Eds.; Routledge: London, UK, 2014; Chapter 15; pp. 146–159.
29. Giradet, H. *Creating Regenerative Cities*; Routledge: London, UK, 2014.
30. Newman, P.; Jennings, I. *Cities as Sustainable Ecosystems*; Island Press: Washington, DC, USA, 2008.
31. United Nations Department of Economic and Social Affairs (UN DESA). Proceedings of United Nations Conference on Sustainable Development, Rio + 20, Rio de Janeiro, Brazil, 20–22 June 2012.
32. United Nations Environment Programme (UNEP). *Towards a Green Economy: Pathways to Sustainable Development and Poverty Eradication*; UNEP: Paris, France, 2011.
33. Organization for Economic Co-operation and Development (OECD). *Fostering Innovation for Green Growth*; OECD: Paris, France, 2011.
34. United Nations Economical Commission for Asia and the Pacific (UN ESCAP). *Low Carbon Green Growth Roadmap for Asia and the Pacific*; UN ESCAP: Bangkok, Thailand, 2012.
35. Asia-Pacific Economic Cooperation (APEC). *The Green Initiative: Second Cycle Daegu Initiative*; APEC Small and Medium Enterprises Working Group: Singapore, Singapore, 2012.
36. Asian Development Bank (ADB). *Green Urbanization in Asia: Key Indicators for Asia and the Pacific 2012 Special Chapter*; ADB: Mandaluyong, Phillppines, 2012.
37. Government of the Republic of Korea. *Road to Our Future: Green Growth National Strategy and Five Year Plan 2009–2013*; Presidential Comission on Green Growth Republic of Korea: Seoul, Korea, 2009.
38. Mazza, L.; Tn Brink, P. *Green Economy in the European Union*; United Nations Environment Programme (UNEP): Nairobi, Kenya, 2012.
39. Bailey, I.; Caprotti, F. The green economy: Functional domains and theoretical directions of enquiry. *Environ. Plan. A* **2014**, *46*, 1797–1813. [CrossRef]
40. Caprotti, F. *Eco-Cities and the Transition to Low Carbon Economies*; Palgrave: London, UK, 2015.
41. Joss, S. *Sustainable Cities: Governing for Urban Innovation*; Palgrave: London, UK, 2015.
42. Smith, A.; Kern, F.; Raven, R.; Verhees, B. Spaces for sustainable innovation: Solar photovoltaic electricity in the UK. *Technol. Forecast. Soc. Chang.* **2014**, *81*, 115–130. [CrossRef]

43. Markard, J.; Raven, R.; Truffer, B. Sustainability transitions: An emerging field of research and its prospects. *Res. Policy* **2012**, *41*, 955–967. [CrossRef]
44. Spash, C. Green economy, red herring. *Environ. Values* **2012**, *21*, 95–99. [CrossRef]
45. Barbier, E.B.; Markandy, A. *A New Blueprint for a Green Economy*; Earthscan: London, UK, 2012.
46. Moretti, E. *The New Geography of Jobs*; Houghton Mifflin Harcourt: New York, NY, USA, 2012.
47. Elder, J. Preparing americans for a green economy. *Sustainability* **2009**, *2*, 240–242. [CrossRef]
48. Milani, B. *Designing the Green Economy*; Roman and Littlefield: Lanham, MD, USA, 2000.
49. Brown, L.R. *Eco-Economy: Building an Economy for the Earth*; Norton and Company: New York, NY, USA, 2001.
50. Hatfield-Dodds, S.; Turner, G.; Schandl, H. *Growing the Green Collar Economy: Skills and Labour Challenges in Reducing Our Greenhouse Emissions and National Environmental Footprint (Report to the Dusseldorp Skills Forum*; Commonwealth Scientific and Industrial Research Organisation (CSIRO): Canberra, Australia, 2008.
51. Haanaes, K.; Reeves, M.; von Strengvelken, I.; Audretsch, M.; Kiron, D.; Kruschwitz, N. *Sustainability nears a Tipping Point: Findings from the 2011 mit and Boston Consulting Group Sustainability and Innovation Global Study*; Massachusetts Institute of Technology (MIT): Cambridge, MA, USA; Boston Consulting Group (BCG): Boston, MA, USA, 2012.
52. Newsweek. Green Rankings 2010: U.S. Companies. Available online: http://www.newsweek.com/2010/10/18/green-rankings-us-companies.html (accessed on 12 February 2015).
53. Australian CleanTech (ACT). January 2015 Cleantech Index Result. Available online: http://www.auscleantech.com.au/PDF/index/January2015PerformanceReport.pdf (accessed on 12 February 2015).
54. Grey, F. Sustainable corporations. In *Transitions: Pathways towards Sustainable Urban Development in Australia*; Newton, P., Ed.; Springer: Dordrecht, The Netherlands, 2008; Chapter 42; pp. 641–650.
55. Neugent, M.; Hughes, P. Government and sustainability reporting. In *Transitions: Pathways toward Sustainable Urban Development in Australia*; Newton, P., Ed.; Springer: Dordrecht, The Netherlands, 2008; Chapter 41; pp. 629–640.
56. United Nations Environment Programme (UNEP). *Decoupling Natural Resource Use and Environment Impacts from Economic Growth*; UNEP: Geneva, Swizerland, 2011.
57. Phillmore, J. Schumpeter, schumacher and the greening of technology. *Technol. Anal. Strateg. Manag.* **2001**, *13*, 23–37. [CrossRef]
58. Davies, A.R. Cleantech clusters:Transformational assembalages for a just, green economy or just business as usual? *Globval Environ. Chang.* **2013**, *23*, 1285–1295. [CrossRef]
59. Cutler, T. *Venturous Australia: Building Strength in Innovation (Report Prepared for the Commonwealth Government)*; Cutler and Company Pty Ltd.: North Melbourne, Australia, 2008.
60. Gilding, M. The tyranny of distance: Biotechnology networks and clusters in the antipodes. *Res. Policy* **2008**, *37*, 1132–1144. [CrossRef]
61. Victor, P.A. *Managing without Growth: Slower by Design, Not Disaster*; Edward Elgar: Cheltenham, UK, 2008.
62. Romano, L. Voices of power: Adviser sets sights on developing green-colar jobs. *Washington Post* **2009**.
63. Friedman, T. As the West Fiddles, China Makes the Most of Green Revolution. Available online: http://newsstore.fairfax.com.au/apps/viewDocument.ac;jsessionid=2F184300761C048BBCBF9FBEA31C0755?sy=afr&pb=all_ffx&dt=selectRange&dr=1month&so=relevance&sf=text&sf=headline&rc=10&rm=200&sp=brs&cls=57&clsPage=1&docID=AGE100112NV1EGVF420M (accessed on 13 Feburary 2015).
64. Friedman, T. *Hot, Flat and Crowded: Why We Need a Green Revolution—and How It Can Renew America*; Picador: New York, NY, USA, 2009.
65. Osnos, E. Green Giant: Beijing's crash program for clean energy. Available online: http://www.newyorker.com/magazine/2009/12/21/green-giant (accessed on 13 Feburuary 2015).
66. Sheehan, P. Achieving Sustained Change in Energy Structure. In Proceedings of the 1st Abrupt Change in China's Energy Path: Implications for China, Australia and the Global Climate Conference, Melbourne, Australia, 26 June 2014.
67. Hargroves, K.C. *Low Carbon Inclusions in Commonwealth and Nsw Government Built Environment Sector Procurement*; Cooperative Research Centre for low Carbon Living (CRCLCL): Sydney, Australia, 2015.
68. Sull, D. Closing the gap between strategy and execution. *MIT Sloan Manag. Rev.* **2007**, *48*, 30–38.

69. Haanaes, K.; Arthur, D.; Balagopal, B.; Kong, M.T.; Reeves, M.; Velkin, I.; Hopkins, M.S.; Kruschwitz, N. *Sustainability the "Embracers" Seize Advantage: How Fast Are Businesses Adopting Sustainability-Driven Management?* Massachusetts Institute of Technology (MIT): Cambridge, MA, USA; Boston Consulting Group (BSG): Boston, MA, USA, 2011.
70. McKinsey and Company. *Sustainability's Strategic Worth: Mckinsey Global Survey Results*; McKinsey and Company: New York, NY, USA, 2014.
71. Kruschwitz, N.; Haanaes, K. First look: Highlights from the third annual global executive survey. *Sloan Manag. Rev.* **2011**, *53*, 86–89.
72. Cox, L. Business seeks action on climate. Available online: www.aldersgategroup.org.uk/asset/download/135/business_seeks_action_on_climate_change.pdf (accessed on 10 July 2015).
73. Economist Intelligence Unit (EIU). *Cleaning up: Australia's Readiness for a Low Carbon Future*; EIU: London, UK, 2011.
74. World Bank. *Turn down the Heat: Why a 4 Degree C Warmer World Must Be Avoided*; World Bank: Washington, DC, USA, 2012.
75. Dixon, T. *Hotting up? An Analysis of Low Carbon Plans and Strategies for UK Cities*; Royal Institute of Chartered Surveyors (RICS): London, UK, 2011.

sustainability

MDPI

Article

Sustainable Urban Transport in the Developing World: Beyond Megacities

Dorina Pojani [1,*] and Dominic Stead [2,*]

1 School of Geography, Planning and Environmental Management, The University of Queensland, Brisbane 4072 QLD, Australia

2 Faculty of Architecture and the Built Environment, Delft University of Technology, Delft 2628 BL, The Netherlands

* Authors to whom correspondence should be addressed; d.pojani@uq.edu.au (D.P.); d.stead@tudelft.nl (D.S.); Tel.: +61-(07)-3365-6091 (D.P.); +31-(0)-15-278-3005 (D.S.).

Academic Editors: Tan Yigitcanlar and Md. Kamruzzaman

Received: 3 March 2015; Accepted: 12 June 2015; Published: 17 June 2015

Abstract: Megacities have frequently received a disproportionate amount of attention over other sizes of cities in recent discourse on urban sustainability. In this article, the authors argue that a focus on smaller and medium-sized cities is crucial to achieving substantial progress towards more sustainable urban development, not only because they are home to at least a quarter of the world's population but because they also offer great potential for sustainable transformations. In principle, their size allows for flexibility in terms of urban expansion, adoption of "green" travel modes, and environmental protection. At the same time, smaller and medium-sized cities often have fewer resources to implement new transport measures and can be more vulnerable to fluctuations in the world economy. This article critically reviews the potential role and impact of nine commonly considered options for sustainable urban transport in cities in developing countries: (1) road infrastructure; (2) rail-based public transport; (3) road-based public transport; (4) support for non-motorized travel modes; (5) technological solutions; (6) awareness-raising campaigns; (7) pricing mechanisms; (8) vehicle access restrictions; and (9) control of land-uses. Drawing on international research and examples of policies to reduce the environmental impacts of transport in urban areas, this article identifies some key lessons for sustainable urban transport in smaller and medium-sized cities in developing countries. These lessons are certainly not always identical to those for megacities in the global south.

Keywords: urban transport; sustainable transport; developing cities; medium-sized cities

1. Introduction

In the space of just a few decades, urban areas across the world, in both developed and developing countries, have become increasingly automobile-dominated and less sustainable. In developing countries in particular, cities have experienced a rapid growth in transport-related challenges, including pollution, congestion, accidents, public transport decline, environmental degradation, climate change, energy depletion, visual intrusion, and lack of accessibility for the urban poor. In more developed countries, particularly in Northern Europe, some cities have witnessed a trend of reclaiming urban space from the automobile and prohibiting cars from major parts of downtown areas and/or confining them in other ways. Today, these places are often considered as leading examples of sustainable urban development, as cities across the world strive to meet urban sustainability standards by improving public transport, encouraging non-motorized modes, creating pedestrian zones, limiting the use of private cars, and otherwise trying to undo the transformation of cities caused by automobile dominance. Concepts of automobile restraint that were unthinkable just a few decades ago are now being considered or even adopted in many urban areas around the globe, both north and south, with the encouragement and support of major international organizations (see [1–5]).

This article critically reviews the potential role and impact of nine commonly considered options for sustainable urban transport in medium-sized cities located in developing countries: (1) road infrastructure; (2) rail-based public transport; (3) road-based public transport; (4) support for non-motorized travel modes; (5) technological solutions; (6) awareness-raising campaigns; (7) pricing mechanisms; (8) vehicle access restrictions; and (9) control of land uses. These options for action are overlapping and interconnected. They cover both the demand and the supply side of urban transport with a focus on the latter. The article is organized around "interventions" rather than according to geographic regions, implementation stages, or types of impact.

A focus on smaller developing cities (*i.e.*, with fewer than one million inhabitants) is important in current urban sustainability discourses since nearly half of the world's 3.9 billion urban dwellers reside in relatively urban settlements with fewer than 500,000 inhabitants, while only around one in eight live in the 28 megacities of 10 million inhabitants or more [6]. Overall, the urban population in developing countries is set to double from 2010 to 2050 while remaining stable in developed countries [7]. At the moment, cities with less than 100,000 inhabitants represent a third of world's urban population, a figure which is predicted to grow to 40% in 2050. The fastest-growing urban agglomerations are medium-sized cities with less than 1 million inhabitants, located in Asia and Africa [7,8]. The term "developing cities" used in this article refers to the cities of all countries included in the World Bank's 2015 list of developing countries [8]. It should be noted that, since this categorization of countries is based on average Gross National Income (GNI) statistics, the wealth of different cities within the same country (and its distribution across society) can vary substantially.

Emerging megacities have received a disproportionate amount of attention, while smaller developing cities are extensively underserved with respect to basic services and lack the necessary institutional capacity to be able to manage their rapidly growing populations [9]. The authors argue that medium-sized cities in the developing world can offer greater potential for more sustainable transformations than megacities. They generally have a smaller ecological footprint, and in principle, their size allows for flexibility in terms of urban expansion, adoption of "green" travel modes, and environmental protection. At the same time, smaller developing cities might have fewer resources to implement necessary transport measures and might be more vulnerable to fluctuations in the world economy and climate. Also, due to their size and density, they are often characterized by less efficient public transport systems, lower modal shares of public transport, and higher transport-related energy consumption per capita than larger developing cities [2].

This article draws on examples of international experience with policies to reduce car use in urban areas. The examples are indicative given the complicated and complex nature of urban transport issues; the authors do not attempt to quantify a particular mix of policies (see, for example, [10]). While most examples are from developing countries, a few come from developed contexts when examples from developing countries are less available. This article targets two main audiences: (a) urban transport sustainability researchers, who are seeking a review of academic literature and state-of-the-art practice; and (b) policy-makers (and politicians) in developing countries, seeking an overview of practice that can be used to inform the development of new urban transport strategies.

The authors strike a cautionary note about the potential for international policy transfer in the urban transport arena. Contrary to a common belief amongst some development agencies that policy solutions already exist and simply need to be implemented more widely, the search, analysis, and uptake of urban transport policy ideas, concepts, or instruments from elsewhere are subject to a range of different influences, including political, professional, institutional, economic, and social. Research to date indicates that there is little evidence or prospect of "copying" of one policy from one area to another, certainly outside national boundaries. The potential for replication of "best practices" is questionable. City size is certainly not the only variable that determines the transferability of policy options.

Cultural penchants and historic trajectories of transport demand and supply (implying path dependence) prevent cities (including smaller ones) from applying the same solutions to apparently similar problems. For example, the public transport trip share is 19% in Latin American cities compared to 5% in African cities; the average non-motorized trip distance is 1.0 km in Latin American cities and 2.1 km in Southeast Asian cities [11]. Where contexts are quite dissimilar (e.g., north to south), caution is suggested both in terms of the appropriateness and effectiveness of standard policy solutions being exported from one place to another [12–14]. Developing cities are advised to consider examples of transport solutions from both developing and developed contexts and to keep in mind that not all innovation originates from the global north. After all, some of the most efficient and cost-effective public transport systems have been developed in Latin America.

2. Nine Options for Promoting Sustainable Urban Transport in Medium-Sized Developing Cities

2.1. Road Infrastructure

In the immediate post-WWII decades, increasing the size and number of roads (including flyovers and tunnels) was a commonly-used approach for addressing congestion and other urban travel issues. In more recent years an understanding has emerged that increasing capacity can lead to greater demand as a result of "induced travel" (also referred to as "latent demand" or "generated traffic"). Induced travel is due to diversion of travel from: (1) other lower volume hours of the day to more peak hour use of improved facilities; (2) parallel commuting routes; and (3) public transportation. The consequence is that congestion levels are soon restored to almost pre-expansion levels and little travel time savings are realized. Road investments also have adverse long-term effects on traffic congestion. They spawn new trips due to the land-use development (and possibly sprawl), which improved car access induces. Moreover, road construction itself is disruptive for densely-built urban areas. In many cases, the demolition of buildings and/or open space is necessary [15].

While induced demand claims have elicited strong reactions and polarized political factions, academic studies have confirmed the induced travel demand theory (see [16–18]). While the degree (*i.e.*, the travel demand elasticity) and the circumstances (a single road facility or a metropolitan area) in which induced travel occurs remain a matter of debate, there is no question that road improvements prompt traffic increases and provide little congestion relief. Since this is the case in developed cities with saturated or nearly saturated car ownership markets, it follows that the findings would certainly apply to developing cities, in which incomes and car ownership are still growing. However, there have been no academic studies testing the induced travel demand in these settings.

The prospect of induced travel lends credence to a transportation policy based on alternative modes. Policy makers in Europe have been sensitive to claims of induced demand and have taken steps to jettison the traditional "predict and provide" policy of trying to accommodate traffic growth (see for example [16,19]). In North America, funding incentives and political inertia have made major change difficult [19]. In developing countries, where new road construction is often seen as a yardstick of modernization, governments have often allocated public expenditures in favor of new road construction at the expense of other urban transport investments and the maintenance of existing infrastructure. In terms of road investment, evidence to date suggests that developing cities need to focus their resources on existing road maintenance rather than new road construction.

Estimates suggest that $45 billion worth of road infrastructure (urban, interurban, and rural) was lost in 85 developing countries in Asia, Africa, and Latin America between 1970 and 1990 due to inadequate maintenance. This loss could have been averted with preventive maintenance costing less than $12 billion [20]. There are multiple reasons why the problem is widespread. They include economic adversity and mistaken investment choices (*i.e.*, policymakers cut maintenance if donors raise funding for new construction) as well as institutional failures (*i.e.*, separation of responsibility and control between the providers and users of roads). Without corrective action, poor roads–as well as inadequate road-based (public and private) transport–can become an insurmountable obstacle to the economic recovery and growth of developing cities. Current evidence suggests that, in developing cities, maintenance expenditures have a positive effect on economic output whereas the construction of new highly-visible road infrastructure is less beneficial for economic development [21,22].

2.2. Rail-Based Public Transport

Increased investments in all types of public transport promise to help boost the urban economy of developing countries. The case of Japan is illustrative in this respect. In the post-WWII period, the Japanese government played a key role in urban transport by adopting public policies that discouraged automobile use and channeled investment into public transport. This urban transportation system contributed to rapid economic growth by minimizing aggregate transportation costs, constraining the consumption of private automobiles, and encouraging savings, albeit with a decided lack of attention to passenger comfort [23].

It is debatable whether rail-based or road-based public transport should be emphasized in medium-sized developing cities. Mobility patterns are influenced by both population size and population density, especially the latter. Urban sprawl has a significant effect on travel distances and hinders public transport supply. In urban areas with a small but dispersed population, the provision of either rail- or road-based public transport might not be economically viable [24]. In particular, rail should have a definite advantage over road-based systems to justify implementation in smaller and/or dispersed cities, since new rail systems are very expensive to construct and operate. A full cost-benefit analysis of both options should guide decision-making. A brief overview of the two main rail-based urban transport systems, Light Rail Transit (LRT), and metro (suburban or heavy rail), is provided below.

LRT ranges from the historical tramways, trolleys, and streetcars of Eastern Europe, which run along other traffic in urban streets, to the sophisticated elevated and completely segregated systems of Singapore. LRT vehicles can be developed on urban streets and run alongside urban traffic because they are fed electricity from overhead wires. This is an advantage over metro systems, which require fully segregated rights-of-way because they have an electrified third rail that increases speeds but is fatal on contact. LRT is expanding rapidly in developed cities with low corridor volumes, sometimes feeding heavy rail systems. In developing countries, LRTs exist only in larger cities such as Tunis, Alexandria, Manila, Buenos Aires, and São Paulo. The cost of building and operating LRT varies widely but it is considerably higher than the cost of alternative public transport forms, such as busways (see below) [25,26]. If LRT operates at grade without priority or protection from obstruction by other traffic, it has little or no performance (speed) advantage over busways [27].

In the past, LRT advantages over busways were the lower local air pollution impact and possibly smother rides for urban travelers. Older LRT vehicles generally had higher carrying capacity than most buses [27]. Evolving technologies (e.g., electric buses, see later) have minimized the differences between bus and rail in terms emissions, capacity, and comfort. However, LRT is generally more appealing to middle class passengers, and investment in this mode is seen as a signal of a more permanent commitment to public transport on a government's part [26]. Surprisingly, a few quantitative analyses employing data from the U.S.-a country in which aversion to bus transport is perhaps the strongest-have shown that riders' preferences for rail travel over bus is null when service characteristics such as travel time and cost are equal. A preference only arises when one mode offers a higher quality service (*i.e.*, fewer transfers and higher frequency) [28].

In view of the evidence, investments in new LRT systems in medium-sized developing cities may have limited economic and practical value. Due their high costs, developing cities often can only construct such systems over a few kilometers in a few limited corridors, which do not meet the broader transport needs of the population. Nevertheless, the public sector may end up with a long-term debt that can affect investment in more pressing policy areas [26]. A secondary argument is that, in many developing cities, power shortages are common which means that a public transport system which relies on grid electricity may not always be a feasible or desirable option. At the same time, full reliance on fossil fuels for public transport is also controversial, especially in countries which lack oil resources.

Metro systems are usually the most expensive form of public transport in terms of construction and operation, but, as fully segregated systems, have the best performance. Since most metros are designed for capacities around 30,000 to 40,000 passengers per hour in the peak direction, usually, only cities with a population of 2–3 million have at least one corridor, which requires this type of facility [29]. In the developing world, only some of the larger cities, such as Beijing, Mexico City, Bombay, and Cairo have metro or suburban rail systems.

As with LRT, the capital costs of building metro systems can vary substantially between cities, between metro systems, and between metro lines within the same city and system. However, they are taxing even for developed nations: $50–$150 million per kilometer, with cost over-runs being the rule rather than the exception [30]. The reasons for high cost variations include differences between projects in terms of the ratio of underground to above-ground construction, ground conditions, station spacing, type of rolling stock, environmental and safety constraints, and labor costs. Underground metro systems cannot be easily integrated into existing urban physical structures, without disrupting building foundations and utility lines [30]. In addition to high capital costs, metro systems have high operating costs and usually require operating subsidies; otherwise the price of the tickets would be prohibitive even in developed cities. While in principle public transport operations do not need to be profitable, given the valuable service that they provide to society, the high capital and operation cost of metros makes them less economically viable in medium-sized developing cities than in megacities.

Ropeways or air gondolas are a special case of rail transport. With capacities comparable to small or medium-sized tramways on rail tracks and with moderate costs, ropeways have evolved to become a reasonable and attractive proposition for mainstream urban public transport in a number of medium-sized developing cities, including Algiers and Oran (Algeria), Medellin (Colombia), and Caracas (Venezuela). They can provide suitable urban transport in hilly terrains, and over rivers, harbors, motorways, as well as over dense housing and historic buildings, and can complement other public transport options [31]. Due to the limited amount of literature available on this mode, policy options in this area are not reviewed in detail here.

2.3. Road-Based Public Transport

Effective road-based public transport is central to economic growth of developing cities. For the majority of residents, road-based public transport (bus and paratransit) is the only means to access employment, education, and public services. In medium and large developing cities, such destinations are beyond viable walking and cycling distances while vast numbers of individuals have limited access to automobiles. Unfortunately, the current state of road-based public transport services in many developing cities does not serve the mobility needs of the population adequately. Formal bus services are often unreliable, inconvenient, uncomfortable, or even dangerous. Informal paratransit services, while providing benefits including on-demand mobility for the transit-dependent, jobs for low-skilled workers, and service coverage in areas devoid of formal transit supply, carry major costs, such as increased traffic congestion, air and noise pollution, traffic accidents, and even violence among route cartels [32]. In addition to regulating paratransit systems through measures in between the extremes of acceptance and outright prohibition, several options are available to medium-size developing cities that wish to improve the quality of formal bus services.

In recent decades, the creation of bus lanes on existing roads (painting of a lane in a different color from the rest of the asphalt) has been a common low-cost strategy for improving the quality of bus systems throughout the world. In some cases, they are shared with high-occupancy vehicles, taxis, and/or non-motorized vehicles, and even with vehicles near turning points. New technologies allow vehicles in bus lanes to gain priority at intersections, with lights automatically turning red for cars and green for buses whenever the later approach shared intersections. Despite their advantages, in developing cities, unsegregated bus lanes alone, particularly those situated in the curb lane, do little to enhance the effectiveness of public transport. Temporary parking by taxis and delivery vehicles, low levels of respect for traffic rules, the unavoidable conflicts with turning vehicles, and limitations in narrow street configurations degrade the usefulness of bus lanes in these contexts. Here, a more effective intervention in favor of public transport is the construction of busways that are physically segregated from other traffic by means of barriers, cones, or other well defined physical features. Located on the curb or in the median of a roadway, they are permanently and exclusively for the use of public transport vehicles-although emergency vehicles are often allowed to use the lane [26].

Bus Rapid Transit (BRT) is a recently developed bus-based mass transit which emulates the performance and amenities of rail transit. "Full BRT" (Box 1) is often more appropriate for large cities as it can transport up to 45,000 passengers per hour per direction, surpassing the capacity of many rail systems. To date, full BRT has been developed only in a few large cities (including Bogotá, Curitiba, and Guangzhou) with very high levels of political commitment and charismatic political leadership in support of quality public transport [26,33]. More standard forms of BRT include segregated busways over the majority of the length of the system's trunk/city center corridors and at least two of the full BRT characteristics. These serve up to around 13,000 passengers per hour per direction, and may be more suitable for medium-sized cities [26].

Box 1. Characteristics of a "full BRT system" (based on [26]).

- metro-quality services
- location of busways in the median of the roadway rather than the curb
- integrated network of routes and fares
- closed high-quality stations that provide level access between the platform and vehicle floor
- pre-board fare payment/verification
- system management through a centralized computerized control center
- clear route maps, signage, and/or real time displays that are visibly placed within the stations/vehicles
- frequent and rapid service
- modern, clean vehicles
- special physical provisions to ease access for the physically disabled
- marketing identity
- clean vehicle technology
- superior images and customer service (*i.e.*, clean buses and uniformed staff)
- entry to system restricted to prescribed operators and a restricted number of vehicles ("closed system")

Prior experience indicates that the best BRT results are achieved when private sector competition is combined with strong public sector oversight [26]. This type of business plan is generally desirable for public transport in developing cities, even those without BRT systems, because in these contexts both fully regulated sectors and completely deregulated sectors have failed [34]. Mainline services can be complemented by existing conventional bus and paratransit systems, which can provide feeder connections and serve remote areas [32]. To avoid overreliance on fossil fuels, BRT vehicles can run on natural gas, electricity, or biofuels (see below). Once a developing city has secured the right of way for a BRT system, it can later upgrade to light rail should funding become available.

To date more than 150 cities worldwide (at least 70 in Asia, Africa, and Latin America) have implemented BRT systems. Comparative assessments of BRTs throughout the world have found that most systems have greatly improved their local travel conditions and the quality and performance of public transport, especially in travel time savings and enhanced reliability. BRT systems have also reduced energy consumption and emissions. They have generally been well-received by the users leading to massive bus ridership increases. Urban enhancements are also evident [26,33,35,36].

BRT can be developed at substantially lower costs than rail transit. BRT systems typically cost between $1 m and $8 m per kilometer depending on the complexity and size of the project (the need for overpasses or underpasses and the need for property acquisition). Even in developed cities with higher labor costs, BRT costs less than $10 m per kilometer. If property designed, BRT can operate at affordable fares ($1/ride) without subsidies. Rapid implementation times (1–5 years) and flexibility to adapt to spatially-constrained historical centers and business districts with narrow roadway segments are other attractive features [26].

Despite the advantages, BRT systems in various developing countries suffer from a range of problems. These include rushed implementation (e.g., several components incomplete at the time of commissioning); very tight financial planning as systems usually do not receive operational subsidies; excessive occupancy levels; early deterioration of infrastructure; fare collection systems requiring very tight supervision; and insufficient user education for initial implementation and system changes. While many of these problems are associated with financial restrictions and institutional constraints, they are not intrinsic BRT issues. Nevertheless, these difficulties affect public perception which means that BRT is often regarded as a "second-best" mode compared to rail, and that politicians frequently offer rail alternatives as part of their electoral proposals [37].

2.4. Support for Non-Motorized Modes

Non-motorized transport—walking and cycling, but also pedicabs and other human-operated vehicles—is the dominant transport mode in many developing cities, especially in Asia and Africa. The smaller the city size, the higher the percentage of non-motorized transport use. Generally, bicycles are used more extensively in developing cities than in developed cities. The urban poor, who cannot afford motorized transport, most often walk or cycle to reach work, school, and other services. Pedicabs and other non-motorized taxi services provide employment for young and poor urban men. While non-motorized trips may last longer than vehicular trips, for many developing city residents a higher time cost is preferable to a higher financial cost for transport [38]. Moreover, increases in non-motorized transport improve traffic safety in cities. Research (employing data from developed cities) has shown that the likelihood of collision between a pedestrian or cyclist and a motor vehicle is inversely related to the amount of foot or bicycle traffic [39]. Therefore, in medium-size developing cities, investment and support for walking and cycling are paramount [38].

Despite its importance, non-motorized transport policy (Box 2) and its related infrastructure are often neglected in policy-making in developing cities. On one hand, the political climate is not favorable to soft modes. Politicians consider walking and cycling as a sign of backwardness and not commensurate with their goals and aspirations. Urban elites distort transport planning in favor of motorized modes, which they are more likely to use themselves. Major international lenders often invest minuscule amounts in non-motorized improvements. The groups which most heavily rely on non-motorized transport are poorly organized and unable to articulate their needs. On the other hand, urban street use in developing cities often has a conflicting nature, with a complex pattern of coexistence between pedestrians, vehicles, vendors, and even animals, which also makes interventions more difficult [38,40].

Box 2. Examples of non-motorized transport policies mainly from Northern European cities (based on [41,42]).

- auto-free zones in city centers
- sidewalks on both sides of the street
- pedestrian refuge islands for crossing wide streets
- zebra crosswalks, often raised and with special lighting for visibility
- pedestrian-activated crossing signals, both at intersections and at midblock crosswalks
- extensive networks of bike paths and lanes
- "bicycle streets" where cars are permitted but cyclists have strict right-of-way
- special bike turn lanes leading directly to intersections
- separate bike traffic signals with advance green lights for cyclists
- bike parking (ranging from secured boxes to simple racks)
- integration of bicycles with public transport (*i.e.*, allowing bicycles on public transport vehicles, at least outside peak hours)
- bike-activated traffic signals at key intersections
- modifications of street networks to create deliberate dead ends and slow, circuitous routing for cars but direct, fast routing for bikes
- traffic calming of residential neighborhoods through speed limits and physical barriers such as raised intersections and crosswalks, traffic circles, road narrowing, zigzag or chicane routes, curves, speed humps, and artificial dead ends created by midblock street closures
- lower speed limits for motor vehicles in cities (e.g., 50 km per hour)
- prohibitions of truck traffic and through traffic of any kind in residential neighborhoods
- driver training focused on avoiding collisions with pedestrians and cyclists
- traffic education of children
- traffic regulations that favor pedestrians and bicyclists, placing the burden of proof on drivers in case of collisions
- strict ticketing and high penalties for motorists, pedestrians, and cyclists who violate traffic regulations

Various recent examples of improvements and innovation in non-motorized transport policy in developing countries can be identified. For example, Chinese cities have rapidly transitioned from human-powered bicycles and gasoline-powered scooters to electric bicycles. Improvements in e-bike technology, growing incomes, and falling e-bike prices, as well as government policies including e-bike national standards and safety regulation, local sale restrictions for gasoline scooters, and banning of gasoline vehicles from city centers, have encouraged this modal shift [43]. The Chinese government has also mandated the reconversion of motorcycle lanes into bicycle lanes [38]. Cycle networks have been developed in several Latin American cities, including Bogotá, Colombia, and Sao Paolo, Brazil. As with buses, physically segregated bicycle paths are more effective than bicycle lanes. Bogotá's Ciclovia-Recreovia and Cicloruta programs are community-wide programs that promote the use of bicycle paths. In Chile, the Vida-Chile is a national program that uses a variety of strategies to promote physical activity [44]. Bicycle-sharing schemes have been introduced in a growing number of developing cities, including Rio de Janeiro (Brazil), Daejon (Korea), and Hangzhou (China). Evidence to date suggests that bicycle-sharing has led to growing bike use but no reductions in car use. The growth of this mode has been concentrated in medium- to small-sized towns with systems of 50 bicycles [45].

2.5. Technological Solutions

The main urban transport-related technological solutions that cities worldwide are currently pursuing (with North American and West European cities leading the way) include alternative-fuel vehicles and intelligent transportation systems (ITS). New technologies may help to tackle certain transport-related problems, such as air and noise pollution, oil dependency, traffic congestion, and accidents [46,47]. Their applicability in developing cities is considered below. The sustainability of these types of technological solutions is subject to three caveats. First, transport technology improvements cannot help tackle reduced physical activity due to car dependence. Second, in most developing countries, where car ownership is growing, the benefits of technological advances will often be offset by the rapid increase in the amount of car travel. Third, many new technologies are outside the financial reach of many residents of developing cities. Consequently, the pursuit of technology-based measures alone is not cost-effective in these contexts [48]. Nonetheless, technological optimism prevails in both developed and developing countries, which may be seen as an expression of people's reluctance to more fundamental changes in lifestyle [49].

A wide range of vehicles–passenger cars, heavy-duty trucks, garbage trucks, three-wheelers (primarily in Asian countries), and buses–can run on alternative fuels including natural gas, electricity, and biofuels. These fuels, which can be produced from any primary energy source, including biomass, wind and solar energy, nuclear energy and decarbonized fossil fuels, constitute a cleaner alternative to diesel and gasoline. Biodiesel can be used in any diesel engine without modification while ethanol, gas, hydrogen, and electricity can be used only in specially-produced or modified cars [50].

In some countries, especially in South America, the availability of natural gas resources and existing pipeline and delivery infrastructure are incentives to encourage natural gas use for transport (compressed natural gas and liquefied petroleum gas). Brazil and Argentina combined have more than half of the world's total natural gas vehicles, but several Asian countries, notably India, China, and Pakistan, have also had significant natural gas vehicle growth beginning in the late 1990s. In other countries, the adoption of natural gas for urban transport requires the co-existence of fuel supply, refueling stations, and appropriate vehicles–a classic "chicken and egg" dilemma [51].

Some governments in developing countries specifically mandate the use of natural gas for transport in highly polluted areas (e.g., for taxis in Buenos Aires, and more recently for buses in Beijing and all pre-1990 taxis in New Delhi). Other developing country governments offer financial incentives to consumers of natural gas fuels and vehicles and to equipment suppliers of alternative fuel vehicles. A pump price of at least 40%–60% below the gasoline price–attributed to government incentives such as favorable taxation, tax breaks for natural gas, or higher taxes on gasoline and/or diesel fuel–is common in most countries that have had successful natural gas vehicle penetration. Experiences in Argentina, Brazil, China, India, and Pakistan indicate that marketing and subsidy programs must be sustained for long periods before diffusion crosses the tipping point. This is due to the long life of a vehicle fleet and social and economic penetration barriers [51].

Electric cars (fuel-cell, battery, or plug-in) have a low range (e.g., 100 km on a full charge in city traffic conditions), and are therefore attractive for use in small and medium-sized urban areas. However, their widespread adoption in developing countries faces a range of barriers. The capital costs of electric vehicles are significantly higher than the costs of conventional cars. In order to amortize the acquisition costs through energy savings, an electric car has to be used to travel significant distances, typically more than 20,000 km a year [52]. While some estimates indicate that in a few decades capital costs could drop significantly, conventional cars will remain cheaper. In terms of fuel costs, accurate predictions are not possible. Another significant challenge is the bulkiness and cost of batteries or hydrogen tanks. In the recent past, recharging battery-powered vehicles was very time-consuming but fast-charging stations are now becoming available. While electric vehicles are non-polluting at the point of use, their overall environmental impact depends on the way in which electricity is generated, stored, and distributed and the problems associated with recycling expired batteries. The availability of lithium in the case of batteries and the availability of platinum in the case of fuel cells constitute barriers as well [50]. Overall, the market penetration of electric vehicles in developing cities is far from favorable, at least in the short term. For example, modelling estimates for Colombia predict that, even by 2050, electricity will not have surpassed gasoline in the Colombian passenger car fleet [53]. India aims to have 100,000 electric vehicles on the roads by 2020-a small share considering its population size-while China's modest target is for the annual sales of "new energy" vehicles (electric, hybrid, *etc.*) to reach 5% in the short term [52].

Biofuels (mainly ethanol, biodiesel, and blends) could provide a significant reduction in urban greenhouse-gas emissions, particulate matter, and volatile organic chemicals. They are attractive to many developing countries where surplus land (e.g., marginal land) can be used for biofuel production [54]. Brazil, China, and India are world leaders in ethanol production, surpassed only by the US [55]. Warm-climate countries such as Argentina, Colombia, Indonesia, India, Malaysia, Thailand, and the Philippines which produce large amounts of palm, coconut, soybean, and jatropha oil show promise for the adoption of biodiesel [56]. However, biofuels present various challenges depending on the type of feedstock. For corn-based biofuel, in particular, life-cycle assessment studies have highlighted a low or negative net contribution to emission reductions [57] Biofuels, especially biodiesel, generate up to 70% higher NO_X emissions depending on feedstock. Moreover, the greenhouse-gas savings potential may disappear once the full impacts of fuel production are taken into account (*i.e.*, the release of carbon stored in forests or grasslands during land conversion to crop production; the frequent burning of cleared vegetation for biofuel production; and the fugitive emission of methane from palm-oil production). Another key concern is the close link between biofuels and food consumption. A switch to second-generation biofuels (manufactured from non-food feedstock) might alleviate these concerns. However, second-generation biofuels still compete with food supply through land-use and are currently constrained by many technical and economic barriers [54].

In addition to alternative fuels, intelligent transportation systems (ITS) have the potential to address urban transport problems in a variety of applications (Box 3). In the industrialized world, ITS have only been adopted at a moderate pace and, in developing countries, adoption has been even slower. In East Asia, Eastern Europe, and Latin America, the most common forms of ITS that have been introduced to date include traffic signal systems, traffic surveillance systems using CCTV, commercial vehicle (e.g., taxi) tracking systems using GPS, electronic ticketing services, electronic toll collection and fare payment systems, bus management systems, and traveler information systems. Further ITS deployment is needed in these settings to improve road safety conditions and mitigate traffic congestion, especially in large, polluted, and congested cities and in harsh climates with hazardous driving scenarios [58].

Box 3. Potential applications of intelligent transportation systems (ITS) (based on [59]).

- advanced traffic management systems, which predict traffic congestion and provide alternative routing instructions to vehicles in real time to improve the efficiency of the road network and maintain priorities for high-occupancy and transit vehicles
- advanced traveler information systems, which provide data to travelers in their vehicles, homes, or workplaces about the location of incidents, weather problems, road conditions, parking availability, and optimal routings, to help them decide what route and mode they should use
- advanced vehicle control systems, which enhance driver's control of the vehicle to make travel both safer and more efficient, and range from collision warning systems to self-driving cars
- automated vehicles, including private cars and automated taxis with fares similar to public transport
- automated road systems, which rely on infrastructure information and control to automatically control the movements of high-occupancy or transit vehicles in special lanes
- on-demand transport and parking (aided by cell phone technology)

Although developing cities are often at a disadvantage in applying ITS relative to developed cities, they also have some advantages. For example, some developing cities can install electronic infrastructure at the same time that physical infrastructure is being constructed, which is far less expensive than retrofitting existing physical infrastructure. Developing cities are also not generally burdened with outdated IT infrastructure that has to be updated. They can take advantage of ITS products and applications, which have already been tested and deployed in developed cities and which are now mature and stable. In theory, they can then leapfrog to an ITS-enabled transportation infrastructure far more rapidly and far less expensively than developed countries [58]. Nevertheless, the viability of ITS in developing countries remains contentious due to a lack of financial resources, basic infrastructure, and institutional capacity. Substantial funding is required in order to implement high-level ITS at a large scale. The indifferent attitude of many local professionals and a lack of user trust in new technologies have also undermined wide-scale ITS acceptance [58,60]. Another concern is that the adoption of "hi-tech" transport solutions in developing cities might eliminate low-skilled jobs, which are desperately needed by the population. In view of these experiences and concerns, ITS needs to be introduced cautiously in smaller developing cities.

2.6. Awareness-Raising Campaigns

Developed and developing countries have used information, education, persuasion, and awareness-raising campaigns in favor of sustainable urban transport with various, but generally limited, degrees of success. Typically, the more effective a measure is, the more resistance it evokes [61]. Social mechanisms and processes, such as status seeking (*i.e.*, the automobile as a status symbol), freedom seeking, or lack of trust in others' cooperativeness, are often at play, especially in the developing world, and perpetuate urban transport problems [49]. Moreover, the publicity generated by the car industry is well ahead of sustainable urban transport promotion. In the collective consciousness, private motorized vehicles have been long associated with pleasure, comfort, speed, convenience, power, protection, superiority, individuality, hedonism, and freedom [62].

Several general strategies to raise awareness on sustainable transport policy can be employed (Box 4). A few specific techniques seeking to reduce car use settings have also been developed and tested but generally only in developed countries. These techniques involve the provision of tailored feedback to individuals and households about travel patterns, costs, and the alternatives to current behavior [63,64]. The main aim of these exercises has been to motivate people to consider the consequences of their travel behavior. While the pilots have yielded promising results (up to 10% reduction in car travel), their effects cannot be fully generalized yet. Although cost-effective, they often require large upfront investments [63].

Box 4. General strategies to raise awareness on sustainable transport (based on [49]).

- provision of information, education, communication about risk generation, types and levels of risk resulting from one's transport choices, others' perceptions and intentions, and risk reduction strategies
- social modeling and support, *i.e.*, demonstrating cooperative behavior and the efficacy of others
- changing values and morality (*i.e.*, appealing to conscience, enhancing "altruism" towards others and future generations, and reducing selfishness)

In developing countries, lower-cost activities are likely to be more feasible and constructive (e.g., car-free days, bicycle-to-work days, free-vehicle-inspection days, bicycle film festivals, car-pooling days, free-public-transport days, and media attention). Effective public awareness activities require novel approaches to capture the audience's attention. Also, campaigns must advance specific ideas (e.g., the creation of cycle tracks) rather than vague notions (e.g., transport sustainability in general). Experience suggests that public awareness campaigns need to be targeted and "integrated" (presenting all the urgent urban transport concerns as interconnected and interdependent). Parking policy must be correctly framed as not merely a public order issue but as a crucial tool in restricting demand for car travel and in raising revenues. Public awareness activities must encourage shifts in existing paradigms. For example, bicycles must be presented as the healthy vehicles of the future (*i.e.*, a new status symbol rather than a vehicle for the poor); cars as imposers of high costs on the community (*i.e.*, an antisocial mode); buses as modern and comfortable (*i.e.*, a choice mode rather than a mode of last resort); walkways as a measure of democratization (*i.e.*, pedestrians as part of the transport system).

It is important to select relevant and context-sensitive advocacy activities as locals might not be able to relate to best practice examples from developed cities. Campaign organizers must project an image of professionalism, expertize, vision, creativity, and persistence. Messages sent by groups which appear amateurish are often ignored. Whether awareness campaigns are organized by government and/or civil society, the support of charismatic political leaders is often crucial [65]. Because many urban transport problems are perpetrated and perpetuated by people with higher education, sustainability concepts must be included in university teaching curricula. Successful courses in sustainability engage participants' heads (cognitive domain; academic study and understanding of transport sustainability), hands (psychomotor domain; enactment of theoretical learning through practicing transport sustainability in real life), and hearts (affective domain; enablement of values and attitudes to be translated into travel behavior) [66].

2.7. Pricing Mechanisms

Even in contexts where drivers are well aware of the adverse impacts of car driving in urban areas, the choice of mode is distorted in favor of road transport, particularly private cars, if drivers are not charged the full costs of motorization [27]. The availability of free or underpriced parking also fuels car ownership and use and increases "search-for-parking" traffic [67]. Keeping fuel prices artificially low-through price control, export or quantity restrictions, or political pressure put on oil companies-produces another set of adverse effects, particularly in developing countries. These include flourishing black markets, smuggling, fuel adulteration, illegal diversion of subsidy funds, large financial losses suffered by fuel suppliers, deteriorating refining and other infrastructure, and acute fuel shortages causing economy-wide damage [68].

The idea of financially penalizing drivers by using coercive pricing mechanisms (Box 5) has long been proposed by transport economists as an effective mechanism to contain car use in urban areas. Notwithstanding their theoretical value, all real-world pricing schemes have limitations and there is no guarantee that their benefits will exceed their setup and operating costs. For example, while fuel taxes are administratively simple and discourage the utilization of vehicles in the short-run, in the long run they alter consumers' purchasing behavior, thereby causing them to switch to more fuel-efficient methods (e.g., smaller vehicles, which ultimately do little to alleviate congestion and can increase safety risk). Even in the short-term, most studies find that fuel taxes lead to welfare loss among lower-income drivers, who lack alternative travel options (*i.e.*, in a city where the public transport system is weak).

Box 5. Examples of coercive pricing mechanisms.

- fuel taxes (based on emissions)
- vehicle import/purchase/registration taxes (based on emissions)
- experience rating of car insurance premiums; tradable mobility/emission credits/quotas
- direct road charges (urban drivers are charged by distance travelled based on information collected through electronic plates installed in cars)
- cordon area pricing (charges apply for the right to access or circulate within limited geographical areas, generally city centers, with some degree of time differentiation)
- parking fees (generally paid hourly in city centers and monthly in residential neighborhoods).

Cordon pricing and direct road charges, which outside the developed world have only seen application in Singapore, employ cameras or other electronic devices that observe the license plates of vehicles entering or moving within the cordon, and charge the driver remotely. While cordon pricing has positive impacts on peak-congestion delays, air pollution, and accidents in city centers, it often has few effects on the overall amount of commuter traffic [69].

While megacities with large numbers of private vehicles and severe congestion problems may prefer congestion charges, smaller developing cities might consider fuel taxes. Generally, in developing countries and cities with low administrative capacities, instruments with smaller or no monitoring costs (e.g., fuel taxes and emission-based vehicle taxes) are more effective than those requiring large monitoring or administrative and compliance costs. No single policy fits all conditions. The policy options presented in this article can be enacted at the local, regional, or national level, depending on the governance arrangements that are already in place [70].

The major barriers to charging drivers the full cost of car use are often related to public acceptability and political feasibility. Even the most sophisticated, equitable, efficient, and sensitive policy designs will create losers and generate opposition. This necessitates approaches that can increase public acceptance while maintaining efficiency. For example, earmarking revenues for local public transportation, and possibly for infrastructure for pedestrians and bicyclists and public space improvements, has public support (more so than placing revenues in general funds). These revenues can then be used to reduce other taxes or could be returned uniformly to adult residents or owners of registered vehicles in a designated area. Overall, studies indicate that attitudes toward road pricing improve after programs are implemented–once residents experience their benefits [61,67,69].

Pricing mechanisms also include subsidies for public transport fares (e.g., limited to vulnerable groups), tax subsidies (or exemptions) for the purchase of clean vehicles, and incentives for scrapping old vehicles. They have all been implemented in various developing cities but in a limited way due to their cost.

2.8. Vehicle Access Restrictions

Pricing mechanisms are generally considered to be more effective than regulatory approaches because they offer car users more choice, raise revenues, and can be adjusted according to different conditions [69]. However, blanket command-and-control policies have a role to play as well. Laws and regulations related to driving include limits on car use based on certain criteria, such as emission levels, noise levels, vehicle weight, fuel consumption, occupancy (*i.e.*, bans of single-occupancy vehicles), days of the week, time of the day, area (usually a city center), and license plate number (in pollution-emergency days or permanently), and quotas for distance travelled or number of motorized trips within a given urban area. Other regulatory options include parking restrictions and speed limits.

Several developing cities have experimented with selective car rationing or banning, often reaching for higher achievement than developed cities. These measures are considered politically easier to implement than pricing mechanisms because of the perception that all sections of the population are treated equally [71]. For example, Bangkok made efforts to restrict all newly registered cars to use exclusively in non-rush hours. In Guangzhou only locally registered motorcycles are allowed to circulate. Many other Chinese cities have limited the operation of commercial vehicles in unprecedentedly detailed ways (in terms of days, hours, and localities). During the 2008 Olympics, the City of Beijing imposed a temporary restriction on car owners based on license plate numbers. In Latin America, larger cities including Mexico City, Santiago, São Paulo, and Bogotá have attempted the same approach for some time. While there are no reported examples of large-scale car restrictions implemented in smaller developing cities, many examples of small-scale pedestrianization schemes in historic or commercial centers can be found, especially in some parts of Latin America and Asia.

In the past, some vehicle restriction measures have had unintended consequences. In Mexico City, for example, a vehicle restriction backfired when more than one fifth of the households (the higher income ones) purchased additional cars with alternating plates (usually cheaper, older, and more polluting) in order to circumvent the restriction. In Santiago, the car ban schedule is changed every few months to prevent this possibility, while in Bogotá, the high price of used cars prevented the problem from arising. While traffic restrictions have received public support in Latin America, they have faced opposition from the auto industry and vehicle owners. To be successful, these types of command-and-control measures must be reinforced by other complementary transport policies and promotional measures [71]. Some types of car restrictions, such as speed limits, are not effective without the traffic law enforcement resources to ensure that limits are followed [72].

An indirect way to alleviate peak-hour congestion through regulations is to mandate employers to implement telecommuting, flexible work, and staggered work shift programs, so that employees shift their commute at different times of the day. Reviews that recount the experience of developing cities with work-related policies and their impact on urban travel are still to be assembled. In developed countries, the evidence suggests that, without supportive policies, telecommuting is unlikely to be enough to affect employee commuting patterns [73].

2.9. Control of Land Uses

Generally, public transport and non-motorized modes require high densities and mixed uses in order to be practically and financially feasible. Compact urban development is also often associated with shorter distances and lower use of motorized transport. Therefore, land-use controls have important implications for travel behavior. In smaller cities in particular, the manipulation of urban form (shape, size, density, compactness, intensification, decentralization, land-use type and mix, building layout and type, and green and open spaces) can help to overcome city problems [74]. However, there are many complexities in the relationship between transport and land-use. The desirable degree of compaction of existing settlements is far from clearly understood. Moreover, a wide range of variation in terms of urban form, density, governance, economy, zoning controls, and enforcement capacity, exists in the developing world. Sustainability choices depend on local characteristics, which are briefly reviewed here.

Developing cities often have higher densities than developed cities, especially in the urban core (the densest urban areas in the world are found in developing countries—e.g., Mumbai and Hong Kong). This is due to higher urbanization rates, smaller dwellings, a prevalence of high-rise housing, later advent of the automobile, and lax regulations that allow land and housing subdivisions. However, there is disparity among regions. While Asian cities are often extremely dense, cities in sub-Saharan Africa are some of the world's most spread out, with large squatter settlements. In contrast to the "planned sprawl" of developed cities, sprawl here is mostly unplanned and poses a different set of challenges. Densities are high in North African cities but low in Latin American ones. Many cities are dual, with dense inner areas and peripheral sprawl. Climate variations, as well as cultural factors, play a role in the level of acceptable space consumption and proximity [74]. Evidence to date suggests that there is no single sustainable urban form but uncontrolled low-density sprawl is never the best option.

Some developing cities have strong economies which enable transport investments and land use control, while other are poor and/or have a laissez-faire approach to development and are dominated by the informal sector. However, in most developing-city contexts, land use intensification often occurs in the absence of land use controls. Strict enforcement to avoid sprawl and high public investment to purchase land or development rights (e.g., for the creation of urban green belts, green corridors, or ecological reserve areas) appear to be unlikely here given a past history of loose or patchy enforcement.

In contrast with de-industrialized European cities, the potential for conversion of derelict urban land (brownfield sites) is limited in many developing cities where industrialization has been minimal. In developing cities which are built in naturally hazardous areas (floodplains, seismically active zones, foot of volcanoes, *etc.*), densification has implications for disaster mitigation and management. The densification is a very contentious issue in both overcrowded inner-city shanty towns and low-density peri-urban squatter settlements with large plot sizes but with maximum lot coverage [74].

In terms of functional mix, developing cities are generally characterized by high levels of mixed use, as well as vitality and vibrancy (*i.e.*, the "urban village" notion). This is due to a limited penetration of modernist concepts and loose land use controls rather than specific policies. However, in higher income developing cities, shopping centers, strip malls, wholesale supermarkets, and sometimes high-tech industry clusters, have started to appear in the peripheries, causing traffic congestion. Interest in linear city models, often at a very large scale, is high in Southeast Asia and South Africa. Attempts to manipulate urban form (e.g., towards a classic radio-circular form for smaller cities and a polycentric "concentrated deconcentration" form for larger cities) have been limited due to the shift from comprehensive planning to piecemeal project-based and strategic planning, which has also occurred in developed cities [74].

Given the great diversity of land use approaches, preferences, and constraints in developing cities, it is difficult to come up with a set of recommendations that can apply to all. Experience to date suggests that some degree of success can be achieved if a pragmatic rather than idealistic approach is taken. If an overall dense and compact development cannot be achieved or if densification is not desirable in a given context (*i.e.*, already hyper-dense inner city areas), densification and intensification of land uses can be encouraged around transport nodes and along transport corridors (the transit-oriented development or TOD model at a regional scale) in order to increase access for larger portions of the population. TOD has been successful in a variety of settings, including some developing cities [75]. Compaction efforts can be concentrated on the development of new neighborhoods rather than on modifying existing ones, although this approach could lead to urban sprawl and consequently, the degradation of inner cities. Incentive schemes involving land-sharing arrangements, the transfer of development rights, and public/private partnerships, show some promise. In some cases, small yet significant interventions, either through planning discourse or symbolic development on the ground, can help change public perceptions of sustainable urban form [74].

3. Conclusions

Smaller and medium-sized developing cities, especially ones which are dense and compact, have great potential to develop sustainable transport systems. Low-cost investments and the imposition of modest fees on road users can deliver substantial environmental and lifestyle benefits for these cities. However, no single type of strategy or policy is effective or sufficient to promote more sustainable urban transport. Moreover, different types of measures may be more appropriate for smaller and medium-sized developing cities than megacities (e.g., fuel taxes rather than congestion charges).

Some of the key strategies to be considered in these developing cities include: (1) street conditions conducive to green modes via low-cost interventions such as sidewalk maintenance and speed restraint; (2) pedestrian-only zones in areas with heavy pedestrian traffic; (3) exclusive lanes for busses and bicycles, which are adequately protected from car traffic; (4) reasonable parking fees; (5) more attention to road infrastructure maintenance rather than the construction of new infrastructure; and (6) awareness-raising and education campaigns.

When considering investments in public transportation in medium-sized developing cities, a key priority should be to improve existing bus systems. BRT is more affordable and cost-effective in these cities than many other types of public transportation systems, including LRT. The high capital and operation cost of metros makes them less economically viable in medium-sized developing cities than in megacities. Promoting more sustainable patterns of urban development is also crucial for reducing the environmental impacts of cities but the appropriateness of different forms of development is context-dependent. Uncontrolled low-density sprawl is, however, rarely appropriate. Technological improvements can help to address urban environmental problems but they cannot address all transport-related problems. Moreover, the benefits of technological advances may be offset by rapid transport growth in developing cities. However, inexpensive technologies such as new mobility services via cell phones (*i.e.*, on-demand transport or parking payments) already exist in developing countries and could be utilized more extensively to promote new innovative forms of urban transport services. The social, economic, and environmental impacts of large-scale alternative fuel adoption in developing cities are often uncertain.

There is increasing recognition that combinations (or packages) of measures are necessary [5,76]. Certain combinations of policies can work together and give rise to synergies, leading to impacts greater than the sum of their individual parts. The identification of policy packages is a crucial issue for promoting more sustainable urban transport: packages should maximize potential synergies. It is crucial to consider local factors such as costs, feasibility, and barriers. Finally, caution is advised both in terms of the appropriateness and effectiveness of policy solutions being transferred to smaller and medium-sized cities in developing countries from larger cities and/or from more developed countries.

Author Contributions: Dorina Pojani performed 80% of the work for this article (including research design, analysis, and writing) and Dominic Stead performed 20%. Both authors have read and approved the final manuscript.

Conflicts of Interest: The authors declare no conflict of interest.

References

1. European Commission. Green Paper: Towards a New Culture for Urban Mobility. Available online: http://eur-lex.europa.eu/legal-content/EN/TXT/PDF/?uri=CELEX:52007DC0551&from=EN (accessed on 23 April 2015).
2. Newman, P.; Kenworthy, J. *Sustainability and Cities: Overcoming Automobile Dependence*; Island Press: Washington, DC, USA, 1999.
3. OECD. *Green Growth in Cities*; Organisation for Economic Co-Operation and Development: Paris, France, 2013; Available online: http://www.oecd-ilibrary.org/docserver/download/0413051e.pdf?expires=1429755160&id=id&accname=ocid177546&checksum=AFEF3E4F87F7D859F4F181885F10EF40 (accessed on 23 April 2015).
4. UNECE. Climate Neutral Cities. How to Make Cities less Energy and Carbon Intensive and more Resilient to Climatic Challenges. Available online: http://www.unece.org/fileadmin/DAM/hlm/documents/Publications/climate.neutral.cities_e.pdf (accessed on 23 April 2015).
5. OECD. *Policy Instruments for Achieving Environmentally Sustainable Transport*; Organisation for Economic Co-Operation and Development: Paris, France, 2002.
6. United Nations. *World Urbanization Prospects*; United Nations: New York, NY, USA, 2011.
7. United Nations. World Urbanization Prospects. Available online: http://esa.un.org/unpd/wup/ (accessed on 28 February 2015).
8. World Bank. Country and Lending Groups. Available online: http://data.worldbank.org/about/country-and-lending-groups (accessed on 23 April 2015).
9. Cohen, B. Urbanization in Developing Countries: Current Trends, Future Projections, and Key Challenges for Sustainability. *Technol. Soc.* **2006**, *28*, 63–80. [CrossRef]
10. Acharya, S. Motorization and Urban Mobility in Developing Countries: Exploring Policy Options through Dynamic Simulation. *J. East. Asia Soc. Transp. Stud.* **2005**, *6*, 4113–4128.
11. Union Internationale des Transports Publics. *Mobility in Cities Database*; UIPT: Brussels, Belgium, 2006.
12. Marsden, G.; Stead, D. Policy Transfer and Learning in the Field of Transport: A Review of Concepts and Evidence. *Transp. Policy* **2011**, *18*, 492–500. [CrossRef]
13. Pojani, D.; Stead, D. Going Dutch? The Export of Sustainable Land-Use and Transport Planning Concepts from The Netherlands. *Urban Stud.* **2014**. [CrossRef]
14. Stead, D.; de Jong, M.; Reinholde, I. Urban Transport Policy Transfer in Central and Eastern Europe. *disP Plan. Rev.* **2008**, *44*, 62–73. [CrossRef]
15. Downs, A. *Stuck in Traffic*; The Brookings Institution: Washington, DC, USA, 1992.
16. Cervero, R. Induced Travel Demand: Research Design, Empirical Evidence, and Normative Policies. *J. Plan. Lit.* **2002**, *17*, 3–20. [CrossRef]
17. Litman, T. Generated Traffic and Induced Travel: Implications for Transport Planning. *ITE J.* **2001**, *71*, 38–47.
18. Noland, R.; Lem, L. A Review of the Evidence for Induced Travel and Changes in Transportation and Environmental Policy in the US and the UK. *Transp. Res. Part D* **2002**, *7*, 1–26. [CrossRef]
19. Banister, D.; Pucher, J.; Lee-Gosselin, M. Making Sustainable Transport Politically and Publicly Acceptable. In *Institutions and Sustainable Transport: Regulatory Reform in Advanced Economies*; Rietveld, P., Stough, R., Eds.; Edward Elgar Publishing: Cheltenham, UK, 2007; pp. 17–50.
20. World Bank. *Road Deterioration in Developing Countries: Causes and Remedies*; Policy Study; World Bank: Washington, DC, USA, 1988.
21. Devarajan, S.; Swaroop, V.; Zou, H. The Composition of Public Expenditure and Economic Growth. *J. Monet. Econom.* **1996**, *37*, 313–344. [CrossRef]
22. Rioja, F. Filling Potholes: Macroeconomic Effects of Maintenance *vs.* New Investments in Public Infrastructure. *J. Public Econom.* **2003**, *87*, 2281–2304. [CrossRef]

23. Hook, W. Role of Nonmotorized Transportation and Public Transport in Japan's Economic Success. *Transp. Res. Record* **1994**, *1441*, 108–115.
24. Haghshenas, H.; Vaziri, M. Urban Sustainable Transportation Indicators for Global Comparison. *Ecol. Indic.* **2012**, *15*, 115–121. [CrossRef]
25. Kühn, F. Bus Rapid or Light Rail Transit for Intermediate Cities? Available online: http://www.codatu.org/wp-content/uploads/Bus-rapid-or-light-rail-transit-for-intermediate-cities-F.-KUHN.pdf (accessed on 13 June 2015).
26. Wright, L.; Hook, W. *Bus Rapid Transit Planning Guide*; Institute for Transportation and Development Policy: New York, NY, USA, 2007.
27. Gwilliam, K. *Cities on the Move: A World Bank Urban Strategy Review*; The World Bank: Washington, DC, USA, 2002.
28. Ben-Akiva, M.; Morikawa, T. Comparing Ridership Attraction of Rail and Bus. *Transp. Policy* **2002**, *9*, 107–116. [CrossRef]
29. UN-Habitat. *Provision of Travelway Space for Urban Public Transport in Developing Countries*; United Nation Centre for Human Settlements (Habitat): Nairobi, Kenya, 1993.
30. Flyvbjerg, B.; Bruzelius, N.; van Wee, B. Comparison of Capital Costs per Route-Kilometre in Urban Rail. *Eur. J. Transp. Infrastruct. Res.* **2008**, *8*, 17–30.
31. Bergerhoff, J.; Perschon, J. The Role of Ropeways to Reshape Urban Mobility in Developing Countries. *Journeys* **2013**, *10*, 13–21.
32. Cervero, R.; Golub, A. Informal Transport: A Global Perspective. *Transp. Policy* **2007**, *14*, 445–457. [CrossRef]
33. Gilbert, A. Bus Rapid Transit: Is Transmilenio a Miracle Cure? *Transp. Rev.* **2008**, *28*, 439–467. [CrossRef]
34. Sohail, M.; Maunder, D.; Cavill, S. Effective Regulation for Sustainable Public Transport in Developing Countries. *Transp. Policy* **2006**, *13*, 177–190. [CrossRef]
35. Hensher, D.; Golob, T. Bus Rapid Transit Systems: A Comparative Assessment. *Transportation* **2008**, *35*, 501–518. [CrossRef]
36. Hidalgo, D.; Graftieaux, P. Bus Rapid Transit Systems in Latin America and Asia: Results and Difficulties in 11 Cities. *Transp. Res. Record* **2008**, *2072*, 77–88. [CrossRef]
37. Hidalgo, D.; Gutiérrez, L. BRT and BHLS around the World: Explosive Growth, Large Positive Impacts and Many Issues Outstanding. *Res. Transp. Econom.* **2013**, *39*, 8–13. [CrossRef]
38. Dimitriou, H.; Gakenheimer, R. (Eds.) *Urban Transport in the Developing World: A Handbook of Policy and Practice*; Edward Elgar Publishing: London, UK, 2011.
39. Jacobsen, P. Safety in Numbers: More Walkers and Bicyclists, Safer Walking and Bicycling. *Inj. Prev.* **2003**, *9*, 205–209. [CrossRef] [PubMed]
40. Vasconcellos, E. *Urban Transport, Environment, and Equity: The Case for Developing Countries*, 2nd ed.; Earthscan: New York, NY, USA, 2013.
41. Pucher, J.; Dijkstra, L. Promoting Safe Walking and Cycling to Improve Public Health: Lessons From The Netherlands and Germany. *Am. J. Public Health* **2003**, *93*, 1509–1516. [CrossRef] [PubMed]
42. Pucher, J.; Buehler, R. Making Cycling Irresistible: Lessons from The Netherlands, Denmark and Germany. *Transp. Rev.* **2008**, *28*, 495–528. [CrossRef]
43. Weinert, J.; Ma, C.; Cherry, C. The Transition to Electric Bikes in China: History and Key Reasons for Rapid Growth. *Transportation* **2007**, *34*, 301–318. [CrossRef]
44. Hoehner, C.; Soares, J.; Perez, D.; Ribeiro, I.; Joshu, C.; Pratt, M.; Legetic, B.; Malta, D.; Matsudo, V.; Ramos, L.; *et al.* Physical Activity Interventions in Latin America: A Systematic Review. *Am. J. Prev. Med.* **2008**, *34*, 224–233. [CrossRef] [PubMed]
45. Midgley, P. Bicycle-Sharing Schemes: Enhancing Sustainable Mobility in Urban Areas. In *Paper for the United Nations*; Commission on Sustainable Development: New York, NY, USA, 2011.
46. Chowdhury, M.; Sadek, A. *Fundamentals of Intelligent Transportation Systems Planning*; Artech House: Norwood, MA, USA, 2003.
47. Kulmala, R. Ex-Ante Assessment of the Safety Effects of Intelligent Transport Systems. *Accid. Anal. Prev.* **2010**, *42*, 1359–1369. [CrossRef] [PubMed]
48. Wright, L.; Fulton, L. Climate Change Mitigation and Transport in Developing Nations. *Transp. Rev.* **2005**, *25*, 691–717. [CrossRef]
49. Vlek, C. Essential Psychology for Environmental Policy Making. *Int. J. Psychol.* **2000**, *35*, 153–167. [CrossRef]

50. Offer, G.; Howey, D.; Contestabile, M.; Clague, R.; Brandon, N. Comparative Analysis of Battery Electric, Hydrogen Fuel Cell and Hybrid Vehicles in a Future Sustainable Road Transport System. *Energy Policy* **2010**, *38*, 24–29. [CrossRef]
51. Yeh, S. An Empirical Analysis on the Adoption of Alternative Fuel Vehicles: The Case of Natural Gas Vehicles. *Energy Policy* **2007**, *35*, 5865–5875. [CrossRef]
52. Leurent, F.; Windish, E. Triggering the Development of Electric Mobility: A Review of Public Policies. *Eur. Transp. Res. Rev.* **2011**, *3*, 221–235. [CrossRef]
53. González-Palencia, J.; Furubayashi, T.; Nakata, T. Techno-Economic Assessment of Lightweight and Zero Emission Vehicles Deployment in the Passenger Car Fleet of Developing Countries. *Appl. Energy* **2014**, *123*, 129–142. [CrossRef]
54. Timilsina, G.; Shrestha, A. How Much Hope Should We Have for Biofuels? *Energy* **2011**, *36*, 2055–2069. [CrossRef]
55. Demirbas, A. Political, Economic and Environmental Impacts of Biofuels: A Review. *Appl. Energy* **2009**, *86*, S108–S117. [CrossRef]
56. Liaquat, A.; Kalam, A.; Masjuki, H.; Jayed, H. Potential Emissions Reduction in Road Transport Sector Using Biofuel in Developing Countries. *Atmos. Environ.* **2010**, *44*, 3869–3877. [CrossRef]
57. Sorda, G.; Banse, M.; Kemfert, C. An Overview of Biofuel Policies across the World. *Energy Policy* **2010**, *38*, 6977–6988. [CrossRef]
58. Shah, A.; Dal, L. Intelligent Transportation Systems in Transitional and Developing Countries. *IEEE Aerosp. Electron. Syst. Mag.* **2007**, *22*, 27–33. [CrossRef]
59. Sussman, J. *Perspectives on Intelligent Transportation Systems*; Springer: New York, NY, USA, 2005.
60. Shah, A.; Ahn, B. Intelligent Transportation Systems: Is it a Compatible Tool for Developing Countries? *J. Adv. Transp.* **2006**, *40*, 289–294.
61. Stead, D. Effectiveness and Acceptability of Urban Transport Policies in Europe. *Int. J. Sustain. Transp.* **2008**, *2*, 3–18. [CrossRef]
62. Diekstra, R.; Kroon, M. Cars and Behavior: Psychological Barriers to Car Restraint and Sustainable Urban Transport. In *The Greening of Urban Transport*; Tolley, R., Ed.; Wiley & Sons: Sussex, UK, 1997; pp. 147–157.
63. Brög, W.; Erl, E.; Mense, N. Individualised Marketing: Changing Travel Behaviour for a Better Environment. In *Communicating Environmentally Sustainable Transport: The Role of Soft Measures*; OECD: Paris, France, 2002; pp. 83–97.
64. Rose, G.; Ampt, E. Travel Blending: An Australian Travel Awareness Initiative. *Transp. Res. Part D* **2001**, *6*, 95–110. [CrossRef]
65. Pardo, C. *Raising Public Awareness about Sustainable Urban Transport*; GTZ: Eschborn, Germany, 2006.
66. Sipos, Y.; Battisti, B.; Grimm, K. Achieving Transformative Sustainability Learning: Engaging Head, Hands and Heart. *Int. J. Sustain. High. Educ.* **2008**, *9*, 68–86. [CrossRef]
67. Shoup, D. *The High Cost of Free Parking*; Planners Press: Chicago, IL, USA, 2005.
68. Kojima, M. *Petroleum Product Pricing and Complementary Policies Experience of 65 Developing Countries Since 2009*; Research Paper 6396; Word Bank: Washington, DC, USA, 2013.
69. Anas, A.; Lindsey, R. Reducing Urban Road Transportation Externalities: Road Pricing in Theory and in Practice. *Rev. Environ. Econom. Policy* **2011**, *5*, 66–88. [CrossRef]
70. Timilsina, G.; Dulal, H. *Fiscal Policy Instruments for Reducing Congestion and Atmospheric Emissions in the Transport Sector: A Review*; Research Paper 4652; Word Bank: Washington, DC, USA, 2008.
71. Mahendra, A. Vehicle Restrictions in Four Latin American Cities: Is Congestion Pricing Possible? *Transp. Rev.* **2008**, *28*, 105–133. [CrossRef]
72. Afukaar, F. Speed Control in Developing Countries: Issues, Challenges and Opportunities in Reducing Road Traffic Injuries. *Inj. Control Saf. Promot.* **2003**, *10*, 77–81. [CrossRef] [PubMed]
73. Cohen-Blankshtain, G.; Rotem-Mindali, O. Key Research Themes on ICT and Sustainable Urban Mobility. *Int. J. Sustain. Transp.* **2015**. [CrossRef]
74. Jenks, M.; Burgess, R. (Eds.) *Compact Cities: Sustainable Urban Forms for Developing Countries*; Spon Press: London, UK, 2000.

75. Curtis, C.; Renne, J.; Bertolini, L. (Eds.) *Transit-Oriented Development: Making it Happen*; Ashgate: Surrey, UK, 2009.
76. Banister, D.; Stead, D. Transport Policy Scenario Building. *Transp. Plan. Technol.* **2003**, *26*, 79–102.

sustainability

MDPI

Article

The Bumpy Road toward Low-Energy Urban Mobility: Case Studies from Two UK Cities

Tim Schwanen

Transport Studies Unit, School of Geography and the Environment, University of Oxford, South Parks Road, Oxford OX1 3QY, UK; tim.schwanen@ouce.ox.ac.uk; Tel.: +44-(0)-1865-285503

Academic Editors: Tan Yigitcanlar and Md. Kamruzzaman

Received: 1 April 2015; Accepted: 26 May 2015; Published: 2 June 2015

Abstract: Cities are increasingly seen as the places where innovations that can trigger a sociotechnical transition toward urban mobility are emerging and maturing. Processes such as peak car, rail renaissance and cycling boom manifest themselves particularly in cities, and success stories of cities experimenting with specific types of low-energy mobility abound in the academic literature. Nonetheless, innovation is known to be a precarious process requiring favorable circumstances. Using document analysis and in-depth interviews, this study examines the nature of low-energy innovation in the everyday mobility of people in two UK cities with favorable conditions for a transition away from fossil fuels—Brighton and Oxford. It shows that clear differences exist between the two cities in the sorts of innovation that emerge and diffuse as a result of path dependencies, local politics, and financial support from supra-local governments and agencies. While low-energy mobility currently has substantial momentum in both cities, the majority of low-carbon innovations in urban mobility are incremental rather than radical in nature, and their future is often imbued with uncertainty. The autonomy of small- and medium-sized cities as agents in bringing about transformational change toward low-energy urban mobility should not be overestimated.

Keywords: low-energy transport; carbon; energy consumption; sociotechnical transition; innovation; cities; United Kingdom

---

1. Introduction

Current transport systems for people and freight are environmentally unsustainable [1–4]. In 2011, transport was responsible for 23% of global energy consumption [5] and 22% of CO_2 emissions [6]. Both energy consumption and emissions continue to grow due to a range of mutually reinforcing factors and processes, including a near total—94% [5]—reliance on oil; population growth; economic development; urban sprawl; the globalization of cultural ideas that tie car use and ownership to social progress, freedom and individuality; and governments' commitment to road building and investment in other infrastructures for carbon-intensive forms of mobility, such as maritime ports and airports [4,7].

Greenhouse gas (GHG) emissions from transport need to be reduced drastically, and the IPCC unambiguously calls for "aggressive and sustained" [4] (p. 602) measures yet also appreciates the monumentality of the challenges ahead: the growth of transport volumes in non-OECD countries and of intercontinental movements of people and goods risks cancelling out the benefits from technological advances and behavior change occurring across parts of Europe, North America and Australasia. In those regions, the growth of car use and ownership may have peaked [8,9] and a rail renaissance, both within cities (metro, light rail) and between cities (high speed rail) [10–12], and cycling boom [13,14] can be witnessed. These developments seem to take place first and foremost in urban areas. Indeed, across the global North, cities lead the way in moving toward low-energy mobility—here used as shorthand for types of everyday mobility that consume less fossil fuels and emit smaller quantities of GHGs than do conventional internal combustion personal vehicles—for various reasons [15–17]: higher population densities make public transport, cycling and walking more practical and attractive compared to car use and ownership; populations are on balance relatively young, highly educated, environmentally conscious and willing to experiment with—or at least support—various forms of low-energy living; and governments are more likely to have the political, institutional and financial capacities to support low-energy mobility initiatives and experiments than are their suburban and rural counterparts.

Whilst encouraging, the currently observable gradual shifts across the global North are unlikely to be sufficient if transport's contribution to anthropogenic climate change is to be minimized. Across academics from various disciplinary backgrounds a consensus appears to be emerging around two key ideas. The first of these revolves around the need for systemic change, whereby prevailing systems of mobility—*i.e.*, not only (vehicle) technologies, physical infrastructures and user practices, but also the cultural meanings associated with various forms of movement, markets, maintenance and repair, forms of regulation and policy making, and formal expertise about transport [18]—become durably reconfigured. The second is that such change can be triggered by the diffusion of one or more low-energy innovations—new technologies, institutional arrangements or user practices that differ to greater or lesser extent from prevailing mobility systems; examples include IT-supported mobility services (e.g., car/bike sharing), bio-fueled buses and smartcard-based integrated ticketing systems. In accordance with these views, a literature on innovations in transport is emerging [19,20], most of which is informed by the theoretical perspectives of strategic niche management (SNM) [21–23] and the multi-level perspective (MLP) [18,20,24,25], although a range of studies have also drawn on cultural approaches like practice theory [26–28] and the energy cultures framework [29,30].

The theoretical lens of SNM/MLP is particularly useful because of its comprehensive system orientation but also tends to privilege the temporal over the spatial dimensions of systemic change. Whilst offering useful conceptualizations of the moments at which innovations can break through and the temporalities involved in systemic change, it has less to say on where such processes will happen, why there, and which geographical circumstances will facilitate or complicate them (see also [31–39]). This can be considered a shortcoming given the urban character of the processes of peak car, rail renaissance and cycle boom across the global North.

The current paper seeks to contribute to a better understanding of the geographical dimensions of innovations in transport by focusing on the everyday mobility of people and adopting a comparative case study approach. It considers the UK cities of Brighton and Oxford as what Flyvbjerg [40] calls "critical cases" that enable conclusions which may also apply to other cities with less favorable conditions for systemic change. The UK is an appropriate setting because it was the first country to adopt legally binding CO_2 emission reduction targets, which are linked to the 2008 Climate Change Act, and has allocated £738.5 m to the Local Sustainable Transport Fund (2011–2016), which funds a program of locally designed and delivered policy packages aimed at stimulating sustainable forms of passenger transport [41]. Within this national context, Brighton and Oxford are two medium-sized cities with strong reputations in sustainable transport: bus usage is amongst the highest in the UK, and whilst Oxford is widely known as a city of cyclists, Brighton—used henceforth as synonym for the city's official name of Brighton and Hove—was among the first to adopt comprehensive cycling promotion policies in the UK. Both cities can be expected to be ahead of the curve in terms of transitioning to low-energy mobility systems and offer potentially fertile grounds for the emergence and flourishing of low-energy innovation in urban mobility. This makes these cities useful sites to examine which local processes and circumstances facilitate, complicate or obstruct systemic transitions toward low-energy urban mobility.

2. Sociotechnical Transitions and Innovation

Theoretically, the starting point for this paper is the multi-level perspective (MLP), which holds that systemic change requires processes at the sociotechnical landscape, regime and niche levels to somehow coincide and become interlinked [18,20,24,25]. A sociotechnical regime is a set of rules—cognitive routines, shared beliefs, social norms and conventions, regulations, industry standards, protocols, contracts, laws and so forth—that fulfill a societal function (e.g., everyday mobility) and thereby condition the practices through which the technology, infrastructure, markets, cultural values, user practices, maintenance and repair, regulation and formal knowledge that make up socio-technical systems are reproduced [18]. Innovation and experimentation occur within a regime but are normally incremental: they tend not to challenge the basic architecture of, and power relations undergirding, the rules associated with a prevailing sociotechnical system, but rather seek to minimize an existing system's negative externalities. In a transport context they are best exemplified by end-of-pipe innovations [3,42], such as fuel economy improvements in internal combustion personal vehicles. Such innovations are typically led by incumbent regime actors.

In contrast, innovations in sociotechnical niches are more radical: they often involve new, non-incumbent actors and potentially challenge—or at least reconfigure—the rules of prevailing systems. This does not mean that regime actors are uninterested or uninvolved; they may participate to learn about new developments and future competition, but their commitment to a given niche's development may ultimately be restricted [43]. The sociotechnical landscape is to some extent a residual category [44]; it gathers all wider contexts and developments over which regime actors have little influence, from anthropogenic climate change to economic crises, demographic shifts and the rise of the Information Age, as well as the transition from welfare to workfare [45] according to which all adults in advanced liberal democracies should, in principle, be in paid employment (and which becomes easier if they can afford to own a private car). According to the MLP, landscape developments can put pressure on, and eventually destabilize, existing regimes, opening up opportunities for innovations in niches to diffuse and reconfigure regimes [25].

The number of studies using the MLP in a passenger transport context is increasing rapidly (e.g., [19,20,39,46–48]), although most focus on specific niches, such as electric vehicles [43,48,49]. Nonetheless, at least two studies have discussed a broader range of niches in contemporary passenger mobility. Based on research in the UK and Sweden, Nykvist and Whitmarsh [19] identify three types: radical vehicle technologies (e.g., battery and hybrid electric vehicles), mobility services (e.g., car/bike sharing, public transport) and mobility management programs seeking to reduce vehicular travel (e.g., promotion of active travel or substitution of physical travel by ICT—telecommuting, e-commerce, *etcetera*). Geels [20] has expanded this classification into six categories, two of which—green propulsion technologies and demand management—are closely aligned with Nykvist and Whitmarsh's radical vehicle technologies and mobility management, respectively. However, the other four categories refine and extend those by Nykvist and Whitmarsh:

- IT-based innovations not only encompass telecommuting programs and the like, but also intelligent transport systems seeking to increase transport's efficiency through real-time information provision and smartcard-based integrated ticketing;
- Public transport innovations, such as light rail systems, and improved information provision and ticketing systems;
- Intermodal travel programs that seek to integrate different systems such as car and bus travel (Park and Ride) or car and public transport with public bike sharing schemes; and
- Cultural and socio-spatial niches that challenge the normality of private car use through land use developments (e.g., compact cities, smart growth, Transit Oriented Development (TOD), Complete/Livable Streets and home zones), or collective/commercial ownership (car/bike sharing).

However useful these classifications of niches may be, they are complicated by three sets of issues. One is that regimes are non-singular in transport [20], especially in cities and more so in continental Europe than in the UK and certainly the USA, Canada and Australia. In many Danish, Dutch and German cities, for instance, there is now a decade-long tradition and sometimes dominance of the provision, regulation and consumption of multiple forms of transport deemed environmentally sustainable—rail, bus, also cycling—which complicate the distinction between regime and niche [13,15]. Many cycling-related, TOD or intermodal transport innovations cannot be considered radical in those cities, although they might be labeled as such in other geographical contexts [3]. Moreover, what counts as an incremental or radical innovation depends on how place-specific regimes interact with each other (see also [50]). In geographical contexts where mobility systems configured around car, public transport and cycling compete with each other—or are encouraged to do so because of wider, modally organized institutional set-ups—, publicly financed bike sharing and/or commercial car sharing schemes may be seen as radical innovations and often face an up-hill struggle to assert themselves. Yet, in places where multiple regimes are more symbiotic than competitive, publicly funded bike sharing and/or commercial car sharing schemes are more incremental innovations, enabled by incumbent actors seeking to exploit synergies.

Additionally, the dynamics of niche innovation cannot be reduced to merely replacing or reconfiguring one or more regimes because of their purportedly unsustainable character. Many niches are not only about creating and expanding social networks, creating shared expectations and visions, and collective learning about technologies or regulation, as much of the strategic niche management literature would have it [22]. They are often political, in one of two senses. Niche innovations can be grassroots innovations [51,52] or urban experiments [35] whose very point is to challenge the ways in which regime actors—and others closely aligned to them—go about addressing the climate change challenge. In the urban mobility context these innovations may be exemplified by community-led initiatives configured around shared cars [53] or focused on cycling. Niche innovations and experiments can also be politically in a strategic sense. This is because they are being framed and used by political and economic elites to enhance the global competitiveness of cities and city-regions in a (post)neoliberal world. This, for instance, was Hodson and Marvin's [31] interpretation of the attempts to create a hydrogen economy in London under Mayor Ken Livingstone in the 1990s and the subsequent demonstration projects with hydrogen-fueled buses.

Thirdly, there is a tendency in the recent transport-oriented literature to focus on specific niches and to do so in isolation from other innovations. As in the MLP literature more widely, the emphasis is often on new technologies, such as electric vehicles (EVs), or innovations in which new technologies play a significant role, such as most recent bike and car sharing schemes. Whilst the latter two can also be seen as socio-cultural innovations insofar as they promote a shift away from individual ownership, there has not been much attention for innovations in which technology plays a very limited role, such as walk/cycle-to-school initiatives or cycling cafes and hubs for cycling training, maintenance and/or storage. Such innovations can nonetheless be of critical importance to low-energy transitions, because they directly seek to reconfigure elements of sociotechnical systems other than technologies and transport infrastructures as conventionally understood, such as cultural values, user practices and maintenance and repair. As protagonists of cultural approaches [26–30] would rightfully argue, those innovations can play an important role in changing the meanings associated with particular ways of moving around the city, enhancing users' skills and confidence and ultimately normalizing low-energy mobility in the city. Moreover, most studies so far have looked at individual or at most several innovations rather than on larger numbers as Nykvist and Whitmarsh [19] and Geels [20] have done. This focus on single or at most several innovations is problematic insofar that innovations may reinforce one another (symbiosis) or rather out-compete each other, and such relations may be place-specific. Moreover, innovations may be "entwined becomings", meaning that one could not obtain its distinctiveness without the realization and evolution of others. Consider, for instance, a smartcard-based integrated ticketing system facilitating not merely the transfer between rail and bus, but also the storage of privately owned bikes at the train station of origin and the use of a shared car at the destination side to overcome the first and last mile challenges, respectively. Focusing on many initiatives and their interdependencies can thus offer a more systemic perspective on innovation in passenger mobility.

This paper addresses the above issues through an analysis of a large number and diverse range of innovations and their linkages in geographical contexts where multiple regimes—auto-, bus- and velomobility—compete, coexist and intermingle. It moves beyond classifications used previously [19,20] to consider not only the geographical scale but also the involvement of specific types of actors and political entanglements. It also explores how the last two dimensions are critical to the mobilization of such resources as funding.

3. Research Design

3.1. Innovations Defined

Innovation is a polysemic and contested term: it means different things to different people in different circumstances (see also [54]). It must therefore be defined carefully, and here the term will be used to denote a new socio-technical configuration. This definition has several implications. For one, it means that an innovation is *a combination of heterogeneous elements that link together and reach out across space and time*. Those elements may include technical artifacts and infrastructures (e.g., the bikes, docking stations and GPS technologies in the case of recent bike sharing schemes), cultural values and norms, standards, regulations and forms of knowledge (e.g., pricing structures, subscription systems, such ideas as "cycling is healthy", the rules used to make decisions on the location of docking stations), as well as user practices. The novelty of an innovation lies in the fact that either existing elements are put together in a novel way, or that one or more of the elements are new. The words "new" and "novel" should be seen as relative and scale dependent. A distinction can be made between genuinely new (*i.e.*, the first manifestation anywhere), and new in a particular bounded territory but possibly existing in a similar format or manner elsewhere. Borrowing from Upham and colleagues [55], we can call the former an original innovation and the latter a diffusion-based innovation. "New" is also relative to particular time scales. For passenger transport systems periods of 5–10 years seem to offer a reasonable time scale to consider something novel.

The above definition mean that at any point in time the number of innovations in urban mobility in a medium-sized city is potentially huge; any inventory or examination of innovations that can be expected to reduce energy consumption will be selective rather than exhaustive. Yet, this is not necessarily problematic, as long as those innovations that are likely to have the greatest impact on energy consumption are taken into consideration. These have been identified for the cities of Brighton and Oxford through an extensive analysis of existing documents and fieldwork (site visits, interviews). Attention has only been directed toward *place-dependent low-energy innovations* with a reasonable degree of organizational embedding in the territory associated with each city. Generic innovations that relate to the national scale—e.g., national government's attempts to subsidize the purchase of EVs or charging equipment by individuals or organizations throughout England and Wales, or peer-to-peer car rental schemes run exclusively at the national level—have been left out of consideration. It should also be appreciated that the focus is on innovations that can be *expected* to reduce GHG emissions and energy consumption per kilometer traveled; whether they do when in use is beyond the current study.

3.2. Data and Analysis

For both cities, data collection started with systematic and detailed internet searches for relevant policy documents, reports by transport providers (e.g., bus and car club operators) and other organizations (e.g., advocacy and community groups), information provided on websites, newspaper articles and blog posts. In this way information from many different types of stakeholders—incumbents, new actors and citizens—has been gathered. For both cities the focus of the search has been on materials since 2005; however, the majority of assembled documents relate to the period since 2010. This seems to reflect a combination of greater innovation activity in urban mobility, increased reporting on such innovations, and a greater inclination among organizations and (local) newspapers to make materials available online in recent years. The initial searches were conducted in the winter and summer of 2014 for Oxford and Brighton, respectively. The compiled archives have since been updated continually; at present (March 2015) 400+ documents are available for Oxford and 500+ for Brighton. Within the broader research project to which this paper belongs, a selection of the assembled documents has been subjected to in-depth textual analysis, although for this particular piece the emphasis has generally been on extracting information on the size and origins of financial budgets for particular initiatives.

The document analysis has been complemented by site visits by the author to experience specific innovations, and by in-depth interviews conducted by the author and/or co-workers. To date, 16 interviews have been conducted in Oxford and Brighton with mid-range and senior officials in local transport policymaking, local politicians, and representatives of bus/rail operators and community organizations. Interviews lasted between 30 minutes and two hours and were audio-recorded. Relevant parts have been transcribed and subjected to a largely inductive form of thematic interview analysis (*cf.* [56]).

3.3. Comparative Case Study Approach

This study adopts a case-study approach on the grounds that a specific city as case—if chosen carefully, and with appropriate caveats—can offer valuable insight in processes that exceed the city in question. This is especially so if a "critical case" with strategic importance in relation to a more general set of issues is examined [40]. However, this paper does not look at a single case-study but at two, which reflects its understanding of cities as "open, embedded and relational" [57] (p. 481). Cities are not sealed off, independent territorial units; they are articulations or nodes within wider—national and increasingly globalized—networks through which ideas, capital, objects and people circulate. Cities shape those circulations and are shaped by them; they are at once relational and territorial [58] as wider flows and dynamics are refracted through place-specific historically sedimented amalgams of social practices and customs, institutions and built environments, including transport infrastructures [59]. In this context considering two cities, rather than one, sensitizes us toward appreciating "interconnected trajectories [and] moves us away from searching for similarities and differences between two mutually exclusive contexts and instead toward relational comparisons that uses different cities to pose questions of one another" [57] (p. 480).

For Flyvbjerg [40] (p. 307) "no universal methodological principles exist by which one can with certainty identify a critical case". Hence, the decision to select Brighton and Oxford as case studies was based on what Bertilsson [60] calls an under-coded abduction. This is an "on trial" inference, or "the best, or the most economical at hand in order to solve [the] puzzle" [60] (p. 379) of identifying something—two critical cases—that can actually only be known through induction afterwards, once large numbers of materials have been assembled and scrutinized in a time-intensive process. In the current study the on trial inference was based on prior understanding of both cities':

- Transport system—both cities are known to have very high levels of bus usage, in the UK only surpassed by London;
- Land use configuration—both are compact cities with significant planning constraints on expansion: Oxford is surrounded by a green belt, Brighton by the sea and nature reserves;
- Local economy—in both cities based on IT, knowledge intensive services, higher education, high tech manufacturing and tourism; and
- Population composition—large shares of highly educated young adults and professionals, often with progressive political beliefs (Brighton and Hove is the only local authority in the UK that in 2011–2015 had a local administration led by the Green Party), in combination with

considerable pockets of social deprivation in South East Oxford and toward the east and north-east of Brighton's city center. Both sets of conditions depress car ownership and usage levels [61] and are conducive to bus travel and walking.

Nonetheless, there are also important differences between the two cities, of which three deserve particular mention. Brighton and Hove's population is almost twice as large, with 273,400 inhabitants against 150,200 in Oxford (2011 Census). Additionally, Brighton is a unitary authority with a single governmental organization—Brighton and Hove City Council (BHCC)—responsible for local transport planning, whereas local transport policy in Oxford involves both Oxfordshire County Council, which is the statutory regional transport authority, and Oxford City Council. This governmental complexity may influence local government support for innovations in urban mobility in Oxford. Finally, whilst Oxford has a long-standing reputation for high levels of cycling (within a UK context), Brighton's hilly topography is widely seen as making cycling less attractive. This is to some extent borne out by Table 1, although this also demonstrates that cycling for commuting has roughly doubled over the 2000s in Brighton.

Table 1. Modal split and increase in cycling for commuting (derived from [62,63]).

		Brighton and Hove	Oxford City
Modal split (2011)	Car/van driver	39.9%	36.2%
	Car/van passenger	3.9%	3.5%
	Train	11.3%	2.7%
	Bus	15.2%	17.8%
	Cycling	5.4%	19.1%
	Walking	22.2%	18.8%
	Other	2.1%	2.0%
Increase in the number of people cycling to work (2001–2011)		109%	33%

For ease of comparison the analysis below concentrates on low-energy innovations within the administrative boundaries of each city, the territories of BHCC and Oxford City Council.

4. Innovations in Brighton and Oxford

4.1. Caveats

In recent years place-dependent low-energy innovation activity with regard to the everyday mobility of people has been extensive in both Brighton and Oxford (Tables 2 and 3). It is even tempting to infer that both cities are transitioning toward reduced emissions and energy consumption in mobility. However, the information in Tables 2 and 3 is inevitably partial and in some ways indicative. That partiality follows from the fact that many small-scale and more or less informal initiatives by local community members (e.g., neighborhood/street-level peer-to-peer car sharing or awareness raising events, mobility management schemes by local employers) could not be included due to non-availability or paucity of information and documentation. Moreover, many minor innovations may well go unnoticed to the individuals involved, irrespective of their social role (community member, entrepreneur, policymaker, and so forth), and are therefore not documented in a publicly available format.

The contents of Tables 2 and 3 is indicative in regards to timing and budgets. As far as timings are concerned, it is usually very difficult to pinpoint the moment at which a particular innovation starts, so the emphasis in Tables 2 and 3 has been placed on the time at which initiatives became available for use, came in operation, or physical (re)construction took place. Information on available budgets in policy documents or press releases is often imprecise and based on a priori intentions; actual expenses can differ from that information in various ways. Budget overruns may occur, and based on the work of Flyvbjerg and colleagues [64,65]—even if those authors' interest was in larger projects than those considered here—their possibility should not be discarded easily. Consider one example from Brighton. Press releases and reports in the local media habitually refer to £850k or £900k as the budget for Brighton Station's Cycle Hub, a building that has recently been erected next to the train station with Dutch-style cycling parking for 500 bikes, as well as a bike repair and cycle shop, a café, changing/shower facilities and a fitness facility. However, during one of the interviews it became clear that in reality the project would cost at least £1.2m because of various contingencies during the construction phase.

Moreover, information on the budgets for particular projects in policy documents or media reports often excludes a range of actual costs, such as the labor costs of involved personnel, money spent on preparatory research (e.g., feasibility studies), and the time needed to prepare bids to funding agencies. On top of this, information on actual expenses is often not publicly available. This makes it particularly difficult to gauge the budgets and costs of initiatives that are not led by local governments; among the latter BHCC tends to disclose more information on budgets than public authorities in Oxford. In light of these considerations, the budgets in Tables 2 and 3 have been reduced to six broad categories—<£100k, £100k–<500k, £500k–<1m, £1m–<3m, £3m–<10m, and >£10m (without correction for inflation)—and should be seen as conservative estimates of actual expenses. Still, a focus on budgets is potentially more revealing of stakeholders' priorities and orientations than the rhetoric that is sometimes used in policy documents, press releases, brochures and the like.

4.2. General Observations

Apart from pointing toward extensive innovation activity in recent years in both cities, the materials on which Tables 2 and 3 have been based highlight the usefulness of thinking about innovations as ensembles of new and old elements (Section 3.1). Consider as an example Oxonbike, the public bike sharing scheme that Oxfordshire County Council has set up for commuters in East Oxford with financial support from the Local Sustainable Transport Fund (LSTF). The novelty of this scheme does most certainly not lie in the idea of sharing bikes, thinking about mobility as a service, or its framing as enhancing Oxford's competitiveness in a globalizing economy; numerous schemes in other cities in the UK, and indeed across the world, had long since exhausted the originality of this set-up by 2010. This scheme is innovative in its symbiotic links to other innovations seeking to reduce car use in East Oxford, such as the expansion of the Thornhill Park and Ride (P&R) site on the eastern fringe of the city, the improvement to the bus lane and services connecting Thornhill with key employment sites in East Oxford and the city center, and mobility management initiatives encouraging employees to change their commuting behavior. This, then, is a clear example of an incremental innovation with the involved actors exploiting synergies between regimes.

Oxonbike also highlights the relativity of innovations. New to Oxford and decidedly different from the completely commercial Brompton bike hire scheme that offers folding bikes to subscribers at the city's train station, it is a singular innovation yet also exemplary of the diffusion of an ensemble of elements tried and implemented (with varying levels of success) in numerous other cities across Europe, North America, Australasia, and increasingly in the global South. The Oxonbike example thus allows a second general observation to be made: the vast majority of innovations are diffusion-based rather than genuinely new (Section 3.1). The EV charging points, railway station design, information provision through signage posts to assist pedestrian way finding, smart ticketing for bus travel, and internet-based journey planners provide only part of the evidence to substantiate this claim. Even the aforementioned Cycle Hub at Brighton Station bears clear resemblance to the Cycling Point at Leeds Station, which opened in 2010, was also developed by a consortium of stakeholders led by a train operating company (Abellio in Leeds, Southern in Brighton) and was itself inspired by similar developments at city-center railway stations in the Netherlands. This is one reason why cities have to be conceptualized as open, embedded and relational [48]. Initiatives move between locations; while they are new to *a*, this observation is relativized once *b*, *c*, ... are brought into the frame. And yet, as the geographical literature on innovation diffusion [66,67] and more recently policy mobilities [68–70] has suggested, diffusion—and particularly effective diffusion—entails adaptation and hence novelty; it is rare that something works if it is simply transplanted from site *x* to *y*. So, whilst Upham *et al.*'s [55] distinction of original *versus* diffusion-based innovations is useful, this cannot be seen as absolute but should be considered in terms of degrees of originality or "diffusionness" (given a particular spatial and temporal scale).

Seen thus, there are some innovations in Tables 2 and 3 that qualify as relatively original, at least in the UK context. The retrofitting of Brighton's Lewes Road—the main connection between Brighton's city center and the city's two universities (Sussex, Brighton), its football stadium and the suburb of Lewes—provides useful examples because of its floating bus stops and the construction of a bus lane to separate a segregated cycle path from car travel. The idea of the latter is that the bus lane offers additional protection to cyclists because it is used much less intensely than the car lane. Regarding the floating bus stops, which are commonly observed in parts of continental Europe, a senior transport planner highlighted the originality of the design and the learning processes involved: "it's one of the first times it's been tried in the UK, we spent quite a bit of time working out how that would actually work here. It's not proven design and there was nothing you pull out of the guidance, the technical guidance saying this is how you should do that. So there had to be a slight willingness [among policymakers] for designers to go out on a bit of a limb".

Table 2. Selected place-dependent low-energy innovations in urban mobility, Brighton.

Category	Innovation	When	Spatial scope	Lead actor(s)	Budget and main source(s)
Green propulsion (*cf.* [20])	Biofueled bus services	2007	Region	nfp bus operator	Unknown—savings, shares, revenue
	EV charging infrastructure	2009/10	City	LG, fp operator	£500k–<1m—LG, CA
	Euro 5 compliant int. combustion buses	2011	Region	fp bus operator	>£10m—bus operators
	E-bike demonstration project	2011/14	Neighborhood	University, LG	£1m–<3m—LSTF, research councils
	HEV buses	2012/13	Region	fp bus operator	£1m–<3m—NG, fp bus operator
Intermodal transport (*cf.* [20])	Bus service at railway station	2011/14	Site, Axis	LG, fp bus operator	£500k–<1m—LSTF
	Railway station redesign	2012/13	Site	LG, NR	£3m–<10m—NG, NR, LSTF
	Integrated ticketing bus and car club	2013	City	fp operators	Unknown—fp operators
	Cycle Hub, Brighton railway station	2014/15	Site (facility), City (catchment area)	Train operator, NR, LG	£1m–<3m—NG, LG, NR, train operator
Public transport (*cf.* [20])	Talking bus stops for vis. impaired	2007	City	LG, fp bus operator	Unknown—fp bus operator
	Real time information at bus stops	2009	Region	LG, bus operators	£1m–<3m—LSTF, bus operators
	Eco-driving technology and training for bus drivers	2009	Region	fp bus operator	Unknown—fp bus operator
	Smartcard ticketing buses	2011	Region	fp bus operator	Unknown—fp bus operator
	Bus stop improvement	2012/13	City	LG, fp bus operator	£1m–<3m—LSTF
	Bus ticketing on smartphone	2013	Region	fp bus operator	Unknown—bus operator
Mobility services (*cf.* [19])	City Car Club	2003	Neighborhood/City	fp operator	Unknown—fp operator
	Streetcar/Zipcar	2005 ‡	Neighborhood/City	fp operator	Unknown—fp operator
	Car clubs to low-income areas	2009	Neighborhood	LG	<£100k—EU
Mobility management and support (*cf.* [19])	Cycling training at schools	2005	City	LG, Sustrans	£500k–<1m—NG, CDT, CT, LG, Sustrans
	Cycling awareness events	2005	City	LG, CO's	£500k–<1m—CDT, CT, LG
	Travel planning, door-to-door	2006	City	LG	£3m–<10m—CDT, CT, LSTF, EU, LG
	Travel planning at schools, universities and employers	2006	City	LG	£500k–<1m—CDT, CT, NG, EU, Sustrans
	Internet-based journey planner	2007	City	LG	£100k–<500k—LG, EU, LSTF
	Bike-Off cycle security project	2008/10	City	LG	<£100k—EU, LG
	Pedestrian way finding info	2011/14	City center, Axes	LG	£500k–<1m—LSTF, LG
	Circus Street Bike Hub, incl. bike maintenance and cycling training	2014	Site (facility), City (catchment area)	CO, LG	<£100k—LSTF, loans & revenues

Table 2. *Cont.*

Category	Innovation	When	Spatial scope	Lead actor(s)	Budget and main source(s)
Physical infrastructure retrofit	Cycling retrofit Grand Avenue/Drive	2005/08	Axis	LG	£500k–<1m—CDT
	Cycling retrofit Old Shoreham Rd	2012	Axis	LG	£500k–<1m—Sustrans, NG, LG
	Minor cycling retrofits (e.g.. ASLs, priority and signage)	2005 onwards	City	LG	£500k–<1m—CDT, CT, EU, LG
	Cycling retrofit other thoroughfares and cycling links/roads	2008 onwards	Axes	LG	£3m–<10m—LG, NG, CT, EU, LSTF
	Cycling parking	2009 onwards	City	LG	£100k–<500k—CT, EU, LG, train operator
	Redesign Seven Dials roundabout	2012	Site	LG	£100k–<500k—NG, LG
	Cycle/bus retrofit Lewes Road	12/14	Axis	LG	£1m–<10m—LSTF
Other	Cycling monitoring (incl. interactive cycling counter)	2005	City	LG	£100k–<500k—CDT, CT, EU, LSTF, LG
	Smart traffic signaling	2011/14	Axis	LG	£500k–<1m—LSTF
	20mph speed limit	2012/14	City	LG	–
	Controlled parking zones	2012	City	LG	–

Notes: Axis = covers a transport a specific road; C(D)T = Cycling (Demonstration) Town grant; City = coverage (most of) the city; City center = covers city center only; CO = community organization; EU = EU CiViTAS grant; (n)fp = (not) for profit; LG = local government; LSTF = Local Sustainable Transport Fund grant; Neighborhood = covers a specific neighborhood outside the city center; NG = national government; NR = Network Rail; Region = exceeds city to cover (most of) the urban region; Site = restricted to a single location; ‡ Stopped operations in Brighton in early 2010s.

Table 3. Selected place-dependent low-energy innovations in urban mobility, Oxford.

Category	Innovation	When	Spatial scope	Lead actor(s)	Budget and main source(s)
Green propulsion (*cf.* [20])	EV charging infrastructure	2009 onwards	City	LG, operator	£1m–<3m—LG, NG
	Mini E field trial	2009/11	City	CM, university, LG	£5m–<10m—CM, NG
	Euro 5 compliant int. comb. buses	2009	Region	Bus operators	£5m–<10m—bus operators
	HEV buses	2009/13	Region	Bus operators, LG	>£10m—NG, bus operators
	Retrofit buses to Euro 5 standard	2012	City centre	Bus operators	£500k–<1m—bus operators, NG
Intermodal transport (*cf.* [20])	Thornhill P&R intelligent access	2012	Site	LG	<£100k—LSTF
	Thornhill P&R expansion	2013/14	Site (facility), Region (catchment)	LG	£1m–<3m—LSTF
	Railway station redesign	2014	Site	NR, LG, train operators	>£10m—NG, LG, rail sector
Public transport (*cf.* [20])	Eco-driving technology and training for bus drivers	2008	Region	bus operator	£100k–<500k—fp bus operator
	RTI at bus stops	2009/14	City	LG, bus operators	£100k–<500k—LSTF, bus operators
	East West rail (Oxford, Bedford), incl. new station Oxford Parkway	2011	Region	NR, train operators, LG	>£10m—NG, LG, rail sector
	Smartcard ticketing buses	2011	Region	Bus operators, LG	Unknown—bus operators, LSTF
	New bus services Headington-city	2013	City	LG, bus operator	£1m–<3m—LSTF, bus operator
Mobility services (*cf.* [19])	Streetcar/zipcar	Mid-2000s	Neighborhood/City	fp operator	Unknown—fp operator
	Co-wheels/ Commonwheels	Mid-2000s	Neighborhood/City	fp operator	Unknown—fp operator
	Oxford Liftshare	2009	Region	LG, fp company	<£100k (start)—LG
	Oxford Brookes Uni car club	2013	Neighborhood	University, fp operator	Unknown—university
	North Oxford e-car club	2013	Neighborhood	CO, nfp operator	Unknown—CO
	Brompton bike share	2013	Site	fp operator	Unknown—fp operator
	Oxonbike bike share	2013, 2014	Neighborhoods	LG, fp operator	£100k–<500k—LSTF
Mobility management and support (*cf.* [19])	Travel hub and easitOxford travel information	2011	City	LG	£100k–<500k—LSTF
	Pedestrian way finding info city center	2011/12	City center	LG	Unknown—LG
	Internet-based journey planner	2012	Region	LG	<£100k—LSTF, LG
	Broken Spoke bike maintenance, cycling training	2012	Site (facility), City (catchment area)	CO	<£100k—CO, LG
Physical infrastructure retrofit	Various cycling paths/routes	2010	Axes	LG	£100k–<500k—LG, LSTF
	Plains roundabout retrofit	2014/15	Site	LG	£1m–<3m—NG, LG
	Bus retrofit/bus lane London Rd	2014/15	Axis	LG	£500k–<1m—LSTF
	Shared space retrofit Frideswide Square	2015	Site	LG	£3m–<10m—LG
Other	Low emission zone, city center	2014	City center	LG, bus operators	–

Notes: Axis = covers a transport a specific road; City = coverage (most of) the city; City center = covers city center only; CM = car manufacturer; CO = community organization; Neighborhood = covers a specific neighborhood outside the city center; (n)fp = (not) for profit; LG = local government; LSTF = Local Sustainable Transport Fund; NG = national government; NR = Network Rail; Region = exceeds city to cover (most of) the urban region; Site = restricted to a single location.

A third observation relates to the difficulty of straightforwardly applying Nykvist and Whitmarsh's [19] and Geels's [20] classifications of niche innovations to the cities of Brighton and Oxford. The categories those authors propose overlap quite strongly. For instance, the IT-based category that Geels distinguishes is closely interwoven with those for public transport, mobility management and mobility services. Once those are discounted, there is little that remains other than a residual smart traffic signaling initiative in Brighton (classified as "Other" in Table 3). Geels's [20] public transport category could also have been merged with others, but has been retained because both cities have a well developed busmobility regime with (local) bus companies as powerful players capable of developing unique innovations that are uni- rather than intermodal in the sense of only seeking to make bus more attractive to potential users. His category of cultural and socio-spatial niches has been traded for Nykvist and Whitmarsh's [19] more specific class of mobility services, and their mobility management category has been expanded slightly to also encompass forms of support, such as cycle training and maintenance facilities. Moreover, both Nykvist and Whitmarsh [19] and Geels [20] seem to under-appreciate the significance of retrofits to existing physical infrastructure (see below for discussion on their status as niche) in attempts to transition toward a more cycling oriented urban mobility system. Physical infrastructure retrofits have therefore been added as a separate category.

A further conclusion to be drawn from Tables 2 and 3 is the dominance of incumbent actors. These include first and foremost local governments, but also the Department for Transport (DfT) at the national level; bus companies that have served Brighton and Oxford for decades and more; and key actors in the rail sector, such as Network Rail—the organization that manages and looks after rail tracks and associated infrastructure as well as many stations on behalf of national government—and well-established train operating companies such as Southern (Brighton) and Chiltern Railways (Oxford). The question can be raised whether this dominance is problematic, and the answer is not immediately obvious. However, as Raven [71] (p. 2392) suggests, dominance of incumbents can enhance entry barriers for new actors with alternative views and practices and result in the "[d]anger of getting stuck into [the] existing regime without radical transformation". This is potentially a significant issue in relation to urban mobility, not least because incumbent transport service providers in UK cities are often focused on a single mode—the one they provide—and, despite significant and ongoing changes in orientation and practice, are conditioned by routines accumulated over decades to a greater extent than new actors tend to be.

The dominance of incumbents can be explained with reference to two sets of factors. One is that urban transport is a tightly regulated domain of society, at least in advanced liberal democracies like the UK. Local government, and because of the UK state's strongly centralized character in regard to transport despite 15 years of devolution [72] also the DfT, holds strong power over urban mobility initiatives. In both Brighton and Oxford these manifest themselves in at least three roles for local government:

- Regulator—for instance, being formally in charge of road/street space they play a pivotal role in the allocation of parking bays to all forms of car club, and by having a decisive say in the granting of franchises to bus operators, they have considerable influence over the practices of the latter. They also provide permits for many grassroots initiatives to provide and raise awareness about low-energy forms of urban mobility;
- Financier—they support many innovations by both private sector and community organizations with grants. Even grassroots innovations with socially and politically progressive identities, such as Oxford's Broken Spoke bike cooperative and the Circus Street cycle hub in Brighton, are to some extent dependent on small-scale financial support from local government. Financial support can also come in the form of officer time, as exemplified by the time council staff spent on the preparation of bids to DfT funding streams to subsidize the purchase of hybrid electric vehicles (HEV) by the main bus operators in Oxford and Brighton; and
- Initiator—they develop, lead, coordinate and are ultimately responsible for many of the innovations listed in Tables 2 and 3. This reflects both the changing configurations of finance in

English transport planning (see below), and the enthusiasm and determination of many policy makers—at all levels of organizational hierarchy—and some local politicians in encouraging a transition toward low-energy urban mobility systems.

The other explanation is related to the first: incumbent actors, such as long-standing bus and train operators and local government, are capable of commanding financial resources for investments in innovations that new actors in the realm of low-energy mobility—community organizations, start-up companies, universities, and so forth—cannot access. As already implied, a very large share of those financial resources comes from national government but also important is capital within incumbent firms, such as Brighton and Hove Bus Company (BHBC) and Oxford Bus Company (OBC). Both these firms are part of the Go-Ahead Group, a national conglomerate of formerly independent local/regional public transport companies, and their membership has facilitated investment in HEV buses in multiple ways. The capacity to command funding is particularly important in the transport sector, where a long-standing culture prevails of addressing perceived problems through techno-fixes and interventions in the physical infrastructures that enable particular forms of transport.

The above discussions also have ramifications for the literature on transitions and innovations discussed in Section 2. There are clear examples of grassroots innovations that explicitly challenge regime actors' ways of reducing GHG emissions from urban mobility; Broken Spoke and the Circus Street Bike Hub have already been mentioned and other examples are North Oxford's e-car club and The Big Lemon, a small social enterprise in Brighton that operates buses and coaches running on bio-fuel based on cooking oil collected from restaurants around the region. Nonetheless, and notwithstanding the earlier caveat about the partiality of Tables 2 and 3, it seems that even in the progressive cities of Brighton and Oxford such grassroots innovations play a fairly limited role within innovation activity at the city level. They are critically important in servicing specific user groups whose mobility needs are not adequately catered for by incumbents and in offering paid/voluntary work to specific individuals who might otherwise struggle in a neoliberal labor market, but they tend to be marginal—and in some ways marginalized—actors within transition trajectories in urban mobility in the two cities studied here.

Indeed, the more strategic sense in which low-energy innovations are political seems to be more significant, particularly in Oxford and especially in recent years. Funding is again important here: under the Conservatives-dominated Coalition Government of David Cameron funding streams from national to local levels have increasingly been reframed and repurposed. Local initiatives seeking to reduce emissions and energy consumption in urban mobility have to make a clearly identifiable—*i.e.*, with conventional transport appraisal methods quantifiable—contribution to economic growth and city-regions' competitiveness in the global marketplace. This even applies to the LSTF and has played a significant part in DfT's decision-making about which local governments to allocate funding.

4.3. Geographies of "Radicalness"

The general observations above also raise questions about the usefulness of the regime/niche distinction in relation to urban mobility in Brighton and Oxford. Given the dominance of incumbent actors in initiating, coordinating and funding low-energy innovations, dualistic opposition—regime *versus* niche—is to be avoided; differences are at best gradual. There are only degrees of "radicalness" of innovations in terms of differences *vis-à-vis* existing regimes in both a technological and a social sense, and those degrees are geographically differentiated. Three examples can clarify this point.

Firstly, car clubs—the English term for formal car sharing schemes operated by a for-profit company (e.g., City Car Club and the North-American company Zipcar which took over UK-based Streetcar in 2010) or a not-for-profit organization (e.g., e-car club which runs the North Oxford community scheme that at the moment of writing was still organized around a single car)—are much better integrated into the regimes of auto- and busmobility in Brighton than in Oxford. This is partly a scale issue: according to City Car Club (CCC), the sole commercial operator in Brighton in 2014, that city has the largest use of car clubs in South East England outside London. Indeed, Oxford's

36 cars across four operators in mid-2014 are clearly outnumbered by CCC's 100+ vehicles, which are also more differentiated in terms of size and type than Oxford's and distributed over more than 60 pick-up locations in Brighton. This is one reason why BHBC collaborate with the car club and bus card holders enjoy a reduced first-year subscription rate to CCC (Table 2). Yet, political support is also important: Brighton's minority Green administration has actively promoted Brighton's car club and in some documents prides itself as championing car sharing (even if some policymakers' support is more ambiguous, in part because CCC was one of several car clubs that were reluctant to get involved in an EU-funded attempt by policymakers to extend car sharing to some of the city's more deprived areas). An important consequence of the Greens' actions has been to facilitate the emergence of symbiosis between auto- and busmobility in Brighton. This stands in sharp contrast to Oxford where car sharing has received much less active support from politicians and policymakers, and relationships with the bus sector are non-symbiotic. Car sharing innovations are more radical in the sense of differing more strongly from prevailing regimes in Oxford than in Brighton.

Secondly, juxtaposing Tables 2 and 3 demonstrates considerably greater investment in HEV buses by bus companies in Oxford than in Brighton. At the same time, BHBC has invested much more in Euro 5 standard compliant (*i.e.*, emitting no more than 2.0 grams of NO_x per kilowatt hour of engine power) buses than OBC, its Oxford sister-company with a smaller bus fleet. This difference is to some extent linked to Oxford City Council's air quality strategy and the collaboration of county and city councils with all bus companies serving the city center in order to create a Low Emission Zone (LEZ). Effective since January 2014, the LEZ concentrates on NO_x levels but still has important GHG emission reduction co-benefits. In Oxford and beyond, the LEZ is seen as a concrete success in tackling long-standing problems with poor air quality. It is one of the first in the UK outside London and stipulates that buses should be at least Euro 5 compliant (although there are many exceptions watering down what at first glance appears to be a very solid measure). However, by ordering many HEV buses, OBC has gone beyond what has been required, which seems to reflect both its wider philosophy—it positions itself "one of the greenest bus companies in the UK" ([73], no pagination)—and anticipation of more ambitious plans by the city to create a zero-emission zone in several years. As they have been embraced more fully and at an earlier point in time in Oxford, HEV bus technologies are more radical in Brighton (albeit considerably less radical than those of The Big Lemon, which is also socially radical because it is a social enterprise that is more deeply committed to environmental and social sustainability). All in all, green propulsion technologies are more radical in the Brighton than the Oxford context.

Finally, there are differences in the position of cycling. Not only has cycling long since been more common in Oxford; that city has also had for some time now a subaltern regime [20] of velomobility characterized, amongst others, by dedicated physical infrastructure along many of the city's main roads, a substantial number of bike shops, local government offering cycling training in primary schools, support for cycling by two universities and—in the UK context—high usage levels among the general public. This regime has certainly expanded and gained in complexity and texture over the past decade, but the situation in Oxford is quite different from that in Brighton. There the local government has been very proactive in trying to create a cycling regime, partly in response to autonomous developments within the local community but also because of strong support from some local politicians, especially from the Green Party (although cycling continues to be a subject of political contention in Brighton). Since 2005, local government has concentrated on two sets of initiatives that can, following convention in transport studies [74,75], be classified as:

- "Soft"—cycling training in schools and more recently also the city's two universities, as well as travel planning for households, employers and businesses; and
- "Hard"—the creation of a network of retrofitted major thoroughfares, including Lewes Road, as well as the provision of bike racks and other minor interventions in the materiality of public spaces to make cycling more convenient. The idea has been, in the words of one senior transport

planner, to create the quality of infrastructure "so you don't have to be a kind of hardcore cyclist, it's designed for all levels of ability".

It has taken about a decade to realize and implement these initiatives and activity is still ongoing. As innovations, they were never very radical—not least because of the strong support and involvement of local policymakers—but over the past ten years they have been "mainstreamed" and become increasingly incremental. As a subaltern regime of velomobility started to emerge within Brighton, more recent innovations lost the greater degrees of radicalness characterizing earlier initiatives (even if they can still be considered radical relative to the national scale, as the earlier discussion of Lewes Road suggests).

4.4. Further Geographical Differentiations

There are three other geographical variations at the city level that merit attention further to the place-specificity of what constitutes a radical innovation. One relates to subtle differences in the geographical scope of innovations (Tables 2 and 3). Partly as a result of the greater expansion of car clubs and emphasis on cycling-related interventions in Brighton, mobility services are more localized in Oxford and there are more axially oriented innovations, which collectively offer citywide coverage, in Brighton.

Secondly, trajectories of path-dependent innovation—and hence transition—are city-specific. The previously discussed developments regarding cycling exemplify one such a trajectory, into which Brighton has gradually become "locked in" due to several interacting processes:

- The positive reinforcement that follows from the perception that earlier initiatives have been successful.
- The emergence of specialized expertise and a (small) workforce of travel planners within BHCC.
- Continued success with raising money from dedicated funding programs run by DfT and the EU. In 2005 Brighton became one of six Cycling Demonstration Towns—national test bed sites allowing governments to learn about what can be achieved in the UK with continental-European level of investment per capita in cycling—and subsequently benefited from the follow-up Cycling Town program (2008–2011) and more recently the LSTF (2011–2016). The city also obtained EU funding (CiviTAS) to bolster its activities.
- The strongly territorialized nature of most cycling-related innovations: partly in response to restrictions imposed by external funders, most travel planning projects and road retrofits have been—and still are—limited to specific neighborhoods in the city, meaning there has been at least until now a supply of areas for retrofit and travel planning.

A similar cycling oriented trajectory cannot be discerned in Oxford. This is partly a consequence of its pre-existing subaltern regime of velomobility yet also reflects a more neoliberal and less interventionist style of governing mobility by local government that has existed for several decades (*cf.* [76]) but seems to be changing. Since 2012 the city has had the Oxford Cycle City strategy, aiming to make "Oxford ... one of the truly great cycling cities of Europe" [77] in response to initiatives in other UK cities (and the desire not to lag behind) and the buzz around cycling safety across the UK more generally.

Oxford has also witnessed something of an electric mobility trajectory, although this is not nearly as coherent as Brighton's string of cycling-related innovations. The electric mobility trajectory ties different modes and actors together as it comprises not only OBC's substantial investment in HEV buses (see Section 4.3) and the community-led e-car club in North Oxford. Also included are the city's network of charging points and the Mini-E trial ran by BMW and Oxford Brookes University (OBU). As part of the latter, 20 Minis were provided to professional organizations, including Oxford City Council, and to two waves of 20 householders in Oxford (and London) for several weeks, so that drivers could learn about EVs and BMW and OBU could study the cars' performance and dynamics in user practices, perceptions and attitudes. The project has so far not had a follow-up but it did make the city council amenable to procuring several EVs and creating some non-publicly available charging points for its own fleet. Oxford also has a network of commercially operated charging points. At the moment of writing, there are eight locations, against four in Brighton; half are located in various parts of the city, and the other can be found at four of the five P&R locations that surround the city (one is equipped with rapid charging technology). As before, an important driver behind this electric mobility trajectory is the city's air quality problem and strategy to tackle this. Yet, this is also where a discrepancy can be noticed between city and county council: the former is focused more strongly on air quality, whilst the latter is formally responsible for reducing road congestion; the consequence is a slight ambivalence in support for EV-related innovations in Oxford.

The third variation concerns funding: a comparison of Tables 2 and 3 reveals greater reliance on supra-local—DfT and EU—funding streams in Brighton than in Oxford. This difference is due to a combination of strategy and differential success rates: over the past decade Brighton's local government has been more proactive in raising external funds to finance its aspirations regarding innovation, and early successes in obtaining grants and the reputational value created by actual delivery on promises seem to have played a role in subsequent awards. However, actors within BHCC also realize that success cannot be taken for granted in the near future. As a consequence of the austerity politics of Cameron's Coalition Government, the annual allocations of DfT funds for transport planning to local authorities will be reduced substantially in the coming years. This will probably mean that competition from other local governments for future funding programs—the successor to LSTF, for instance—will become fiercer, and that less staff capacity within BHCC will be available to prepare future bids.

Oxford may well be a strong competitor in future funding programs. Both city and council governments seem to have become more interventionist in recent years and are now more proactive in attempts to raise external funds from national government and the EU's current funding program Horizon 2020. Moreover, local government is reaching out more actively than before to actors in the private sector and also to the city's two universities. These more complex arrangements for governing innovations in low-energy mobility are partly a response to funder requirements but also reflect local actors' shared aspiration to become a "smart city". Transport in general and low-energy mobility in particular have been identified as an important domain for experimentation with new (IT-based) technologies, and local government appreciates that the city's universities and IT firms can make critical contributions to the realization of further innovations in electric mobility, intermodal mobility, mobility services and mobility management. These potential innovations—like many of those included in Tables 2 and 3—cannot be considered as being simply or even predominantly local. They transcend administrative boundaries and hierarchically ordered, nested geographical scales (*cf.* [34]) but without being placeless: they also internalize and embody the historically sedimented specificities of the territory where they become part of existing mobility systems.

5. Conclusions

Using a case study approach and informed by the sociotechnical transitions literature, this paper has demonstrated that many innovations contributing to reduced GHG emissions and energy consumption from everyday mobility have been realized recently in Brighton and Oxford. Activity has been particularly pronounced since 2010, focusing on different transport modes—and sometimes at intermodal mobility—and involving many different actors, both incumbent and new. Examples of the latter include start-up companies specializing in low-energy mobility, not-for-profit organizations and collectives, and local universities. It is tempting to conclude that in both cities an innovation-based systemic change toward low-energy mobility futures is underway. Yet, the admittedly very basic statistics in Table 1 also suggest that, at least for commuting, private car use (in internal combustion vehicles) remains by far the most important form of transport. There is still a long way to go toward urban mobility that is genuinely environmentally sustainable.

There is also a clear geography to the innovation patterns characterizing both cities, in a double sense. Place-specific and path-dependent trajectories of innovation in each city can be identified. There is considerably greater emphasis on commercial car sharing; cycling oriented retrofitting of road infrastructure; and travel planning focused on residences, workplaces and educational institutions in Brighton. In contrast, there has been greater activity in electric mobility in Oxford, partly because of greater air quality issues and more concerted action from local government on these. The second sense relates to how innovation activity in each city is relational and embedded in wider geographical configurations. In both cities most activity is funded by money that comes from national government sources, but Brighton's local government has so far been more proactive and more successful in raising money from specialized funding programs oriented at cycling and/or sustainable transport run by national government and the EU. This appears to be changing, however, and Oxford has recently become more proactive in obtaining supra-local government funding. Whilst this is to some extent the consequence of wider processes of state restructuring in the UK (*cf.* [78]), the findings also demonstrate the importance of understanding cities as at once relational and territorial [58] rather than as bounded, sealed-off and independent containers in which innovations unfold.

In fact, the financial arrangements discussed in this paper are one source of vulnerability and uncertainty when it comes to future innovation activity and the continuation of transition trajectories. This is because national and EU funding is to be obtained via competitive bidding processes, the outcomes of which cannot be guaranteed in advance. Whilst this study suggests that success in obtaining grants and project delivery seems to breed success, the consequences of failure in obtaining the next tranche of funding are not insignificant. Not only will particular lines of innovation activity have to be cut short, local governments will also lose staff and hence expertise and human capital. This may reduce the speed of sociotechnical transition as well as the trajectories followed, even if the broader direction of travel towards reducing energy consumption and GHG emissions may remain unaltered. Another vulnerability that has surfaced relates to fluctuations in political support for particular types of innovations. As has been suggested, support from local politicians has been important in driving forward cycling projects and the expansion of car sharing in Brighton and to a lesser extent cycling-related innovations in Oxford. Whilst very important, such support is also precarious: it is tied to electoral cycles and can also shift in-between election periods (for instance, under influence of local media campaigns or the actions of pressure groups). It is beyond this paper to explore the temporalities of political support in both case-study cities; suffice to say that there have been significant fluctuations in Brighton in particular, and that these have at times threatened the survival of specific innovations.

The above reflections on vulnerabilities and cities' situatedness in wider spatial configurations of finance—as well as regulation and expertise—raise another issue. As previously discussed, both Brighton and Oxford are in many ways privileged when it comes to the emergence and flourishing of low-energy innovations in the everyday mobility of people. If low-energy transition trajectories and (future) innovations are already prone to precariousness, vulnerability and uncertainty in those cities, what can be expected for others where conditions are less favorable? This question obviously demands empirical scrutiny but, as far as the pace of transition processes is concerned, the findings for Brighton and Oxford do not bode particularly well for smaller and medium-sized cities (up to ± 0.5 M inhabitants) at the heart of urban areas with much weaker bus- and velomobility systems in terms of physical infrastructure, regulation, formal expertise, financial resources, cultural values and user practices. It would appear, then, that cities have a significant role to play in reducing GHG emissions and energy consumption in transport in advanced liberal democracies like the UK but expectations about what they can achieve should be realistic. By no means should cities be heralded as actors who can bring about significant change semi-autonomously. This is to deny both the social, political, cultural, technological and other struggles that characterize cities, and the critically important role that national and supranational (EU) level actors play in shaping their innovation and low-energy transition trajectories.

Acknowledgments: The research described in this paper has been supported by the Centre on Innovation and Energy Demand (www.cied.ac.uk), funded by the EPSCR/ESCR (grant number EP/KO11790/1). The comments of three anonymous reviewers on an earlier version of the paper are gratefully acknowledged.

Conflicts of Interest: The author declares no conflict of interest.

References

1. Banister, D. *Unsustainable Transport: City Transport in the New Century*, 1st ed.; Routledge: Abingdon, UK, 2005.
2. Chapman, L. Transport and climate change: A review. *J. Transp. Geogr.* **2007**, *15*, 354–367. [CrossRef]
3. Banister, D.; Anderton, K.; Bonilla, D.; Givoni, M.; Schwanen, T. Transportation and the environment. *Annu. Rev. Environ. Resour.* **2011**, *36*, 247–270. [CrossRef]
4. Sims, R.; Schaeffer, R.; Creutzig, F.; Cruz-Núñez, X.; D'Agosto, M.; Dimitriu, D.; Figueroa Meza, M.J.; Fulton, L.; Kobayashi, S.; Lah, O.; *et al.* Transport. In *Climate Change 2014: Mitigation of Climate Change. Contribution of Working Group III to the Fifth Assessment Report of the Intergovernmental Panel on Climate Change*, 1st ed.; Edenhofer, O., Pichs-Madruga, R., Sokona, Y., Farahani, E., Kadner, S., Seyboth, K., Adler, A., Baum, I., Brunner, S., Eickemeier, P., *et al.*, Eds.; Cambridge University Press: Cambridge, UK, 2014; pp. 599–670.
5. International Energy Agency. *World Energy Outlook 2012*, 1st ed.; International Energy Agency: Paris, France, 2012.
6. International Energy Agency. *CO_2 Emissions from Fuel Combustion: Highlights*; International Energy Agency: Paris, France, 2013.
7. Urry, J. *Climate Change & Society*, 1st ed.; Polity Press: Cambridge, UK, 2011.
8. Millard-Ball, A.; Schipper, L. Are we reaching peak travel? Trends in passenger transport in eight industrialized countries. *Transp. Rev.* **2011**, *31*, 357–378.
9. Newman, P.; Kenworthy, J. 'Peak car use': Understanding the demise of automobile dependence. *World Transp. Policy Pract.* **2011**, *17*, 31–42.
10. Babalik-Sutcliffe, E. Urban rail systems: Analysis of the factors behind success. *Transp. Rev.* **2002**, *22*, 415–447. [CrossRef]
11. Lane, B. Significant characteristics of the urban rail renaissance in the United States: A discriminant analysis. *Transp. Res. Part A* **2008**, *42*, 279–295. [CrossRef]
12. Garmendia, M.; Ribalaygua, C.; Ureña, J.M. High speed rail: Implication for cities. *Cities* **2012**, *29* (Suppl. 2), S26–S31. [CrossRef]
13. Buehler, R.; Pucher, J. *City Cycling*, 1st ed.; MIT Press: Cambridge, MA, USA, 2012.
14. Lanzendorf, M.; Busch-Geertsema, A. The cycling boom in large German cities—empirical evidence for successful cycling campaigns. *Transp. Policy* **2014**, *36*, 26–33. [CrossRef]
15. Newman, P.; Kenworthy, J. *Sustainability and Cities: Overcoming Automobile Dependence*, 1st ed.; Island Press: Washington, DC, USA, 1999.
16. Banister, D. Cities, mobility and climate change. *J. Transp. Geogr.* **2011**, *19*, 1538–1546. [CrossRef]
17. Castán Broto, V.; Bulkeley, H. A survey of urban climate change experiments in 100 cities. *Glob. Environ. Chang.* **2013**, *23*, 92–102. [CrossRef] [PubMed]
18. Geels, F.W. From sectoral systems of innovation to socio-technical systems: Insights about dynamics and change from sociology and institutional theory. *Res. Policy* **2004**, *33*, 897–920. [CrossRef]
19. Nykvist, B.; Whitmarsh, L. A multi-level analysis of sustainable mobility transitions: Niche development in the UK and Sweden. *Technol. Forecast. Soc. Chang.* **2008**, *75*, 1373–1387. [CrossRef]
20. Geels, F.W. A socio-technical analysis of low-carbon transitions: Introducing the multi-level perspective into transport studies. *J. Transp. Geogr.* **2011**, *24*, 445–467. [CrossRef]
21. Kemp, R.; Schot, J.; Hoogma, R. Regime shifts to sustainability through processes of niche formation: The approach of strategic niche management. *Technol. Anal. Strateg. Manag.* **1998**, *10*, 175–195. [CrossRef]
22. Schot, J.; Geels, F.W. Strategic niche management and sustainable innovation journeys: Theory, findings, research agenda, and policy. *Technol. Anal. Strateg. Manag.* **2008**, *20*, 537–554. [CrossRef]
23. Smith, A.; Raven, R. What is protective space? Reconsidering niches in transitions to sustainability. *Res. Policy* **2012**, *41*, 1025–1036.

24. Rip, A.; Kemp, R. Technological change. In *Human Change and Climate Change—Resources and Technology 2*, 1st ed.; Rayner, S., Malone, E.L., Eds.; Batelle: Columbus, OH, USA, 1998; pp. 327–399.
25. Geels, F.W. Technological transitions as evolutionary reconfiguration processes: A multi-level perspective and case-study. *Res. Policy* **2002**, *31*, 1257–1274. [CrossRef]
26. Shove, E.; Walker, G. Governing transitions in the sustainability of everyday life. *Res. Policy* **2010**, *39*, 471–476. [CrossRef]
27. Watson, M. How theories of practices can inform transition to a decarbonised transport system. *J. Transp. Geogr.* **2014**, *24*, 488–496. [CrossRef]
28. Aldred, R.; Jungnickel, K. Why culture matters for transport policy: The case of cycling in the UK. *J. Transp. Geogr.* **2014**, *34*, 78–87. [CrossRef]
29. Hopkins, D.; Stephenson, S. Generation Y mobilities through the lens of energy cultures: A preliminary exploration of mobility cultures. *J. Transp. Geogr.* **2014**, *38*, 88–91. [CrossRef]
30. Stephenson, J.; Hopkins, D.; Doering, A. Conceptualising transport transitions: Energy Cultures as an organising framework. *Wiley Interdiscip. Rev. Energy Environ.* **2015**. [CrossRef]
31. Hodson, M.; Marvin, S. Cities mediating technological transitions: Understanding vision, intermediation and consequences. *Technol. Anal. Strateg. Manag.* **2009**, *21*, 515–534. [CrossRef]
32. Coenen, L.; Raven, R.; Verbong, G. Local niche experimentation in energy transitions: A theoretical and empirical exploration of proximity advantages and disadvantages. *Technol. Soc.* **2010**, *32*, 295–302. [CrossRef]
33. Coenen, L.; Benneworth, P.; Truffer, B. Toward a spatial perspective on sustainability transitions. *Res. Policy* **2012**, *41*, 968–979. [CrossRef]
34. Raven, R.; Schot, J.; Berkhout, F. Space and scale in socio-technical transitions. *Environ. Innov. Soc. Transit.* **2012**, *4*, 63–78. [CrossRef]
35. Bulkeley, H.; Castán Broto, V. Government by experiment? Global cities and the governing of climate change. *Trans. Inst. Br. Geogr.* **2013**, *38*, 361–375. [CrossRef]
36. McCormick, K.; Anderberg, S.; Coenen, L.; Neij, L. Advancing sustainable urban transformation. *J. Clean. Prod.* **2013**, *50*, 1–11. [CrossRef]
37. Longhurst, N. Towards and 'alternative' geography of innovation: Alternative milieu, socio-cognitive protection and sustainability experimentation. *Environ. Innov. Soc. Innov.* **2015**, in press. [CrossRef]
38. Murphy, J.T. Human geography and socio-technical transition studies: Promising intersections. *Environ. Innov. Soc. Innov.* **2015**, in press. [CrossRef]
39. Sengers, F.; Raven, R. Toward a spatial perspective on niche development: The case of Bus Rapid Transit. *Environ. Innov. Soc. Innov.* **2015**, in press. [CrossRef]
40. Flyvbjerg, B. Case study. In *The Sage Handbook of Qualitative Research*, 4th ed.; Denzin, N.K., Lincoln, Y.S., Eds.; Sage: Thousand Oaks, CA, USA, 2011; pp. 301–316.
41. Department for Transport. Local Sustainable Transport Fund. Available online: https://www.gov.uk/government/collections/local-sustainable-transport-fund (accessed on 30 March 2015).
42. Unruh, G.C. Escaping carbon lock-in. *Energy Policy* **2002**, *30*, 317–325. [CrossRef]
43. Bakker, S.; Maat, K.; van Wee, B. Stakeholders interests, expectations, and strategies regarding the development and implementation of electric vehicles: The case of the Netherlands. *Transp. Res. Part A* **2014**, *66*, 54–64. [CrossRef]
44. Geels, F.W. The multi-level perspective on sustainability transitions: Responses to seven criticisms. *Environ. Innov. Soc. Transit.* **2011**, *1*, 24–40. [CrossRef]
45. McDowell, L. Work, workfare, work/life balance and an ethic of care. *Progr. Hum. Geogr.* **2004**, *28*, 145–163. [CrossRef]
46. Geels, F.W.; Kemp, R.; Dudley, G.; Lyons, G. *Automobility in Transition? A Socio-technical Analysis of Sustainable Transport*, 1st ed.; Routledge: New York, NY, USA, 2012.
47. Sengers, F.; Raven, R. Metering motorbike mobility: Informal transport in transition? *Technol. Anal. Strateg. Manag.* **2014**, *26*, 453–468. [CrossRef]
48. Van Bree, B.; Verbong, G.P.J.; Kramer, G.J. A multi-level perspective on the introduction of hydrogen and battery-electric vehicles. *Technol. Forecast. Soc. Chang.* **2010**, *77*, 529–540. [CrossRef]
49. Bakker, S.; Farla, J. Electrification of the car—Will the momentum last?: Introduction to the special issue. *Environ. Innov. Soc. Innov.* **2015**, *14*, 1–4. [CrossRef]

50. Raven, R.; Verbong, G. Multi-regime interactions in the Dutch energy sector: The case of combined heat and power technologies in the Netherlands 1970–2000. *Technol. Anal. Strateg. Manag.* **2007**, *19*, 491–507. [CrossRef]
51. Seyfang, G.; Smith, A. Grassroots innovations for sustainable development: Towards a new research and policy agenda. *Environ. Polit.* **2007**, *16*, 584–603. [CrossRef]
52. Smith, A.; Seyfang, G. Constructing grassroots innovations for sustainability. *Glob. Environ. Chang.* **2013**, *23*, 827–829. [CrossRef]
53. Ornetzeder, M.; Rohracher, H. Of solar collectors, wind power, and car sharing: Comparing and understanding successful cases of grassroots innovations. *Glob. Environ. Chang.* **2013**, *23*, 856–857. [CrossRef]
54. Perren, L.; Sapsed, J. Innovation as politics: The rise and reshaping of innovation in UK parliamentary discourse 1960–2005. *Res. Policy* **2013**, *42*, 1815–1828. [CrossRef]
55. Upham, P.; Kivimaa, P.; Mickwitz, P.; Åstrand, K. Climate policy innovation: A sociotechnical transitions perspective. *Environ. Polit.* **2014**, *23*, 774–794. [CrossRef]
56. Boyatzis, R.E. *Transforming Qualitative Information: Thematic Analysis and Code Development*, 1st ed.; Sage: Thousand Oaks, CA, USA, 1998.
57. Ward, K. Towards a relational comparative approach to the study of cities. *Progr. Hum. Geogr.* **2010**, *34*, 471–487. [CrossRef]
58. McCann, E.; Ward, K. Relationality/territoriality: Toward a conceptualization of cities in the world. *Geoforum* **2010**, *41*, 175–184. [CrossRef]
59. Massey, D. *For Space*, 1st ed.; Sage: London, UK, 2004.
60. Bertilsson, T.M. The elementary forms of pragmatism: On different types of abduction. *Eur. J. Soc. Theory* **2004**, *7*, 371–389. [CrossRef]
61. Lucas, K.; Jones, P. *The Car in British Society*, 1st ed.; RAC Foundation: London, UK, 2009.
62. Office for National Statistics. 2011 Census Analysis, Method of Travel to Work in England and Wales. Available online: http://www.ons.gov.uk/ons/publications/re-reference-tables.html?edition=tcm%3A77-295663 (accessed on 30 March 2015).
63. Office for National Statistics. Cycling to Work. Available online: http://www.ons.gov.uk/ons/publications/re-reference-tables.html?edition=tcm%3A77-295663 (accessed on 30 March 2015).
64. Flyvbjerg, B.; Skamris Holm, M.; Buhl, S. Underestimating costs in public works projects: Error or lie? *J. Am. Plan. Assoc.* **2002**, *68*, 279–295. [CrossRef]
65. Flyvbjerg, B. Cost overruns and demand shortfalls in urban Rail and Other Infrastructure. *Transp. Plan. Technol.* **2007**, *30*, 9–30. [CrossRef]
66. Ormrod, R.K. Local context and innovation diffusion in a well-connected world. *Econ. Geogr.* **1990**, *66*, 109–122. [CrossRef]
67. Ormrod, R.K. Adaptation and cultural diffusion. *J. Geogr.* **1992**, *91*, 258–262. [CrossRef]
68. McCann, E. Urban policy mobilities and global circuits of knowledge: Toward a research agenda. *Ann. Assoc. Am. Geogr.* **2011**, *101*, 107–130. [CrossRef]
69. Peck, J. Geographies of policy: From transfer-diffusion to mobility-mutation. *Progr. Hum. Geogr.* **2011**, *35*, 773–797. [CrossRef]
70. McCann, E.; Ward, K. *Mobile Urbanism: Cities and Policymaking in the Global Age*, 1st ed.; The University of Minnesota Press: Minneapolis, MN, USA, 2012.
71. Raven, R. Niche accumulation and hybridisation strategies in transition processes towards a sustainable energy system: An assessment of differences and pitfalls. *Energy Policy* **2007**, *35*, 2390–2400. [CrossRef]
72. Shaw, J.; MacKinnon, D.; Docherty, I. Divergence or convergence? Devolution and transport policy in the United Kingdom. *Environ. Plan. C* **2009**, *27*, 546–567. [CrossRef]
73. Oxford Bus Company. *Annual Corporate and Social Responsibility Report 2014*; Oxford Bus Company: Oxford, UK; Available online: http://assets.goaheadbus.com/media/cms_page_media/70/CSR%20Report%20WEB.pdf (accessed on 28 March 2014).
74. Enoch, M. *Sustainable Transport, Mobility Management and Travel Plans*, 1st ed.; Ashgate: Aldershot, UK, 2012.
75. Sebastian Bamberg, S.; Fujii, S.; Friman, M.; Gärling, T. Behaviour theory and soft transport policy measures. *Transp. Policy* **2011**, *18*, 228–235. [CrossRef]
76. Lawton Smith, H. Local innovation assemblages and institutional capacity in local high-tech economic development: The case of Oxfordshire. *Urban Stud.* **2003**, *40*, 1353–1369. [CrossRef]

77. City of Oxford, Cycling and Walking. Available online: http://www.oxford.gov.uk/PageRender/decTS/Cycling_and_Walking_occw.htm (accessed on 13 May 2015).
78. MacKinnon, D.; Shaw, J. New state spaces, agency and scale: Devolution and the regionalisation of transport governance in Scotland. *Antipode* **2011**, *42*, 1226–1252. [CrossRef]

sustainability

Article

Managing Knowledge to Promote Sustainability in Australian Transport Infrastructure Projects

Jay Yang [1,*], Mei Yuan [1], Tan Yigitcanlar [1], Peter Newman [2] and Frank Schultmann [3]

1 School of Civil Engineering and Built Environment, Queensland University of Technology, G.P.O. Box 2434, Brisbane 4001 QLD, Australia; mei.yuan01@gmail.com (M.Y.); tan.yigitcanlar@qut.edu.au (T.Y.)
2 Curtin University Sustainability Policy Institute (CUSP), Curtin University, G.P.O. Box U1987, Perth 6845 WA, Australia; p.newman@curtin.edu.au
3 Institute for Industrial Production (IIP), Karlsruhe Institute of Technology, Hertzstrasse 16, 76187 Karlsruhe, Germany; frank.schultmann@kit.edu

* Author to whom correspondence should be addressed; j.yang@qut.edu.au; Tel.: +61-7-3138-1028; Fax: +61-7-3138-1170.

Academic Editor: Marc A. Rosen

Received: 30 March 2015; Accepted: 18 June 2015; Published: 24 June 2015

Abstract: To deliver tangible sustainability outcomes, the infrastructure sector of the construction industry needs to build capacities for the creation, application and management of ever increasing knowledge. This paper intends to establish the importance and key issues of promoting sustainability through knowledge management (KM). It presents a new conceptual framework for managing sustainability knowledge to raise the awareness and direct future research in the field of transport infrastructure, one of the fast growing sectors in Australia. A holistic KM approach is adopted in this research to consider the potential to "deliver the right information to the right person at the right time" in the context of sustainable development of infrastructure. A questionnaire survey among practitioners across the nation confirmed the necessity and identified priority issues of managing knowledge for sustainability. During infrastructure development, KM can help build much needed industry consensus, develop capacity, communicate decisions, and promote specific measures for the pursuit of sustainability. Six essential elements of the KM approach and their priority issues informed the establishment of a conceptual KM framework. The transport infrastructure sector has come to realise that development must not come at the expense of environmental and social objectives. In practice however, it is facing extensive challenges to deliver what has been promised in the sustainability agenda. This research demonstrates the importance of managing sustainability knowledge, integration of various stakeholders, facilitation of plans and actions and delivery of tangible benefits in real projects, as a positive step towards meeting these challenges.

Keywords: infrastructure; transport projects; sustainability; knowledge management; practitioner perceptions

1. Introduction

Infrastructure in the Australian context typically includes utilities such as roads, ports, railways, power lines, water pipes, power generation buildings, sewer plants, and other tangible structures [1]. Transport infrastructure systems, consisting of highways, roads, railways and ancillary facilities such as bridges, are often considered as the backbone of a country's economy. They support basic urban activity and play a fundamental role in determining the nation's economic efficiency and productivity [2,3]. Over the recent years, the growth of Australia's dispersed population in emerging regional centres, the boom of resource exports, growing social expectations, and international competition have been the driving forces of infrastructure development and regeneration [4–6]. Most projects are in the transport segment, including road networks, railway links and airport expansions.

While responding to the growth with real action plans, local, state and federal governments have recognised that infrastructure development can cause significant economic, environmental and social concerns [7,8], which are the foundational pillars of the commonly recognised triple bottom lines (TBL) of sustainability. Infrastructure development, particularly transport-oriented, is closely related to all facets of sustainability, as it often requires significant levels of resources and finance, causes disturbance to the natural environment and impacts upon the local community [9]. With depleting resources, increasing public awareness on wellbeing, stronger evidence of climate change and the global economic slowdown, it is not difficult to build a logical link between sustainability and transport infrastructure. On the positive side, expanding and regenerating Australia's infrastructure system has been seen as the principal means of enhancing economic growth and national productivity [3,8,9].

Increasingly, stakeholders accept that the development of infrastructure must not come at the expense of the environment and social objectives. As an example, in Queensland, the Department of State Development, Infrastructure and Planning replaced the Integrated Planning Act with the Sustainable Planning Act 2009 [10], to reflect a stronger focus on ecological sustainability as the outcome of planning decisions. Within only a few years, the Act was changed to the Sustainable Planning (Infrastructure Charges) and Other Legislation Amendment Act 2014, to embrace new requirements [11].

Practitioners of the infrastructure sector generally agree that infrastructure sustainability can be delivered through incorporating sustainable solutions into design, construction and operation of the transport facility. In practice however, they are facing extensive challenges to apply sustainability principles into their professional domains and real projects to deliver tangible outcomes [2,12]. According to past research, the challenges may be manifold. Immaturity and uncertainty of the body of ever increasing sustainability knowledge, different interpretation of starting and ending points, varying professional perspectives and priorities, the lack of exemplar projects, and shortage of knowledgeable practitioners are just a few examples [13–16]. Compared to other fields such as buildings, sustainability in infrastructure development has not received as much research focus as it requires [15,17].

Pathriage *et al.* [18] argue that the construction industry needs to intensify its efforts to move to a knowledge intensive mode. It is also believed that better decision-making towards sustainability goals can only be achieved when stakeholders are informed of and continuously updated with new concepts, knowledge and expertise across organisational, professional and hierarchical boundaries [19]. Furthermore, as stated by Wallace [20], sustainability needs multi-disciplinary collaboration and innovative solutions, which are highly dependent on a knowledgeable workforce, good communication and a culture of care and innovation. There is a clear and urgent message to the infrastructure sector to not only build and expand the body of sustainability knowledge among practitioners, but also find a better way to manage it at individual, project, organization and industry sector levels.

From a functional perspective, knowledge management (KM) can be seen as a "systematic approach to manage the use of information in order to provide a continuous flow of right knowledge to the right people at the right time, enabling efficient and effective decision making in their everyday business" [21]. Despite being a relatively new concept, KM is regarded as a broad and expanding tool to manage new information in diverse contexts in industries and businesses [22,23]. KM has shown its worth through existing applications in the building construction sector [24–26]. Various attempts have been made to develop strategies and mechanisms to manage construction knowledge [24,27]. However as shown by earlier examples, few have incorporated sustainability issues and requirements in infrastructure works [18,28–31].

With reference to literature and findings of related work in other construction sectors, this paper discusses results of a recent questionnaire survey among transport infrastructure practitioners which helped verify the necessity, determined priority issues and outlined focus areas for managing sustainability knowledge. They responded to six aspects of KM issues from individual professional experiences as well as project and organizational circumstances. Stemming from this finding, a conceptual framework of managing sustainability knowledge is developed for practitioners to promote a holistic KM approach to facilitate sustainability knowledge creation, capture, share and application in the Australian transport infrastructure sector. It will also be used as the foundation for on-going research work on the processes and guidelines of knowledge management specific to this industry sector.

2. Research Method and Data Collection

2.1. Research Focus

Knowledge needs to be managed for effective applications. Knowledge management is an expanding topic contributed by a diverse range of disciplines such as cognitive science, information technology, organisational theory, library and information sciences, and document and information management, with a multi-fold mix of strategies, tools, and techniques [22,23,32]. In this research context, sustainability knowledge can be defined as the type of knowledge that improves the understanding, uptake, and end results of sustainability during projects' life cycles. For transport infrastructure development, sustainability knowledge may come in the form of past experiences, design ideas, guidelines and policies, rating tools and indicators and technology innovations (Table 1).

Table 1. Types, locations and characteristics of sustainability knowledge examples.

-	Project Experiences	Government Guidelines/Polices	Assessment Tools/Indicators	Underpinning Theory/Technology
Location	Individuals	External resources	External resources	External
Type	Explicit/tacit	Explicit	Explicit	Explicit
Characteristic	- Mostly reside in people minds; - Hard to track, record and widely share.	- Vague and general. - More about environment protection	- Mainly developed from building assessment tools - More for environment performance assessment - Evolving in nature	- Complex and should consider extra systems - Many of them do simply not exist currently

Embedding sustainability principles for project delivery presents many challenges [17]. Wallace [20] summarised five important conditions to be met in order to improve project sustainability performance: (1) a knowledgeable and committed project owner; (2) a high-performance project team; (3) alternative procurement and contracting mechanisms; (4) high but achievable sustainability goals and objectives; and (5) access to and willingness to share knowledge and achievements. With reference to previous studies on managing knowledge in the construction industry, the authors considered how transport infrastructure development processes relate to these conditions and decided on a survey study focusing on six key KM elements as listed below:

- The body of sustainability knowledge
- KM strategy
- Stakeholder integration
- KM process
- KM enablers
- Outcomes of managing sustainability knowledge

2.2. Survey Distribution

A structured questionnaire, consisting of nine sections with 37 questions, was developed to collect practitioners' opinions on the abovementioned key elements. Research objectives, methods and approach, and literature review findings guided the formulation of the survey questions. As the majority of the survey questions are concerned with the prioritisation of perspectives rather than dealing with relationships among factors, they were framed on a five-point Likert scale with the mean, frequency and standard deviation of the grouped data assessed to determine the rank order. Because of this, the analysis used simple statistics to consider Means and Standard Deviations only, rather than engaging non-parametric studies.

Respondents were selected from the main stakeholder categories involved in large transport infrastructure projects in Australia, representing both public and private sectors, e.g., general contractors, subcontractors, suppliers, government agencies, consulting firms, and clients.

The survey was conducted through an online survey tool named Survey Monkey, which helps publish surveys on a website and generate pass code and filters to collect responses.

The first distribution approach was through email. Initially, 120 potential respondents were selected from leading organisations that worked on transport development projects across Australia, using a combination of existing industry databases, online search, and recommendations from industry-based research collaborators. They were each given a login name and password to the online survey by email. The initial responses were monitored. Non-respondents were followed up during the course of the survey with email reminders. In the end, out of the 120 questionnaires mailed to the potential respondents, 18 invitations were returned undelivered and 39 received responses. A total of 31 out of the 39 responses were fully completed and considered valid for data analysis, which gives an effective response rate of 30% (31/102) in this first part.

The second distribution approach was via professional associations. AIPM (Australian Institute of Project Management) and Engineers Australia (EA) offered their assistance by including the survey information in their regular newsletters. A specific online collector was set up to collect these responses. Twenty-eight were collected during the course of the survey, out of which 18 were valid for full completion of the survey as well as fitting the required respondent profile. However, it was not possible for the researchers to identify how many potential respondents were reached through this method, and potentially, only a small percentage of AIPM and EA members may have fulfilled the requirements of this survey.

In the end, 49 valid responses were used for data analysis. This study considers the prevailing opinions and common practices of managing knowledge in order to identify priorities and portray a general picture, as the foundation for further research.

2.3. Survey Respondents

Figures 1–3 shows the characteristics of the respondents. Figure 1 shows that about 70% of the respondents have more than 10 years of professional experience. Figure 2 shows that 86% of them are from engineering and project management professional backgrounds directly required and involved in infrastructure projects. This confirms the suitability and authenticity of the data collected. This also made the researchers decide not to use non-parametric analysis to consider group dynamics.

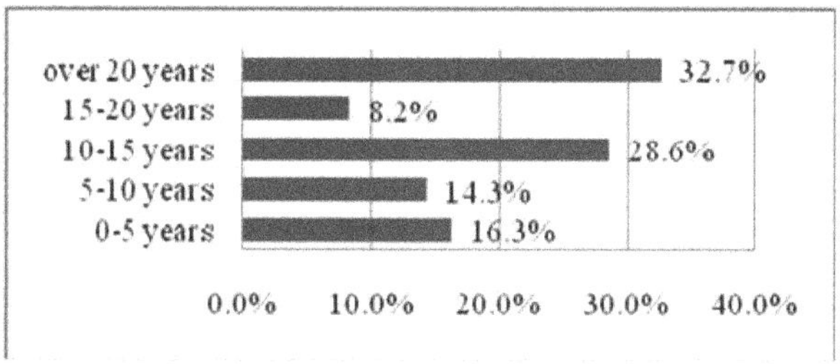

Figure 1. Respondents' professional experience in the infrastructure sector.

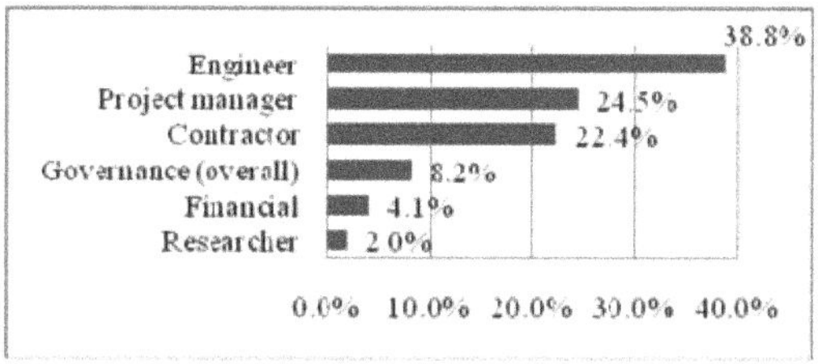

Figure 2. Professional role of respondents.

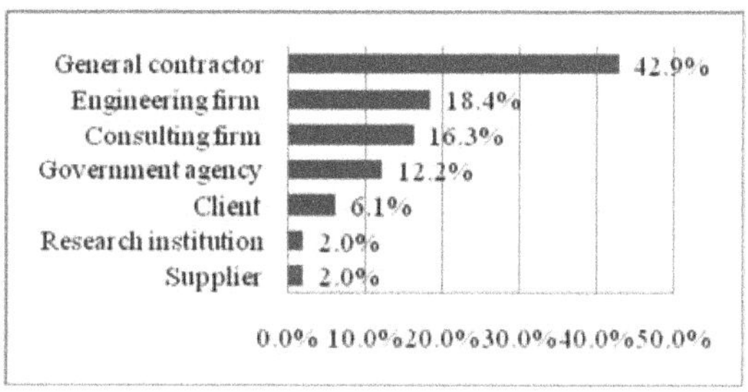

Figure 3. Organisation's type.

Respondents are from both public and private organisations, and together represent every stakeholder type in the industry (Figure 3). The main portion of the respondents is from the general contractor (42.9%), which is the main stakeholder in the transport infrastructure sector. Others are from engineering firms, consulting firms and government agencies (18.4%, 16.3% and 12.2%, respectively). The category of organisation is based on self-description by the respondents, thus there might be a slight overlap between engineering firms and consulting firms as some of the consulting firms provide specialised engineering services. Only 2% represented clients of infrastructure projects. Part of the reason is that it is hard to recognise or target clients of infrastructure markets. The other is because many infrastructures are state owned or temporally owned by the private sector (e.g., BOT/PPP projects), thus some of the real clients are hidden in the contractor and government agency category.

3. Research Findings

The authors analysed valid questionnaires using simple statistics to portray a general picture of KM application in the transport infrastructure sector. Findings are categorised into the six key KM elements described in the last section. Accordingly they are discussed in the following six sections.

3.1. KM Strategy

Knowledge management encompasses a systematic approach to managing the use of information. In this way, it provides a continuous flow of knowledge to enable the efficient and effective decision making by key decision makers. The approach is underpinned by a KM strategy, which enables an aim and focus for KM activities [22].

According to the results in Table 2, the logical link and necessity of adopting KM to promote sustainablity application in infrastructure developement has again been confirmed. It is widely accepted that sustainability considerations are very important (4.47/0.58) and managing related knowledge are considered greatly helpful (4.22/0.65). Respondents also believe that achievements in sustainability contribute to organisational performance.

Table 2. Linkage between infrastructure, sustainability and KM.

Statements	Mean	SD
Sustainability issues need to be considered when developing infrastructure projects.	4.47	0.58
Consideration of sustainability issues can help my organisation's performance.	4.08	0.95
Managing related knowledge will help in promoting the sustainability of infrastructure projects.	4.22	0.65
My organisation currently has a KM strategy or is willing to have a KM strategy.	3.47	0.92
In my organisation, there are specific KM criteria to manage sustainability knowledge.	3.14	0.89

Note: 1-Strongly Disagree, 2-Disagree, 3-Neutral, 4-Agree, 5-Srongly Agree.

With strong competition over the volatile economic conditions in the last decade, some construction organisations have begun using KM to obtain and maintain advantages [33]. However, according to this survey, most of the infrastructure practitioners hold reservations on the current usage of KM according to their experience. In their opinion, KM is still a new concept. For the infrastructure sector, there is no general KM process or framework to follow.

Table 3 shows the main impetus ranked by the respondents in promoting sustainability pursuit and application in the organisation. The biggest drivers in this regard are client awareness and requirement, business benefit, government regulation/legislation and social responsibility.

Table 3. Main Impetus for promoting sustainability pursuit and applications.

Rank	Impetus	Mean	SD
1	Client's awareness and requirement	4.45	0.77
2	Business benefits	4.10	0.87
3	Government regulations and legislation	4.04	0.96
4	Organisational reputation	3.88	0.88
5	Social responsibility	3.63	1.23
6	Community awareness	3.50	0.96
7	Improved competitiveness through labels, such as "Green Firm"	3.39	1.24
8	Threat of climate change and/or other global crisis	3.33	1.30
9	Problem solving	3.24	1.22
10	Use of new procurement method in which the developer is responsible for maintaining the project for a period (e.g., PPP, BOT)	3.10	1.25

Note: Level of Importance: 1 Low → 5 High.

Client awareness and requirement are the most important impetuses (highest mean, lowest standard deviation), something which is also frequently suggested by the literature. In real projects, if the client is not particularly interested in sustainability, the contractor usually is unlikely to actively pursue it, as the contracting industry will typically do just enough to get the project built and pushes the community or social issues back to the client. It is common that the whole development is driven by client's documentation. To truly progress sustainability, clients need to be adamant in the project proposal and be clear about the requirements.

The main barriers to the application of sustainability knowledge are listed in Table 4.

Table 4. Key Barriers to pursue and manage sustainability knowledge.

Rank	Barrie	Mean	SD
1	Difficult to measure the return on investment	4.04	1.02
2	Lack of co-ordination and consensus between stakeholders	3.73	0.95
3	Sustainability concepts are not well understood	3.73	1.00
4	Hard to connect sustainability target with organisational business strategy	3.61	0.81
5	Highly fragmentary nature of the industry	3.55	1.04
6	Will increase project budget/cost	3.47	1.26
7	No standardisation of key processes to follow	3.43	1.15
8	Project-oriented nature of the business development type	3.35	1.28
9	Labour-incentive nature of the infrastructure construction industry	3.16	1.39
10	Lack of top management's commitment	3.12	1.03
11	Complex nature of infrastructure project development	2.98	1.15
12	Poor financial resources	2.90	1.19
13	Poor non-financial resources	2.82	0.97
14	Long duration of infrastructure project development	2.74	1.22
15	Will extend project completion period	2.73	1.22
16	Will generate higher risk	2.67	1.39
17	High staff turnover	2.56	1.15
18	Intellectual Property protection issues	2.51	1.24

Note: Level of influence: 1 Low → 5 High.

"Business benefits" is ranked second in Table 3. This shows that sustainability will be strongly promoted if organisations can see tangible results from its application. However, as shown in Table 4, the biggest barrier is the difficulty in measuring investment returns in regards to sustainability (4.02/1.02). As sustainability applications in transport infrastructure are still being explored, industry best practices and exemplar projects are rare. The body of knowledge provides limited underpinning for the industry to tangibly measure the input and output of sustainability. Easy-to-follow frameworks and examples should be put on top of the development agenda.

The limited understanding of the sustainability concept and the lack of consensus between stakeholders are the second big hurdles which may strongly limit people's absorptive ability and differentiating ability towards more sustainable designs and activities. Literature suggested that the lack of general and professional education is one of the reasons [17,34]. Better decisions can only be made if all the stakeholders have abundant and up-to-date knowledge to tap into. It also makes a big difference if they deeply care about sustainability through internal volition. For example, it is common in the construction industry that people see sustainability as another expression of "environmental protection". Such views narrow the scope of professional responsibility, as well as the opportunities to perform beyond compliance.

3.2. The Body of Sustainability Knowledge

Table 5 describes the current quality of the body of sustainability knowledge. Respondents agree that sustainability knowledge is very subjective (4.12/0.63) and challenging to acquire (3.73/0.95). As suggested by the literature, sustainability can only be achieved though a multi-disciplinary approach, which is especially important for infrastructure development, as the knowledge is fragmentary and evolving constantly [20,33].

Table 5. Characteristics of sustainability knowledge for infrastructure development.

Rank	Statement	Mean	SD
1	It is subjective, means different things to different people.	4.12	0.63
2	It is challenging to acquire.	3.73	0.95
3	It is dynamic and evolving constantly.	3.61	0.95
4	It is fragmentary.	3.59	0.86
5	It is challenging to articulate and comprehend.	3.59	0.81
6	It is embedded in personal mind, hard to be codified and transferred.	3.33	0.88
7	It is challenging to be adjusted and used in other project.	3.22	0.94
8	It is immature to be implemented.	3.14	0.91
9	It is contextually/culturally sensitive.	2.94	0.94

Note: 1-Strongly Disagree, 2-Disagree, 3-Neutral, 4-Agree, 5-Srongly Agree.

Table 6 lists the main sources through which industry practitioners can obtain sustainability knowledge. People tend to find help within the organisation as colleagues and internal experts are ranked the highest (4.2/0.84, 3.92/0.84, respectively).

Table 6. Primary sources for sustainability knowledge.

Rank	Knowledge Resource	Mean	SD
1	Colleagues	4.2	0.84
2	Internal expert	3.92	0.82
3	Industry best practice	3.59	1.10
4	Deliverables from previous stages	3.49	0.84
5	External consultant	3.31	1.16
6	The construction project team	3.31	0.99
7	Internal training/workshop/seminar	3.29	1.08
8	Government agency	3.00	1.10
9	Quality Assurance Process	2.90	1.14
10	Research institution	2.86	0.96
11	Internal database	2.90	1.03
12	Industry association	2.73	1.25
13	Other industry	2.55	1.08
14	Other organisation	2.45	1.26
15	Local communities	2.49	1.06

Note: Level of Importance: 1 Low → 5 High.

It is common nowadays for organisations to use internal databases for information storage and applications. Many also invest heavily in computer tools that can generate various reports. However, respondents of this survey do not currently see internal databases as a highly valued knowledge resource. Reasons for this can vary. While specialised computer packages are important, better-structured catalogue systems and more user-friendly and controllable databases can be just as helpful. Furthermore, an index of expert and knowledge maps can be an efficient and applicable tool for locating sustainability knowledge.

Sustainability knowledge is people- and projects-based. Thus a "subject" KM approach should be developed which puts emphasis on ways to promote, motivate, encourage, nurture or guide the process of knowing, and abolishes the idea of trying to capture, then distribute knowledge. This view mainly understands KM as a social communication process, which can be improved by collaboration and cooperation support tools. In this approach, knowledge is closely tied to the person who developed it and is shared mainly through person-to-person contacts. The main role of Information and Communication Technology (ICT) in this case is to help people to communicate knowledge, not store it.

The authors also attempted to categorise sustainability knowledge according to respondents, however taxonomy could not be built by using the survey results. The top three ranked items are lessons learned from internal projects (4.24/0.85), staff personal capabilities/skills/experience (4.06/1.05) and industry best practice (4.00/0.89). This finding echoes the observation that the main carriers of sustainability knowledge are exemplar projects, as well as experienced and knowledgeable industry professionals.

3.3. Infrastructure Stakeholder Integration

Construction professionals, from both private and public sectors, are under pressure to maximise economically feasible, socially viable and environmentally accountable project outcomes. They will require the adoption of sustainability principles during project conception, design and planning, and innovative technologies and products during construction and operation. However, due to their different level of input and their professional influence, the implementation of sustainability in a certain project always highly relies on the stances of the key stakeholders and their priorities.

Table 7 lists the ranking of the influence of the main stakeholders in regard to the pursuit and management of sustainability knowledge, while Table 8 shows the current status of willingness of different stakeholders regarding the same.

The client is considered the most powerful stakeholder (4.49/0.82). This is consistent to findings in Table 5 that client's requirement is the most important driver for sustainability. Project sustainability needs to be driven from the very beginning, from the project conception and proposal stage. It also needs to involve the end-user early. Designers are the key people to embed sustainability concepts and principles into the project design while the project manager is the key person to transfer the design into the actual project by managing resources, time and providing funding. As the main stakeholders for the project, clients and designers together may source sustainability consultants for extra support. The consultant is ranked as comparatively the most enthusiastic stakeholder in sustainability knowledge pursuit and application. This is their core competitive advantage. However, the attendance and influence of the consultant in a project still depends on the endorsement of the key stakeholders, especially the client.

Table 7. Influence of main stakeholders to pursue and manage sustainability knowledge.

Rank	Stakeholder	Mean	SD
1	Client	4.49	0.82
2	Designer	4.02	0.72
3	Project manager	3.67	0.99
4	Consultant	3.57	0.76
5	Government agency	3.53	1.12
6	Engineer	3.49	1.02
7	Contractor	3.12	1.07
8	Local community	2.73	1.2
9	Research institution	2.27	0.97
10	Sub-contractor	2.18	0.95
11	Quantity surveyor	2.12	0.99

Note: Level of influence: 1 Low → 5 High.

Table 8. Willingness of main stakeholders to pursue and manage sustainability knowledge.

Rank	Stakeholder	Mean	SD
1	Consultant	3.65	0.67
2	Designer	3.59	0.76
3	Project Manager	3.43	0.68
4	Client	3.41	0.91
5	Engineer	3.41	0.64
6	Government Agency	3.31	0.89
7	Research Institution	3.29	0.96
8	Contractor	3.16	0.80
9	Local Community	3.08	1.00
10	Quantity Surveyor	2.37	0.88
11	Sub-contractor	2.20	0.84

Note: Level of influence: 1 Low → 5 High.

3.4. KM for Infrastructure Sustainability

Table 9 shows the main stages of managing sustainability knowledge in a typical KM cycle while Table 10 indicates the relative difficulties of these KM tasks. Knowledge application is seen as the most important (4.65/0.60) while also being the most challenging task ranked by the respondents (3.86/1.00). To promote infrastructure sustainability and deliver tangible results, knowledge application at the project level is the critical stage, which 'links talks to actions' while other stages may happen at the organisational level.

It is interesting to see that knowledge creation is seen as the second most challenging task (3.54/1.11), however as the lowest in regards to its importance to the respondents (3.55/0.94). According to the findings of a 2002 survey by the Australian Construction Industry Forum (ACIF), the Australian building and construction industry in general is a very slow innovator, lagging behind many other sectors [35]. Today, the industry still has not changed much, as most of the respondents did not recognise the importance of knowledge creation, or at least as not high on the agenda. From the survey, a brief KM cycle for sustainability knowledge management can be proposed, which integrates "identify knowledge, obtain knowledge, share knowledge, maintain knowledge and apply knowledge", while knowledge application in the actual projects should be especially emphasised. Researchers and practitioners around the globe also consider knowledge *using* ("application", "execution", "processing", "utilisation" were used by different authors as well) as the most essential activity within KM.

Table 9. Main KM process for infrastructure sustainability.

Rank	KM Stage	Mean	SD
1	Apply knowledge	4.65	0.60
2	Share knowledge	4.39	0.79
3	Maintain knowledge	4.00	0.71
4	Obtain knowledge	3.92	0.98
5	Identify knowledge	3.86	0.96
6	Contextualise knowledge	3.65	1.20
7	Measure knowledge	3.57	0.98
8	Search knowledge	3.57	1.04
9	Store knowledge	3.55	1.06
10	Create knowledge	3.55	0.94

Note: Level of Importance: 1 Low → 5 High.

Table 10. Challenging KM tasks.

Rank	KM Stage	Mean	SD
1	Apply knowledge	3.86	1.00
2	Create knowledge	3.54	1.11
3	Measure knowledge	3.51	0.96
4	Contextualise knowledge	3.41	1.22
5	Share knowledge	3.37	0.81
6	Maintain knowledge	3.22	0.87
7	Identify knowledge	3.19	1.08
8	Obtain knowledge	3.08	0.96
9	Search knowledge	2.71	1.08
10	Store knowledge	2.63	1.05

Note: Level of Difficulty: 1 Low → 5 High.

3.5. KM Enablers

Many factors may affect the success of KM initiatives within a project team or organization, by presenting a positive environment for knowledge management. Enablers for sustainability knowledge management were investigated and are shown in Table 11 in the order of importance.

Out of the 43 listed enablers, 35 are scored over 3.5 and 13 over 4.0 on the Likert scale. Leadership, communication skill and organisation culture are the top three enablers for sustainability knowledge management.

Due to limited space, individual enablers will not be specifically discussed here. However the survey results depict that "soft" enablers are much valued than the "hard" facilities. Managing sustainability knowledge should be driven by forming leadership, establishing supportive culture and promoting personal capability, rather than focusing on IT technology and systems.

Table 11. Main KM enablers (Level of Importance: 1 Low → 5 High).

Rank	KM Enabler	Mean	SD
1	Leadership	4.71	0.50
2	Communication skill	4.43	0.58
3	Organisational culture	4.41	0.76
4	Project target	4.22	0.55
5	Team work	4.16	0.80
6	Organisational mission, vision and strategy	4.13	0.82
7	Technical expertise	4.12	0.67
8	Training and education	4.08	0.67
9	Multi-discipline integration	4.04	0.99
10	Sense of social responsibility	4.04	0.96
11	T-shape skill (having knowledge and skills that are both deep and broad; multi-disciplinary)	4.04	0.84
12	Capability to codify thoughts, conceptions and experiences *etc.* into written document	4.00	0.58
13	Stakeholder integration	4.00	0.87
14	Financial support	3.98	0.88
15	Sense of social responsibility	3.98	0.97
16	Project budget	3.90	1.01
17	Creativity	3.88	0.88
18	Innovation capability	3.88	0.78
19	Trust	3.82	0.91
20	Trust between colleagues	3.78	0.85
21	Networks of subject matter expertise	3.78	0.80
22	Financial support	3.73	0.86
23	Project risk management	3.71	0.87
24	Government administration	3.71	1.35
25	Professional education	3.69	0.82
26	Research and design	3.67	0.90
27	Change management	3.67	0.83
28	Process and organisational structure	3.67	0.69
29	Time frame	3.55	0.96
30	Industry Codes of Conduct	3.55	0.91
31	Collaboration of academics and industry practice	3.49	0.79
32	Professional association action	3.35	1.18
33	Team structure	3.31	0.85
34	Document management	3.17	0.84
35	Time management	3.16	0.75
36	Administrative support	3.14	0.96
37	IT technology and infrastructure	3.14	0.91
38	Staff incentive schemes	3.04	0.96
39	Technology and IT support	3.00	0.84
40	Increasing average profit rate	2.94	1.23
41	Ambition	2.86	1.35
42	Human resource turnover	2.86	1.04
43	Loyalty to the organisation	2.69	1.12

The listed enablers can be divided into four categories: external environment, organisational environment, project environment and personal capability. In order to make the infrastructure sector more knowledgeable and active in pursuit of sustainability, organisations should establish appropriate culture, integrate sustainability to their organisational mission and strategy and provide relevant training. Furthermore, in the project team, leadership formulation, target setting and multidiscipline and stakeholders integrations are indispensible. Industry practitioners should demonstrate communication skills and improve technical expertise while developing cooperation with other disciplines and arousing a sense of social responsibility. The industry as a whole should improve its innovation ability and promote the liaison between relevant experts.

3.6. Outcomes of Managing Sustainability Knowledge

As the infrastructure sector is highly project-oriented, outcomes are divided into two parts in this study: intermediate outcome, which is mainly achievable at the project level, and organisation's performance. Multi-choices questions were used to identify what outcomes could be improved by managing sustainability knowledge. As shown in Table 12, project reputation is the top ranked project outcome (75.3%) in this regard. It is followed by employee's sense of responsibility, value delivery, organisation's ability to exploit market opportunity and customer satisfaction, which were chosen by more than half of the respondents. The top selected area in organisational performance is corporate reputation (83.3%) while customer recognition, intellectual assets and profit come next.

Many businesses have public relations departments dedicated to managing their reputation. The findings in this study explained why some leading construction contractors publish a regular Sustainability Report as part of the marketing scheme, as well as for consultancy purposes.

Although the direct link between promoting sustainability and profit has not been clearly built, half of the respondents believe profit levels can be increased through better usage of sustainability knowledge.

Table 12. Outcomes of managing sustainability knowledge.

Rank	Intermediate Outcomes	%	Performance	%
1	Project's reputation	73.5%	Corporate reputation	83.3%
2	Employee's sense of social responsibility	69.4%	Customer recognition	68.8%
3	Value delivery	65.3%	Intellectual asset	60.4%
4	Organisation's ability to exploit market opportunity	59.2%	Profit	50.0%
5	Customer satisfaction	53.1%	Market share	47.9%
6	Knowledge leadership	46.9%		
7	Organisational adaptability	42.9%		
8	Organisational creativity	40.8%		
9	Reusable content created	36.7%		

4. A Conceptual KM Framework

The survey results brought out a number of important issues with regards to managing sustainability knowledge in the infrastructure sector. The key messages are highlighted as follows.

- Internal enablers are more important that external enablers, especially at the project level.
- Enablers are mainly soft in nature. IT tools are not considered very important.
- The first important driver of sustainability knowledge pursuit and application is the Awareness and Requirement of clients, who traditionally do not actively pursue sustainability knowledge.
- The monitoring process for sustainability application should be entwined with project management processes throughout the project and should involve the client.
- Sustainability knowledge is highly people centred. Indexing expert and best practice index can be applicable. A "subjective" KM approach should be adopted which focuses on communication and networking among industry practitioners and subject-matter experts.
- Promoting industry best practice should be at the top of the action plan for the transport sector.
- Sustainability knowledge application is the most important phase of the whole KM cycle; however, it is also the most challenging.
- The role of knowledge creation warrants more study as the respondents ranked the importance of knowledge creation the lowest, despite they believe it is highly challenging.

- According to the characteristics of sustainability knowledge, seeking consensus among stakeholders is a first priority task.
- It is not often possible to increase profit through managing knowledge for sustainability, as project sustainability itself is not often directly linked to profits. However, other important aspects of organizational performance can be improved to realize profit gains indirectly.
- Governments can impact on the KM of transport projects in the following ways:
    - Enable KM through administering the transport infrastructure sector;
    - Be a knowledge provider;
    - Promote sustainability considerations through regulations and legislation;
    - Be the most influential stakeholder and the most important driver of sustainability applications (as the client of the infrastructure projects).

The above analyses provide important clues as to how to formulate strategies on managing knowledge in the transport industry sector. Accordingly, a preliminary conceptual KM framework is proposed and shown in Figure 4.

The conceptual KM framework is organised by input, processes, and outcomes. With a focus on KM processes, it highlights the "what-to-do" aspects of an integrated knowledge management approach. Combining findings of this study and previously developed KM theories, a new KM process is proposed with recommendations on five key aspects of managing Sustainable Knowledge (SK): identify, obtain, share, apply and maintain. The framework recommends the use of specific KM strategies to facilitate KM activities such as communication of knowledgeable practitioners, recordkeeping, use and transfer of good industry practices, and stakeholder integration.

Future research will focus on "how-to-do-it" aspects according to the framework, for its potential application in construction organisations and actual infrastructure projects. The next stage of work will focus on the establishment of recommendations and action guidance for transport infrastructure practitioners. To reach this goal, interviews and case studies are considered the most appropriate mechanism. They are being carried out to elicit information.

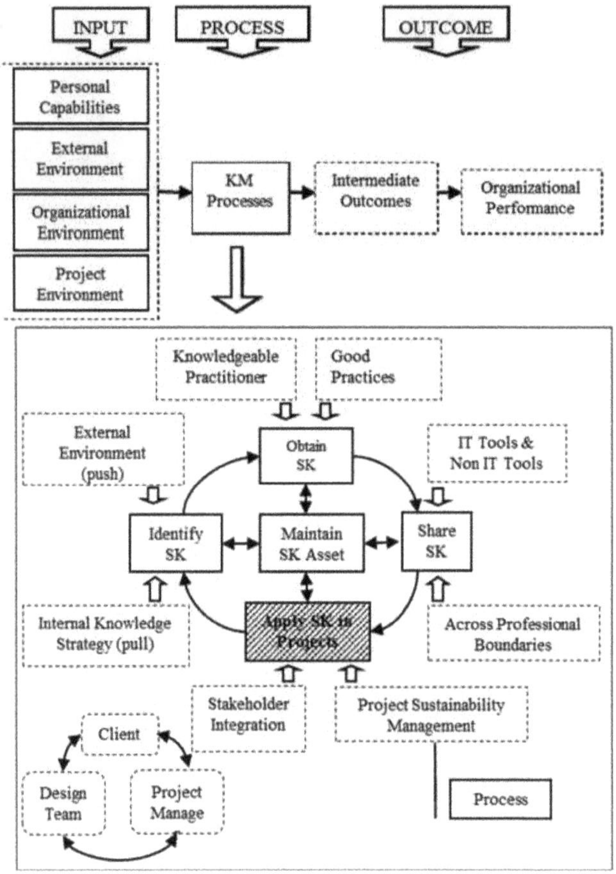

Figure 4. A conceptual KM framework.

5. Conclusions

The transport infrastructure sector is characterised by its highly project-oriented nature, a relatively conservative culture, long and arduous development processes, enormous financial undertakings, and significant potential impact on the natural environment. Therefore, it is imperative for infrastructure development to respond to the sustainability challenge. While few would argue against such principles, the perspectives, priorities and indeed the actual approaches adopted by the practitioners can largely vary, thus putting project end sustainability deliverables at risk.

Knowledge management (KM) has been proven as a useful tool to promote innovation and change in many industry sectors including building construction. However, very limited studies have been done on transport infrastructure sustainability from a KM perspective. No existing KM framework provides a holistic approach to address sustainability issues in transport projects. Through a comprehensive questionnaire survey, this research explored how sustainability knowledge is currently considered in the Australian infrastructure sector and identified the main challenges for its management. The holistic conceptual KM framework proposed here outlines the important aspects and possible strategies of managing knowledge to promote sustainability.

Interviews with practitioners and case studies are now underway to explore appropriate KM enablers, their inherent relationships and implementation strategies. This will lead to the specific procedures and guidelines of KM practice suited for this sector. By investigating the various forms of sustainability knowledge, illustrating main knowledge activities and priority requirements, and suggesting effective strategies for managing sustainability knowledge, research such as this contributes to new dimensions of disciplinary knowledge and promotes infrastructure sustainability outcomes.

Within a relatively short timeframe, this research confines itself to the issues affecting the management of sustainability knowledge at the macro level of transport infrastructure projects. Future research should consider the unique characteristics of each type of transport project, as the adopted approaches in highway projects may be different to those in bridge construction. Detailed relationships between key issues may also be examined in future. For example, contrary to the notion that innovation is critical for the success of sustainable development [36,37], this survey found that interviewees, as transport infrastructure practitioners, see creating knowledge as the least important but the second most challenging KM activity. There may be specific reasons behind this that warrant further investigation.

Author Contributions: The research reported in this paper was conducted by Mei Yuan and Jay Yang. Tan Yigitcanlar, Peter Newman, and Frank Schultmann contributed to knowledge management, infrastructure sustainability and decision making aspects of this paper respectively. All authors have read and approved the final manuscript.

Conflicts of Interest: The authors declare no conflict of interest.

References

1. AGIC. AGIC Charter, the Australian Green Infrastructure Council (AGIC). Available online: http://www.agic.net.au/final_agic_charter_for_website_18_january_2010-1.pdf (accessed on 18 June 2015).
2. Sahely, H.R.; Kennedy, C.A.; Adams, B.J. Developing sustainability criteria for urban infrastructure systems. *Can. J. Civ. Eng.* **2005**, *32*, 72–85. [CrossRef]
3. Goh, K.C.; Yang, J. Importance of sustainability related cost components in highway infrastructure: Perspective of stakeholders in Australia. *ASCE J. Infrastruct. Syst.* **2014**, *20*, 04013002. [CrossRef]
4. Byrnes, J.; Dollery, B.; Crase, L.; Simmons, P. Resolving the infrastructure funding crisis in Australian local government: A bond market issue approach based on local council income. *Australas. J. Reg. Stud.* **2008**, *14*, 167–175.
5. Newton, P.W. Regenerating cities: Technological and design innovation for Australian suburbs. *Build. Res. Inf.* **2013**, *41*, 575–588. [CrossRef]
6. Newman, P. Density, the Sustainability Multiplier: Some Myths and Truths with Application to Perth, Australia. *Sustainability* **2014**, *6*, 6467–6487. [CrossRef]
7. Mallon, K.; Burton, D. *Synthesis Report: Climate Change and Infrastructure Expert Summit*; Climate Risk Pty Limited: Sydney, Australia, 2009.
8. Dollery, B.; Kortt, M.A.; Grant, B. Harnessing Private Funds to Alleviate the Australian Local Government Infrastructure Backlog. *Econ. Papers J. Appl. Econ. Policy* **2012**, *31*, 114–122. [CrossRef]
9. Yigitcanlar, T.; Teriman, S. Rethinking sustainable urban development: Towards an integrated planning and development process. *Int. J. Environ. Sci. Technol.* **2015**, *12*, 341–352. [CrossRef]
10. SPA2009 (2009) Sustainable Planning Act, Queensland State Government. Available online: https://www.legislation.qld.gov.au/LEGISLTN/CURRENT/S/SustPlanA09.pdf (accessed on 18 June 2015).
11. SPICOLA (2014) Sustainable Planning (Infrastructure Charges) and Other Legislation Amendment Act, Queensland State Government. Available online: http://www.dsdip.qld.gov.au/planning-and-development/sustainable-planning-act-2009.html#SPICOLA (accessed on 18 June 2015).
12. Markard, J. Transformation of Infrastructures: Sector Characteristics and Implications for Fundamental Change. *J. Infrastruct. Syst.* **2011**, *17*, 107–117. [CrossRef]
13. Kibert, C.J.; Rinker, M.E. The next generation of sustainable construction. *Build. Res. Inf.* **2007**, *35*, 595–601. [CrossRef]
14. Ugwu, O.; Haupt, T.C. Key performance indicators and assessment methods for infrastructure sustainability—A South African construction industry perspective. *Build. Environ.* **2007**, *42*, 665–680. [CrossRef]

15. Yang, J.; Yuan, M. Managing knowledge to promote sustainability for infrastructure development. In Proceedings of the CIB 3rd International Conference on Smart and Sustainable Built Environment (SASBE2009), Delft, The Netherlands, 15 June 2009; p. 66.
16. Rusly, F.; Sun, P.Y.T.; Corner, J.L. The impact of change readiness on the knowledge sharing process for professional service firms. *J. Knowl. Manag.* **2014**, *18*, 687–709. [CrossRef]
17. Lim, S.K.; Yang, J. Developing frameworks and processes to enhance sustainability deliverables in infrastructure projects. In *Clients Driving Construction Innovation: Benefiting from Innovation*; Icon. Net Pty Ltd.: Brisbane, Australia, 2008; pp. 233–240.
18. Pathirage, C.P.; Amaratunga, D.G.; Haigh, R.P. Tacit knowledge and organisational performance: Construction industry perspective. *J. Knowl. Manag.* **2007**, *11*, 115–126. [CrossRef]
19. Yigitcanlar, T.; Dur, D.; Dizdaroglu, D. Towards prosperous sustainable cities: A multiscalar urban sustainability assessment approach. *Habitat Int.* **2015**, *45*, 36–46. [CrossRef]
20. Wallace, B. *Becoming Part of the Solution: The Engineer's Guide to Sustainable Development*; ACEC/American Council of Engineering Companies: Washington, DC, USA, 2005.
21. Teece, D. Strategies for Managing Knowledge Assets: The Role of Firm Structure and Industrial Context. *Long Range Plan.* **2010**, *33*, 35–54. [CrossRef]
22. Dalkir, K. *Knowledge Management in Theory and Practice*; Elsevier Inc.: Oxford, UK, 2005.
23. Sandhawalia, B.S.; Dalcher, D. Developing knowledge management capabilities: A structured approach. *J. Knowl. Manag.* **2011**, *15*, 313–328. [CrossRef]
24. Kamara, J.M.; Augenbroe, G.; Anumba, C.J.; Carrillo, P.M. Knowledge management in the architecture, engineering and construction industry. *Constr. Innov. Inf. Process Manag.* **2002**, *2*, 53–67. [CrossRef]
25. Jafari, M.; Fesharaki, M.N.; Akhavan, P. Establishing an integrated KM system in Iran aerospace industries organization. *J. Knowl. Manag.* **2007**, *11*, 127–142. [CrossRef]
26. Robinson, H. *Governance and Knowledge-Management for Public-Private Partnerships*; John Wiley and Sons: New York, NY, USA, 2010.
27. Anumba, C.J.; Egbu, C.; Carrillo, P. *Knowledge Management in Construction*; Wiley-Blackwell: Oxford, UK, 2005.
28. Venters, W.; Cornford, T.; Cushman, M. Knowledge about sustainability: SSM as a method for conceptualising the UK construction industry's knowledge environment. *J. Comput. Inf. Technol.* **2004**, *13*, 137–148. [CrossRef]
29. Shelbourn, M.A.; Bouchlaghem, D.M.; Anumba, C.J.; Carillo, P.; Khalfan, M.M.K.; Glass, J. Managing knowledge in the context of sustainable construction. *ITcon* **2006**, *11*, 57–71.
30. Maqsood, T.; Walker, D.H.T. Facilitating knowledge pull to deliver innovation through knowledge management. *Eng. Constr. Archit. Manag.* **2007**, *14*, 94–109. [CrossRef]
31. Tan, H.C.; Carrillo, C.J.; Bouchlaghem, D.; Kamara, J.; Udeaja, C. *Capture and Reuse of Project Knowledge in Construction*; Wiley-Blackwell: Hoboken, NJ, USA, 2010.
32. Witherspoon, C.L.; Bergner, J.; Cockrell, C.; Stone, D.N. Antecedents of organizational knowledge sharing: A meta-analysis and critique. *J. Knowl. Manag.* **2013**, *17*, 250–277. [CrossRef]
33. Yigitcanlar, T.; Bulu, M. Dubaization of Istanbul: Insights from the knowledge-based urban development journey of an emerging local economy. *Environ. Plan. A.* **2015**, *47*, 89–107. [CrossRef]
34. Bartholomew, D. *Building on Knowledge: Developing Expertise, Creativity and Intellectual Capital in the Construction Professions*; Wiley-Blackwell: Oxford, UK, 2008.
35. PricewaterhouseCoopers. *Innovation in the Australian Building and Construction Industry—Survey Report*; Australian Construction Industry Forum (ACIF): Canberra, Australia, 2002.
36. Vollenbroek, F.A. Sustainable development and the challenge of innovation. *J. Clean. Prod.* **2002**, *10*, 215–223. [CrossRef]
37. Newman, L. Uncertainty, innovation, and dynamic sustainable development. *Sustain. Sci. Pract. Policy* **2005**, *1*, 25–31.

sustainability

MDPI

Review

Trees in Canadian Cities: Indispensable Life Form for Urban Sustainability

Peter N. Duinker [1,*], Camilo Ordóñez [1], James W. N. Steenberg [2], Kyle H. Miller [3], Sydney A. Toni [1] and Sophie A. Nitoslawski [1]

1. School for Resource and Environmental Studies, Dalhousie University, 6100 University Ave., Halifax, NS B3H 4R2, Canada; camilo.ordonez@dal.ca (C.O.); sydney.toni@dal.ca (S.A.T.); sophie.nitoslawski@dal.ca (S.A.N.)
2. Environmental Applied Science and Management, Ryerson University, Toronto, ON M5B 2K3, Canada; james.steenberg@ryerson.ca
3. Geography and Program in Planning, University of Toronto, Toronto, ON M5S 3G3, Canada; kyle.miller@mail.utoronto.ca

* Author to whom correspondence should be addressed; peter.duinker@dal.ca; Tel.: +1-902-494-7100; Fax: +1-902-494-3728.

Academic Editors: Tan Yigitcanlar and Md. Kamruzzaman

Received: 25 March 2015; Accepted: 3 June 2015; Published: 9 June 2015

Abstract: We argue that a healthy urban forest contributes immensely to the sustainability of cities. The argument is based on a comprehensive array of values elicited from Canadians in several cities. To begin, we define the urban forest as inclusive of all the trees in the city and thus representing the predominant contributor to a city's green infrastructure. Then we enumerate and explain the broad diversity of ways in which urban people value trees in the city. We, thus, show the myriad pathways by which trees contribute positively to any city's social, economic, and ecological sustainability. Following a short summary of the ways in which trees may detract from people's quality of life, we present promising management directions for urban-forest improvement, as we understand the situation in Canada. We conclude that all cities can enhance their sustainability by improving the urban forest.

Keywords: sustainability; tree; urban forest; forest values

1. Introduction

The world is decidedly urban today, given that more than half of all humans live in cities [1]. The proportion is much higher in countries like Canada, where it is well over 80% [2]. The global trend toward urban living will continue for some decades. Clearly, the sustainability of human civilization depends to a great degree on determining sustainable pathways for city living and urban development.

Propositions for how to make urban living more sustainable are numerous and diverse and address a wide range of themes (Table 1). Such themes are, of course, not independent of each other; indeed, they are often deeply intertwined. In this paper, we argue that one particular type of plant, implicit in the "greenspace and biodiversity" theme in Table 1, has the potential to make immense contributions to city sustainability, not only in their own right but also as strong influences on most of the other themes. Those plants are trees, and together they define the urban forest.

Table 1. A selection of themes in the discourse of urban sustainability.

Theme	Reference(s)
Community/Social Cohesion	[3,4]
Education	[5]
Employment	[6]
Energy	[7,8]
Food Security	[9,10]
Greenspace and Biodiversity	[11,12]
Human Health	[13,14]
Justice and Human Rights	[15–18]
Transportation	[19–21]
Urban Form and Design	[22,23]
Water and Air Quality	[24–26]

Our objective is to demonstrate, through the lens of values, the potency and promise of trees and their associated organisms to enhance sustainability in any urban setting. As Table 1 shows, urban sustainability consists of a broad range of elements. Without doubt, all these elements need to be carefully and diligently addressed in the pursuit of city sustainability. In the context of Canadian cities, our position on the centrality of trees is this: because people value trees so strongly and for such a wide range of benefits and services, and because both urban residents [27] and professionals [28] agree that more trees are highly desirable in the city, trees therefore represent a superlative opportunity for advancing urban sustainability.

By focusing on trees in this article, we do not imply that other plants are irrelevant. Indeed, shrubs and non-woody annual and perennial plants, even manicured grass, can be considered vital components of a city's green infrastructure. Some of the values we describe later in the paper can well be satisfied by these other types of plants, especially when they exist in association with trees. Trees are special and unique in the plant community for many reasons (see below), perhaps most profoundly because they predominate in contributing to the vertical dimension of the plant community on account of their height.

We begin with a contextual and conceptual discussion on sustainability and urban forests so as to lay a foundation under our claims of trees' contributions to sustainability. The main part of the paper highlights the diverse suite of values people ascribe to trees in the city (these are sometimes also called urban-forest benefits or services). To be fair to a comprehensive treatment of how trees and people in the city interact, we identify several ways in which trees can be detrimental to urban dwellers. Finally, we explain the main avenues for improving urban forests, particularly in the Canadian setting in which we, as urban-forest scholars, have most of our personal and professional experiences. The paper draws heavily on our research on these topics across Canada, as well as our ongoing work in support of planning and implementation of new directions for the urban forest in Halifax, Canada [29].

2. Concepts of Sustainability in Relation to Cities

As the places where most people on the planet live and work, cities have become central to both global socioeconomic development and environmental sustainability [30]. The concepts of development and sustainability are intertwined; entities such as the Organisation for Economic Co-operation and Development and the World Bank are now promoting urban sustainability for its value in attracting investment [31]. With virtually all worldwide population growth until 2050 expected to take place in urban areas, strategies for managing resource use and environmental degradation have become essential [32].

A city is, by itself, unsustainable. While calling them "parasites on the biosphere" may be extreme [33] (p. 290, as cited in [32]), it is true that cities depend on vast hinterlands for both essential resources and waste disposal [34]. Thus, rather than in absolute terms, urban sustainability is commonly conceptualized as a city's ability to maximize relative socioeconomic and environmental benefits while minimizing harm [35]. Within these broad categories, considerations of urban sustainability include inter- and intra-generational equity, protection of the natural environment, minimization of natural resource use, and community and individual well-being [36]. Considerable ambiguity remains in the definition of urban sustainability, but ambiguity is arguably desirable, because it allows communities to adopt individual understandings of sustainability that incorporate local values, circumstances, and environmental considerations [32].

Because cities concentrate people and wealth, they play a crucial role in global efforts to address climate change, but they also concentrate societal and infrastructural vulnerability to natural disasters [37]. However, while city living in developed countries generally remains more environmentally friendly than the alternatives [38], the reverse may hold in developing countries, although that is slowly changing [1]. Many cities, particularly in middle-income countries, have begun to lead by example, promoting compact urban forms, encouraging non-motorized modes of transportation, and (re)introducing efficiency concepts, such as passive solar design [39]. For true urban sustainability, cities must implement a wide range of interventions, from high-level policy to new technological solutions, to adjustments in building materials and streetscape [32].

The last two decades have also witnessed increasing focus on urban sustainability reporting, from Maclaren [36] identifying a dearth of clear reporting methods to the recent compilation of many new urban sustainability indicator methods [35]. Many urban sustainability indicator frameworks now explicitly address the urban forest canopy, including ISO standard #37120, "Sustainable Development of Communities" [40]. It is clear that urban sustainability is a complex phenomenon requiring concerted attention to a host of economic, social, and environmental issues.

3. Conceptions of Urban Forests

In dialogues and policy forums across the globe, from sub-national to global levels, there is a wide range of definitions and conceptions of forest and tree. A forest is usually conceived as a tree-dominated ecosystem. A key question, though, is what "dominated" means. When are the trees of insufficient size or density such that the ecosystem in question, rural or urban, is not considered a forest? In the urban setting, there are abundant sites and locations where the land cover is dominated by built infrastructure. Sometimes there are vast expanses of mowed grass with a few scattered trees. A conception of urban forest needs to account for such situations.

Defining a tree is somewhat less arbitrary, but there are still difficulties. One is the distinction between a shrub and a tree, both of which are woody plants but, at maturity, small ones are defined as shrubs whereas large ones are trees. For some species, the size threshold is arbitrary. For this paper, a tree is a living stem of a species determined by local, regional, or national authorities to be a tree species. Even minuscule seedlings of tree species are still trees.

Given the above, what is an urban forest? We find the most helpful conception to be an inclusion of all trees in the town or city [41]. In our own setting of the city of Halifax, the municipal government has jurisdiction over an area greater than a half-million hectares. The municipality has some 400 km of Atlantic coastline, and is largely wooded beyond the urban core. The Urban Forest Master Plan [29] applies only to what is called the urban core, which was defined as the territory in which all homes are serviced with municipal sewers and water.

Thus, once the urban boundary has been defined, the urban forest is defined as all trees within that boundary. The circumstances under which any urban tree might find itself are varied indeed. The definition is blind to land ownership. Whether trees are on private land (*i.e.*, homes and businesses) or government land (*i.e.*, municipal, provincial, national), they are all part of the urban forest. Whether a tree is standing alone in a grassy park or a tree pit on a busy commercial street, in a row of trees along a residential street, grouped in a back yard, or in a dense stand in an urban wooded park, it is part of the urban forest. The discussion below on values associated with urban trees makes it clear that each and every tree in the city has the potential to make a significant contribution to sustainability of the urban environment. Each and every tree is a contributor to the city's green infrastructure.

4. How Trees Contribute to Sustainability of Cities

The premise of this paper is that trees make strong contributions to desirable ecological, social, and economic conditions in any city, which can be interpreted as enhancing the city's sustainability. While the literature is replete with accounts of forest benefits, services, and values, the treatments of these in individual accounts is either broad and shallow—as in bare lists—or narrow and deep, as when one forest value, e.g., carbon sequestration, is examined at length. Our presentation below aims at a compromise between the two—our list is long, and we have presented a short explanation for each item presented.

We prefer the term "value" in our research because it helps citizens give considered responses to questions about what they deem to be important about trees in the city (see [42] for further justification). Our research group has recently engaged in studies of urban-forest values in selected Canadian cities (e.g., [43,44]), and has produced materials promoting trees in cities through enumeration of diverse forest values [45]. We draw upon this work in the following summary account of more than twenty ways in which trees contribute to urban sustainability. The list could be much longer–we regard the following ones as the leading forest values.

We acknowledge potential shortcomings of such a list. For example, some values (e.g., shade) are pre-requisites for or important influences on other values (e.g., extension of road-surface life, and enhancement of human heath). A reader may judge some of the values as substantially more important than others. The list is not logically established based on deductive reasoning–it arises from empirical research based mainly on what urban citizens say in answer to the importance question. Perhaps the values could be clustered into meaningful classes such as economic, social, and ecological values (e.g., [46]). All that said, we stand by the list as a broad suite of values identified, using the very words found below, in our survey, interview, and focus-group elicitations involving hundreds of people across Canada [27,43,44].

A Suite of Urban-Forest Values

Trees enhance the aesthetic beauty of the city. The idea that trees are pretty may be self-evident to any urban dweller as visions of green usually inspire calm and pleasant feelings. Indeed, city trees provide a wide set of visual pleasures, from the different colours of the leaves to the size and texture of tree trunks [47]. Trees also have sounds, such as when leaves rustle in the wind or a carpeted floor of fallen leaves is walked upon. Trees give people joy through the senses.

Trees provide shade. Shade is at the root of so many of the social, economic, and environmental benefits provided by trees in the city. It is a vital service in promoting people's health, as shade trees block harmful ultraviolet radiation [48]. Shade trees, especially ones that are strategically situated where people congregate outdoors like in parks and on sidewalks, can help to mitigate the risk of skin cancer. Children are especially vulnerable to skin cancer, and thus trees in playgrounds and school yards have the potential of significantly reducing their exposure and risk of skin cancer. Through shade, trees sustain human health.

Trees help conserve fuel by reducing emissions from parked vehicles. Outdoor parking lots can be considered miniature urban heat islands with extensive impervious surfaces and low albedo. Vehicles parked in sunlight heat up and emit hydrocarbons into the air from the fuel in their tanks. Well-placed trees can reduce the amount of fuel that evaporates into the atmosphere from vehicles [49]. Thus, trees conserve fuel while concurrently helping to sustain the urban atmosphere.

Trees cool the city environment. Due to their built infrastructure, cities frequently experience higher temperatures than the surrounding countryside—this is the urban heat-island effect [50]. Trees reduce ambient air temperatures by altering wind speeds, shading surfaces, and blocking solar radiation [51,52]. Trees also transpire water vapour into the air and, thus, cool it [53,54]. Consequently, the more trees there are in the city, the greater the cooling effect will be. This is especially relevant given expected climatic change and the associated warmer temperatures.

Trees clean the air. Air pollution is an issue for most cities, and urban trees help improve air quality. Gaseous pollutants like ground-level ozone, sulphur dioxide, nitrogen dioxide, and carbon monoxide are removed from the atmosphere by trees by absorbing them through their leaves [55]. Particulate matter is also removed from the air and stored temporarily on the plant surfaces until it washes off in the rain [56]. Overall air quality is better with more trees in the city.

Trees foster health and healing. Trees help purify the air and reduce rates of asthma and other respiratory illnesses in urban populations [57,58]. When we fall ill, or are convalescing from illness, trees can play a pivotal role in healing. Landmark research by Ulrich [59] showed that surgery patients recovered faster and better when the view through their hospital bedroom window was dominated by trees rather than another building. Research in Japan has shown clear health benefits from exposure to forest landscapes [60].

Trees enhance community safety. Security and safety are serious concerns in a city. Among the psychological causes of insecurity and feelings of danger, we find mental fatigue and elevated levels of stress. Although earlier research suggested that dense and naturally vegetated areas were perceived as insecure and threatening (e.g., [61]), these perceptions have been changing through the years with our understanding of how mental fatigue and stress are mitigated by green space and its most dominant feature–trees. More recent research shows that residents living in greener surroundings report lower levels of fear, fewer incivilities, and less aggressive and violent behaviour [62].

Trees increase property values. Trees in the city are a major component of a neighbourhood's aesthetic appeal and benefit homeowners by adding monetary value to properties. For Portland, Oregon, Donovan and Butry [63] found that a large tree on a residential property can add some $9000 to the sale price of a house. Land owners are also helping others by having trees on their residential properties, because adjacent homes and even entire neighbourhoods benefit from the increased property value. This is an important message to communicate to homeowners, since there is frequently more available space to plant trees on private residential properties than next to publicly owned streets.

Trees reduce energy costs. In some places, roughly 5%–10% of urban electricity demand is spent on cooling/heating buildings. Trees around buildings and houses can act as heat insulators and heat absorbers, shielding buildings from a high-temperature environment through shade, or keeping buildings from losing their heat in winter by increasing the humidity of the surrounding area and slowing down wind [64]. Simulations in Canadian cities have shown that an increase in a neighbourhood's tree cover by about three trees per house reduces the heating energy of that house by up to 10% and the cooling energy by up to 40% [65]. The annual savings in heating and cooling costs can reach the hundreds of dollars every year depending on house size.

Trees prolong the life of infrastructure. Trees help reduce the amount of maintenance and repair required for city streets, thus reducing costs against the city budget. The asphalt used to pave streets is made up of aggregate held together by asphalt cement. The asphalt cement is a petroleum product, which breaks down and evaporates in the sunlight, causing streets to crack and eventually crumble into potholes, which need to be repaired, or the whole street repaved, at great cost. McPherson and Muchnik [66] found that just a 20% shading of streets in Modesto, California, could save 60% of resurfacing costs over a 30-year period. This service provided by trees is a huge incentive for engineers and indeed all municipal managers to increase tree canopy over asphalt surfaces.

Trees capture and store carbon. Atmospheric carbon dioxide is one of the main drivers of climate change [67]. Its concentration in the air is rising largely because of the burning of fossil fuels like coal, oil, and gas [67]. Anything we can do to slow down emissions of carbon dioxide and increase the rate of its removal from the air will be good for the future of cities [68]. Trees capture carbon dioxide from the air and store the carbon in their trunks, roots, and branches. As long as a tree is healthy and growing, it stores increasing amounts of carbon. The more trees we have in the city, and the larger and longer they grow, the more carbon dioxide will be taken out of the atmosphere [69].

Trees slow down stormwater flow and consequently improve water quality. Urban land is covered mainly by roads, sidewalks, rooftops, and parking lots. Most of these surfaces are impervious and prevent rainwater from being absorbed directly into the ground. Consequently, stormwater and wastewater systems, as well as natural water bodies, are strained during heavy rains as runoff flows into them off impervious surfaces. Excessive runoff can lead to flooding, sewage spillover, and aquatic pollution. This is especially the case in cities like Halifax, where the older parts of the city have combined storm and sanitary sewers. City trees intercept some amounts of rainfall and retain it in their foliage for a period of time [70]. Trees, thus, provide a critical economic service in stormwater management.

Trees provide employment opportunities. As they grow up into overhead wires, shed their leaves, grow new branches in undesirable directions, drop dead branches, or die, trees are cared for by municipal workers as well as private landscape contractors and other specialized tree caretakers. Salaries make up large proportions of the budget of tree-care organizations. The more trees there are in a city, the greater the amount of economic activity associated with their maintenance [71]. Trees represent smart societal investments because they demand expenditures on caring for vital urban green infrastructure.

Trees support business activity. The services provided by trees provide tangible financial benefits to business owners. Research has shown that consumers perceive business districts with trees as better places to shop [72]. Moreover, consumers say they are willing to pay higher prices, travel further and longer, and shop longer and more frequently in areas with green streetscapes [73]. This not only benefits business owners, but also provides incentive for them to become more actively involved in the stewardship of urban trees.

Trees enhance recreational opportunities. City residents frequently visit treed areas for recreation. Recreation in these areas can be passive or active, ranging from gentle activities such as cultural events, walking, picnics, or tree climbing, to active sports such as running or biking [74,75]. The many types of urban forest formats, ranging from treed streets to dense and naturalized forest remnants, serve diverse recreational uses [76].

Trees enhance tourism. Among the few studies linking urban forests and tourism, Majumdar *et al.* [77] concluded that for Savannah, Georgia, the better the urban forest, the more attractive the city is for tourists. This seems a reasonable conclusion considering that most city residents would like more trees and better urban-forest management in their own cities [27]. All other things being equal, it seems fair to say that tourists would prefer to visit a well-wooded city as opposed to one with few trees.

Trees provide diverse foods. Trees have been a source of food for people throughout the ages. Urban settings are highly suitable for growing the full range of fruit and nut trees [78]. Additionally, there may be opportunities to pick edible berries and mushrooms that grow on the forest floor of treed parks and other naturalized areas. Thus, trees can contribute, even if in a small way, to food security in the city.

Trees conserve biodiversity. Biodiversity, in the simplest terms, refers to the full diversity of life on earth and includes the diversity of gene pools, species, communities and ecosystems. Trees themselves represent important elements of biodiversity, but they also serve as host and habitat for a wide range of other organisms. The ability of trees to contribute to urban biodiversity increases as one moves from single isolated trees to lines of trees along streets and lanes, and further to stands of trees in parks and other areas [79]. Urban forests can contribute immensely to biodiversity conservation through inclusion of the full range of native tree species in their full spectrum of ages and community associations [80].

Trees promote learning opportunities. Trees provide habitat for many kinds of wild organisms, including fungi, insects, lichens, birds, mammals, and other vascular plants. An excellent focus to start learning about terrestrial nature in the city is the trees. Indeed, there is no better place to learn about nature than to be in it [81]. Getting away from the city and out into natural forests can be costly and may even be impossible for some people. The alternative is to study nature in the city. Trees in the city can provide excellent opportunities to learn about the kinds of species and natural ecosystems there are in the countryside and the wilderness. Research has also shown that urban trees can enhance the learning capacity of learners [82].

Trees impart a sense of place. Feeling a sense of belonging in the city is important to its citizens. The vegetation of a locale can contribute strongly to this sense of place. The presence of trees transforms barren areas into pleasant, welcoming spaces that infuse the city environment with a positive sense of self [83]. It has been shown that well-kept treed neighbourhoods serve to strengthen the ties among residents, generating a sense of place and stewardship among neighbours [84]. This in turn generates important civic values such as a greater sense of safety and adjustment, more use of neighbourhood common spaces, and fewer incivilities [85].

Trees contribute to a sense of well-being. The main argument is that the more that people can experience nature, the better they feel, emotionally, mentally, and physically [11,86,87]. The lack of nature in cities means that many people cannot benefit from it as directly as they might. In many cities around the world, trees dominate the natural ecosystems. If we are to bring nature to the people in an urban environment, that means more trees, not just in total but also more trees in naturalized conditions (see [88]). A healthy urban forest contributes to a healthy and happy people.

To sum up, we conclude that trees boast an extensive and diverse array of values to cities. We suggest that they are the greatest contributors to urban sustainability of all forms of plants. In absolute terms, the contributions of trees to urban sustainability are substantial. The more trees there are in the city, the better the city can serve as a good place to live.

5. Can Trees Detract from Urban Sustainability?

Unfortunately, trees can be detrimental to urban dwellers in several ways. We believe that the benefits of trees outweigh the costs, but these potential inconveniences or risks should be taken into consideration when planning a sustainable city. Some costs may be unavoidable; some stem from the location of the tree in relation to infrastructure, others from tree species selection. Yet other costs relate to how well or how badly the built infrastructure accommodates trees.

Trees can damage infrastructure. Roots may crack sidewalks, and branches or whole trees can fall and cause damage during extreme weather events such as windstorms [89,90]. Roots are also know to invade underground drainage pipes [91], although this normally happens in the context of leaky pipes. Trees may require frequent maintenance to ensure they do not damage infrastructure, such as pruning programs to keep tree branches out of overhead wires [92]. Moreover, natural or disturbance-related mortality and the presence of standing deadwood both present a similar hazard or significant costs for removal, as is the case currently with widespread ash (*Fraxinus* spp.) mortality caused by the emerald ash borer (*Agrilus planipennis*) [93]. Similarly, trees can create unwanted debris, such as sap and leaves [93]. Lastly, infrastructure and people may be put at risk by the increase in wildfire potential that comes with an increase in urban trees [29].

Humans can have allergic reactions to the pollen produced by trees and plants in the urban forest [94,95]. Treed areas can create feelings of unease due to real or perceived danger [96]. Urban residents occasionally identify heightened feelings of unease and fear of crime in some treed urban areas due to the potential of concealed individuals or activities [85,97,98]. Trees provide habitat for other species, but not all these species may be desirable. Urban wildlife can annoy or threaten pets and humans, or damage plants and structures (e.g., birds and fruit trees) [99]. For example, in Helsinki, Finland, the excrement on sidewalks from aphids in linden trees produced a vomit-like odour during warm spells [100].

Tree shade may be valued by some but be seen as an inconvenience by others. Trees located over flower or food gardens can provide undesirable shade that limits the growth of more-desired plants, such as vegetables [101]. They may also hinder the growth of garden plants in the understory when their needles and leaves decompose. Thus, the location of a tree in its ecological and social contexts influences whether the tree's services are of value.

Similarly, tree shade that lowers summer cooling costs could increase winter heating costs [102]. Trees might be positioned in a way that limits winter sunlight, turning shade into a disservice [103]. While we generally think of street trees as cooling agents, they also have the potential locally to increase air temperatures. For example, a canopy of immature trees can block airflow but still allow solar radiation to strike and heat the ground [51].

The touted improvements in quality of life that trees bring can result in increased financial costs to nearby residents. The increase in property values from trees may result in an increased property tax for the owner. Dwyer *et al.* [104] conservatively estimated that, in the United States, $1.5 billion/year in property taxes might be attributed to the value trees add to property. As trees increase the value of the spaces and buildings around them, the increased values benefit owners wishing to sell but represent higher real-estate costs for would-be buyers.

Trees and parks are differentially distributed in cities. Ethnic/racial minorities and low-income residents have lower access to green space [105,106]. While city planning that incorporates tree planting and park development in lower-income areas aims to rectify this social injustice, it also increases the desirability of these areas and can contribute to their gentrification, pushing out the original residents [107].

More trees means more maintenance. The equipment used for tree maintenance is usually powered by fossil fuels, so engine emissions need to be factored against trees' carbon-storage capabilities [102]. These costs might start to outweigh the benefits of trees if urban vegetation is short-lived or stressed, and requires constant attention or removal and replacement [108]. Lastly, some trees may not always be the optimal choice for a sustainable city. The water demands of many tree species may make them unsuitable for cities in arid and semi-arid climates [109].

Despite the above drawbacks of trees in the city, we reiterate our own views about the net benefits of trees, as well as what we perceive to be the trend in Canada based on our research [27,28]: the benefits of trees in the city far outweigh the drawbacks, and hence both professionals and citizens favour increases in urban tree canopy. When urban forests are managed well, as outlined below, the net benefits can be significantly increased.

6. Enhancing Urban Forests for City Sustainability

Shaping the urban forest so it delivers on citizens' value expectations requires active and careful intervention—there is no passage without fare. Investment is needed to improve the supply of urban-forest services while also mitigating potential disservices. The majority of trees in the city require substantial management actions to maintain function and ensure tree establishment and survival. Cities represent rather different conditions from those under which tree species have evolved. An abundance of stressors and disturbances afflict city trees (e.g., poor soils, pollution, invasive species, vandalism, poor management), resulting in urban forests suffering relatively high rates of tree decline and mortality [110]. Ensuring that city dwellers live well among trees, and the challenges associated with doing so, has therefore spawned a wide array of policies and practises for maintaining and enhancing urban forests.

Urban forestry in North America, as a practice and profession that separated from traditional forestry in the 1960s, represents an amalgamation of forestry with horticulture, arboriculture, and landscape architecture [111]. Arguably, broad establishment and acceptance of urban forestry was cemented by the Dutch elm disease (*Ophiostoma novo-ulmi*), which ravaged the extensively planted elm (*Ulmus* spp.) populations of North American cities and first exposed the public to the consequences of widespread urban canopy loss [112]. While urban forest management in many ways remains reactive and practice-oriented, being driven by immediate disturbance and threats like the Dutch elm disease, more advanced and progressive models of urban forest governance are emerging [113]. Urban forest management today continues to grow beyond operational necessities like tree planting, maintenance, and removal.

Across Canada, the urban forest is increasingly becoming an item on municipal planning agendas and many cities are creating policies and strategic plans that address their urban forest [28,114]. The design and implementation of tree protection regulations for public and private properties are becoming commonplace in larger cities [115]. Many of these regulations focus on the formal planning process, since development practices and land-use change are among the largest contributors to urban tree mortality and canopy loss [116]. Trees also continue to garner attention in the urban design process as vital pieces of green infrastructure, where the realities of growing trees that survive to maturity is no longer an afterthought in landscape architecture and engineering [117]. Experiments are underway to explore how to grow trees on buildings—examples include the famous Hundertwasserhaus in Vienna and the Bosco Verticale (Vertical Forest) in Milan. The use of underground structural soil cells in high-density streetscapes is a prime example of designing cities for trees rather than adapting trees to cities [118].

Another growing trend in urban forest enhancement is incentive-based programs that offer financial or non-financial motivation for residents, businesses, and organizations to take part in urban forest stewardship activities (e.g., free-tree programs) with the long-term goal of changing citizen behaviours and fostering stewardship [119,120]. Lastly, many cities are adopting comprehensive and strategic urban forest management plans [28]. These plans generally consist of guidelines for tree planting and species selection, tree maintenance (e.g., pruning-cycle establishment), pest management, conservation goals, and performance standards [121]. Management plans are an important step for communities to acknowledge urban forests as a public good and a key stage of policy development for ensuring explicit and consistent goals for long-term sustainable urban forest management [122].

The complexity and heterogeneity of the urban forest and the growing importance of cities have demanded a more enlightened and interdisciplinary approach to understanding and managing them. This requires not only maintenance operations and municipal policies, but also ongoing research from the social, natural, and applied sciences, as well as partnerships involving governments, industry, academia, and communities [123]. In fact, today it is argued that urban forest management and governance have become particularly innovative as they often involve partnerships with a variety of non-government stakeholders, such as environmental non-governmental organizations, citizen associations, landowners, and industry [113]. Indeed, these partnerships are arguably necessary given the complex and fragmented ownership regimes of cities and their urban forests. Ultimately, it is the partnership between humans and trees that contributes to the overall sustainability of the modern city.

7. Conclusions

Municipal managers across Canada seek to increase the proportion of the urban landscape dominated by trees. This objective rests firmly on the understanding that trees are, on balance, extremely good for people and the city environment. We have shown above that trees offer a relatively much stronger and broader array of benefits to people compared to their negative aspects, suggesting that they are strong net contributors to urban sustainability. Careful and sensitive management of a city's tree population, with an eye to reducing tree disservices and increasing tree services, as well as catering to people's tree-related values, can substantially enhance those contributions.

Urban sustainability is a journey of improving city conditions economically, socially, and environmentally. It aims to improve the human condition in respect of people's health and welfare. City development is most often dominated by construction and maintenance of grey infrastructure, mostly buildings and transportation networks. Without question, such infrastructure is needed for a city to function. However, at one end of the spectrum, grey infrastructure can be developed with no regard to the quantity and quality of green infrastructure, *i.e.*, trees and other plants. At the other end, it can be designed and installed to cater well to, and even celebrate, the green infrastructure. Sustainable urban development will proceed well when cities are reconceived from concentrations of concrete, steel, and asphalt sprinkled about with a few amenity trees, to a natural or semi-natural landscape into which concrete, steel, and asphalt are judiciously introduced. Where the former predominates, urban redevelopment is warranted with emphasis on renewal and expansion of green infrastructure. If the latter exists anywhere in the world, it is to be emulated and repeated. Trees are indeed indispensable for city sustainability.

Acknowledgments: Our urban-forest research has been funded by the Social Sciences and Humanities Research Council of Canada as well as Halifax Regional Municipality. We are grateful for the comments of two anonymous reviewers on an earlier version of the paper.

Author Contributions: Peter Duinker conceived the idea for the paper, served as principal investigator of all research grants and contracts pertaining to the topic, wrote several sections, submitted the manuscript, and implemented revisions. Camilo Ordóñez, James Steenberg, Kyle Miller, and Sydney Toni each drafted specific sections of the manuscript, and Sophie Nitoslawski edited the entire paper and contributed to early revisions. Ordóñez, Steenberg, Toni, and Nitoslawski are students or former students of Duinker and contributed to the many discussions that lie behind the concepts of urban-forest sustainability as documented herein.

Conflicts of Interest: The authors declare no conflict of interest. The research sponsors had no role in: the design of the study; the collection, analyses, and interpretation of data; the writing of the manuscript, and the decision to publish the results.

References

1. UN-HABITAT. *The State of the World's Cities 12/13—Prosperity for Cities*; United Nations HABITAT Programme, United Nations Human Settlements Programme (UN-HABITAT): Nairobi, Kenya, 2012; Available online: http://www.unhabitat.org/pmss/ (accessed on 1 March 2015).
2. Statistics Canada. Population, urban and rural, by province and territory (Canada). Available online: http://www.statcan.gc.ca/tables-tableaux/sum-som/l01/cst01/demo62a-eng.htm (accessed on 1 March 2015).
3. Dempsey, N.; Bramley, G.; Power, S.; Brown, C. The social dimensions of sustainable development: Defining urban social sustainability. *Sustain. Dev.* **2011**, *19*, 289–300. [CrossRef]
4. Vojnovic, I.; Darden, J.T. Class/racial conflict, intolerance, and distortions in urban form: Lessons for sustainability from the Detroit region. *Ecol. Econ.* **2013**, *96*, 88–98. [CrossRef]
5. Trencher, G.P.; Yarime, M.; Kharrazi, A. Co-creating sustainability: Cross-sector university collaborations for driving sustainable urban transformations. *J. Clean Prod.* **2013**, *50*, 40–55. [CrossRef]
6. Bugliarello, G. Urban sustainability: Dilemmas, challenges and paradigms. *Technol. Soc.* **2006**, *28*, 19–26. [CrossRef]
7. Hawkey, D.; Webb, J.; Winskel, M. Organisation and governance of urban energy systems: District heating and cooling in the UK. *J. Clean Prod.* **2013**, *50*, 22–31. [CrossRef]
8. Uyarra, E.; Gee, S. Transforming urban waste into sustainable material and energy usage: The case of Greater Manchester (UK). *J. Clean Prod.* **2013**, *50*, 101–110. [CrossRef]
9. Koc, M.; MacRae, R.; Mougeot, L.J.A.; Welsh, J. (Eds.) *For Hunger-Proof Cities: Sustainable Urban Food Systems*; International Development Research Centre (IDRC): Ottawa, ON, Canada, 1999.
10. Barthel, S.; Isendahl, C. Urban gardens, agriculture, and water management: Sources of resilience for long-term food security in cities. *Ecol. Econ.* **2013**, *86*, 224–234. [CrossRef]
11. Chiesura, A. The role of urban parks for the sustainable city. *Landsc. Urban. Plan.* **2004**, *68*, 129–138. [CrossRef]
12. Dearborn, D.C.; Kark, S. Motivations for conserving urban biodiversity. *Conserv. Biol.* **2010**, *24*, 432–440. [CrossRef] [PubMed]
13. Jackson, L.E. The relationship of urban design to human health and condition. *Landsc. Urban. Plan.* **2003**, *64*, 191–200. [CrossRef]
14. Grahn, P.; Stigsdotter, U.K. Landscape planning and stress. *Urban. For. Urban. Green.* **2003**, *2*, 1–18. [CrossRef]
15. Stephens, C. Healthy cities or unhealthy islands? The health and social implications of urban inequality. *Environ. Urban.* **1996**, *8*, 9–30. [CrossRef]
16. Di Chiro, G.D. Nature as Community: The Convergence of Environment and Social Justice. In *The Struggle for Environmental Democracy: Environmental Justice Movements in the United States*; Faber, D.R., Ed.; The Guilford Press: New York, NY, USA, 1998.

17. Marcuse, P.; Connolly, J.; Novy, J.; Olivo, I.; Potter, C.; Steil, J. *Searching for the Just City: Debates in Urban Theory and Practice*; Routledge: New York, NY, USA, 2009.
18. Pearsall, H.; Pierce, J. Urban sustainability and environmental justice: Evaluating the linkages in public planning/policy discourse. *Local Environ.* **2010**, *15*, 569–580. [CrossRef]
19. Kenworthy, J.R.; Laube, F.B. Automobile dependence in cities: An international comparison of urban transport and land use patterns with implications for sustainability. *Environ. Impact Assess.* **1996**, *16*, 279–308. [CrossRef]
20. Himanen, V.; Lee-Gosselin, M.; Perrels, A. Sustainability and the interactions between external effects of transport. *J. Transp. Geogr.* **2005**, *13*, 23–28. [CrossRef]
21. Richardson, B.C. Sustainable transport: Analysis frameworks. *J. Transp. Geogr.* **2005**, *13*, 29–39. [CrossRef]
22. Badland, H.; Schofield, G. Transport, urban design, and physical activity: An evidence-based update. *Transport. Res. D* **2005**, *10*, 177–196. [CrossRef]
23. Ahern, J. Urban landscape sustainability and resilience: The promise and challenges of integrating ecology with urban planning and design. *Landsc. Ecol.* **2013**, *28*, 1203–1212. [CrossRef]
24. Hellström, D.; Jeppsson, U.; Kärrman, E. A framework for systems analysis of sustainable urban water management. *Environ. Impact Assess.* **2000**, *20*, 311–321. [CrossRef]
25. Fenger, J. Urban air quality. *Atmos. Environ.* **1999**, *33*, 4877–4900. [CrossRef]
26. Milman, A.; Short, A. Incorporating resilience into sustainability indicators: An example for the urban water sector. *Global Environ. Chang.* **2008**, *18*, 758–767. [CrossRef]
27. Ordóñez, C.; Duinker, P.N.; Sinclair, J.A.; Beckley, T.M. Public values of urban forests in Canada using a sidewalk-interception survey in Fredericton, Halifax, and Winnipeg. *Arboricult. Urban. For.* 2015; in press.
28. Ordóñez, C.; Duinker, P.N. An analysis of urban forest management plans in Canada: Implications for urban forest management. *Landsc. Urban. Plan.* **2013**, *116*, 36–47. [CrossRef]
29. HRM Urban Forest Planning Team. *Halifax Regional Municipality (HRM) Urban Forest Master Plan*; Halifax Regional Municipality (HRM): Halifax, NS, Canada, 2013; Available online: https://www.halifax.ca/Property/UFMP/index.php (accessed on 1 March 2015).
30. Matusitz, J. Glurbanization theory: An analysis of global cities. *Int. Rev. Sociol.* **2010**, *20*, 1–14. [CrossRef]
31. Davidson, K.; Gleeson, B. The sustainability of an entrepreneurial city? *Int. Plan. Stud.* **2014**, *19*, 173–191. [CrossRef]
32. Vojnovic, I. Urban sustainability: Research, politics, policy and practice. *Cities* **2014**, *41*, S30–S44. [CrossRef]
33. Odum, E.P. *Ecology: Bridge between Science and Society*; Sinauer Associates Incorporated: Sunderland, MA, USA, 1997.
34. Rees, W.; Wackernagel, M. Urban ecological footprints: Why cities cannot be sustainable—And why they are a key to sustainability. *Environ. Impact Assess.* **1996**, *16*, 223–248. [CrossRef]
35. Mori, K.; Yamashita, T. Methodological framework of sustainability assessment in City Sustainability Index (CSI): A concept of constraint and maximisation indicators. *Habitat Int.* **2015**, *45*, 10–14. [CrossRef]
36. Maclaren, V.W. Urban sustainability reporting. *J. Am. Plann. Assoc.* **1996**, *62*, 184–202. [CrossRef]
37. Hoornweg, D.; Freire, M.; Lee, M.J.; Bhada-Tata, P.; Yuen, B. (Eds.) *Cities and Climate Change: Responding to An Urgent Agenda*; The World Bank: Washington, DC, USA, 2011.
38. Van Oers, R.; Roders, A.P. Editorial: Historic cities as model of sustainability. *J. Cult. Herit. Manag. Sustain. Dev.* **2012**, *2*, 4–14. [CrossRef]
39. Jabareen, Y. Planning the resilient city: Concepts and strategies for coping with climate change and environmental risk. *Cities* **2013**, *31*, 220–229. [CrossRef]
40. ISO. *Sustainable Development of Communities—Indicators for City Services and Quality of Life*; ISO 37120:2014; International Standards Organization (ISO): Geneva, Switzerland, 2014; Available online: http://www.iso.org/iso/catalogue_detail?csnumber=62436 (accessed on 1 March 2015).
41. Konijnendijk, C.C.; Ricard, R.M.; Kenney, A.; Randrup, T.B. Defining urban forestry: A comparative perspective of North America and Europe. *Urban. For. Urban. Green.* **2006**, *4*, 93–103. [CrossRef]
42. Ordóñez, C.; Duinker, P.N. Interpreting sustainability for urban forests. *Sustainability* **2010**, *2*, 1510–1522. [CrossRef]
43. Peckham, S.; Duinker, P.N.; Ordóñez, C. Urban Forest Values in Canada: Views of citizens in Calgary and Halifax. *Urban. For. Urban. Green.* **2013**, *12*, 154–162. [CrossRef]

44. Sinclair, J.A.; Diduck, J.; Duinker, P.N. Elicitation of urban forest values from residents of Winnipeg, Canada. *Can. J. For. Res.* **2014**, *44*, 922–930. [CrossRef]
45. CUFRG. *In Support of Trees in the City: A Message for Municipal Councillors, Developers, and NGOs*; Canadian Urban Forest Research Group (CUFRG), School for Resource and Environmental Studies, Dalhousie University: Halifax, NS, Canada, 2013; Available online: http://www.canadianurbanforest.ca (accessed on 1 March 2015).
46. Moyer, J.N.; Owen, R.J.; Duinker, P.N. Forest Values: A framework for old-growth forest with implications for other forest conditions. *Open For. Sci. J.* **2008**, *1*, 27–36. [CrossRef]
47. Smardon, R.C. Perception and aesthetics of the urban environment: Review of the role of vegetation. *Landsc. Urban. Plan.* **1988**, *15*, 85–106. [CrossRef]
48. Heisler, G.M.; Grant, R.H. Ultraviolet radiation in urban ecosystems with consideration of effects on human health. *Urban. Ecosyst.* **2000**, *4*, 193–229. [CrossRef]
49. Scott, K.I.; Simpson, J.R.; McPherson, E.G. Effects of tree cover on parking lot microclimate and vehicle emissions. *J. Arboricult.* **1999**, *25*, 129–142.
50. Solecki, W.D.; Rosenzweig, C.; Parshall, L.; Pope, G.; Clark, M.; Cox, J.; Wiencke, M. Mitigation of the heat island effect in urban New Jersey. *Global Environ. Chang.* **2005**, *6*, 39–49. [CrossRef]
51. Heisler, G.M. Effects of individual trees on the solar radiation climate of small buildings. *Urban. Ecol.* **1986**, *9*, 337–359. [CrossRef]
52. Rosenfeld, A.H.; Akbari, H.; Romm, J.J.; Pomerantz, M. Cool communities: Strategies for heat island mitigation and smog reduction. *Energy Build.* **1998**, *28*, 51–62. [CrossRef]
53. Streiling, S.; Matzarakis, A. Influence of single and small clusters of trees on the bio-climate of a city: A case study. *J. Arboricult.* **2003**, *29*, 309–316.
54. Armson, D.; Stringer, P.; Ennos, A.R. The effect of tree shade and grass on surface and globe temperatures in an urban area. *Urban. For. Urban. Green.* **2012**, *11*, 245–255. [CrossRef]
55. Nowak, D.J.; Crane, D.E.; Stevens, J.C. Air pollution removal by urban trees and shrubs in the United States. *Urban. For. Urban. Green.* **2006**, *4*, 115–123. [CrossRef]
56. Freer-Smith, P.H.; Holloway, S.; Goodman, A. The uptake of particulates by an urban woodland: Site description and particulate composition. *Environ. Pollut.* **1997**, *95*, 27–35. [CrossRef] [PubMed]
57. Tiwary, A.; Sinnett, D.; Peachey, C.; Chalabi, Z.; Vardoulakis, S.; Fletcher, T.; Leonardi, G.; Grundy, C.; Azapagic, A.; Hutchings, T.R. An integrated tool to assess the role of new planting in PM10 capture and the human health benefits: A case study in London. *Environ. Pollut.* **2009**, *157*, 2645–2653. [CrossRef] [PubMed]
58. Donovan, G.H.; Butry, D.T.; Michael, Y.L.; Prestemon, J.P.; Liebhold, A.M.; Gatziolis, D.; Mao, M.Y. The relationship between trees and human health: Evidence from the spread of the emerald ash borer. *Am. J. Prev. Med.* **2013**, *44*, 139–145. [CrossRef] [PubMed]
59. Ulrich, R. View through a window may influence recovery from surgery. *Science* **1984**, *224*, 420–421. [CrossRef] [PubMed]
60. Lee, A.C.K.; Maheswaran, R. The health benefits of urban green spaces: A review of the evidence. *J. Public Health* **2011**, *33*, 212–222. [CrossRef] [PubMed]
61. Schroeder, H.W. Environmental perception rating scales: A case for simple methods of analysis. *Environ. Behav.* **1984**, *16*, 573–598. [CrossRef]
62. Kuo, F.E.; Sullivan, W.C. Environment and crime in the inner city: Does vegetation reduce crime? *Environ. Behav.* **2001**, *33*, 343–367. [CrossRef]
63. Donovan, G.H.; Butry, D.T. Trees in the city: Valuing street trees in Portland, Oregon. *Landscape Urban. Plan.* **2010**, *94*, 77–83. [CrossRef]
64. Akbari, H.; Pomerantz, M.; Taha, H. Cool surfaces and shade trees to reduce energy use and improve air quality in urban areas. *Solar Energy* **2001**, *70*, 295–310. [CrossRef]
65. Akbari, H.; Taha, H. The impact of trees and white surfaces on residential heating and cooling energy use in four Canadian cities. *Energy* **1992**, *17*, 141–149. [CrossRef]
66. McPherson, E.G.; Muchnick, J. Effects of street tree shade on asphalt concrete pavement performance. *J. Arboricult.* **2005**, *31*, 303–310.
67. IPCC. *Climate Change 2014: Synthesis Report*; Cambridge University Press: Cambridge, UK; New York, NY, USA, 2014.

68. Hoornweg, D. Cities and climate change: An urgent agenda. In *Sustainable Low-Carbon City Development in China*; Baeumler, A., Ijjasz-Vasquez, E., Mehndiratta, S., Eds.; The World Bank: Washington, DC, USA, 2012; pp. 3–32.
69. Nowak, D.J.; Crane, D.E. Carbon storage and sequestration by urban trees in the USA. *Environ. Pollut.* **2002**, *116*, 381–389. [CrossRef]
70. Xiao, Q.; McPherson, E.G.; Ustin, S.L.; Grismer, M.E.; Simpson, J.R. Winter rainfall interception by two mature open-grown trees in Davis, California. *Hydrol. Process.* **2000**, *14*, 763–784. [CrossRef]
71. City of Oakville. *Urban Forest Strategic Management Plan, Town of Oakville: 2008–2027*; Kenney, A., Ed.; Urban Forest Innovations Inc. (UFII), City of Oakville: Oakville, ON, Canada, 2008; Available online: http://www.oakville.ca/forestry.htm (accessed on 1 March 2015).
72. Wolf, K.L. Trees and business district preferences: A case study of Athens, Georgia, US. *J. Arboricult.* **2004**, *30*, 336–346.
73. Wolf, K.L. Business district streetscapes, trees, and consumer response. *J. For.* **2005**, *103*, 396–400.
74. Giles-Corti, B.; Donovan, R.J. The relative influence of individual, social and physical environment determinants of physical activity. *Soc. Sci. Med.* **2002**, *54*, 1793–1812. [CrossRef]
75. Gobster, P.H.; Westphal, L.M. The human dimensions of urban greenways: Planning for recreation and related experiences. *Landsc. Urban. Plan.* **2004**, *68*, 147–165. [CrossRef]
76. Nowak, D.J.; Noble, M.H.; Sisinni, S.M.; Dwyer, J.F. People and Trees: Assessing the US Urban Forest Resource. *J. Forest* **2001**, *99*, 37–42.
77. Majumdar, S.; Deng, J.; Zhang, Y.; Pierskalla, C. Using contingent valuation to estimate the willingness of tourists to pay for urban forests: A study in Savannah, Georgia. *Urban. For. Urban. Green.* **2011**, *10*, 275–280. [CrossRef]
78. Clark, K.H.; Nicholas, K.A. Introducing urban food forestry: A multifunctional approach to increase food security and provide ecosystem services. *Landsc. Ecol.* **2013**, *28*, 1649–1669. [CrossRef]
79. Adams, L.W. In our own backyard: Conserving urban wildlife. *J. For.* **1994**, *92*, 24–25.
80. Alvey, A.A. Promoting and preserving biodiversity in the urban forest. *Urban. For. Urban. Green.* **2006**, *5*, 195–201. [CrossRef]
81. Louv, R. *The Nature Principle: Human Restoration and the End of Nature-Deficit Disorder*; Algonquin Books: Chapel Hill, NC, USA, 2011.
82. Taylor, A.F.; Kuo, F.E.; Sullivan, W.C. Coping with ADD: The surprising connection to green play settings. *Environ. Behav.* **2001**, *33*, 54–77. [CrossRef]
83. Hull, R.B.; Lamb, M.; Vigob, G. Place identity: Symbols of self in the urban fabric. *Landsc. Urban. Plan.* **1994**, *28*, 109–120. [CrossRef]
84. Elmendorf, W. The importance of trees and nature in community: A review of the relative literature. *Arboricult. Urban. For.* **2008**, *34*, 152–156.
85. Kuo, F.E.; Sullivan, W.C. Aggression and violence in the inner city: Effects of environment via mental fatigue. *Environ. Behav.* **2001**, *33*, 543–571. [CrossRef]
86. Jay, M.; Schraml, U. Understanding the role of urban forests for migrants–uses, perception and integrative potential. *Urban. For. Urban. Green.* **2009**, *8*, 283–294. [CrossRef]
87. Qin, J.; Zhou, X.; Sun, C.; Leng, H.; Lian, Z. Influence of green spaces on environmental satisfaction and physiological status of urban residents. *Urban. For. Urban. Green.* **2013**, *12*, 490–497. [CrossRef]
88. Toni, S.; Duinker, P.N. A framework for urban-woodland naturalization in Canada. *Environ. Rev.* **2015**, in press.
89. Lopes, A.; Oliveira, S.; Fragoso, M.; Andrade, J.A.; Pedro, P. Wind risk assessment in urban environments: The case of falling trees during windstorm events in Lisbon. In *Bioclimatology and Natural Hazards*; Střelcová, K., Mátyás, C., Kleidon, A., Lapin, M., Matejka, F., Blaženec, M., Škvarenina, J., Holécy, J., Eds.; Springer Netherlands: Berlin, Germany, 2009; pp. 55–74.
90. McPherson, E.G. Expenditures associated with conflicts between street tree root growth and hardscape in California, United States. *J. Arboricult.* **2000**, *26*, 289–297.
91. Fisher, D.R.; Svendsen, E.S.; Connolly, J. *Urban. Environmental Stewardship and Civic Engagement: How Planting Trees Strengthens the Roots of Democracy*; Routledge: Oxon, UK, 2015.
92. Powell, A.S.; Lindquist, E.S. Effects of power-line maintenance on forest structure in a fragmented urban forest, Raleigh, NC. *Southeast. Nat.* **2011**, *10*, 25–38. [CrossRef]

93. Kovacs, K.F.; Haight, R.G.; McCullough, D.G.; Mercader, R.J.; Siegert, N.W.; Liebhold, A.M. Cost of potential emerald ash borer damage in U.S. communities, 2009–2019. *Ecol. Econ.* **2010**, *69*, 569–578. [CrossRef]
94. Cariñanos, P.; Casares-Porcel, M. Urban green zones and related pollen allergy: A review. Some guidelines for designing spaces for low allergy impact. *Landsc. Urban. Plan.* **2011**, *101*, 205–214. [CrossRef]
95. D'Amato, G.; Cecchi, L.; Bonini, S.; Nunes, C.; Annesi-Maesano, I.; Behrendt, H.; Liccardi, G.; Popov, T.; van Cauwenberge, P. Allergic pollen and pollen allergy in Europe. *Eur. J. Allergy Clin. Immunol.* **2007**, *62*, 976–990. [CrossRef] [PubMed]
96. Talbot, J.F.; Kaplan, R. Needs and fears: The response to trees and nature in the inner city. *J. Arboricult.* **1984**, *10*, 222–228.
97. Koskela, H.; Pain, R. Revisiting fear and place: Women's fear of attack and the built environment. *Geoforum* **2000**, *31*, 269–280. [CrossRef]
98. Jorgensen, A.; Anthopoulou, A. Enjoyment and fear in urban woodlands—Does age make a difference? *Urban. For. Urban. Green.* **2007**, *6*, 267–278. [CrossRef]
99. Clucas, B.; Marzluff, J.M. Coupled relationships between humans and other organisms in urban areas. In *Urban Ecology: Patterns, Processes, and Applications*; Niemelä, J., Ed.; Oxford University Press: New York, NY, USA, 2011; pp. 135–147.
100. Lyytimaki, J.; Petersen, L.K.; Normander, B.; Bezak, P. Nature as a nuisance? Ecosystem services and disservices to urban lifestyle. *J. Integr. Environ. Sci.* **2008**, *5*, 161–172. [CrossRef]
101. Fraser, E.D.G.; Kenney, W.A. Cultural background and landscape history as factors affecting perceptions of the urban forest. *J. Arboricult.* **2000**, *26*, 106–113.
102. Nowak, D.J.; Dwyer, J.F. Understanding the benefits and costs of urban forest ecosystems. In *Urban and Community Forestry in the Northeast*; Kuser, J.E., Ed.; Springer: New York, NY, USA, 2007; pp. 25–46.
103. Tyrväinen, L. Economic valuation of urban forest benefits in Finland. *J. Environ. Manag.* **2001**, *62*, 75–92. [CrossRef] [PubMed]
104. Dwyer, J.F.; McPherson, E.G.; Schroeder, H.W.; Rowntree, R.A. Assessing the benefits and costs of the urban forest. *J. Arboricult.* **1992**, *18*, 227–234.
105. Pham, T.; Apparicio, P.; Séguin, A.; Landry, S.; Gagnon, M. Spatial distribution of vegetation in Montreal: An uneven distribution or environmental inequity? *Landsc. Urban. Plan.* **2012**, *107*, 214–224. [CrossRef]
106. Landry, S.M.; Chakraborty, J. Street trees and equity: Evaluating the spatial distribution of an urban amenity. *Environ. Plan. A* **2009**, *41*, 2651–2670. [CrossRef]
107. Wolch, J.R.; Byrne, J.; Newell, J.P. Urban green space, public health, and environmental justice: The challenge of making cities "just green enough". *Landsc. Urban. Plan.* **2014**, *125*, 234–244. [CrossRef]
108. Nowak, D.J.; Stevens, J.C.; Sisinni, S.M.; Luley, C.J. Effects of urban tree management and species selection on atmospheric carbon dioxide. *J. Arboricult.* **2002**, *28*, 113–122.
109. Roloff, A.; Korn, S.; Gillner, S. The climate-species-matrix to select tree species for urban habitats considering climate change. *Urban. For. Urban. Green.* **2009**, *8*, 295–308. [CrossRef]
110. Nowak, D.J.; Kuroda, M.; Crane, D.E. Tree mortality rates and tree population projections in Baltimore, Maryland, USA. *Urban. For. Urban. Green.* **2004**, *2*, 139–147. [CrossRef]
111. Miller, R.W. *Urban Forestry: Planning and Managing Urban Greenspaces*, 2nd ed.; Waveland Press Inc.: Long Grove, IL, USA, 1997.
112. Johnston, M. A brief history of urban forestry in the United States. *Arboricult. J.* **1996**, *20*, 257–278. [CrossRef]
113. Lawrence, A.; de Vreese, R.; Johnston, M.; van den Bosch, C.K.; Sanesi, G. Urban forest governance: Towards a framework for comparing approaches. *Urban. For. Urban. Green.* **2013**, *12*, 464–473. [CrossRef]
114. Steenberg, J.W.N.; Duinker, P.N.; Charles, J.D. The neighbourhood approach to urban forest management: The case of Halifax, Canada. *Landsc. Urban. Plan.* **2013**, *117*, 135–144. [CrossRef]
115. Conway, T.M.; Urbani, L. Variations in municipal urban forestry policies: A case study of Toronto, Canada. *Urban. For. Urban. Green.* **2007**, *6*, 181–192. [CrossRef]
116. Kenney, W.A.; Idziak, C. The state of Canada's municipal forests—1996 to 1998. *For. Chron.* **2000**, *76*, 231–234. [CrossRef]
117. Jim, C.Y. Sustainable urban greening strategies for compact cities in developing and developed economies. *Urban. Ecosyst.* **2013**, *16*, 741–761. [CrossRef]
118. Urban, J. *Up by Roots: Healthy Soils and Trees in the Built Environment*; International Society of Arboriculture (ISA): Champaign, IL, USA, 2008.

119. Bengston, D.N.; Fletcher, J.O.; Nelson, K.C. Public policies for managing urban growth and protecting open space: Policy instruments and lessons learned in the United States. *Landsc. Urban. Plan.* **2004**, *69*, 271–286. [CrossRef]
120. Randrup, T.B.; McPherson, E.G.; Costello, L.R. Tree root intrusion in sewer systems: Review of extent and costs. *J. Infrastruct. Syst.* **2001**, *7*, 26–31. [CrossRef]
121. Van Wassenaer, P.J.E.; Schaeffer, L.; Kenney, W.A. Strategic planning in urban forestry: A 21st century paradigm shift for small town Canada. *For. Chron.* **2000**, *76*, 241–246. [CrossRef]
122. Clark, J.R.; Matheny, N.P.; Cross, G.; Wake, V. A model of urban forest sustainability. *J. Arboricult.* **1997**, *23*, 17–30.
123. Konijnendijk, C.C. Enhancing the forest science-policy interface in Europe: Urban forestry showing the way. *Scand. J. For. Res.* **2004**, *19*, 123–128. [CrossRef]

sustainability

MDPI

Article

Framing Processes in the Envisioning of Low-Carbon, Resilient Cities: Results from Two Visioning Exercises

Stephen McGrail [1,*], A. Idil Gaziulusoy [2] and Paul Twomey [3]

1 Swinburne Institute for Social Research, Swinburne University, Hawthorn VIC 3122, Australia
2 Victorian Eco-Innovation Lab (VEIL), Faculty of Architecture, Building and Environment, University of Melbourne, Parkville VIC 3010, Australia; idil.gaziulusoy@unimelb.edu.au
3 Faculty of Built Environment and Centre for Energy and Environmental Markets, University of New South Wales, Sydney NSW 2052, Australia; p.twomey@unsw.edu.au

* Author to whom correspondence should be addressed; smcgrail@swin.edu.au or stephen.mcgrail@gmail.com; Tel.: +61-0-419-533-712.

Academic Editors: Tan Yigitcanlar and Md. Kamruzzaman

Received: 10 May 2015; Accepted: 24 June 2015; Published: 3 July 2015

Abstract: Visioning exercises were convened in Melbourne and Sydney, Australia, to explore how these cities could become low-carbon and maintain resilience over the next 25 years. Drawing on the concept of frames—in particular Schon and Rein's conceptualisation of a frame as a "diagnostic-prescriptive story" that is based on an underlying structure of beliefs, perceptions and appreciation—this paper seeks to: Attend to the ways that workshop participants framed the problems (of emissions reduction and maintaining resilience); surface framing processes and potential related sources of political contention; and discuss the role of visioning exercises in sustainability transitions. Five frames are identified, along with the interpretive orientations underpinning each frame, framing processes and the potential for frame conflict and alignment. The study suggests that the designers and facilitators of visioning exercises need to be attentive to framing processes, potential framing contests, and related social processes during a visioning exercise. Key implications are identified, with a focus on whether an exercise seeks to "open up" a complex issue or to agree upon a singular, *i.e.*, consensual, agenda.

Keywords: frames; framing processes; low carbon cities; visioning; urban resilience

---

1. Introduction

Over the past decade or so, anticipatory knowledge and visions of the future have been the focus of increasing social scientific research. For example, scholars have considered these in the context of the state such as their use to generate support for policy initiatives, in science and technology policy, and in relation to socio-technical change processes (e.g., [1–4]). In particular researchers in the field of Science and Technology Studies (STS) have paid increasing attention to how knowledge claims about the future are produced and mobilised for a range of purposes (e.g., [5,6]). This research has revealed the diverse roles played by expectations such as the role of technological expectations and expectations about market potential in investment decisions, and the role of normative expectations in coalition-formation. It has also demonstrated that how such knowledge is made and used matters.

Participatory "foresight" exercises are one such activity where future expectations are elicited and, to some extent, shaped through discussion and analysis [7]. This paper reports on forward-looking workshops that were run as part of an Australian project called Visions and Pathways 2040 (hereafter, VP2040). Through research and engagement activities the VP2040 project is examining visions, scenarios, and pathways for low-carbon resilient futures in Australian cities. The VP2040 project is one of an increasing number which view cities as a significant intervention point for climate change action and views the social challenge of rapidly reducing emissions and generating resilience in urban contexts as primarily a system innovation challenge. The project is funded by the Cooperative Research Centre (CRC) for Low Carbon Living, a multi-university and cross-sectoral research initiative established in 2012 to advance low-carbon innovation and urban transitions.

The first phase of the VP2040 project sought to elicit *existing* actors' expectations about the future focussed on what a desirable, low-carbon, resilient future for major Australian cities could look like. This was primarily accomplished through engagement exercises (which are outlined in more detail in below), in which participants shared ideas and were encouraged to imagine "business-as-unusual" futures. Designers and architects were also engaged to help with visualising these concepts. This paper describes, and presents an analysis of, the key outputs of this phase and other data gathered via a pre-workshop survey. The primary aims of this paper are to: (1) describe the different perspectives on a desirable, low-carbon, resilient future that were articulated (primarily focussed on major Australian cities); (2) identify the related problem framings and understandings that shaped these perspectives; (3) assess whether, and in what ways, framing processes may have influenced these exercises; and (4) consider the implications of these diverse framings and framing processes, such as regarding the potential for frame conflict or supportive alignments. Drawing on this analysis, we also discuss the role of visioning-style exercises and framing processes in sustainability transitions.

A related central argument is developed. We argue that the results of the study show that in participatory visioning exercises diverse institutionally-related and culturally-grounded framings of the complex social issues are voiced which are informed and shaped by *contextual* assumptions and *already circulating* understandings (e.g., about urban planning and city futures). These framings influence consideration of alternative solutions. Proponents of alternative framings often compete to influence the resulting visions through a complex social process that can be termed "framing contests" [8]. Surprisingly, this is also true for the relatively homogenous groups which participated in these workshops. From a practitioner perspective the paper concludes that the organisers and facilitators of visioning exercises need to be more attentive to framing contests and related social processes, along with whether the exercise seeks to open up debate or to achieve closure on a consensual agenda.

Introduction to the Visioning Exercises

In the first half of 2014, visioning exercises were held in Melbourne and Sydney, the two largest cities in Australia. A diverse mix of participants from government, business, academic, and the broader community attended—103 in total (see Table 1 below), 66 in Melbourne, 37 in Sydney—who shared a concern about climate change and related sustainability issues and interest in the future of cities. The attendees were the result of an invitation process based on an active search for relevant experts (e.g., web research, review of conference presenters to identify suitable experts) and the existing professional networks of project staff. One consequence of this approach is that most participants also had a strong knowledge of, and/or relevant interest in, the built environment along with related areas of public policy and innovation. Other consequences of this approach are possible, such as via sampling effects, and these are considered in the analysis and discussion sections.

Table 1. Breakdown of sectoral representation in the workshops.

Sector	Number of Participants
Business sector (built environment sector: e.g., urban design firm, engineering/construction; renewable energy/clean energy, *etc.*)	27
Research (university-based researcher; and non-university based researcher)	22
Local Government	15
NGO (e.g., WWF Australia)	11
State Government	10
Peak bodies (e.g., the Green Building Council of Australia)	6
Media/creative arts	4
Utilities	3
Social enterprises	3
Financial bodies/institutions	2

Participants were asked to complete a pre-workshop survey (please refer to the Supplementary Information for full details) that sought to both prompt some future-oriented thinking and help the organisers to better understand participants' expectations for, and interest in, the workshops. A total of 43 participants from the Melbourne workshop completed the survey (response rate of 65%); 22 participants in the Sydney workshop completed the survey (59%).

A two-stage workshop process was then conducted and used the following procedure:

- An initial half-day workshop was conducted in which participants shared and discussed their long-term visions for a low-carbon, highly resilient Melbourne or Sydney (depending on the workshop) in the year 2040. In small groups they moved through five themed discussion tables—shelter, energy, transport, food and water, and lifestyle and behavior—with the intention of gradually enabling a more holistic view of urban life and related systems. This workshop was also attended by designers who played central roles in the next two steps. The designers were asked to treat the workshop dialogue and workshop outputs as a design brief;
- The designers were then tasked with developing rough visualisations of the future(s) that were discussed in the initial workshop. These visualisations were termed glimpses as they are partial snapshots of possible futures (not a blueprint or a complete vision); and
- A second focus group style workshop was then held in which the designers presented the rough visualisations, followed by table-based small group dialogue on the glimpses. Participants responded to the glimpses, such as commenting on their desirability and plausibility.

Participants were also instructed to consider two core normative targets. The future city had to achieve a significant reduction in greenhouse gas emissions, specifically an 80% reduction by 2040 on 2014 levels (which is consistent with recommendations made by the Climate Change Authority, the relevant statutory authority in Australia). Secondly, the imagined future city must also be one that is resilient such as with respect to the future challenges of climate change (e.g., adjusting to new climate patterns) and having the ability to cope with related environmental shocks.

The analysis presented in Section 4 reports on the pre-workshop survey which was completed by the majority of participants, visioning workshop dialogue and associated process outputs.

2. Conceptual Framework and Theoretical Background

Concepts like low carbon living and urban resilience, and successful forms of adaptation to climate change, have different meanings for different people. Moreover, in important ways these meanings and associated judgements (e.g., regarding the question of climate adaptation "success" or "failure") are historically contingent and context-specific [9]. Whilst aspects can be determined or measured in a traditional scientific manner, many aspects are subjective and/or normative.

As Snow *et al.* [10] outline, the concept of a frame refers to the schemata of interpretation that are used when individuals perceive, identify, and label events (or "occurrences") in their life and in the world at large. This conceptualisation draws on Erving Goffman's original theorisation of frames (e.g., see his book *Frame Analysis*). Benford and Snow [11] further define a frame as "constituting a broader, interpretive answer or definition to 'what is going on' or 'should be going on'", emphasising the interpretive dimensions. Related framing processes can produce *collective action frames* which are more than an individual construct; they are shared action-oriented sets of beliefs and meanings, involving the negotiation of share dmeanings. In all these ways frames and framing processes are about reality construction and the production and maintenance of associated meanings [11].

In their influential book *Frame Reflection* Schon and Rein [12] make a broader related argument about how actors engage with problematic complex situations such as the challenge of decarbonising a city or an economy. They argue that actors must cognitively process situations that are ambiguous, indeterminate, and/or vague in ways that make them more intelligible. Through a process termed "naming and framing" they argue that actors construct *selective* stories: "From a problematic situation that is vague, ambiguous and indeterminate, each story selects and names different features and relations that become the "things" of the story—what the story is about. Each story places the features it has selected within the frame of a particular context" [12]. Actors select out the features and relations seen as relevant and important. According to this perspective, complex problem situations can be framed in multiple, sometimes incompatible, ways and there is no "correct" description.

Schon and Rein further argue that frames have normative implications and execute a "normative leap". That is, frames should be understood as "diagnostic-prescriptive stories". For example, this is seen in the construction of policy frames and "generative metaphors that underlie them" [12].

Recently the concept of a frame has gained currency in many social sciences and it has especially been used to help to underscore the culturally constructed dimensions of social action [11,13]. For instance, social movement scholars argue that collective action frames are constructed and used by movement actors, and others, to simplify and condense aspects of the social world in strategic ways. An example is the way issues get defined in ways that are intended to mobilise adherents, garner bystander support, and demonise antagonists. By resonating with varying groups and being open to interpretation and modification, frames can help to enable collective action [13]. Framing processes can also generate or draw on cultural understandings that offer actors motivating identities.

Some sustainability-focussed scholars also emphasise framing. In *Dynamic Sustainabilities* Leach *et al.* [14] emphasise the framing processes which they argue shape the analysis of complex systems and influence how complex sustainability problems are analysed and understood. For them, framing refers to "the particular contextual assumptions, methods, forms of interpretation and values that different groups bring to a problem, shaping how it is bounded and understood". Such processes involve selective perception (*i.e.*, choices about what elements to pay attention to and emphasise), and subjective value judgements, and are also shaped by the contexts in which people think and act; as summarised in Table 2 below. Leach *et al.* [14] further argue that "framings are produced by particular actors and co-constituted with their particular institutional, political and life settings". Drawing on a social constructivist epistemological perspective, they further argue that diverse understandings of sustainability problems are inevitable and call for approaches that aim to enhance reflexivity.

Table 2. Key dimensions of framing [14].

Choice of Elements	Subjective Judgements
• Scale	• Interests
• Boundaries	• Values
• Key elements and relationships	• Goals
• Dynamics in play	• Perspectives
• Outputs	• Notions of relevant experience

A number of related arguments are made. Firstly, that greater attention needs to be given to the multiple, diverse framing of complex sustainability issues in order to open up related debates and policy-making processes. Secondly, framings—such as problem framings, or system framings—often become part of larger cultural *narratives* that are constructed about a problem, its causes, and the solutions that are seen as required to bring about a set of preferred/desired outcomes [14].

Finally, Vogel [15] has developed the related concept of anticipatory frames. For example, Vogel [16] demonstrated the existence of competing framings of biotechnology in assessments of biosecurity threats and related policies. In Vogel's analysis, anticipatory frames have three main elements: Core *assumptions* (e.g., made about the future); particular pieces of *information* that are privileged or fixated on; and frames are related to *practices* (e.g., organisational practices). In the case of frames about emerging technologies (e.g., biotechnology) a frame also incorporates actors' beliefs about technological trajectories, temporal and spatial orientations, and specific foci of attention [16].

In the analysis that follows here, a similar approach to Vogel's analysis will be used to explicate the frames of low-carbon, resilient cities that were expressed by workshop participants. In addition, following Schon and Rein's conceptualisation, we consider a frame to be a diagnostic-prescriptive story which is based on underlying structures of beliefs, perceptions, and selective appreciation.

Framing Processes, Low-Carbon Transitions and Broader Sustainability Transitions

Some transition theorists such as Geels have considered the role of framing and related cultural understandings and factors in transition processes. Geels has considered two different aspects. First, building on Snow and Bedford's work (which was cited earlier) he argues that framing strategies are used by incumbent actors to try to avoid or forestall threatening changes [17]. As an example of problem reframing, he argues incumbents in existing energy industries and regimes benefitted from the redefining of energy policy problems to also include maintaining energy security and energy affordability (*i.e.*, not only reducing of greenhouse gas emissions). For example, "coal was increasingly (re)positioned as an answer to energy security and affordability" [17]. Second, other forms of framing include motivational and prognostic framing. Some historical cases of regime destabilisation and major transitions indicate that cultural understandings and frames can play roles in achieving such transitions. Drawing on the history of the British coal industry, Turnheim and Geels [18] argue that "public support for purposive transitions depends not only on a perceived urgency of problems" but also requires a broader "cultural package" which, in the case they examined, included "new visions of modern, clean, convenient, smokeless households". In other words, diagnostic and prognostic framings can contribute to public support for policy change and associated transitions.

Additionally, it is important to note that, in the sustainability transitions field, transitions towards sustainability are typically "framed from a systems perspective" [19]. That is, changes are seen as being required and as unfolding at the broader system-level, such as for entire mobility systems or energy systems. System changes are envisaged to occur through interconnected technological, organisational, and social and institutional changes, often termed socio-technical transitions.

3. Materials and Methods

Three key sets of data were collected during the visioning process and reviewed for this study: (1) pre-workshop survey responses (n = 65); (2) notes from the table-based dialogue that occurred during the visioning workshops (the authors were also co-facilitators of these workshops); and (3) the visualisations that we produced based on the dialogue at the workshops, which are referred to below as glimpses. The visioning exercises (described in Section 1) and survey (see Supplementary Information) provided opportunities for participants to articulate their views on preferred urban and low-carbon futures, including views on their subjective preferences in relation to these futures and their diagnosis of problems that need to be addressed in order to move towards such a preferred future. Consequently, these statements provide rich sources of data that can be reviewed in order to consider what frames were articulated and drawn on and related framing processes.

Responses to the following three pre-workshop survey questions were focussed on:

- What have been the main forces and processes of change that have shaped *[specific city]* life over the past 30 years? *[Specific city = Melbourne or Sydney]*
- Imagine you are living in *[specific city]* in 2040. It is now a global model city of a super-low-carbon city that can bounce back from extreme weather events and other shocks. You receive a visit from friends (via a time machine) from the year 2014. What are the first things you would want to show these time tourists that would "prove" that an extraordinary 26 years have passed?
- What "disruptive" forces do you think will affect *[specific city's]* pathway to a low carbon city over the next 30 years?

Survey respondents were also asked to provide details on their profession and work (e.g., the organisation they work for). This allowed for high-level consideration of the institutional settings of those articulating frames, e.g., whether there were such patterns as predicted by framing theory.

By review we mean that an informed assessment of the data was conducted which considered relevant framing theory. For example, we asked: Did participants focus on different scales (e.g., focus on local-level decarbonisation activities and localised resilience at the suburb/precinct scale or, alternatively, adopt a multi-scale perspective on decarbonisation and resilience)? Did participants tend to "bound" the problems in unique ways (e.g., primarily considering greenhouse gas emissions produced within the city such as from transportation or, alternatively, adopting a broader problem boundary such as also considering the greenhouse gas emissions that are embedded in the consumption choices of city residents)? How did participants think such problems ought to be addressed—was there wide agreement or distinct preferences/positions? This analysis progressed through three main stages: (1) the survey responses were initially reviewed to identify statements and beliefs indicative of potential frames and related elements (e.g., institutional settings); (2) the visualisations (workshop outputs) were reviewed to further "capture" each frame and consider their consistency with each frame (that is, was the prognostic aspect of the frames consistent with outputs?); (3) the uniqueness of each identified frame was systematically considered as per frame theory and framing elements discussed above. Following the visioning workshops the project team also critically discussed the perspectives that were voiced at the workshop, considering the main areas of divergence/convergence and any related conflicts. These team discussions provided a further way to review, clarify and assess the frame analysis.

Informed by the literature review, six major dimensions were considered when defining and comparing each frame, allowing for a more systematic analysis of each frame:

- *Diagnostic framing:* e.g., diagnosis of the focal problem(s); main aspects emphasised;
- *Choice of elements:* e.g., focal scale/spatial orientation (e.g., suburb, city-scale), boundaries, *etc.*;
- *Assumptions:* e.g., assumptions about technological options and trajectories;
- *Primary actors;*
- *Subjective judgements:* e.g., more or less important social/policy goals and outcomes; and
- *Institutional settings* (*i.e.*, of those who articulated the frame).

Five different, and partially competing, frames were identified, and associated framing processes (that influenced these frames) were examined. These frames and framing processes are described in the next section, and are subsequently critically discussed in Sections 4.8 and 5.

4. Results and Discussion

4.1. Frame 1: Localised Urban Sustainability; Re-imagining the Suburb, Neighbourhood and Home

Many participants focussed on envisioned changes at the suburb and/or precinct scale. The focus is on new types of suburbs, neighbourhoods, and homes within them, which are significantly more self-sufficient or self-contained, especially with respect to energy, water, and food. One participant termed such future suburbs "micro cities". Keywords for this frame include local, decentralised, and distributed (e.g., distributed local energy production systems). Urban agriculture is also embraced. Use of distributed and decentralised energy generation—especially via renewable and other lower-emissions technologies, combined with new energy storage technologies—is envisaged to be the best way to reduced greenhouse gas emission. Resilience is interpreted as requiring greater self-reliance and self-sufficiency, with some participants viewing existing long supply chains and associated infrastructure as a source of vulnerability and arguing that "re-localisation" is necessary for maintaining resilience. These changes are combined with other priority social goals such as community development, and re-balancing private and public space (see Figures 1 and 2).

Participants expressing this framing of low-carbon, highly-resilient cities stated that they would show time travellers (who visited this future city in 2040) the following key features:

> *"Self-sufficient and interconnected suburbs and precincts—with regard to:*
>
> *- energy consumption and production*
>
> *- fresh food production*
>
> *- a different structure to the common working week so that if one is able and chooses, one can manage one's life re: food production or producing wares or developing oneself personally or giving time to things that are important like family and children.*
>
> *[Sector: Peak body (green building sector); Workshop location: Melbourne]*

> *"Transformation of suburbs to double or triple (or more) the local housing yields and populations. Tangible focus on local neighbourhoods as self-sustaining, with local jobs, clear local centres and production loops"*
>
> *[Sector: Business—urban design/urban planning; Workshop location: Melbourne]*

> *"I'd take them out to the parts of Melbourne that were formerly car dominated, homogenously residential, consumption based suburbs, to observe a distinctly urban phenomena: People of all ages and gender cycling; local employment, services and production; complete utilisation of urban water resources; urban forests and food production. And the crowning glory—"living buildings" that are internally homeostatic, run on sunlight and take carbon out of the atmosphere."*
>
> *[Sector: Local Government; Workshop location: Melbourne]*

> *"The urban villages where people live, work, create and grow without needing to travel far ... [the] solar panels coating surface after surface. The roads that have been handed over to pedestrians"*
>
> *[Sector: Research; Workshop location: Sydney]*

> *"[I'd show the "time traveler" that] suburbs are micro cities [and] they are self-contained"*
>
> *[Sector: Business; Workshop location: Melbourne]*

> *"The elimination of the car and the rise of urban agriculture in the spaces the car left behind."*
>
> *[Sector: Business—urban design/urban planning; Workshop location: Sydney]*

In some cases this frame also included reference to development of "polycentric" cities, e.g.,

> *"There are six other business districts (hubs) in the Melbourne metro area other than the CBD"*
>
> *[Sector: Social enterprise; Workshop location: Melbourne]*

> *"[I'd show the "time traveler"] the green building and tree-lined avenue of Parramatta Rd, the shady, water features of Penrith and Liverpool CBDs, the agricultural businesses thriving in Sydney's green spaces and on its fringe."*
>
> *[Sector: Business—built environment/professional services; Workshop location: Sydney]*

Others envisioned related changes such as: Development of "new, dense, diverse urban places" *[Sector: urban design; workshop location: Sydney]* and "rectification of urban sprawl" *[Sector: urban design; workshop location: Melbourne]*, and a future in which "nearly everything, that is practicable, is local: Food, energy, water, jobs, education, *etc.*" *[Sector: Local government; workshop location: Melbourne]*. Another participant from local government stated that: "We'd visit a once sprawling outer suburban area where (at great expense) public transport had been added and a dense walkable and vibrant hub had been created" *[Sector: local government; workshop location: Melbourne]*.

This framing also informs some of the visualisations that were developed. Some of the glimpses portray a future city with more communal land (such as the glimpses shown in Figures 1 and 2 below) and urban "villages" that support more localised living (see Figure 3).

Figure 1. Less emphasis on owning a private backyard: Residents work together to maintain communal and/or shared land (that was formerly private backyards) such as for food production.

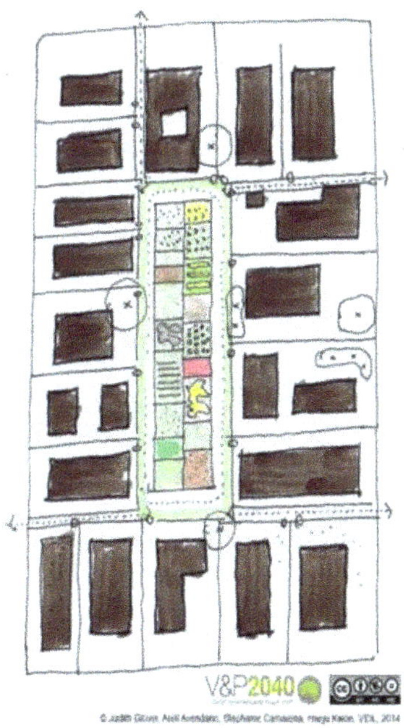

Figure 2. Growth of communal spaces in new housing estates and urban precincts.

Figure 3. Local shopping and produce: a mosaic of urban "villages" now exists including localised hospitals, employment, schools, and self-contained water and food production. Supermarkets do not disappear but incorporate local produce. Technology improves the flow of information and introduces new models of communication, e.g., an app informing residents which trees are currently fruiting.

Additional glimpses conveyed the concept of local community "hubs". The concept of new flexible civic/educational/commercial community spaces is shown in Figure 4 below.

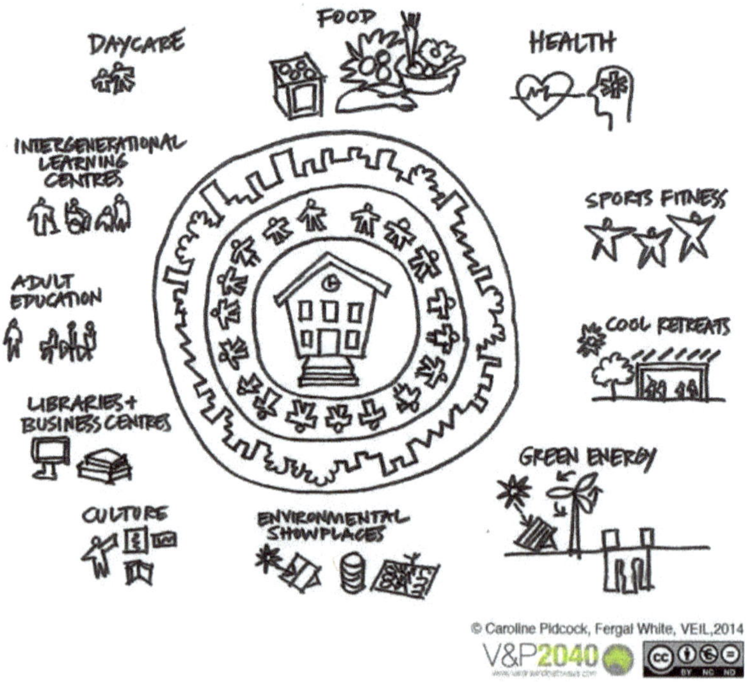

Figure 4. Local "hubs": flexible community spaces that support and empower local people and businesses, and can also serve as "cool retreats" during major heat events.

Major elements of this frame are summarised in Table 3 below.

Table 3. Summary of the "localised urban sustainability" frame.

Element	Details (in Brief)
Diagnostic framing	• Poor government/urban planning and related market-driven urban change "locking-in" consumption and high emissions-intensity • Car-dependence related to poor planning and continued urban sprawl • "Introverted", individualistic urban living (cultural problems)
Elements emphasised	• Local-level changes: Suburbs, precinct planning, dwelling types • Urban density • Threats to the resilience of urban communities/environments • Factoring in broader resilience issues into urban planning (e.g., resilient food systems, water supply/systems, natural resource limits)
Assumptions	• A high degree of localised self-sufficiency is a realistic possibility • Long supply chains and associated existing infrastructures (e.g., food and energy supplies) make urban communities vulnerable • Limited trade-offs exists between increased urban density and other social, environmental and other policy objectives
Primary actors	• Local citizens and local governments • Urban designers and planners
Subjective judgements	• Preference for localisation and local non-commercial ownership • Related goals of greater decentralisation and local "connectedness"
Institutional settings (main themes)	• Local government • Urban design firms, architecture/design professionals; related peak bodies (e.g., Green Building Council of Australia) • University-based research organisation

4.2. Frame 2: Radical Shifts in Political Economy and Material Consumption

Some workshop participants emphasised much more radical shifts in socio-economic organisation and associated ways of life and patterns in economic activity when asked what they would show a time traveller from 2014. Similarly, some of the glimpses shown below were framed by shifts towards greater sharing behaviours and were characterised as being "slow" and "less products, more experiences", about communalisation and changes in values.

According to this issue framing, low carbon living and associated transformations towards a low carbon economy and resilient communities must focus on reducing material consumption, and associated changes in values and economic activity. Additional aspects that are emphasised by this framing—and not the others described in this paper—are the *indirect* greenhouse gas emissions generated by production processes located elsewhere and by freight/logistics. Consumer culture is critiqued, along with dominant status symbols (e.g., a large home, privately-owned vehicles, other accumulated belongings, *etc.*), and the pursuit of affluence. Simpler living is promoted.

Additionally, participants expressing this framing of low-carbon, highly-resilient cities stated that they would show time travellers (who visited this future city) the following key features:

> *"I would want to show them a city where public transport and bike travel eclipses car use. I would want to show them a city with no bottled water, no takeaway food, no smoking, fit and healthy people, community groups actively practicing yoga and meditative practices, I would like to show them large tracts of land and parks filled with edible gardens tended. I would like to show them new education and learning models that are collaborative and non-competitive."*
>
> *[Sector: Social enterprise; Workshop location: Melbourne]*

> *"We have sophisticated social and economic systems for measuring externalities like carbon, health, wellbeing, and happiness. We have a post-growth, steady-state economy. Nurturing, protecting and valuing the Commons is the key priority for business, government and society. People define wealth through their social relationships. Large amounts of food is grown locally in community gardens. There are no cars in the city. Energy is produced locally through distributed grids using renewable energy. People only work for money by choice and not more than 3 days per week."*
>
> *[Sector: Social enterprise; Workshop location: Melbourne]*

> *"[I'd show the "time traveler"] documents from a variety of flourishing sharing economy initiatives I'm part of: A cooperatively owned electric car sharing club, my food cooperative that makes regular farm visits, and photos of the community energy scheme I've invested in."*
>
> *[Sector: Research (university-based); Workshop location: Sydney]*

> *"The whole place will have (for the 2014 visitor) a strong "whiff" of what he or she would describe, perhaps based on earlier travels to less-developed countries, as poverty, but which seems unremarkable to the 2040 citizens—e.g., obvious repairs to clothing; old objects and tools well maintained; breakdowns encountered with good humour. Obvious use of (and reliance on) "legacy technologies". Lots of people engaged in maintenance activity. People obviously taking the time to help each other with day-to-day tasks. A subtle sense of reduced busy-ness—simultaneously with a lot of engagement in hard physical work."*
>
> *[Sector: Research (non-university based); Workshop: Melbourne]*

> *"[I'd show the "time traveler" the] massive social change that has taken place to enable these things, people are more aware of their place in the ecological balance and values are different. Combined with this, new forms of democracy that involve the community in INFORMED decision making (possibly inspired by Swiss direct democracy model)"*
>
> *[Sector: Research (university-based sustainability institute); Workshop location: Sydney]*

> *"[I'd show the "time traveler"] the lack of cars, what people do with their extra free time"*
>
> *[Sector: Research (university-based); Workshop: Melbourne]*

Similarly, another respondent would show a visitor from 2014 the "palpable culture of sharing", "culture and governance of making decisions bottom-up", and "hyper local pockets of diverse, interdependent economies and social neighbourhoods" among other key changes *[Sector: Business—architecture practice; Workshop location: Melbourne]*.

One way that some aspects of this frame can be further described is as a critique of urbanism as a *way of life* (*i.e.*, urban modes of social life). For example, in his sociological theorisation of urbanism Wirth [20] argued that urban life in large cities has fundamentally altered the nature of social relationships, such as generating a tendency for weaker social bonds. He further termed this "the superficiality, the anonymity and the transitory character of urban-social relations". He further pointed to a loss of "the sense of participation that comes from living in an integrated society". The pace of life in urban contexts is generally much faster than rural areas, along with more insecure. Critiques of urban ways of life are, in this frame, linked to the low-carbon and resilience agendas. Or, put another way, the low carbon cities debate provides a new space in which these issues are given voice.

Another survey question asked participants to nominate positive and negative "disruptive" forces what will shape the city's "pathway to a low carbon city over the next 30 years". Noting responses by the participants whose interpretations fully, or in part, express this frame helps to further explicate this perspective. Disruptive forces that were listed as *negative* include: Aggressive pursuit of "affluence entitlement"; narcissism; and patents. Disruptive forces listed as *positive* included: Voluntary simplicity; sharing and sharing economy initiatives; collaboration; peer production; participatory democracy; peak oil; and tangible ecological collapse and climate change.

This framing also informs some of the visualisations that were developed. Some of the glimpses portray a future in which there is less emphasis on ownership, such as ownership of private vehicles (see Figure 5), and shopping centres are transformed due to reduced consumerism (Figure 6).

Figure 5. Deprivatisation of personal transport and repurposing of land and buildings previously dedicated to private vehicles (e.g., car parks, fewer roads, *etc.*).

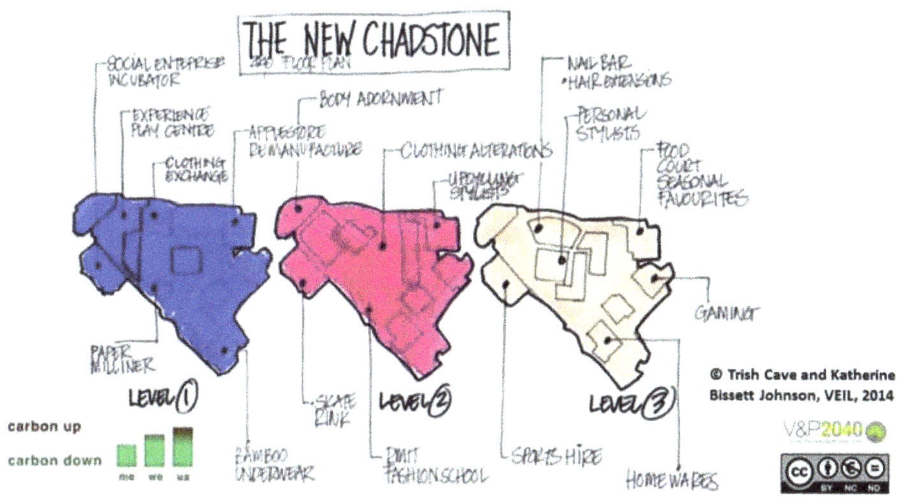

Figure 6. Large shopping centres (such as Chadstone Shopping Centre in Melbourne) are now centres of alternative forms of exchange, product "reuse" and social enterprise.

Some of the glimpses also depicted new forms of exchange (see Figure 7) and slower ways of life with improved work-life balance and more "real-here-now" contact (Figure 8 below).

Figure 7. In Melbourne in 2040 residents routinely trade energy. For example, a household with solar panels may "trade" excess/unneeded energy for other goods/services.

Figure 8. The balance between work and leisure/life activities and the balance between virtual social networks and "real-here-now" contact has shifted.

Major elements of this frame are summarised in the Table 4 below.

Table 4. Summary of "radical shifts in political economy and material consumption" frame.

Element	Details (in Brief)
Diagnostic framing	• The climate change problem is a symptom of deeper issues within socio-economic systems (e.g., economic models, values) • Longer-term sustainability and resilience requires socio-economic transformations (not only technological and regulatory changes)
Elements emphasised	• Consumption (e.g., material consumption, energy consumption) • Economic models • Cultural aspects (e.g., urban ways of life, dominant social values) • Indirect greenhouse gas emissions including those generated by production processes elsewhere (broader spatial boundary)
Assumptions	• Imminent resource and energy supply constraints demand socio-economic changes which form a key part of urban resilience • Climate change is not primarily a technological problem
Primary actors	• Citizens/city residents • Social entrepreneurs • Activists
Subjective judgements	• Low-carbon transitions are an opportunity for socio-economic transitions (*i.e.*, frame emphasises socio-economic goals) • Communitarian values preferred over individualistic values
Institutional settings (main themes)	• Social enterprise (e.g., social entrepreneur) • NGOs/Social activist • Research: e.g., independent researcher

4.3. Frame 3: Achieving Low-carbon Transitions through Digital Lives and "Green Orwellianism"

Other workshop participants expressed somewhat more techno-optimist views and framed the problem in related ways. For example, some discussed the potential development of more "smart infrastructure" systems and related high-tech urban environments and envisaged ways that these could shape transport systems, workplaces and the nature of work, and leisure patterns. Some participants also speculated on the potential for advanced virtual experiences and communications technologies to reduce the need for travel and material consumption. Some participants also broadened the frame from urban resource consumption and local greenhouse gas emissions to the emissions associated with mobile modern lives. Participants who framed low-carbon transitions in this way often emphasised challenges associated with achieving substantial behavioural change. Some suggested that, in the future, high carbon behaviours could become more strongly regulated through greater transparency and real-time data, enabled by new high-tech feedback systems and surveillance systems.

A range of related disruptive forces were also nominated by some workshop participants. These included: Sensor networks; smart city technology; the growth of data, *i.e.*, big data; digital lifestyles and further digitisation; and the adoption of "Gov 3.0" tools, technologies and techniques.

Such approaches have been termed "digital lives" and related potential systems have been termed a kind of Orwellian future, due to the potential for highly surveilled movements and associated regulation [21,22]. The former refers to the potential for the physical travel of people and objects to be substituted by forms of digital communication and virtual experiences. Additionally, more advanced digital networks of control and associated technologies of tracking and tracing could be used to optimise and allocate road space; enable greater automation (e.g., self-navigating vehicles); and enable the tracking, tracing and public reporting of carbon allowances and carbon expenditures [23].

For example, in one glimpse (shown in Figure 9) the carbon footprint of an urban village is made public, along with comparisons with other urban villages with a focus on energy, water, and waste. The idea is that transparency and the ability to compare performance will be a driver of change.

The so-called digital disruption was also connected to other envisaged changes to work and life. Participants expressing this framing of low-carbon, highly-resilient cities stated that they would show time travellers (who visited this future city) the following key features:

> *"[I'd show the "time traveler" that] work, home and play are intertwined activities, and telecommuting is the norm"*
>
> *[Sector: Business—specialist engineering, design and technical services; Workshop location: Sydney]*

> *"[I'd show the "time traveler" that the] technological infrastructure exists to make remote/virtual teams the norm, rather than the exception"*
>
> *[Sector: Business—consultancy (Digital economy/cloud-based data analytics); Workshop location: Sydney]*

> *"Large and small organisations embracing smart work trends that are dispersed throughout Melbourne (no more massive head offices that everyone has to travel to)—creating more co-working and collaborative opportunities, increasing productivity and happiness."*
>
> *[Sector: NGO; Workshop location: Melbourne]*

This framing also informs some of the visualisations. Figures 9 and 10 below convey the expectation that greater transparency and monitoring will be a major driver of social change.

Figure 9. Public carbon footprints (precinct and household footprint data).

Figure 10. Neighborhood energy monitoring: Real-time energy usage and carbon footprint monitoring for each community is freely available to everyone. People can read their community's energy and carbon performance as well as that of their neighbours. Feedback and learning new ways to be more energy and carbon efficient, as well as having pride in the ecological performance of one's community, is a standard part of culture.

Similarly, real-time data analytics and wearable technologies were depicted as providing up-to-date personal footprint data, as shown in Figure 11 below. Some participants viewed related surveillance and regulatory systems as a "dictatorship of carbon" (shown in Figure 12).

Figure 11. Use of smart wearable technologies like new wrist-bands to check a person's community "contribution index" in real-time (*i.e.*, a current, continually updated, measurement for their "contribution" to the city or a community). Contributions could be measured in terms of energy generation, personal consumption, and participation in recycling. Such a system could also provide constantly updated data on personal contributions to greenhouse gas emissions and other impacts of personal choices/consumption.

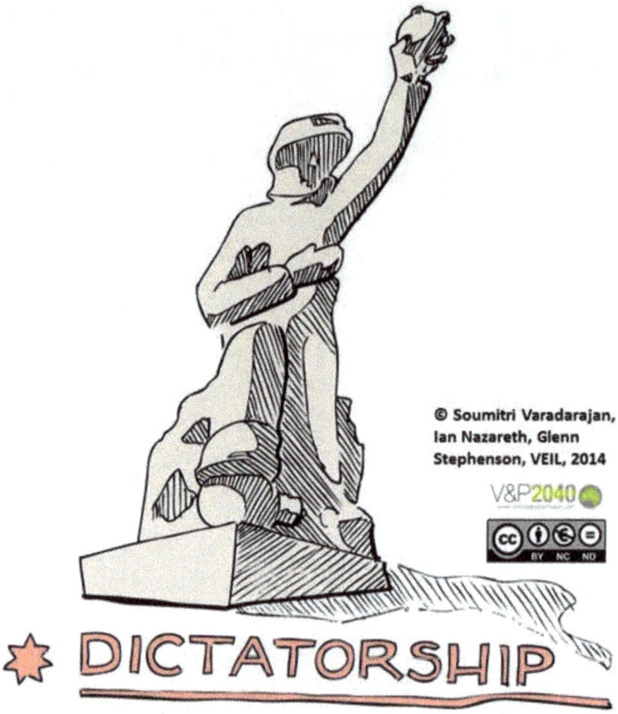

Figure 12. Strict enforcement of carbon allowances or quotas by governments, and related monitoring and regulatory systems, termed the "dictatorship of carbon".

Major elements of this frame are summarised in Table 5 below.

Table 5. Summary of "low-carbon transitions through digital lives and green Orwellianism".

Element	Details (in Brief)
Diagnostic framing	• Entrenched "high-carbon" behaviours which are difficult to change (e.g., transport/mobility, food consumption, leisure) • Contemporary modern lifestyles (including mobility expectations, and related mobility requirements of work, leisure, *etc.*) and their related resource requirements and greenhouse gas emissions
Elements emphasised	• Use of "smart" real-time systems to drive behavioural change • Substitution of physical travel and real-world experiences • Low-carbon urban transitions a way to tackle urban congestion and related pressures (e.g., on transport systems at peak times)
Assumptions	• Substitutability of real-world activities for virtual/digital activities • Public acceptance of new monitoring and surveillance systems • Unsustainability of contemporary mobility

Table 5. *Cont.*

Element	Details (in Brief)
Primary actors	• Planning professionals • Private firms and public sector bodies developing and administering new "smart" city and infrastructure solutions • City residents
Subjective judgements	• Goal of shifting from high-carbon to low-carbon behaviours • Preferences for new modes of work and workplaces; • Arguments about the potential productivity benefits of such changes • Need to revitalise suburban and regional centres
Institutional settings (main themes)	• Policy, and urban and transport planning professionals (see Sustainable Digital Cities Network) • Specialist engineering, design and technology firms

4.4. Frame 4: Advancing Urban Resilience and Climate Change Mitigation by Greening and Radically Adapting the Urban Form

The framing emphasises potential adaptation challenges facing Australian cities, given a changing and warming climate, and the potential for climatic changes to require disruptive forms of adaptation. The potential for existing buildings and urban forms to adapt to a changed climate is questioned and dystopian perspectives were also occasionally voiced. On the other hand, the potential for creatively and effectively adapting or repurposing existing urban infrastructure, buildings and systems to deal with a more extreme climate is central, as depicted in Figures 13 and 14 below. Participants who expressed this framing saw the potential to (partially) address climate and adaptation problems and achieve greater energy efficiency through additional urban greening initiatives.

For example, participants expressing this framing of low-carbon, highly-resilient cities stated that they would show time travellers (who visited this future city) the following key features:

> *"Community resilience networks, adaptive building design and green roofs, walls, and space has eliminated urban heat island"*
>
> *[Sector: State government (NSW Office of Environment and Heritage); Workshop location: Sydney]*

> *"A city in a garden: A beautiful (nature) green city with vast parklands and plants rising up all city buildings to rooftop gardens all doing their best to absorb the CO2 in the air ... [and] trails to everywhere ... A Melbourne Food Bowl: A city producing organic food for residents"*
>
> *[Sector: State government (Parks Victoria); Workshop location: Melbourne]*

> *"There is substantial green tree planting where possible—perhaps 30–40 per cent of the city is covered in greenery such as trees as it is a good foil for reducing heat from heat waves. There are sea walls built around the main city foreshore to prevent storm surges. Storm water harvesting systems have been installed around Sydney to capture and purify all the water."*
>
> *[Sector: Business: Environmental consultant; Workshop location: Melbourne]*

> *"Vegetative cooling technologies; water reuse and recycling and urban food production on rooftops across the city."*
>
> *[Sector: Business: Environmental consultant; Workshop location: Melbourne]*

> *"The greening of concrete corridors, with vine-covered paths that are "air-conditioning" the passage of pedestrians and cyclists"*
> *[Sector: Media/Creative Arts; location: Melbourne]*

Many participants envisaged, in a range of ways, related changes to create far greener cities. For example, in the survey responses the followings sorts of changes were envisaged:

- Protected urban and peri-urban "green lungs";
- Green public spaces; greened inner city precincts; green urbanity;
- Mitigating the urban heat island effect with green roof infrastructure (e.g., comprehensive green roof coverage) and greater urban greening (tree planting, other greenery in the CBD);
- Major urban flora and food planting;
- Permaculture and green spaces all through the city and broader metro area providing food and leisure (recreation) opportunities; and
- Greater urban biodiversity.

Related, future disruptive forces were also nominated by some workshop participants—especially those in the natural environment. These included: Increased frequency of major heatwave events/increased heat; drought and fire; climate change (more broadly); disruption of the seasons and agricultural production; and water scarcity. A related positive disruptive force stated by another participant stated was "more green spaces to deal with heat stress", and a related negative force is the "collapse of keystone species e.g., bees, bats, and birds concerning so need to support biodiversity in our cities". Some participants expected that infrastructure failures will be a disruptive force.

This framing also informs some of the visualisations that were developed. Some glimpses portray a city in which climate change led to the repurposing of abandoned buildings and unusable infrastructure (Figures 13 and 14 below) and urban greening provides a way to cool buildings (Figure 15).

Figure 13. An abandoned building now being used as an urban greenhouse.

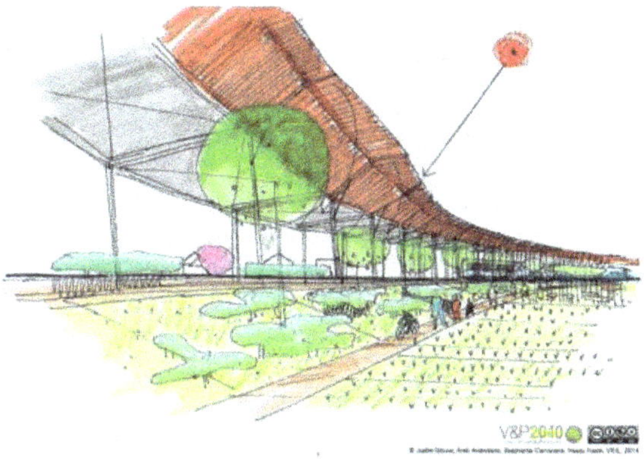

Figure 14. New uses of existing infrastructure to enable new forms of adaptation to extreme climates: Example of reclaimed land under former freeways being used for new purposes due to protection from sun and other climate extremes.

Figure 15. A breeze blows through and cools buildings in Melbourne 2040. Buildings are living systems and use vegetation as permeable "skins".

Additional glimpses conveyed other forms of "extreme" adaptation to climate change. For example, one glimpse depicts a shift to living underground termed "burrowing" (Figure 16).

Figure 16. The concept of "burrowing" and building houses underground is mainstream in 2040. This Melbourne suburb consists of underground houses that escape extremes of heat and bushfires on the urban fringe. The backyard is now the "top yard", and is used for recreation, food production and natural habitats. This provides climate protection, decreases energy costs and provides a greater sense of connection to nature.

Major elements of this frame are summarised in Table 6 below.

Table 6. Summary of the "Advancing resilience and climate change mitigation by greening and adapting the urban form" frame.

Element	Details (in Brief)
Diagnostic framing	• Risks to urban resilience posed by climate change/extreme events • Urban forms and "environments" as providing a constructed barrier between human beings and nature (e.g., extreme weather events) • Role of urban/spatial forms in "locking-in" consumption (e.g., energy)
Elements emphasised	• Local-level adaptive action, *i.e.*, a human settlement focus • Existing urban infrastructure: Repurposing buildings/structures • Links between the built/urban form and experienced climates
Assumptions	• More substantive warming and climate impacts over the next 25 years (by 2040)—e.g., increasing weather-related natural disasters and losses (e.g., droughts, fire threat), heat stress effects, *etc.*—and more extreme forms of "preparedness" and adaptation required • Local climate adaptation actions often also have mitigative functions (e.g., urban greening programs reducing energy consumption)
Primary actors	• Local and state governments • City residents • Urban foresters and urban biodiversity experts • Research organisations
Subjective judgements	• Existing major settled areas should be retained (rather than being abandoned), even in the event of catastrophic climate change • We should maintain existing outdoor lifestyles and leisure activities
Institutional settings (main themes)	• Local Government • State Government Department/Agencies • Consultancy (e.g., environmental; communications e.g., media/creative)

4.5. Frame 5: Low-Carbon Transitions as Changes to City-Scale Infrastructure and Larger-Scale Systems

Other participants, when asked to state what had changed between 2014 and 2040 that they would show to a time traveller from 2014 who visited the city (*i.e.*, Melbourne or Sydney) in 2040, focussed on changes to city-scale hard infrastructure and related changes to large-scale systems such as for regional transport (e.g., inter-state high-speed rail) and changes to the national electricity grid. This was sometimes linked to changes in public policy or changes in behaviour (e.g., increased cycling).

The following responses to the survey questions are also illustrative of this particular frame:

> *"[I'd show the "time traveler" the] extensive punctual and widespread underground railway system, superior to New York, Hong Kong, Paris and London; car free city centre with plenty of bicycle parking facilities; and numerous district renewable energy generation mini grids with storage that have completely replaced the need for brown coal fired power generation from one central grid."*
>
> *[Sector: Business—renewable energy/clean energy sector; Workshop location: Melbourne]*

> *"[I'd show the "time traveler" the] high speed rail station delivering passengers to Australia's major cities in record time, replacing air travel... and all the light rail public transport making so many outlying areas of Melbourne accessible to the city"*
>
> *[Sector: Media/Creative Arts; Workshop location: Melbourne]*

> *"Fast train network from the CBD to our airports, renewable energy generation facilities near to Melbourne, safe and linked access routes for cycling into and out of the city . . . "*
>
> *[Sector: Business: Renewable energy firm; workshop location: Melbourne]*

> *"A reactivated Sydney Harbour used for transport and aquaculture. The Sydney Harbour Tidal Barrage"*
>
> *[Sector: Business—urban systems/built environment consultant: Location: Sydney]*

> *"There are no cars in the city—a second city loop [underground train system/network] and a series of tree-lined walking/cycling paths make it easy to get around . . . [and] there are six new rail lines going out to the suburbs and an outer suburban loop 100% renewable powered public transport network."*
>
> *[Sector: Local Government; Workshop location: Melbourne]*

> *"Mass transport for produce and big things in special freeway zones—possibly underground—which is high speed and automated; large tubes dispersing goods and produce."*
>
> *[Sector: Peak body (Green building industry); workshop location: Melbourne]*

Related, future disruptive forces were also nominated by some workshop participants, such as: High speed rail; clean transport technologies being available right now; cost effective large-scale renewable energy; and increasing interest in and decreasing costs of renewable energy. Another respondent suggested that "current infrastructure not [being] up to scratch" would be a disruptive force.

Some participants pointed to related technical changes and potential innovations that could lead towards low-carbon resilience cities, particularly for buildings, transportation and energy, e.g.,

- Electric cars and busses;
- Driverless vehicles;
- Highly efficient buildings that minimise energy use and the use of other resources;
- Distributed energy generation and new storage technologies;
- Advanced battery-assisted cycles (e-bikes/commuter bikes); and

- New transport infrastructure such as for cycling throughout the city (such as the cycling highway shown in Figure 17) and inter-city high-speed rail services (Figure 18).

Figure 17. A Melbourne train station in 2040, next to an elevated cycling highway. The drones in the sky are part of a delivery system, in which post and other goods are distributed.

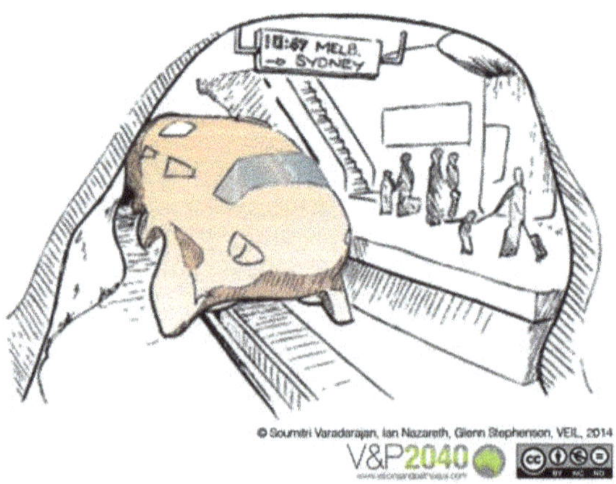

Figure 18. High speed trains link Australian cities in 2040, negating the need to fly for business travel.

Major elements of this frame are summarised in Table 7 below.

Table 7. Summary of "low-carbon transitions as changes to city-scale infrastructure and larger-scale systems" frame.

Element	Details (in Brief)
Diagnostic framing	• Major changes to major public and enabling infrastructure (such as for transport, energy) are required for low-carbon transitions • Lack of funding for alternative lower-emissions infrastructure (e.g., lack of funding for expanding public transport systems)
Elements emphasised	• Larger-scale, hard infrastructure—city-scale, regional, national • Emphasises transport and energy; less focus on consumption and other sources of emissions (e.g., agriculture/food production) • Technical change/technological innovation
Assumptions	• Services such as high-speed rail between State capitals are viable • It is in the public interest to sustain national hard infrastructure over the long-term (e.g., the national grid supplying cities)
Primary actors	• Federal/State governments (infrastructure investment and policy) • Other relevant public sector bodies • Electricity sector and firms • Construction sector and firms
Subjective judgements	• Preference for "big infrastructure" solutions to social problems • Related relevant expertise includes engineering and financing
Institutional settings (main themes)	• Infrastructure provision • Renewable energy companies/clean energy sector • Built environment organisations

Overall, the frames can be summarised by outlining each diagnostic frame (*i.e.*, the way that the problem was described and/or explained) and the associated prognostic framing (*i.e.*, prescriptions for addressing the problem). The five frames are summarised in Table 8 below:

Table 8. Summary of identified frames.

Frame	Diagnostic Framing	Prognostic Framing
Localised urban sustainability	Urban low-carbon transitions and resilience requires a new logic of urban (re)development, which incorporates greater localised self-reliance.	• Self-reliance; distributed systems. • Transformation of precincts and/or suburbs (citizen-led) via localised emergent mobilisation and community engagement. The state acts as an enabler, changing planning frameworks and enabling "polycentric" forms.

Table 8. *Cont.*

Frame	Diagnostic Framing	Prognostic Framing
Radical shifts in political economy and consumption	The climate change problem is a symptom of deeper issues with existing socio-economic systems. Longer-term sustainability and resilience requires socio-economic transformations.	• Radical changes to patterns of consumption and reorganisation of the structure of economies. • Institutional transformation required for urban low-carbon transitions
Digital lives and "green Orwellianism"	Social and technological barriers to changing existing high-carbon behaviours (e.g., travel, meat-intensive diets, material consumption).	• Use of informational and "smart" technologies to enable major shifts from high-carbon to low-carbon behaviours and to drive such changes.
Greening and radically adapting the urban form	Urban environmental risks associated with climate change, are presenting cities with significant, emerging challenges (e.g., adaptation to dynamic new climate risks).	• "Mitigative adaptation": Adapt urban form to address climate change and risks; exploit mitigation co-benefits (e.g., reduced energy use) • Creatively repurposing existing urban structures and existing buildings.
Changes to city-scale "hard" infrastructure and large-scale systems	Urban and broader low-carbon transitions demand major changes to public infrastructure and associated enabling "hard" infrastructure.	• New city-scale, regional and national infrastructure ("technical fixes") to increase the resilience of infrastructure, and support low-carbon living.

4.6. *Dominant and Alternative Frames*

The "localised urban sustainability" frame was the dominant frame in both the Melbourne and Sydney workshops, although adherents to all five frames were present at both workshops. This is reflected in the work of the designers/design teams, with the work of seven (out of nine) strongly expressing the prognosis emphasised by frame 1. Frame 5 was expressed by fewest designers/design teams (two designers/design teams). This finding surprised some project members who had previously encountered resistance to similar concepts emphasising re-localisation and more distributed systems (with energy, food and water production embedded in urban landscapes), such as the vision of a future multi-local "city of short distances" which was developed in the Melbourne 2032 project which was also led by the Victorian Eco-Innovation Lab (VEIL). This result may be reflective of the mainstreaming of such ideas, at least within the built environment sector. In recent years, metropolitan planning strategies in Australia have increasingly emphasised ideas such as "infill" (as per the concept of more compact cities), polycentric urban forms, walkability and living more locally within your neighbourhood (e.g., as per increasing focus on making Melbourne a so-called "20 minute city"), along with the increased investment in photovoltaic solar power technologies in Australian cities. An alternative, or complementary, explanation is that the dominance of the frame is an artifact of the sample, *i.e.*, influenced by who did and didn't participate, their institutional and/or life settings [14], and/or their ideological leanings [11]. These aspects are further discussed below.

4.7. Identified and Potential Framing Processes

4.7.1. Evidence of the Influence of Framing Processes and Related Social Processes

Framing theorists argue that frames are co-constituted by the "particular institutional, political and life settings" of actors [14]. The survey data provides some evidence of this. For example, "localised urban sustainability" was the dominant frame of those working in a specific local setting (e.g., participants who work in local government) or whose work otherwise focusses on specific places (e.g., precincts). "Radical shifts in political economy and consumption" was the dominant frame of participants from social enterprises, NGOs and the research sector (and not those from government and business). The two respondents from State government agencies (Parks Victoria and the New South Wales Office of Environment and Heritage) emphasised "advancing resilience and climate change mitigation by greening and adapting the urban form" (Frame 4). Respondents from specialty built environment and/or large infrastructure-related consulting firms (e.g., AECOM) and utilities emphasised "Changes to city-scale "hard" infrastructure and large-scale systems" (Frame 5). These patterns are indicative of the ways that frames are co-constituted by actors' settings.

Table 9 below compares the frames expressed by some of the categories of participants who responded to the pre-workshop survey (*i.e.*, from different institutional and life settings).

Table 9. Main frames expressed by participant category.

Sector	Category	Dominant Frame(s) in Survey Responses	Other Frames Evident in Survey Responses
Business	Urban design/related practice (*n* = 8)	Frame 1: "Localised urban sustainability" (7 of 8 responses)	• Frame 5: "Changes to city-scale "hard" infrastructure and large-scale systems" (3 of 7 responses) • Frame 2: "Radical shifts in political economy and consumption (3 of 7)
	Specialty built environment and/or large infrastructure-related consulting firm, e.g., engineering and/or construction services (*n* = 6)	Frame 5: "Changes to city-scale "hard" infrastructure and large-scale systems" (5 of 6 responses)	• Frame 1: "Localised urban sustainability" (2 of 6 responses) • Frame 3: Digital lives and "green Orwellianism (1 of 6 responses")
	Consultancy—general sustainability-related or environmental (*n* = 5)	Frame 1: "Localised urban sustainability" (4 of 5 responses)	• Frame 4: Greening and radically adapting urban form (3 of 5) • Frame 2: "Radical shifts in political economy and consumption (2 of 5) • Frame 5: "Changes to city-scale "hard" infrastructure and large-scale systems" (2 of 5 responses)

Table 9. *Cont.*

Sector	Category	Dominant Frame(s) in Survey Responses	Other Frames Evident in Survey Responses
Government	Local government (n = 7)	Frame 1: "Localised urban sustainability" (5 of 7 responses)	• Frame 2: "Radical shifts in political economy and consumption (3 of 7) • Frame 5: "Changes to city-scale "hard" infrastructure and large-scale systems" (2 of 7 responses)
Government	State government agency/department (n = 2)	Frame 4: "Advancing resilience and climate change mitigation by greening and adapting the urban form" (2 of 2)	• Frame 1: "Localised urban sustainability" (1 of 2 responses)
Utilities	Energy distribution/generations [Ausgrid, Pacific Hydro] (n = 2)	Frame 5: "Changes to city-scale "hard" infrastructure and large-scale systems" (2 of 2 responses)	• NONE
Other	Research (n = 9 [excluding Ph.D. students])	Frame 2: "Radical shifts in political economy and consumption (6 of 9)	• Frame 1: "Localised urban sustainability" (4 of 9 responses) • Frame 5: "Changes to city-scale "hard" infrastructure and large-scale systems" (2 of 9 responses)
	Social enterprise (n = 3)	Frame 2: "Radical shifts in political economy and consumption (3 of 3)	• Frame 1: "Localised urban sustainability" (2 of 3 responses)
	NGOs (n = 3)	Frame 2: "Radical shifts in political economy and consumption (2 of 3 responses) Frame 1: "Localised urban sustainability" (2 of 3 responses)	• Frame 3: Digital lives and "green Orwellianism" (1 of 3)

The broader city context can also be considered. Although a diversity of frames was evident at both workshops, some alternative frames were more prevalent at particular workshops. More participants in the Melbourne workshop emphasised significant current and emerging urban environmental risks due to climate change and proposed associated extreme adaptation measures (see Frame 4). This difference may be due to the more significant climate disruptions that have recently been experienced in Melbourne, in particular extreme heat events during the summer months. The Sydney workshop produced more glimpses related to reimagined suburban life. This may reflect a context in which concerns about urban sprawl had intensified—following the release of more land for housing on the fringe of the city—and associated additional attention was placed on "urban renewal" strategies for achieving more development within *existing* suburbs and increased density. Additionally, as car dependence and transport infrastructure provision are significant issues in both Melbourne and Sydney it was unsurprising to see this dimension of Frame 5 articulated in both workshops.

Relating the five frames to frames identified by other researchers, and to common framings of green transformations, suggests further insights. Knight [24] contrasts two framings of action to address climate change: Those framing it as a technological problem, and those who centrally see it as a problem generated by "the excesses of modern consumption". This cleavage is apparent in the contrast between Frame 2 and the more technology-centric frames. Scoones, Newell & Leach [25,26] argue that green transformations get framed in four different ways: Technocentric transformations; marketised transformations; state-led transformations; or as citizen-led transformations (e.g., transitioning to solidarity-based sharing or gift economies). They note that some frames focus on "the right combination of technologies" rather than "reorganizing economies or institutions and unsettling power relations" as per a technocentric transformation narrative [26]. The consistency of the research findings with the framings identified in other studies is, perhaps, indicative of deeper sets of competing beliefs and values. Indeed, Benford and Snow [11] argue that frames "function as innovative amplifications and extensions of, or antidotes to, existing ideologies or components of them".

4.7.2. Potential for Frame Conflict, Contests and/or Alignment

Participant deliberation on the glimpses revealed the potential for both frame conflict and alignment processes. On the conflict side, contrasting preferences were articulated and some participants expressed concerns about how the wider community might respond to the concepts. For example, one participant remarked that "I think the output was great for us but I can't see it being engaging for an older citizen, [or] a member of the parliament *etc.*". With respect to conflicting preferences, some participants were attracted to highly localised ways of life that are far more communal (as per some of the ways the "localised urban sustainability" frame was visualised), whereas others reacted strongly to the sense that communalisation and localisation was being presented as *the* "way to go". There were similar differences and tensions in participants' responses to "green Orwellianism".

Building on Snow *et al.'s* [10] analysis of frame alignment processes, there may be opportunities for what they term frame bridging, frame amplification, frame extension, or frame transformation. For example, frame bridging refers to "the linkage of two or more ideologically congruent but structurally unconnected frames regarding a particular issue or problem" [10]. Frame bridging could be attempted by trying to link the more technocentric frames (e.g., "Low-carbon transitions through digital lives and "green Orwellianism" [Frame 3] and "Low-carbon transitions as changes to city-scale "hard" infrastructure and larger-scale systems" [Frame 5]), or to link the frames that more strongly emphasise social and/or cultural change (e.g., "localised urban sustainability" [Frame 1] and "radical shifts in political economy and material consumption" [Frame 2]). Additionally, frame extensions could be attempted by encompassing additional interests or points of view that are "of considerable salience to potential [frame] adherents" [10]. For example, those promoting a "localised urban sustainability" future could consider ways of portraying such urban futures which would be more congruent with potential adherents who have *different* values and interests (e.g., those with less collectivist social values) and that would also encompass additional auxiliary interests.

Sociological research has consistently demonstrated the importance of frame alignment and "micromobilisation" processes [10,11,13,27,28]. Effective frames are often those which have been crafted to "resonate with varying groups" and have a high degree of interpretive flexibility [13].

Linkages between frames evident in some data also indicate such opportunities. Some participants expressed more than one of the identified frames, for example both Frame 1 and 5. The reason for this linkage requires more investigation; however it may be due to the synergies between urban redevelopment (Frame 1) and improved public and active transport systems (Frame 5).

However, some level of frame conflict may be unavoidable. For example, framing contests between those who see radical reorganisation of the structure of contemporary economies (e.g., towards post-growth economies) as an essential part of low-carbon transitions and resilience (as per Frame 2) and others who favour technocentric–style green transformations may be inevitable.

The overall diversity of frames also meant that there was a risk that adherents to the dominant frame would dominate discussion and prevent other framings from being heard, with framing contests resulting in some frames being marginalised. A diversity of frames also raises complications for visioning processes, an issue that we briefly take-up in the discussion section below.

4.8. Discussion

4.8.1. Framing Processes and the Use of Visioning Processes in Sustainability Research

Visioning practices are often centred on the pursuit of a shared vision or, at the very least, defining and agreeing on *common* goals and associated plans. As McCann notes, visioning is typically viewed and practiced as a consensus-oriented approach that, in urban contexts, "develops goals for the future of a city through consensus-based meetings open to all interested parties" [29]. This study, and related theorisation by sustainability-oriented scientists [14,25,30], raises questions about the suitability of this approach for complex prospective changes such as urban low-carbon transitions and resilience-focused urban change. Political contention over alternatives and multiple proposed low-carbon transformations—rather than consensus—may be inevitable and in some cases more desirable [25].

Theorisation of frames and framing processes provides some support for this view. If, as Schon and Rein [12] have argued, the ways actors' cognitively process problematic complex situations (such as the challenge of decarbonising cities) is necessarily both selective and partial then exploration of multiple frames and associated contested pathways may be necessary to address such situations.

The approach taken in VP2040 was to question the common focus on defining a singular blueprint of the desired future—which is typically a single detailed vision—and to, instead, explore glimpses of possible futures. Whilst the process didn't explicitly explore or make space for multiple frames, the emphasis on glimpses (plural)—rather than a single blueprint—allowed for their expression.

Additional considerations emerging from this study relate to the question of how to address the existence of multiple frames and potential framing contests in visioning processes. This study shows that even fairly homogenous groups—such as the groups convened in the visioning processes discussed here—don't necessarily share a single problem frame or collective action frame. Diverse groups are likely to contain participants that adhere to a greater diversity of frames. The study suggests that organisers and facilitators of visioning exercises should be more attentive to framing processes and the potential for framing contests, as well as the ways these shape the outcomes of such exercises.

4.8.2. Additional Reflections on the Design of the Visioning Process and Research

Convenors of visioning processes also need to be mindful that the framing and design of such processes shapes the outputs and outcomes. A few features of the study are discussed below.

The process had a dual focus on decarbonisation and resilience. This decision may have led to a stronger focus on food production (although this is also important for emissions reduction), urban greening (as per the ideas for mitigating the urban heat island effect), and localised resilience and self-reliance (as per Frame 1). This highlights the influence of process framing decisions.

This study also indicates that the convenors of visioning processes need to be mindful of how the institutional and life settings of participants can shape process outcomes. This is especially the case for workshops for which attendance is invitation only. For example, in the present study achieving a better balance of participants from local, State, and Federal government perhaps should have been a priority (although process convenors often have limited control over this). This may have reduced to dominance of a single frame—localised urban sustainability—and thereby encouraged fuller consideration of the highly complex problems being grappled with in the workshops.

The central roles played by the designers (some of whom are architects by profession) may also have influenced the process outcomes, including contributing to the "localised urban sustainability" frame being central to most glimpses. Their usual focus on smaller-scales (such as buildings, precincts, streetscapes, *etc.*), rather than whole cities and associated large-scale systems, may have influenced some of the visualisations/glimpses.

The potential influence of the designers' own preferences also cannot be entirely discounted. Some design teams/designers and participants had also previously worked with VEIL on similar projects. Linked with this, the research focus of the lab leading the VP2040 project (the Victoria Eco-Innovation Lab [VEIL]) on distributed systems, food security, and resilience may also have influenced the results of the study.

Additionally, we can consider the implications of using framing theory and examining framing processes. The key strengths include the following: it is a useful approach for surfacing and (thereby) discussing existing selective understandings of complex problem situations; framing attends to the meaning work done by actors as they engage with—and prescribe solutions to—complex social problems by constructing and using interpretive frames [11]; and the theory recognises cognitive limitations [12]. The positives must be balanced with the potential to reinforce differences and conflict (e.g., by highlighting the existence of competing frames) and participant considerations.

5. Conclusions

This study identified five framings of low-carbon resilient cities and found that *diverse* institutionally-related and culturally-grounded framings of issues are voiced, even when relatively homogenous groups participate. Consistent with Schon and Rein's conceptualisation of frames—as *selective* diagnostic-prescriptive stories based on underlying structures of beliefs, perceptions, and appreciation—the frames can be understood and contrasted according to the problem diagnosis and description (diagnostic framing) and associated prescriptions (prognostic framing). The identified frames varyingly have the potential to contribute to conflicts or actor alignment. The survey data, whilst having significant limitations, provided evidence of related framing processes.

As noted in the discussion that followed, the study also suggests that the convenors and facilitators of visioning processes need to be aware of framing processes and related dynamics (e.g., framing contests). Such processes and dynamics may have contributed to "localised urban sustainability" being the dominant frame in the workshops. At a minimum the organisers and facilitators of visioning exercises need to be attentive to framing contests and related social processes. A further key consideration is whether a visioning exercise seeks to open up a complex issue (such as by surfacing and clarifying frames) or to achieve agreement on a singular, *i.e.*, consensual, agenda.

This study has made two related contributions. First, it demonstrated the *diverse* ways actors frame the problems of urban decarbonisation and resilience. The five frames would be a good starting point for other urban decarbonisation/resilience projects (with appropriate contextualisation). For example, it may be possible to use them as collective action frames for building coalitions. The diverse ways that actors frame such problems also means that if a sustainable city-related visioning process aims to achieve consensus the process will either: need to allow for framing contests (by welcoming or making room for them) and achieve a deeper level of frame reflection and associated reframing [12]; or it must involve participants from similar institutional and life settings (but those actors will necessarily have a selective and partial understanding of the problems being addressed). This has major implications for visioning practices. Second, it demonstrated the need to be aware of framing processes in the context of visioning and action for realising sustainable cities. As noted earlier, such action may be enhanced by the exploration of multiple frames and associated contested pathways such as through further examination of the frames identified in this paper and by enabling related collective action.

Acknowledgments: We would like to acknowledge funding received through the Cooperative Research Centre for Low-carbon Living and its partners that has supported this research (Visions and Pathways 2040 project).

Author Contributions: Stephen McGrail conceived of the paper, led the data analysis and wrote the first draft of the paper. A. Idil Gaziulusoy and Paul Twomey contributed to the survey design and research, data analysis, and reviewed the draft manuscript. All authors were co-facilitators of the visioning workshops reported on in this paper.

Conflicts of Interest: The authors declare no conflict of interest.

References

1. Bakker, S.; van Lente, H.; Meeus, M.T.H. Credible expectations—The US Department of Energy's Hydrogen Program as enactor and selector of hydrogen technologies. *Technol. Forecast. Soc. Change* **2012**, *79*, 1059–1071. [CrossRef]
2. Van Lente, H.; Rip, A. Expectations in technological developments: An example of prospective structures to be filled in by agency. In *Getting New Technologies Together: Studies in Making Sociotechnical Order*; Disco, C., van der Meulen, B., Eds.; Walter de Gruyter: Berlin, Germany, 1998; pp. 203–229.
3. Van Lente, H. Navigating foresight in a sea of expectations: Lessons from the sociology of expectations. *Technol. Anal. Strateg. Manag.* **2012**, *24*, 769–782. [CrossRef]
4. Borup, M.; Brown, N.; Konrad, K.; van Lente, H. The sociology of expectations in science and technology. *Technol. Anal. Strateg. Manag.* **2006**, *18*, 285–298. [CrossRef]
5. Brown, N.; Rappert, B.; Webster, A. *Contested Futures: A Sociology of Prospective Techno-Science*; Ashgate Publishing Ltd.: Aldershot, UK, 2000.
6. Brown, N.; Rappert, B.; Webster, A.; Cabello, C.; Sanz-Menendez, L.; Merkx, F.; van de Meulen, B. *Foresight as a Tool for the Management of Knowledge Flows and Innovation (FORMAKIN)—Final Report*; European Commission: Brussels, Belgium, 2001.
7. Truffer, B.; Voß, J.-P.; Konrad, K. Mapping expectations for system transformations: Lessons from Sustainability Foresight in German utility sectors. *Technol. Forecast. Soc. Change* **2008**, *75*, 1360–1372. [CrossRef]
8. Kaplan, S. Framing contests: Strategy making under uncertainty. *Organ. Sci.* **2008**, *19*, 729–752. [CrossRef]
9. Moser, S.C.; Boykoff, M.T. Climate change and adaptation success: The scope of the challenge. In *Successful Adaptation to Climate Change: Linking Science and Policy in a Rapidly Changing World*; Moser, S.C., Boykoff, M.T., Eds.; Routledge: London, UK, 2013.
10. Snow, D.A.; Burke Rochford, E.; Worden, S.K.; Benford, R.D. Frame alignment processes, micromobilization, and movement participation. *Am. Sociol. Rev.* **1986**, *51*, 464–481. [CrossRef]
11. Benford, R.D.; Snow, D.A. Framing processes and social movements: An overview and assessment. *Annu. Rev. Sociol.* **2000**, *26*, 611–639. [CrossRef]
12. Schon, D.A.; Rein, M. *Frame Reflection: Towards the Resolution of Intractable Policy Controversies*; Basic Books: New York, NY, USA, 1994.
13. Fligstein, N.; McAdam, D. *A Political-Cultural Approach to the Problem of Strategic Action, in Rethinking Power in Organizations, Institutions, and Markets*; Courpasson, D., Golsorkhi, D., Sallaz, J.J., Eds.; Emerald Group Publishing Limited: Bingley, UK, 2012; pp. 287–316.
14. Leach, M.; Scoones, I.; Stirling, A. *Dynamic Sustainabilities: Technology, Environment, Social Justice*; Earthscan: London, UK; Washintgon, DC, USA, 2010.
15. Vogel, K.M. Iraqi Winnebagos™ of death: Imagined and realized futures of US bioweapons threat assessments. *Sci. Public Policy* **2008**, *35*, 561–573. [CrossRef]
16. Vogel, K.M. Framing biosecurity: An alternative to the biotech revolution model? *Sci. Public Policy* **2008**, *35*, 45–54. [CrossRef]
17. Geels, F.W. Regime Resistance against low-carbon transitions: Introducing politics and power into the multi-level perspective. *Theory C. Soc.* **2014**, *31*, 21–40. [CrossRef]
18. Turnheim, B.; Geels, F.W. Regime destabilisation as the flipside of energy transitions: Lessons from the history of the British coal industry (1913–1997). *Energy Policy* **2012**, *50*, 35–49. [CrossRef]
19. Farla, J.C.M.; Markard, J.; Raven, R.; Coenen, L. Sustainability transitions in the making: A closer look at actors, strategies and resources. *Technol. Forecast. Soc. Change* **2012**, *79*, 991–998. [CrossRef]
20. Wirth, L. Urbanism as a way of life. *Am. J. Sociol.* **1938**, *44*, 1–24. [CrossRef]
21. Urry, J. *Societies beyond Oil: Oil Dregs And Social Futures*; Zed Books: London, UK; New York, NY, USA, 2013.
22. Dennis, K.; Urry, J. *After the Car*; Polity Press: Cambridge, UK, 2009.
23. Urry, J. Climate change, travel and complex futures. *Br. J. Sociol.* **2008**, *59*, 261–279. [CrossRef] [PubMed]
24. Knight, E. *Reframe: How to Solve the World"s Trickiest Problems*; Black, Inc.: Collingwood, Australia, 2012.
25. Scoones, I.; Newell, P.; Leach, M. The politics of green transformations. In *The Politics of Green Transformations*; Scoones, I., Leach, M., Newell, P., Eds.; Routledge: New York, NY, USA, 2015.
26. Scoones, I.; Leach, M.; Newell, P. *The Politics of Green Transformations*; Routledge: New York, NY, USA, 2015.

27. Fligstein, N.; McAdam, D. *A Theory of Fields*; Oxford University Press: New York, NY, USA, 2012.
28. Snow, D.A.; Benford, R.D. Ideology, frame resonance, and participant mobilization. *Int. Soc. Mov. Res.* **1988**, *1*, 197–218.
29. McCann, E.J. Collaborative visiong or urban planning as therapy? The politics of public-private policy making. *Prof. Geogr.* **2001**, *53*, 207–218. [CrossRef]
30. Stirling, A. Transforming power: Social science and the politics of energy choices. *Energy Res. Soc. Sci.* **2014**, *1*, 83–95. [CrossRef]

sustainability

MDPI

Article

Aligning Public Participation to Stakeholders' Sustainability Literacy—A Case Study on Sustainable Urban Development in Phoenix, Arizona

Matthew Cohen [1,2,*], Arnim Wiek [1], Braden Kay [1] and John Harlow [1]

1 School of Sustainability, Arizona State University, 800 South Cady Mall, Tempe, AZ 85281, USA; arnim.wiek@asu.edu (A.W.); brkay@asu.edu (B.K.); john.harlow@asu.edu (J.H.)

2 Earth and Environmental Sciences Department, Furman University, 3300 Poinsett Highway, Greenville, SC 29613, USA

* Author to whom correspondence should be addressed; matthew.cohen@furman.edu; Tel.: +1-864-294-2052.

Academic Editor: Tan Yigitcanlar

Received: 15 April 2015; Accepted: 29 June 2015; Published: 3 July 2015

Abstract: In public planning processes for sustainable urban development, planners and experts often face the challenge of engaging a public that is not familiar with sustainability principles or does not subscribe to sustainability values. Although there are calls to build the public's sustainability literacy through social learning, such efforts require sufficient time and other resources that are not always available. Alternatively, public participation processes may be realigned with the sustainability literacy the participants possess, and their capacity can modestly be built during the engagement. Asking what tools might successfully align public participation with participants' sustainability literacy, this article describes and evaluates a public participation process in Phoenix, Arizona, in which researchers, in collaboration with city planners, facilitated sustainability conversations as part of an urban development process. The tool employed for Visually Enhanced Sustainability Conversation (VESC) was specifically designed to better align public participation with stakeholders' sustainability literacy. We tested and evaluated VESC through interviews with participants, city planners, and members of the research team, as well as an analysis of project reports. We found that the use of VESC successfully facilitated discussions on pertinent sustainability issues and embedded sustainability objectives into the project reports. We close with recommendations for strengthening tools like VESC for future public engagements.

Keywords: sustainability; sustainability literacy; urban development; public participation; urban planning; civic engagement

1. Introduction

Both sustainability science and urban planning literature identify public participation as an important decision making procedure [1–5]. Yet, high-quality public participation is not always achieved. Some challenges include overly dominant government agencies, participants who lack capacities to engage, and inequitable distributions of resources [6]. Particularly, in planning for *sustainable* urban development, public participation requires not only engaged participants, but also participants that are literate in sustainability principles, norms, and behaviors [7,8].

Engaging with a public that lacks sustainability literacy is a significant challenge for planners and experts working towards sustainable urban development. Cohen and Wiek [9] identify ten ways in which the public participation process (as designed by planners and experts) is often misaligned with the local context. Participants' sustainability literacy is one key misalignment that can impair the quality of public conversations on urban sustainability and ultimately compromise the robustness of sustainability outcomes in project planning and implementation.

Obstacles to high-quality public participation, including low sustainability literacy, can often be avoided with careful process design [10,11]. However, the body of planning literature on public participation is weak in providing empirically informed directives for designing good processes. The majority of published papers on public participation are theory-based, and there are far fewer examples of empirical studies [12]. The magnitude of empirical research on the topic lags behind both the theoretical and practitioner communities [11,13], due in part to a lack of evaluative studies on public participation and stakeholder engagement processes [14–16].

Therefore, this article describes and evaluates a case in which researchers, in collaboration with city planners, attempted to align the public participation process to stakeholders' sustainability literacy in a sustainable urban development process in Phoenix, Arizona. The project team encountered low sustainability literacy amongst participants and redesigned engagement activities to strengthen sustainability conversations at public visioning workshops. This study asks how public participation processes might be structured to cope with low sustainability literacy, with the objective of generating evidence-supported directives for designing public participation for sustainable urban development.

2. Public Participation Processes in Urban Development Projects

2.1. Defining Public Participation

"Public participation" may describe a variety of processes applied across diverse contexts [10]. For the purposes of this study, we define public participation as a process that is part of an official urban development project. It is situated within the urban development project, which itself is positioned within a specific context. *Strategic Agents* (*i.e.*, elected officials and investors) supervise the process, which is carried out by *Operating Agents* (*i.e.*, city staff and project partners), and *Participating Stakeholders* (*i.e.*, residents, non-profits, businesses) provide input through a structured process. Stakeholders may be involved throughout multiple project phases, including preparing, planning, implementing, and evaluating project outcomes. Stakeholders may be asked to engage through different methods, including public meetings, focus groups, workshops, citizen juries, and other protocols [9,17].

2.2. Misalignments between Public Participation Process and Local Context

The literature on public participation features theoretical debates about whether participation yields positive outcomes or instead produces adverse effects. In planning literature, little empirical evidence supports either argument [12], yet some scholars contend that most shortcomings of participation processes can be mitigated through careful process design [10,11].

To avoid challenges to high-quality public participation, one must first identify these pitfalls. Through a broad review of the literature on public participation in urban development, Cohen and Wiek [9] found common challenges to public participation and organized these issues into ten categories, conceptualized as misalignments between the public participation process and the local context, including policy maker support, community civic engagement, and stakeholders' collaborative capacity, among others. These common challenges to public participation in urban development occur when the public participation process (as designed by experts and planners) does not align with the local context where participation is taking place.

This article is concerned with one specific misalignment between the public participation process and the local context: Stakeholders' sustainability literacy. Generating robust sustainability outcomes through public participation processes requires stakeholders to engage in sustainability-oriented conversations. Many members of the public, however, lack a strong grounding in sustainability principles, and their values and behaviors may be in conflict with sustainability. Meaningful engagement around sustainability is challenging when participants are not, in this sense, "literate" in sustainability.

When a public participation process is not aligned with stakeholders' sustainability literacy, there is a knowledge and/or values gap on sustainability between experts and stakeholders. When this problem persists, participants may feel confused, they may harbor frustrations or distrust, and their input may be incompatible with sustainability goals and objectives [9]. This is not to imply that sustainability-literate participants would guarantee a consensus around sustainability outcomes. Rather, a public participation process aligned to participants' sustainability literacy might yield constructive dialogue, seek compromise, find common understanding, and enable robust sustainability-oriented outcomes to influence subsequent policy decisions.

2.3. *Approaches for Aligning Public Participation Processes to Participants' Sustainability Literacy*

Building stakeholders' capacity for sustainability can strengthen the quality of public participation in urban sustainability governance [7]. Sustainability competence has been defined as a bundle of knowledge, attitudes, and skills about sustainability [18], and we adopt this comprehensive perspective on sustainability competence as the ideal for high sustainability capacity. Stakeholders' capacity for sustainability can be developed through social learning, *i.e.*, experiences in which participants build understanding and shape their values through collaboration with others. Social learning is a common potential benefit of public participation, and there are many cases of facilitating social learning of sustainability [19–21]. Although social learning is popular in sustainability governance literature, it is important to note, that this form of capacity building requires a significant investment of time and other resources that may not be available in all participatory processes [20]. In an empirical study of social learning in public participation for sustainability outcomes, Garmendia and Stagl [22], for example, found that participants need ample time for interaction and deliberation.

As an alternative to enhancing sustainability literacy through separate capacity building events, the planning process itself can be designed as a learning experience, using guidelines from the literature. In the following, we review a set of guidelines for aligning public participation processes to participants' sustainability literacy.

Innes and Booher [6] identify keys to successful public participation, one of which is *dialogue*. Dialogue can be transformative because participants who listen to and inform each other can develop new ideas and shared meanings. They recommend that public agencies promote procedures that foster deliberation between stakeholders and that planners be trained to design and manage collaborative processes.

Other scholars repeat this recommendation. Through case studies of participatory processes, Fung and Wright [23] synthesize key principles of empowered participatory governance. They state that procedures should lead dialogue beyond an abstract discourse on values and instead focus on conversations about practical issues and concrete problems. People affected by the problems being discussed should be given an opportunity to deliberate solutions to the problems. Healey [24] discusses creative urban governance as an alternative to the "established routines (and) rule-bound bureaucratic procedures (p. 88)" that typify planning processes. Under creative urban governance, processes would support informative conversations that are facilitated through experimental practices (*cf.* [5]).

Rowe and Frewer [16] propose four criteria for evaluating a public participation procedure: Criterion of resource accessibility, criterion of task definition, criterion of structured decision making, and criterion of cost-effectiveness. Two of these criterion—resource accessibility and structured decision making are particularly helpful for guiding the construction of participatory procedures to foster social learning and structure participant interaction and deliberation. The first vital criterion, resource accessibility, stipulates the resources that participants need to make a decision. This includes (a) information resources, or the facts needed to make an informed decision; (b) human resources, or access to experts and other individuals that can provide needed information; and (c) material resources, or objects like projectors or whiteboards to facilitate understanding. These resources would all be important for communicating sustainability concepts, problems, and solutions to diverse participants.

The second vital criterion, structured decision making, states that activities should follow clear mechanisms for facilitating decision making. Coping with low participant sustainability literacy in participatory procedures would lend to a need for competent facilitation. Subsequent works [15,25] question whether this criterion should be further broken into more specific criteria, and Rowe *et al.* [25] consider including assessments of adequate and fair elicitation as well as information presentation.

One way to present information is through the use of images. In a study on the use of imagery for public engagement on climate change, O'Neill *et al.* [26] found that images of climate change impacts made participants feel that climate change was important (salience), but reduced their feeling that they could do something to address the problem (self-efficacy). On the other hand, images of energy futures increased participant efficacy. While images of the problem reduced participant self-efficacy, images of potential solutions awoke in participants a sense that they could tackle the challenge. This finding would support Fung and Wright's assertion that participants should be engaged on the topic of solutions, which is doubly relevant to sustainability planning, as sustainability science is framed as a solution-oriented endeavor [27]. The framing of issues significantly impacts participant input [22,28,29], and in the case of sustainability, these examples from the literature indicate that such topics should be framed as solutions to real problems.

In summary, a public participation procedure to support sustainable urban development may be effective if it fosters deliberation about real problems and their potential solutions, and if it supports social learning. Due to common time constraints of participatory planning processes, we focus particularly on information resources and presentation as well as facilitation to enable conversations about sustainability outcomes amongst stakeholders who may not be comfortable or familiar with sustainability. Therefore, facilitators need to be specially trained to lead deliberations on sustainability supported by materials like visuals to aid participant understanding.

3. Research Methods: Evaluation of a Public Participation Procedure

This study was conducted on a participatory visioning process for an urban development project in Phoenix, Arizona. The authors were members of a research team that designed and implemented a participatory process to elicit stakeholder input for sustainability visions for districts along the City of Phoenix's light rail corridor. Through one-on-one interviews, surveys, community organization meetings, and public workshops engaging nearly 300 participants in the first district (Gateway District), researchers observed a challenge of facilitating discussions around sustainability with stakeholders who had little background on the subject. The researchers identified stakeholders' low sustainability literacy as a barrier to quality public participation and redesigned workshop activities and materials for use in other transit districts (including the Midtown District, which serves as the reference district in this study).

This article evaluates the tool of Visually Enhanced Sustainability Conversation (VESC) that was designed to better facilitate deliberation on sustainability options during the public visioning process. We evaluate VESC using select criteria from Rowe and Frewer [16] that would support participant deliberation over sustainability issues (resource accessibility, including information resources, human resources, and material resources; and structured decision making, including elicitation and information presentation), and we apply a fifth criterion that the activity must facilitate public discussion on sustainability. The ultimate research question is whether or not the application of VESC effectively facilitated conversations and decision-making about sustainability options.

To assess VESC in terms of the above criteria, the evaluation includes researchers' direct observations; document analysis of vision reports from two of the transit districts (Gateway and Midtown); and 11 interviews with participant stakeholders, project partners, and members of the research team. Table 1 outlines the data inputs used in the evaluation.

After public participation concluded in each district, a steering committee of stakeholders was formed to support the visions towards implementation. Steering Committee members were co-selected by researchers and staff from the City of Phoenix Neighborhood Services Department and the Planning and Development Department. Selection criteria emphasized a diversity of perspectives, including anchor institutions (*i.e.*, hospitals, universities, and religious institutions), businesses and business coalitions, K-12 schools, developers, landowners, neighborhood associations, and residents. Special attention was given to underrepresented populations, including refugee communities, recent immigrants, public housing residents, and American Indians. Those recommended were contacted and interviewed by researchers to determine their availability and interest in serving on the committees. Over a span of 18–24 months, Steering Committees reviewed and edited planning documents, recommended potential investments (*i.e.*, support for road improvements or development of business incubators), and supported grants for infrastructure projects.

Table 1. Data inputs for evaluating visually enhanced sustainability conversations.

Data Type:
Direct observations
Document Analysis:
• District Vision Report • Midtown District Vision Report
Interviews:
• Gateway District Steering Committee Member #1, interviewed 6 January 2015 • Gateway District Steering Committee Member #2, interviewed 9 January 2015 • Gateway District Steering Committee Member #3, interviewed 14 January 2015 • Midtown District Steering Committee Member #1, interviewed 5 December 2014 • Midtown District Steering Committee Member #2, interviewed 11 December 2014 • Midtown District Steering Committee Member #3, interviewed 12 December 2014 • Midtown District Steering Committee Member #4, interviewed 14 December 2014 • Midtown District Steering Committee Member #5, interviewed 19 December 2014 • City of Phoenix Planner, interviewed 17 December 2014 • Research Team Member #1, interviewed 30 January 2015 • Research Team Member #2, interviewed 3 February 2015

For this study, researchers identified steering committee members as ideal respondents because they were involved at public engagements during the visioning process, they hold a big-picture perspective of development goals in their districts, and they were selected to be representative of significant stakeholder groups in their districts. Researchers conducted interviews with three Gateway Steering Committee members and five Midtown Steering Committee members to compare the workshop experiences in the two districts. A City of Phoenix planner was interviewed to gain insight from a project partner. The planner also recommended the steering committee members to be interviewed for this evaluation. Finally, to reduce bias in the evaluation, interviews with two members of the research team provide feedback from individuals that helped design the VESC tool and facilitate public participation, but are *not* authors of this article. The interviews with the research team members are particularly valuable because both respondents are now practicing urban planners in major metropolitan areas in the U.S.

Respondents participated in semi-structured interviews. A researcher met with the respondent, reviewed copies of workshop activity posters from both districts and VESC posters from the Midtown district. In each interview, respondents compared the experiences from the two districts and provided feedback on the tools that were used. After each interview, responses were coded by evaluative criteria, and the researcher assessed whether feedback was negative, ambivalent, or positive. This approach relies to some extent on the researchers' own judgments (coding and assessment), but respondent quotes provide rich details that support the assessment decisions.

4. Reinvent Phoenix: Aligning Public Participation Process to Stakeholders' Sustainability Literacy

4.1. The Reinvent Phoenix Participatory Visioning Process

Reinvent Phoenix was an urban development project in Phoenix, Arizona. Funded by the U.S. Department of Housing and Urban Development, Reinvent Phoenix was a partnership between the City of Phoenix, Arizona State University, St. Luke's Health Initiative, and other community organizations. The project sought to promote transit-oriented and sustainable urban development along Phoenix's light rail corridor. This goal was to be achieved over multiple phases that included a public participation process to develop sustainability visions for five specific transit districts: Gateway, Eastlake-Garfield, Midtown, Uptown, and Solano (Figure 1). The visions would then inform a zoning process to create form-based codes that support transit-oriented and sustainable urban development.

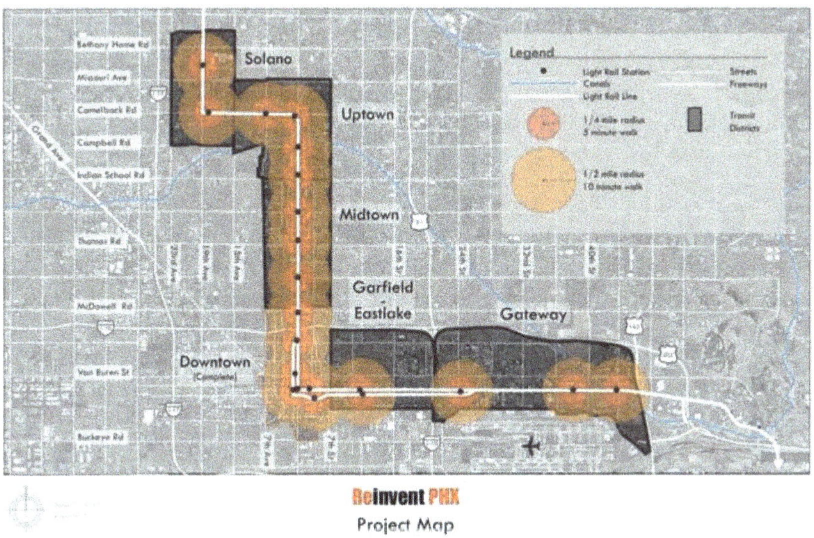

Figure 1. Reinvent Phoenix project map.

The authors and their research team managed the public participation process to create the sustainability visions for each district. This article focuses on a particular aspect of the visioning processes in the Gateway and Midtown Districts. The Gateway District is the farthest east district of Phoenix' light rail corridor. It is one of the most ethnically diverse transit districts, and it features the corridor's highest poverty rates and lowest educational levels. The Midtown District, by contrast, is the most affluent of Phoenix's transit districts. Comparing participation between these two districts is challenging as socioeconomic, educational, and other factors significantly impact participation [30]. However, Gateway's visioning process provides a convenient baseline against which to measure the effectiveness of VESC as this tool was not designed until after public participation in Gateway had concluded.

A pilot process was conducted in the Gateway District from September–December 2012. During this period, researchers engaged with stakeholders, who live, work, do business in, or visit the district, through one-on-one interviews, community organization meetings, two public mapping forums, and two public visioning workshops. Through the engagements leading to the visioning workshops, participants identified areas they would like to see preserved or changed, and they discussed the types of changes they would like to see occur. Researchers identified consensus areas for change (transition areas) and prepared a visioning workshop to enable participants to discuss in detail how each of the transition areas might look in the future.

Accordingly, researchers engaged workshop participants in discussions about specific changes for the identified transition areas in the Gateway District. These conversations revolved around a Visual Preference Survey (VPS) in which participants discussed and voted on preferences for issues including building height, street design, and landscaping. The VPS facilitated form-based discussions that focused primarily on the district's physical form. Sustainability outcomes were tacitly included in certain options. For instance, taller building heights promoted density, complete street designs fostered multi-modal transportation options, and low-water landscaping would use fewer natural resources. Sustainability was not explicitly addressed in these conversations, and function-oriented conversations (*i.e.*, how participants would live and work in the buildings; how participants would travel on streets) occurred organically, if at all.

At the end of the public participation process in the Gateway District, researchers drafted a vision based strictly on stakeholder input. After the report was complete, one member of the research team conducted a criteria-based sustainability appraisal of the Gateway vision. The appraisal showed that the initial vision lacked sustainability substance [31]. Throughout the process, the researchers felt that public discussions about sustainability outcomes could have been stronger and that more targeted discussions would have further infused sustainability into the vision.

As the Gateway District was the first transit district in which visioning activities occurred, researchers had an opportunity to take lessons from that experience and revise the approach for subsequent districts. Identifying low sustainability literacy of participants and seeking to strengthen sustainability conversations at visioning workshops, the researchers devised new workshop activities and materials for facilitating public discussions. The Midtown District is one of the districts in which the revised participation procedures were implemented.

The Midtown visioning process occurred January–May 2013. Researchers employed a slightly altered process for identifying transition areas, including one-on-one interviews, forums at existing neighborhood meetings, and tabling at community events. At the visioning workshops, researchers again led participants through a VPS activity. After concluding the VPS, researchers facilitated a new activity, titled Visually Enhanced Sustainability Conversation (VESC).

4.2. Designing the Visually Enhanced Sustainability Conversation

The intention of VESC was to facilitate a public discussion to prioritize sustainability objectives and identify means (vision elements) for achieving these objectives that would be acceptable to stakeholders. To foster deliberation towards sustainability outcomes, researchers pre-selected the objectives and vision elements prior to the visioning event. Figure 2 shows the hierarchy of a sustainability vision, from the most general component (guiding principle) to the most specific (vision element).

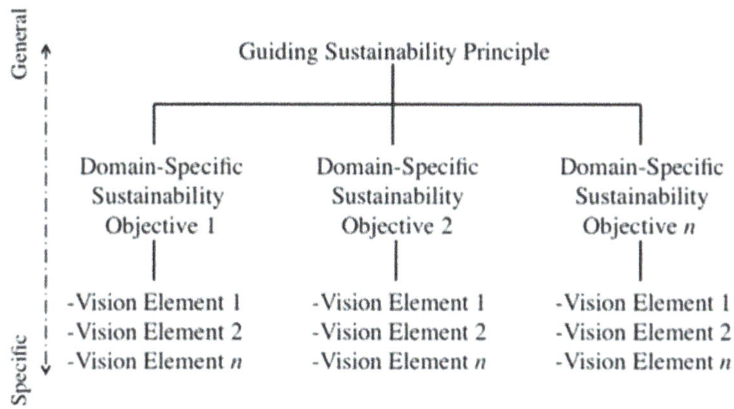

Figure 2. Hierarchy of a sustainability vision.

In the case of Reinvent Phoenix, researchers populated this sustainability vision hierarchy through a transparent process. Guiding sustainability principles were adapted from Gibson's widely accepted synthesized set of six criteria [32]. For each sustainability principle, neighborhood-specific objectives were identified through intensive public engagements via one-on-one stakeholder interviews, surveys, community organization meetings, and public forums [33]. Researchers then identified criteria-based elements that could be operationalized to achieve the neighborhood-specific objectives and the guiding sustainability principles. Researchers then vetted the objectives and vision elements with project partners in the City of Phoenix Planning Department. Table 2 outlines the objectives and their related vision elements for each of Midtown's three transition areas.

Researchers sought local and regional examples of each vision element in order to capture visual evidence of the element's use in the Phoenix area. In some cases, local examples did not exist, and researchers attempted to locate images as relevant to the Phoenix context as possible.

Although the content of the VESC tool is critical, it is not the only success factor for a participatory vision. The facilitators themselves are equally as important as the facilitation tools. In preparation for public visioning workshops, members of the research team were paired as facilitators and note-takers. Facilitators and note-takers underwent multiple trainings. Through a series of dry runs, the facilitators and note-takers practiced the visioning activity that revolved around VESC. The facilitator played a focal role leading the exercise, and note-takers played a support role by recording all conversation onto a laptop, joining the group discussion when the facilitator couldn't manage multiple conversations, and organizing support materials like pens, markers, and post-it notes. In some cases, when group dynamic dictated, the note-taker became a secondary facilitator.

Table 2. Sustainability objectives and vision elements by transition area.

Transition Area	Objective	Vision Element 1	Vision Element 2	Vision Element 3
Park Central Mall	Economic vitality through strong local businesses	Buy-local initiative	Small business support organization	Business in mixed-use building
	Diverse employment and training opportunities	Co-working spaces	University-community partnership	Participant suggestions
	Cool neighborhoods	Cool pavement	Vegetation	Living roof
	Walkable and bikeable neighborhoods	Neighborhood circulator	Pedestrian malls and promenades	Park-and-ride
Central Avenue Corridor	Diverse employment and training opportunities	Co-working spaces	University-community partnership	Participant suggestions
	Cool neighborhoods	Cool pavement	Living roof	Vegetation
	Walkable and bikeable neighborhoods	Neighborhood circulator	Pedestrian malls and promenades	Participant suggestions
	Saving money through conserving natural resources	Adaptive reuse	Energy efficient homes	Solar houses
Third Street Corridor	Economic vitality through strong local businesses	Business incubator	Small business support organization	Buy-local initiative
	Cool neighborhoods	Cool pavement	Living roof	Vegetation
	Walkable and bikeable neighborhoods	Neighborhood circulator	Pedestrian malls and promenades	Participant suggestions
	Saving money through conserving natural resources	Adaptive reuse	Solar houses	Energy efficient homes

4.3. Facilitating the Visually Enhanced Sustainability Conversation

During the visioning activities, the facilitator employed the VESC tool by first showing participants a poster stating a small number (3–6) of sustainability objectives to discuss. Participants voted on the objectives that most interested or resonated with them (top 1–3). This voting served as an input for planners on how to prioritize objectives in the planning and implementation process of the Reinvent Phoenix project. The facilitator then led structured conversations about the most popular objectives, sharing a poster for each objective that illustrated potential vision elements for achieving the objective (Figure 3).

The poster for each objective listed two to three potential vision elements, and some posters prompted participants to suggest additional elements. For each vision element, there was a short description and a photograph providing visual representation. For issues involving building height and street design, we used photo-realistic visuals that depicted the vision elements (*i.e.*, 5–8 story buildings) as they would actually look in a given location. The facilitator described each vision element and fielded questions. The facilitator then guided a pros/cons conversation in which participants provided strengths for each vision element as well as potential obstacles to successful implementation. The facilitator noted participant responses on sticky notes and placed these inputs on the poster.

Participant input was inserted directly into the Midtown District vision. The vision for each transition area was organized around the sustainability objectives that gained the greatest stakeholder interest. The specific vision elements that participants supported were included in the report, and any nuances in participant preferences were addressed. For instance, in discussing rooftop photovoltaics (strategy: *Solar houses*) as a vision element for *saving money through conserving natural resources* along the Central Avenue Corridor, one participant noted that he was not comfortable with photovoltaic panels being visible from the street in the historic neighborhoods near Central Avenue. Through deliberation, he acknowledged that photovoltaic panels on historic homes were acceptable provided they were visible only from backyards. The note-taker recorded this request, and the vision stipulated that photovoltaic installations should not compromise historic character in such neighborhoods [34].

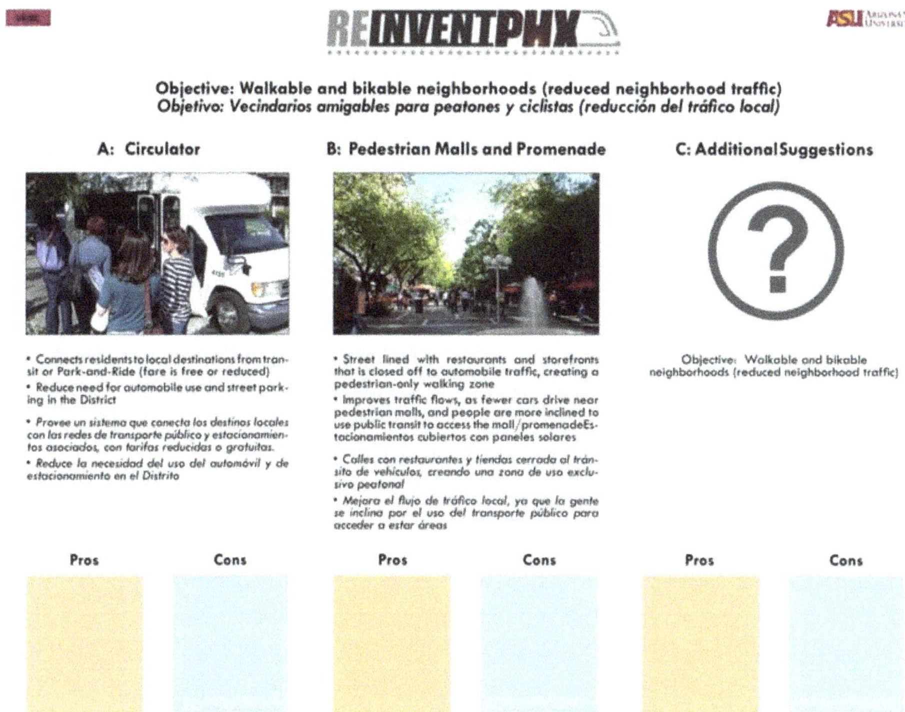

Figure 3. Example poster from a Visually Enhanced Sustainability Conversation.

Because the vision was oriented around pre-selected sustainability objectives, researchers conducted the criteria-based sustainability appraisal during the design of VESC. As described in Section 4.2, sustainability scientists vetted potential vision elements prior to the visioning workshops (based on objectives that aligned criteria identified by stakeholders and expert-based sustainability criteria), ensuring that the public discussions revolved around sustainability-oriented outcomes. A cursory sustainability appraisal of the vision reinforced that the vision does indeed describe a *sustainable* Midtown [34].

5. Evaluating the Alignment between Participatory Visioning and Participants' Sustainability Literacy in Reinvent Phoenix

The central focus of this article is to determine if Visually Enhanced Sustainability Conversation (VESC) aligned Reinvent Phoenix's participatory process to stakeholders' sustainability literacy. We therefore evaluate the tool in terms of how it facilitated participant conversations about sustainability. Respondents provided predominantly positive feedback on VESC for each of the evaluation criteria, and Table 3 shows which inputs (*i.e.*, Steering Committee members, research team members, direct observations) yielded positive or negative assessments for each criterion.

Table 3. Results of evaluation of visually enhanced sustainability conversations.

Evaluation Criteria	Assessment	
	Negative	**Positive**
Information Resources	• 2 Midtown Steering Committee members	• 2 Gateway Steering Committee members • 2 Midtown Steering Committee members • 1 research team member
Human Resources	• 1 Direct observation	• 1 Midtown Steering Committee member • 1 research team member • 1 Direct observation
Material Resources		• 1 Gateway Steering Committee member • 1 Research team member
Elicitation	• 1 Research team member	• City of Phoenix planner • 2 Gateway Steering Committee members • 5 Midtown Steering Committee members • 1 Research team member • 1 Direct observation
Information presentation	• City of Phoenix planner • 1 Gateway Steering Committee member • 2 Midtown Steering Committee members	• 3 Gateway Steering Committee members • 3 Midtown Steering Committee members • 1 Research team member
Sustainability Discussions		• City of Phoenix planner • 2 Gateway Steering Committee members • 5 Midtown Steering Committee members • 2 Research team members • 1 Direct observation • Midtown Vision Report [34]

In general, respondents regarded VESC as a useful tool for facilitating public discussions on sustainability objectives and vision elements. Respondents from Midtown reported that VESC did support the goal of infusing sustainability into both discussions at workshops and the resulting vision. Respondents from Gateway reviewed the VESC materials and felt that the tool would have been helpful in their district as well. The City of Phoenix planner and members of the research team also acknowledged the benefits of VESC. Although respondents supported the use of VESC, some also provided critiques and made recommendations for improving the tool. The most salient remarks and observations are discussed below.

6. Discussion

This section discusses the results for each evaluation criterion and highlights key quotes from respondents that illustrate both the strengths and weaknesses of VESC as it was implemented in Reinvent Phoenix. The following codes are used to cite each evaluation source:

- Direct observations: DO
- Gateway District Vision Report: [31]
- Midtown District Vision Report: [34]
- Three Gateway Steering Committee members: GW01; GW02; GW03
- Five Midtown District Steering Committee members: MT01; MT02; MT03; MT04; MT05

- City of Phoenix Planner: CP01
- Two research team members: RT01; RT02

6.1. Information Resources

The VESC posters, as information resources, were intended to define sustainability-oriented development objectives and present details about potential vision elements for achieving those objectives. The goal here was to foster informative conversations [24]. One Gateway respondent noted, "Everybody needs a starting point to enter the discussion" (GW03). A member of the research team pointed out that by placing objectives and vision elements on the posters, the researchers were reassuring participants "that these options are sustainable" (RT01). By presenting to participants sustainable options to discuss with examples and details, VESC gave participants the starting point they needed to enter the discussion about sustainability options.

To improve VESC posters as information resources, a Midtown respondent recommended providing more examples for each objective and offering more local examples so that participants could understand the examples through a context with which they are familiar (MT02). Researchers attempted to provide local images of vision elements, but not all vision elements had been implemented locally. The research team included external examples so as not to limit the vision to only what had already been accomplished in Phoenix. O'Neill *et al.* [26] reported on the importance of images in facilitating public discussions on sustainability issues, and respondent feedback in this evaluation seems to support this assertion. One Midtown respondent was concerned that participants that were learning of a potential objective or vision element for the first time might walk away from the activity with an overly optimistic view if the group is unable to identify shortcoming and pitfalls (MT04). For a balanced perspective [35], the posters may need to include disadvantages and facilitators might need to be more transparent about the pros and cons of each option. To alleviate this concern, the VESC poster and facilitation included an opportunity to discuss shortcomings of each vision element. Ensuring then that participants have a firm grasp of each option depends on competent and transparent facilitation.

6.2. Human Resources

Rowe and Frewer [16] highlight the importance of human resources to a participatory exercise, and VESC involved two groups of people: (1) the facilitator and note-taker; and (2) the participating stakeholders. A member of the research team was very positive about the quality of facilitation and attributed facilitator competence to the training and preparation of facilitators prior to the event (RT02). Facilitators helped research the sustainability objectives and vision elements and therefore were familiar with each option. A respondent from the Midtown District also noted the human resource value of fellow participants, saying, "It was good to have people with different ideas that can consider something different than their own point of view. There were times that I changed my opinion when I heard other people's ideas" (MT01). This form of social learning was fostered by the facilitated discussions about the benefits and obstacles for each vision element. Furthermore, this outcome from VESC supports the importance of dialogue and deliberation emphasized by Innes and Booher [6] and Fung and Wright [23].

6.3. Material Resources

During the interviews, respondents discussed the VESC posters as a mode for sharing information with participants. One member of the research team felt strongly that posters were resources that aided "people that didn't necessarily have a literacy in sustainability or urban planning" (RT02). A Gateway Respondent said, "I think the materials are fine. I look at this and I see 'here is a priority and here are three strategies to do that.' I think that is great" (GW03). The objectives and vision elements offer solutions to sustainability problems in the districts, and the VESC materials focus deliberation on solutions, as recommended by Fung and Wright [23].

6.4. Elicitation

Interviews showed a favorable assessment of VESC as an elicitation tool. Respondents tended to like the structure of the activity, and they felt that VESC prompted discussion by first providing examples that participants could see and understand. One respondent noted, "If you give them tangible examples, then they can see themselves in it" (MT02). Another respondent liked the structure, saying "It might help with people that aren't as knowledgeable, getting them on the right path" (MT04). The City of Phoenix Planner said VESC "helps because (...) you need to have directions so you can prompt them to think and then go with it. You initiate some conversations and then you elicit additional ideas because you prompted" (CP01). Contrasting VESC against the experience in Gateway, a member of the research team said, "In Gateway, it was hard to facilitate without the material to guide the conversations. We were less able to elicit responses because we didn't have the tools to do that" (RT02). She felt the VESC tool solved this issue. These comments highlight VESC's strength at structuring decision-making, emphasized by Rowe and Frewer [16], and facilitating informative conversations, promoted by Healey [24] and Zint and Wolske [5].

In critique of elicitation under VESC, one Gateway Respondent said, "I think you should always leave an opportunity for people to come up with ideas that you might not have already thought of" (GW03). Some objective posters presented two options (vision elements) and asked for additional ideas, while other posters presented three options and did not elicit additional input. It would be possible to design posters that always ask for additional ideas. A member of the research team also felt that while the structured conversation around pre-selected options was helpful, "there would be more value if (...) there could have been a more organic discussion about, for example, what sustainable land use looks like" (RT01). One solution to this concern might be to initiate the conversation with the objective/element discussion and then facilitate a bigger-picture discussion of general sustainability once the participants have been prepared by first discussing tangible examples. However, such a structure may not be reasonable if an event is facilitated under time constraints.

6.5. Information Presentation

O'Neill *et al.* [26] show that images of solutions inspire participants and increase participants' self-efficacy. The VESC sought to accomplish similar goals by including images of potential vision elements for achieving the sustainability objectives. The images allowed participants to see themselves in the sustainable future (MT02). VESC posters would also help participants who were visual thinkers (GW02), and visuals help move "the discussion along quicker in terms of people comprehending what we're comparing in terms of several options" (GW03). All five respondents that participated in VESC in Midtown District felt that the visuals improved the activity and fostered good conversation.

There were several issues that should be improved so that VESC can better present information. The City of Phoenix Planner felt that "the language was very planner wonky. Why can't it just say "good job choices?" (instead of "diverse employment opportunities") (CP01). Although there is credence to the respondent's opinion, the language on each poster was negotiated between project partners, evaluated in pre-tests, and revised multiple times. Another critique focused on the quality of translation, because some of the Spanish language translation was considered inaccurate, making the activity confusing for Spanish speakers (GW01). This critique is especially troublesome because the Reinvent Phoenix research team included native Spanish speakers who translated the materials. Pre-testing VESC with Spanish speaking participants is one potential solution for checking translation quality. The presentation of vision options would also have been better if images depicted implementations local to the Phoenix area (RT01). The goal was to show local examples, but there was not always local evidence available for the selected vision elements.

6.6. Sustainability Discussions

Respondents overwhelmingly rated VESC positively as a tool for leading discussions about sustainability outcomes, and researcher observations support these conclusions. A member of the research team succinctly justified VESC and highlighted the activity's outcomes: "Our mandate was from a grant which stipulated that the vision had to be something sustainable. We weren't just talking about sustainability in general terms. Because the future has to be sustainable—what options would you support out of this pool of ideas? The VESC guided the conversation in a particular direction" (RT02). One respondent speculating on how VESC would have supported visioning in Gateway said, "I think you have to present some sustainable strategies and put those forth rather than work through 15 ideas people throw out that aren't sustainable. And if it is an opportunity for folks to learn about sustainability by discussing strategies that are based on sustainability, it allows people to meaningfully engage" (GW03). A Midtown respondent said, "If you just give someone a question, their mind goes blanker than anything. These posters were a good way to start" (MT01). Another Midtown respondent praised VESC for supporting his own thought process, saying, "I'm thinking along these lines anyways. I might organize my ideas. But without these objectives, my ideas might not be so formalized" (MT04).

7. Limitations

The evaluation of VESC provides insights about designing tools for supporting public discussion about sustainability, but the evaluation did have some limitations that are discussed below.

The evaluator was a member of the research team: While there is opportunity for bias to cloud the evaluation, the authors' role in the process afforded rich opportunities to collect direct observations of the design of visioning workshop, the VESC tool, workshop implementation, and participant experiences. Because the evaluation occurred up to two years after the public participation events, the authors were also able to distance themselves from the research project and approach the evaluation more objectively.

The evaluation occurred as much as two years after the public participation events: While this gap in time afforded greater objectivity, it might have slightly undermined the quality of participant reflection (*cf.* [36]). Some respondents clearly remembered the participation events and activities. Others were able to recall what occurred after brief conversations about the process. Others had trouble remembering specific conversations and themes. Ideally the research questions and evaluation would have been established prior to the visioning process in each district and conducted during and immediately after. Still, given the circumstances, the authors felt that there was value in collecting feedback to learn from the experience.

No formal assessment of stakeholders' sustainability literacy: The researchers did not formally evaluate stakeholders' sustainability literacy, and the decision to create tools to better align the engagement process with participants' sustainability literacy was based on direct observations and a heuristic process. Were more time available, critically assessing participant capacities could have further informed the design of the engagement tools and procedures. Still, although there was no formal pre-testing of literacy, through an informal appraisal we found sustainability literacy was very low, and in Gateway, this experience was repeated at four separate tables at each of four public events. Facilitators at each table faced the same challenges in engaging participants on issues of sustainability. From this low level of observed literacy, researchers set a baseline from which to communicate through VESC.

Sustainability not publicly defined with participants: While neighborhood-specific sustainability objectives were derived through robust stakeholder engagement, guiding sustainability principles were drawn from the literature, and a shared conceptualization of sustainability was not co-created with participating stakeholders. In-depth discussions about what sustainability means to a community is a way to engage on a vision. Pragmatically, in this project researchers wanted visioning workshops to produce tangible outcomes so that the vision could produce implementations and experiments.

In most cases, VESC fostered discussions that went beyond the preselected options, and the vision elements were often just a starting point, as noted above by respondents MT01 and MT04.

No interviews with participants outside of steering committee members: Because the evaluation took place long after public visioning concluded, the authors chose to interview steering committee members out of convenience. Steering committee members were easy to contact because they were still involved in Reinvent Phoenix. Steering committee members were also representative of certain constituents in each district, and the authors felt that their perspectives would be valuable, and that they could speak on behalf of other participants.

The evaluation lacks quantitative data: Prior to the evaluation, the authors did not establish clear metrics for what would be negative or positive assessments of each criterion. Respondent interviews do, however, provide rich details about VESC and the experience of participants at Reinvent Phoenix visioning workshops.

8. Conclusions

In Reinvent Phoenix, researchers experienced low sustainability literacy amongst participants. Researchers went from a model in which participants couldn't engage with sustainability principles to a model under Visually Enhanced Sustainability Conversations (VESC) in which they could. VESC helped sustainability experts translate abstract, hard to understand sustainability principles into something that is tangible, down to earth, and reasonably easy to understand.

One of the goals of VESC was to align a public participation process with stakeholders' sustainability literacy in order to improve sustainability-oriented discussions at public visioning workshops. To achieve this goal, the tool would have to provide participants with information resources and structure decision making [16]. Through an evaluation of VESC, the study investigated in how far the tool effectively facilitated public deliberation about sustainability outcomes and in how far VESC improved public discussion about sustainability compared to previous engagements.

In general, all respondents had favorable opinions of VESC. Midtown District respondents thought the activity was successful as they experienced it, and Gateway District respondents thought the exercise would have been beneficial to visioning in their own district. Respondents thought VESC was strong because it stimulated conversation. By seeding participants with example ideas, participants were then able to think more creatively. In terms of meeting sustainability goals, presenting vision elements that were already vetted as sustainable steered the conversation towards additional ideas that were more likely to lead to sustainability outcomes.

VESC did have some flaws. For instance, despite the presence of native Spanish speakers on the research team, the Spanish translations were not perfect, and some of the technical language should have been better translated. Terminology in English could have been simplified, yet all language was negotiated between project partners. More local examples of successful strategies could have inspired further support from participants, and additional images of each vision element could have made the options even more tangible. However, despite detailed research, few local examples of vision elements existed. These illustrate clear concerns regarding VESC, but the solutions for improving the tool are not simple.

There are some concerns regarding the evaluation of VESC, but the evaluation is transparent and provides a discussion of the tool's strengths and weaknesses. The evaluation describes the tool and highlights the aspects to be replicated as well as aspects to improve upon. VESC is a tool that was created through academic work. This study describes the intention, design process, implementation, and outcome of the tool to support participatory researchers and planning professionals in utilizing similar engagement tools to align public participation processes to the local context. We fully acknowledge that there was no formal testing. As we are moving forward with VESC, it would be important to include a formal pre/post test.

Misalignments between the public participation process and local context, like low sustainability literacy of participating stakeholders, can undermine sustainability outcomes in public participation processes in urban development projects. Facilitation and deliberation tools can improve discussions amongst members of the public. This study offers insights from which planners and experts can learn when designing their own public participation activities and materials.

Acknowledgments: The authors would like to thank David Manuel-Navarrete and Daniel Schugurensky for providing comments on early drafts of this manuscript. We also thank the participating members of the Gateway and Midtown District Steering Committees for providing their perspectives on the effectiveness of the VESC tool.

Author Contributions: Matthew Cohen and Arnim Wiek conceived of the study. Braden Kay and John Harlow contributed to the research design and implementation. Matthew Cohen and Arnim Wiek wrote the manuscript with contributions from Braden Kay and John Harlow. The authors acknowledge financial support through the Community Challenge Grant Award "Reinvent Phoenix—Cultivating Equity, Engagement, Economic Development and Design Excellence with Transit-Oriented Development," funded by the U.S. Department of Housing and Urban Development (HUD) (2012–2014). The findings and observations contained in this article are those of the authors and do not necessarily reflect the views of the U.S. Department of Housing and Urban Development (HUD).

Conflicts of Interest: The authors declare no conflict of interest.

References

1. Arnstein, S.J. A ladder of citizen participation. *J. Am. Inst. Plan.* **1969**, *35*, 216–224. [CrossRef]
2. Agyeman, J.; Evans, T. Toward just sustainability in urban communities: Building equity rights with sustainable solutions. *Ann. Am. Acad. Polit. Soc. Sci.* **2003**, *590*, 35–53. [CrossRef]
3. Hawkins, C.V.; Wang, X. Sustainable development governance: Citizen participation and support networks in local sustainability initiatives. *Public Works Manag. Policy* **2011**, *17*, 7–29. [CrossRef]
4. Smith, R.; Wiek, A. Achievements and opportunities in initiating governance for urban sustainability. *Environ. Plan. C Gov. Policy* **2012**, *30*, 429–447. [CrossRef]
5. Zint, M.; Wolske, K.S. From information provision to participatory deliberation: Engaging residents in the transition toward sustainable cities. In *The Elgar Companion to Sustainable Cities: Strategies, Methods and Outlook*; Mazmanian, D., Blanco, H., Eds.; Edward Elgar Publishing: Northampton, MA, USA, 2014; pp. 188–209.
6. Innes, J.E.; Booher, D.E. Reframing public participation: Strategies for the 21st century. *Plan. Theory Pract.* **2004**, *5*, 419–436. [CrossRef]
7. Cuthill, M. Exploratory research: Citizen participation, local government and sustainable development in Australia. *Sustain. Dev.* **2002**, *10*, 79–89. [CrossRef]
8. Holden, M.; Owens, C.; Mochrie, C. Lessons from a community-based process in regional sustainability indicator selection: The case of the Regional Vancouver Urban Observatory. In *Community Quality-of-Life Indicators: Best Cases IV*; Sirgy, J.M., Philips, R., Rahtz, D.R., Eds.; Springer: Berlin, Germany, 2009; pp. 59–80.
9. Cohen, M.; Wiek, A. *Avoiding Misalignments between Public Participation Process and Local Context in Urban Development*; School of Sustainability, Arizona State University: Tempe, AZ, USA, 2015.
10. *Public Participation in Environmental Assessment and Decision Making*, 1st ed.; Dietz, T.; Stern, P.C. (Eds.) The National Academies Press: Washington, DC, USA, 2008.
11. Von Korff, Y.; Aquino, P.; Daniell, K.A.; Bijlsma, R. Designing participation processes for water management and beyond. *Ecol. Soc.* **2010**, *15*. Article 1.
12. Shipley, R.; Utz, S. Making it count: A review of the value and techniques for public consultation. *J. Plan. Lit.* **2012**, *27*, 22–42. [CrossRef]
13. Carpini, M.X.D.; Cook, F.L.; Jacobs, L.R. Public deliberation, discursive participation, and citizen engagement: A review of the empirical literature. *Annu. Rev. Polit. Sci.* **2004**, *7*, 315–344. [CrossRef]
14. Bailey, K.; Grossardt, T.; Ripy, J. Toward environmental justice in transportation decision making with structured public involvement. *Transp. Res. Rec. J. Transp. Res. Board* **2012**, *2320*, 102–110. [CrossRef]
15. Walls, J.; Rowe, G.; Frewer, L. Stakeholder engagement in food risk management: Evaluation of an iterated workshop approach. *Public Underst. Sci.* **2010**, *20*, 241–260. [CrossRef]
16. Rowe, G.; Frewer, L. Public participation methods: A framework for evaluation. *Sci. Technol. Hum. Values* **2000**, *25*, 3–29. [CrossRef]

17. Wiek, A.; Kay, B.; Cohen, M.; Golub, A. *Public Participation for Sustainable Urban Development-Conceptual Framework, Evaluative Scheme, and Case Studies*; School of Sustainability, Arizona State University: Tempe, AZ, USA, 2014.
18. Wiek, A.; Withycombe, L.; Redman, C.L. Key competencies in sustainability: A reference framework for academic program development. *Sustain. Sci.* **2011**, *6*, 203–218. [CrossRef]
19. Pahl-Wostl, C. Towards sustainability in the water sector—The importance of human actors and processes of social learning. *Aquat. Sci.* **2002**, *64*, 394–411. [CrossRef]
20. Tippett, J.; Searle, B.; Pahl-Wostl, C.; Rees, Y. Social learning in public participation in river basin management—Early findings from HarmoniCOP European case studies. *Environ. Sci. Policy* **2005**, *8*, 287–299. [CrossRef]
21. Holden, M. Public participation and local sustainability: Questioning a common agenda in urban governance. *Int. J. Urban Reg. Res.* **2011**, *35*, 312–329. [CrossRef]
22. Garmendia, E.; Stagl, S. Public participation for sustainability and social learning: Concepts and lessons from three case studies in Europe. *Ecol. Econ.* **2010**, *69*, 1712–1722. [CrossRef]
23. Fung, A.; Wright, E.O. Deepening democracy: Innovations in empowered participatory governance. *Polit. Soc.* **2001**, *29*, 5–41. [CrossRef]
24. Healey, P. Collaborative planning in perspective. *Plan. Theory* **2003**, *2*, 101–123. [CrossRef]
25. Rowe, G.; Horlick-Jones, T.; Walls, J.; Poortinga, W.; Pidgeon, N.F. Analysis of a normative framework for evaluating public engagement exercises: Reliability, validity and limitations. *Public Underst. Sci.* **2008**, *4*, 419–441. [CrossRef]
26. O'Neill, S.J.; Boykoff, M.; Niemeyer, S.; Day, S.A. On the use of imagery for climate change engagement. *Glob. Environ. Chang.* **2013**, *23*, 413–421. [CrossRef]
27. Miller, T.R.; Wiek, A.; Sarewitz, D.; Robinson, J.; Olsson, L.; Kriebel, D.; Loorbach, D. The future of sustainability science: a solutions-oriented research agenda. *Sustain. Sci.* **2014**, *9*, 239–246. [CrossRef]
28. Pan, Z.; Kosicki, G.M. Framing as a strategic action in public deliberation. In *Framing Public Life: Perspectives on Media and Our Understanding of the Social World*; Reese, S.D., Gandy, O.H., Jr., Grant, A.E., Eds.; Lawrence Erlbaum Associates Inc.: Mahwah, NJ, USA, 2001; pp. 35–66.
29. Kahan, D.; Braman, D. Cultural cognition and public policy. *Yale Law Policy Rev.* **2006**, *24*, 149–172.
30. Fagotto, E.; Fung, A. Empowered participation in urban governance: The Minneapolis neighborhood revitalization program. *Int. J. Urban Reg. Res.* **2006**, *30*, 638–655. [CrossRef]
31. Wiek, A.; Golub, A.; Kay, B.; Harlow, J.; Cohen, M.; Minowitz, A.; Soffel, M.; Avallone, D.; Castaneda, M.; Quinn, J.; *et al. Vision Report for the Gateway District, Phoenix*; School of Sustainability, Arizona State University: Tempe, AZ, USA, 2014.
32. Gibson, R.B. Sustainability assessment: Basic components of a practical approach. *Impact Assess. Proj. Apprais.* **2006**, *24*, 170–182. [CrossRef]
33. Harlow, J.; Wiek, A.; Kay, B. *Initial Stakeholder Engagement for Urban Sustainability Transition Efforts*; Sustainability Transition and Intervention Research Lab, School of Sustainability, Arizona State University: Tempe, AZ, USA, 2015.
34. Wiek, A.; Golub, A.; Kay, B.; Harlow, J.; Cohen, M.; Kao, K.; Champagne, E.; Gehrlich, G.; Gwiszcz, J.; Quinn, J. *Vision Report for the Midtown District, Phoenix*; School of Sustainability, Arizona State University: Tempe, AZ, USA, 2013.
35. Newman, E.; Feigenson, N. The truthiness of visual evidence. *Jury Expert* **2013**, *25*, 1–6.
36. Wiek, A.; Talwar, S.; O'Shea, M.; Robinson, J. Towards a methodological scheme for capturing societal effects of participatory sustainability research. *Res. Eval.* **2014**, *23*, 117–132. [CrossRef]

sustainability

MDPI

Article

Spatio-Temporal Features of China's Urban Fires: An Investigation with Reference to Gross Domestic Product and Humidity

Zhenbo Wang [1,2], Xiaorui Zhang [3,*] and Bo Xu [4]

1 Key Laboratory of Regional Sustainable Development Modeling, Chinese Academy of Sciences, Beijing 100101, China; wangzb@igsnrr.ac.cn
2 Institute of Geographical Sciences and Natural Resources Research, Chinese Academy of Sciences, Beijing 100101, China
3 Department of Urban Planning, Hefei University of Technology, Hefei 230009, China
4 China Academy of Safety Science and Technology, Beijing 100012, China; xub@cassp-pec.com
* Author to whom correspondence should be addressed; rgdhf@hfut.edu.cn; Tel.: +86-10-6488-8147; Fax: +86-10-6488-9301.

Academic Editor: Tan Yigitcanlar

Received: 30 April 2015; Accepted: 8 July 2015; Published: 22 July 2015

Abstract: Frequent fire accidents pose a serious threat to human life and property. The spatio-temporal features of China's urban fires, and their drivers should be investigated. Based on the Spatio-temporal Dynamic panel data Model (SDM), and using fire data gathered from 337 Chinese cities in 2000 to 2009, the influence of spatio-temporal factors on the frequency of urban fires was analyzed. The results show that (1) the overall fire incidence of China increased annually before 2002 and reduced significantly after 2003, and then high fire incidence increased in western China; (2) Spatio-temporal factors play a significant role in the frequency of Chinese urban fires; specifically, the fire assimilation effect, fire inertia effect and fire caution effect. The ratio of fire incidence of China has reduced significantly, and the focus of fire incidence moved towards the western region of China. GDP and humidity have a significant effect on urban fire situation change in China, and these effects may be referred to as "fire assimilation effects", "fire inertia effects" and "fire caution effects".

Keywords: urban fire; spatio-temporal features; SDM, humidity; GDP; the fire assimilation effect; fire inertia effect; fire caution effect

1. Introduction

The process of urbanization and industrialization has undergone rapid growth, leading to changes in fuel amounts, composition, and its configuration [1]. In addition to the change of land use and management activities, cities have become a highly-possible location for fire occurrence. According to statistics, on average more than 500 daily fire accidents occurred in China from 2000 to 2009. Frequent fire accidents pose a serious threat to human life and property. Fires can be classified into forest fires, grass fires, residential fires, building fires, industrial fires and mine fires based on the places where they occur. This paper focuses on the last four types listed above. The fuels, an ignition source, and sufficiently dry weather are the essential conditions of fire occurrence [2,3], whose relative importance at different scales can be difficult to quantify [4]. At landscape scale, the determinants are fuel, weather, human activities and topography [5–9]. At regional scale, climate variability and vegetation distribution are the more sensitive drivers of fire [10,11].

Urban areas belong to the landscape scale. Urban fire patterns are mainly influenced by the distribution of fuels, weather conditions (temperature and humidity) [12], and topography [13,14].

Human activities have a close relationship with the factors above. Humans can have both positive and negative effects on fire. Fuels provide the raw material acted on by fire. Cities are major spaces where substantial wealth is increasingly accumulated by rapid social and economic development [15]. On the one hand, with more and more fuels being used in urban construction and consumption of electricity and fire, there is a potentially high agglomeration of ignition source. On the other hand, social economic growth also promotes the development of firefighting facilities and flame retardant materials [16]. Higher human population density often increases ignition sources, whereas fire prevention activities tend to decrease fire occurrence [17,18]. All the factors abovementioned will undoubtedly impose significant influence on the occurrence of urban fires. Research analysis shows a significant relationship between the level of regional economic development and the rate of fire accidents However, different analysis result may be drawn from the variety of studies that have examined the economic development level. For example, Duncanson *et al.* found that fire accidents tended to decrease with increasing economic development [19]. The Australian research bureau of fire protection showed that an economic downturn coupled with imperfect firefighting equipment increased the fire risk [20]. In Smith *et al.*'s analysis, fire risk was shown to be relatively high in single-parent households and in households where the households members are unemployed, according to the research into fire accidents in Britain from 2002 to 2004 [21]. However, historical fire accident data in China before the year 2000 and after showed that the frequency of fires increases with economic development, except for Shanghai [22].

The relative importance of various drivers of fire activity can vary depending on the scale at which they are measured and the scale of analysis [23]. It is known that a warmer climate will increase fire activity at regional [24] to global scales [25]. At local scale, seasonality and meteorological conditions (temperature, humidity, wind speed and potential fuels) are more relevant. Material space morphology leads to microclimate change, evidenced by the heat island effect increasing the temperature and decreasing the humidity [26] of urban areas. Architectural form [27] in cities will also affect the local weather conditions in terms of wind orientation and speed. In the Western Amazon, drought intensity was the most important predictor of fire occurrence in local scale [24]. In Queensland, Australia, a high fire occurrence rate is related to high temperature [28]. Gunther found that, because there are few fixed heating systems in the rural parts of South America, the death rate from fires increased on cold days owing to the risks from using other types of heating [29]. Moreover, poverty makes this problem even worse. In the two largest cities in Indonesia—Jakarta and Surabaya—the number of fire accidents each month is affected by a change in climate, with fires occurring more frequently in the arid season; with, in Surabaya in particular, the number of fires decreasing with increasing humidity, while the number of fires increasing significantly when the humidity is less than 70% or when the rainfall measures less than 6 mm [30]. Similarly in China, a change in climate and humidity has a significant impact on the frequency of fires. Therefore, meteorological factors are closely related to fire occurrence. Climate and humidity have the largest influence on fire [31]. Specifically, a more arid climate increases the fire incidence rate while a more humid climate decreases the rate.

Climate change and socio-economic development have significant temporal and spatial characteristics; hence the incidence of fires is also associated with temporal and spatial characteristics [12,14,32,33]. Recent studies have focused on the spatio-temporal analysis of grassland and forest fires [34], and natural disasters [35]. A few researchers have tried to describe the geographical characteristics of urban fires. Jennings pointed out the geographical factors of fire accident analysis, and realized the spatial problems are related to fire accidents and the building environment. He argued that both the population and buildings are distributed along a certain density gradient; therefore, the rate of fire accidents is unevenly distributed [36]. Corcoran *et al.* analyzed the temporal and spatial change based on temporal and spatial exploratory analysis by using Queensland as an example [28]. Chang adopted the Spatial Lag Model (SLM) and the Spatial Error Correction Model (SEM) to analyze the relationship between the fire incidence rate and spatial factors, using Tainan City as an example [37].

In general, although progress has been made in related research on urban fires, there are still gaps. Firstly, the research above has examined either economic factors or climatic factors. However, urban fires are affected by both socio-economic and climatic factors, and the occurrence of fires seems to be random and uncertain. With rapid economic development and global climate change, academic research must take into consideration the compounding effects of both economic development and climate change on urban fire accidents. Secondly, most of the existing research on the influential effects of temporal and spatial factors on urban fires is based on exploratory analyses of some certain cities or regions, while little spatial econometric analysis has been carried out on the impact of large-scale temporal and spatial factors on urban fires. Thirdly, by using older data, the current situation regarding urban fires has not been addressed. In this study, we gathered urban fire accident data relating to China in the years 2000 to 2009, and used the Spatio-temporal Dynamic panel data Model (SDM) to demonstrate the effects of spatio-temporal factors on urban fire occurrences under the influence of rapid economic development and climate change. The fire accidents mentioned below all refer to urban fires.

This paper identifies the fire incidence and its change to represent the fire situation and trend in different cities and regions. (Data set of climate data annual value from China Macao and Taiwan region are missing). The statist China Meteorological Administration website shows the average annual temperature and relative humidity from 2000 to 2009 from 188 meteorological stations. By Kriging interpolation, the raster maps of national annual average temperature and relative humidity are formed and then Zonal statistics available for each unit on the annual average relative humidity. The ArcGIS10.1 spatial analysis tools are utilized of Kriging interpolation and Zonal Statistics in this study. According to the common interpolation Kriging value, the average standard deviation of prediction result error of each year is less than 0.01 and the root-mean-square standard prediction deviation ranges between 1 ± 0.1.

2. Study Area and Data

The study is aimed at 337 Chinese prefecture-level (at least) administrative units (data of only Haikou and Sanya of Hainan province are presented but the data of Zhongwei of Ningxia province and Hong kong, Macao and Taiwan region are missing). The statistics on urban fires are obtained from Statistic Yearbook of Fire in China (2001 to 2010). The statistics on consensus and GDP per capita are derived from Statistic Yearbook of Regional Economy in China (2001 to 2010).

One of our objectives was to study the influence of economic development on the macro-statistical characteristics of the urban fire rate in China, with GDP per capita as the representative index. The change in GDP per capita reflects the basic macro-economic situation in one country or region and is highly related to education, investment, household income, *etc*. Existing research showed that climate and humidity had the largest influence [23,29–31], so we chose to principally study the influence of climate and humidity on urban fire, selecting average air temperature and humidity as the index of climate change. The number of fire accidents represents an important index which demonstrates severity. The occurrence index (the number of fire accidents among 1 million people) was chosen to represent the severity of fire occurrence in different areas and cities.

According to the "Meteorological annual data set from the China international exchange station ground", the grid's profile of national average air temperature and humidity could be generated through Kriging interpolation after capturing the annual average humidity in 188 meteorological stations from 2000 to 2009. Furthermore, the annual average relative humidity per unit can be calculated by using zonal statistics. This paper adopts spatial analytical tools to perform Kriging interpolation and calculate zonal statistics. The standardized mean of interpolation result forecasting error is less than 0.01 and the root-mean-square standardized is 1 ± 0.1; therefore it is a precise method.

3. Methods

3.1. Spatial Autocorrelation Model (Moran's I Model)

Tobler (1970) put forward the First Law of Geography that any object is related to other objects with special consideration of distance, which shows the more closely located the objects, the stronger correlation exists between them [38]. It is called spatial autocorrelation which can be measured by Moran's I index. When the observation values are similar within a certain distance (d), the Moran's I is positive in significant level (p-value less than 0.1), otherwise it is negative. If the observation values are arranged randomly, the Moran's I is zero. Moran's I can be classified into Global Moran's I (GMI) and Locan Moran's I (LMI). GMI is used to judge the spatial agglomeration degree of China's urban fires, and LMI is used to explore the spatial distribution of the "hot spots" and "cold spots". Due to the possible local spatial autocorrelation observations existing in the overall spatial random sample distribution, GMI and LMI are both applied in this paper to analyze the agglomeration features of the urban fires. GMI refers to:

$$I = \frac{n}{S_0} \frac{\sum_{i=1}^{n} \sum_{j=1}^{n} w_{ij}(x_i - \overline{x})(x_j - \overline{x})z_j}{\sum_{i=1}^{n} (x_i - \overline{x})^2} \quad (1)$$

in which n is the total number of the sample, w_{ij} is spatial weight matrix ($n \times n$), x_i and x_j are respectively the fire occurrence rates in places of i and j. $\overline{x}$ is the average value, and S_o is the sum of all its elements. LMI is defined as:

$$I_i = \sum w_{ij} Z_i Z_j \quad (2)$$

in which Z_i and Z_j are the standardized values of fire occurrence rate. When I_i and Z_i are both positive at a significant level (p-value less than 0.05), it means that the fire occurrence rate in place i and the units nearby are high. It is termed as High Concentration Area (HH); by contrast, while I_i and Z_i are negative, it shows that the fire occurrence rate in place i and its neighboring units are low which are called Low Concentration Area (LL). If I_i is positive and Z_i is negative, it means that the fire occurrence rate of place i is higher than that of the neighboring units, which are referred to High Low Concentration Area (HL). While if I_i is negative and Z_i is positive, the fire occurrence rate of place i is lower than that of the units nearby, which are termed as Low High Concentration Area (LH).

3.2. Selecting Incident-Inducing Factors of Fire Accidents

The starting point in the Granger causality test of judgment was to examine whether the prophase information of X contributed to the decrease of MSE (Mean Square Error) of Y [39]. Moreover, this paper compared the MSE value with that when not using prophase information of X. If there is no change in MSE value, X has no influence on Y under the Granger causality test, or "Y is not caused by X Granger". On the contrary, there is causality under Granger, or "X is the Granger reason of Y". Gao describes specific methods [40].

The annual average temperature and average relative humidity between the years 2000 and 2009 were recorded based on those of 188 China International Exchange Ground Stations (Figure 1).

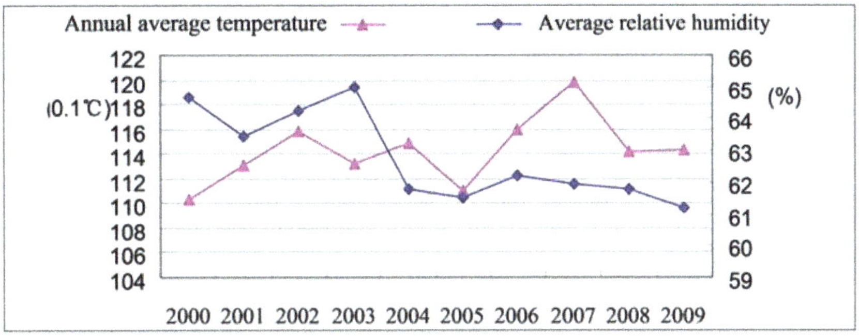

Figure 1. Annual average temperature and annual average relative humidity of China, 2000 to 2009.

Assuming all the disaster-inducing factors (GDP per capita, annual average relative humidity, annual average temperature) are not the Granger cause of the fire occurrence situation, the Granger causality test examines whether the chosen disaster-inducing factors really cause the change in fire occurrence (Table 1).

Table 1. Granger causality test results of fire rate and other factors of China.

Lags	Null Hypotheses	F Stat.	Prob.	Conclusions
1	*H* is not the Granger cause of *F*	9.42	0.01	rejected
	T is not the Granger cause of *F*	0.73	0.41	accepted
	G is not the Granger cause of *F*	9.45	0.00	rejected
2	*H* is not the Granger cause of *F*	2.39	0.16	accepted
	T is not the Granger cause of *F*	0.73	0.52	accepted
	G is not the Granger cause of *F*	2.57	0.15	accepted

Note: *H, T, G, F* represent the logarithmic values of the annual average relative humidity, annual average temperature, GDP per capita, and fire rate, respectively.

When the lag is 1, H and G are both the Granger cause of F, but T is not the Granger cause of F; when the lag is 2, H, T and G are not the Granger cause of F. This means that the annual average relative humidity and GDP per capita are the Granger causes of fire rate; moreover, they also have a significant influence on the fire rate one year later. However, the annual average temperature does not significantly influence the fire rate. Therefore, the annual average relative humidity and GDP per capita can be considered to be incident-inducing factors. This paper only considers annual average relative humidity as the characteristic factor of climate change, and uses GDP per capita to represent economic development and the fire occurrence rate to represent the change in urban fires.

3.3. Fire-GDP-Humidity Model (FGHM)

C.W. Cobb and Paul H. Douglas put forward the Cobb-Douglas production function in the 1930s [39]. Labor L and investment K are explanatory variables, Y is the output, and the model is shown below:

$$Y = AK^{\alpha}L^{\beta}e^{u} \quad (3)$$

As mentioned in Section 3.2, the fire situation and trend has a strong correlation with GDP and humidity. Assuming GDP per capita and annual average relative humidity as inputs, and fire occurrence rate as output, according to the Cobb-Douglas production function, the fire input-output model is:

$$F = AG^{\beta}H^{\gamma}e^{u} \quad (4)$$

where G is GDP per capita, H is annual average relative humidity, F is fire occurrence rate, β,γ are the elasticity of the economic and climate factors.

This model is based on the panel data model through logarithmic transformation:

$$\ln(F_{it}) = \alpha + \alpha_i + \beta \ln(G_{it}) + \gamma \ln(H_{it}) + u_{it} \quad (5)$$

where α is the average intercept value of each section, α_i is the deviation value that each section deviates from the average α, μ_i is the residual value of each section, F_{it} is the fire occurrence rate of unit i in the t phase, G_{it} is GDP per capita (thousand yuan) of unit i in the t phase, H_{it} is the annual average humidity of unit i in the t phase.

3.4. Co-Integration Test of Each Index in FGHM

In order to avoid spurious regression [40], we used the co-integration test of ln (F), ln(G) and ln(H).

The Pedroni (Engle-Granger based) method, Kao (Engle-Granger based) method and Fisher (combined Johansen) method were used to perform the test. The trace test and Max-Eigen test both rejected the null hypothesis at a 1% significance level. There is a co-integration relationship among the variables and there are three co-integration vectors.

After examining the above three test methods, the results show a co-integration relationship among ln(F), ln(G), and ln(H). There is a long-term equilibrium relationship among economic development, climate change and fire situation change in urban China between 2000 and 2009. In the short term, the fire situation may deviate from average value, and it will return to the equilibrium situation determined by the economic development level and climate situation as time goes by. Therefore, it is feasible to build a regression function based on the three variables and the FGHM model is efficient.

3.5. The Fire-Space Dynamic Model

The autocorrelation of a temporal series must be considered when analyzing natural environment and socio-economic problems. The values before and after some points are correlated, which reflects the regularity of temporal series. The spatial panel data model combines both the cross-section and time series panel data. This model is widely used because it takes into consideration both variables' regional difference and time influence and avoids the multi-collinearity towing to the missing independent variables. The SDM is one of the spatial panel data models that has been developed recently, and it can reflect the dynamic change of research sample both in space and time. It not only considers the spatial lag of explained variables, but also considers the time lag and spatial time united lag. Thus, this model has more explanatory strength [41].

The basic form of SDM is as follows:

$$Y_{nt} = \lambda_0 W_n Y_{nt} + \gamma_0 Y_{n,t-p} + \varphi_0 W_n Y_{n,t-p} + X_{nt}\beta_n + c_n + \varepsilon_n \quad (6)$$

where Y_n is the dependent variable, X_n, β_n is the cross-section explanatory variable k-dimensional vector and k-dimensional parameter corresponding to i = 1, 2, . . . , N. W_n is the spatial weight matrix, c_n is the intercept, ε_n is the random error vector, $Y_{n,t-p}$ is the time lag p-order time lag, λ_0, γ_0, φ_0 is the parameter of spatial lag, time lag and spatial and time united lag.

The FGHM does not consider the spatial correlation of the fire situation. The fire occurrence rate of cities tended to increase from 2000 to 2002 and started to decline from 2002 to 2009. The same trend is reflected from the autocorrelation Moran index (Figure 2). Each year's Z(I) value is larger than 1.96 and illustrates the significant autocorrelation of the fire occurrence rate each year. The fire occurrence rate in mainland China has significant spatial accumulative features. Regions with a high fire occurrence rate are located near regions with a high fire occurrence rate; similarly, regions with a low fire occurrence rate are located near areas with a low fire occurrence rate.

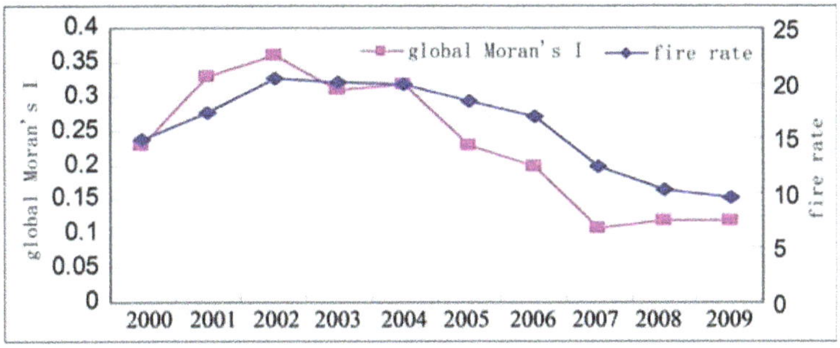

Figure 2. Fire rate and global Moran Index from 2000 to 2009.

Given that the fire distribution has spatial accumulative features, the spatial autocorrelation of fire should be considered in the analysis. The effect of spatial factors on the change in the fire occurrence rate is very important because there are imitative, competitive and cooperative behaviors in terms of the productive level, living standard and fire safety management of the surrounding regions. Moreover, the fire analysis should also consider the dynamic effect of time considering that the fire situation change is continuous and gradual. Therefore, the fire SDM (FSDM) is constructed after an extension of the FGHM based on the SDM:

$$\ln(F_{it}) = \alpha + \alpha_i + \beta_1 \ln(G_{it}) + \beta_2 \ln(H_{it}) + \beta_3 \ln((I_t \otimes W_N)F_{it}) + \beta_4 \ln(F_{i,t-1}) + \beta_5 \ln((I_t \otimes W_N)F_{i,t-1}) + \mu_i \quad (7)$$

where α is the average intercept value of each section, α_i is the deviation value between the intercept of each section and average value of α. μ_i is the residual value of each section, F_{it} is the fire occurrence rate of section *i* of the *t* phase (number/ million people), G_{it} is the GDP per capita (thousand yuan) of section *i* of the *t* phase, H_{it} is the annual relative humidity(%), I_t is the unit matrix, W_N is the spatial weight matrix, β_1, β_2, β_3, β_4, β_5 is the spatial lag parameter, time lag parameter, spatial and time united lag parameter of the fire occurrence rate.

If adopting the spatial weight based on adjacency relation, the spatial lag variable is the average value of K surrounding regions, that is $F_{it}\prime = (I_t \otimes W_N)F_{it} = \frac{1}{k}\sum_{i=1}^{k} F_{it}$, $F_{it}\prime$ is the average fire occurrence value of K surrounding areas in unit *i* of the *t* phase. Then FSDM can be simplified as follows:

$$\ln(F_{it}) = \alpha + \alpha_i + \beta_1 \ln(G_{it}) + \beta_2 \ln(H_{it}) + \beta_3 \ln(F_{it}\prime) + \beta_4 \ln(F_{i,t-1}) + \beta_5 \ln(F_{i,t-1}\prime) + \mu_i \quad (8)$$

where $F_{it}\prime$ represents the 1-order spatial lag parameter of the fire occurrence rate, $F_{i,t-1}$, is the 1-order time lag parameter of the fire occurrence rate, $F_{i,t-1}\prime$ refers to the 1-order time and spatial unit lag parameter of the fire occurrence rate.

The FSDM is as follows based on the estimation of EVIEWS6.0:

$$\ln(F_{it}) = 4.63 + \alpha_i - 0.15\ln(G_{it}) - 0.99\ln(H_{it}) + 0.57\ln(F_{it}\prime) + +0.54\ln(F_{i,t-1}) - 0.20\ln(F_{i,t-1}\prime) + \mu_i \quad (9)$$

The fit of the model is good. All the parameters pass the 1% significance level test. The goodness of fit of this model is 0.94. The adjusted RR reaches to 0.93 and natural logarithm of the likelihood value is −4545.30. All these values illustrate that the model has a high degree of fitting.

4. Result Analysis

4.1. The Spatio-Temporal Features of China Fires

In Equation (1), the GMI result of urban fires each year indicated by Figure 2 shows that Z(I) is more than 1.96 each year. Residuals are examined by spatial autocorrelation, achieving the significant result of their Moran's I less than 0.05 in average in 2003, 2004 and 2009 while insignificant during other years by contrast [42–44]. It illustrates the significant autocorrelation of the fire occurrence rate each year.

Table 2. The statistical tests of the post-fit residuals.

	2001	2002	2003	2004	2005	2006	2007	2008	2009
Moran'I	−0.02	0	0.07	0.03	0	−0.01	−0.02	−0.01	−0.04
Z	−1.44	0.17	4.99 **	2.32 *	0.47	−0.58	−0.97	−0.2	−2.38 *

* and ** represent significant difference at a level of 0.05 and 0.01, respectively.

GIM increased from 0.23 to 0.36 from 2000 to 2002, which represents the higher level of spatial agglomeration of China fires occurrence rate. The following decreasing trend shows the lower level of its spatial agglomeration, taking on the dispersion trend instead. The trend of fire spatial agglomeration is consistent with that of the overall fire occurrence rate of China (Figure 2). Since then governments of all levels have enhanced the fire safety management and its investment, thus decreasing the fire occurrence rate.

LMI index in Equation (2) indicates the fire agglomeration features of some local areas (seen in Figure 3). There are significant fire spatial agglomeration phenomenon in local China. Each year the HH units are mainly distributed in northeast, central-north part and Xinjiang province of China. In early periods, they also gather in Bohai Rim and Yantze River Delta, which are the important areas for fire prevention and protection. LL units mainly spread in southwest part of China with low level of fire occurrence rate due to the local warm and humid climate yet underdeveloped economy. As for LH and HL units, the spatial correlation is not that obvious with little distribution for each year. Table 3 shows the number of cities for each type of area. During the period from 2000 to 2009, the number of HH units gradually declined year by year. The urban spatial agglomeration of high fire occurrence rate has become weaker and tended to be randomly distributed with decentralization. LL units experienced oscillating declines in amount and totally disappeared after 2007. The spatial agglomeration of low fire occurrence rate took on a random and dispersed distribution. After 2007, the adoption of a related national policy contributed to the constant decreasing fire occurrence rate, evidenced by the following regulations. In 2006, the State Council issued The State Council Opinion on Further Strengthening.

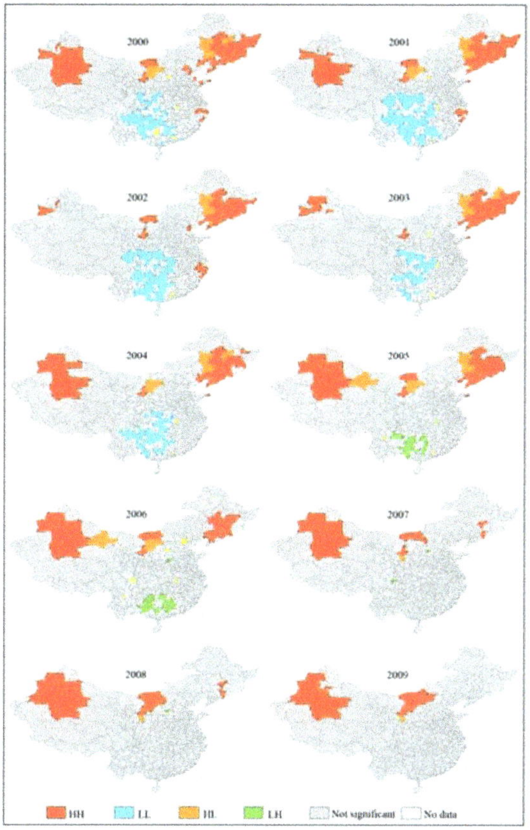

Figure 3. The analysis result on fire occurrence rate local autocorrelation of 2000 to 2009.

The Work of Fire (Guo Fa [2006] No.15), depending on which governments of all levels and public security fire departments launched a nationwide investigation and remediation of fire hazards, further enhancing the fire supervision and management. The Ministry of Public Security, the National Development and Reform Commission and the Ministry of Finance jointly issued the "The 3rd-Phase Promotion Planning of Fire Fighting Equipment" on 29 May 2007, in which a decision was made that from 2007 to 2010, with the total investment of 931.3 million yuan mainly subsidized by central government and partially supported by local government, 120 public security fire secret squadrons would be established in China's main cities and areas, gradually improving the fire aid system and enhancing the capacity to withstand major disasters and accidents. All these measures have effectively curbed the fire occurrence, providing guarantee for social steady and harmonious development.

Table 3. The units number of each type of local autocorrelation, 2000 to 2009.

Year	HH	HL	LH	LL
2000	47	6	4	33
2001	43	5	4	52
2002	43	3	3	49
2003	40	7	3	24
2004	40	4	4	32
2005	41	7	4	13
2006	35	8	2	15
2007	18	5	1	0
2008	21	3	1	0
2009	17	2	1	0

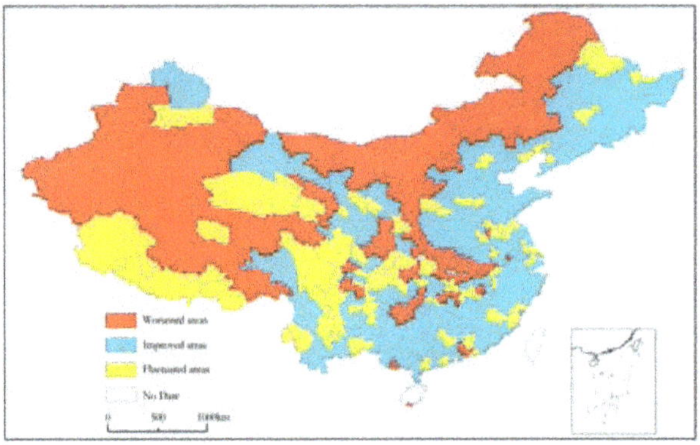

Figure 4. The classification of fire occurrence rate time series trend.

To apply the method of "Three Points Moving Average Algorithm" in calculating the correlation coefficient Pearson(r, t) from 2000 to 2009, of which r refers to the fire change rate and t for time. The result that Pearson(r, t) less than −0.5 represents the negative correlation between t and r, *i.e.*, the fire occurrence rate taking on the declining trend with the time passing by; by contrast, if Pearson(r, t) is over 0.5, the correlation between t and r is positive, which means the fire occurrence rate is increasing year by year. As for the units, we use t as the independent variable and r as the dependent variable to calculate the slope(r, t), which reflects the degree or speed of r. According to the calculation results, the cities around the nation can be classified into three types: Improved areas: Pearson(r, t) < -0.5, decreasing fire occurrence rate. The higher slope(r, t) absolute value shows the accelerated improvement trend; Worsened areas: Pearson(r, t) >0.5, increasing fire occurrence rate. The higher slope(r, t) absolute value means the faster deterioration trend; Fluctuated areas: $-0.5 \leq$ Pearson(r, t) ≤ 0.5, fluctuations or no significant features in fire occurrence rates (Figure 4).

There are in total 186 cities with improved areas, mainly distributed in the northeast, northern, eastern, southern part of China. In recent years, the fire occurrence rate has gradually declined, especially in northeast China. The area has a relatively developed economy and has achieved remarkable results in fire disaster prevention and reduction within the sound development cycle.

There are in all 74 cities that have worsened areas, mainly distributed in the west of Heihe-Tengchong Line, as well as Shaanxi and Hubei provinces. The urban fire occurrence rate increased relatively quickly influenced by development. Economic development has been promoted later in this area, however, fire safety has not developed at the same rate, thus leading to the higher fire occurrence rate.

There are 77 cities in total in the fluctuation area, distributed in southwest China. The area can be described in terms of the following three situations: firstly, the fire occurrence rate initially trended upwards and then downwards, *i.e.*, the fire occurrence rate worsened during the early stages and then improved more recently; secondly, the fire occurrence rate declined at first and then increased, *i.e.*, the fire occurrence rate demonstrated a sound trend but later worsened; thirdly, the fire occurrence rate continually fluctuated. These units are expected to further enhance fire supervision and management in the future to prevent a worsening trend and are expected to achieve gradual improvements.

4.2. The Influence of GDP and Humidity

The distribution of the average fire rate from 2000 to 2009 is shown in Figure 2. The urban fire occurrence rate distribution in China shows a high occurrence rate in the north and a low rate in the south; the eastern area also has a higher occurrence rate than the western area (Figure 5a). This is in accordance with the regional climate conditions and the regional economic development level on a whole. In China, the southern area is warm and wet while the northern area is cold and dry (Figure 5c,d). The eastern area is well developed and the western area is less developed (Figure 5b).

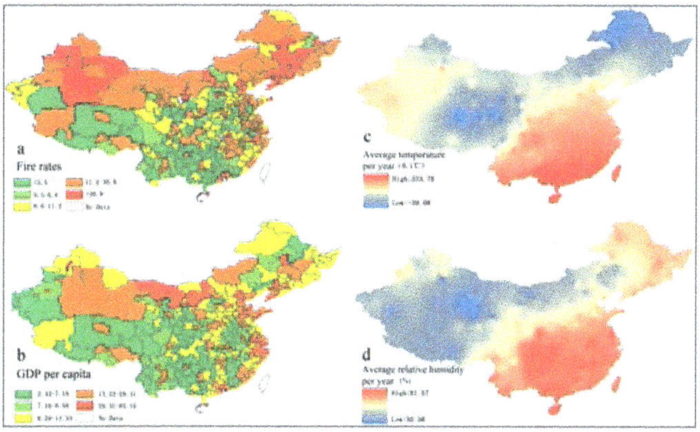

Figure 5. The average fire rate, temperature, GDP per capita and relative humidity of China, 2000 to 2009.

(1) The influence of GDP on the fire situation. The parameter of economic factor (β_1) is negative. This indicates that economic development helped to improve the fire situation in general. This is consistent with research from other countries. However, we obtained different results after analysis because we chose data from the late 1990s to early 2000. During this important period of economic transformation, there was a lack of firefighting investment in each city. The cost of firefighting equipment was very high and the numbers of those fighting the fires was relatively small. The fire safety level and ability to respond to emergencies was affected by a lack of public firefighting facilities. Therefore, the fire risk and fire occurrence rate increased prior to 2002. After 2003, all levels of government increased their fire safety management, investment and prevention, and strengthened their fire response capability. The fire occurrence rate decreased with rapid economic development. From 2000 to 2009, economic development reduced the fire occurrence rate. A 1% increase in GDP per capita led to a 0.15% decrease in the annual average fire occurrence rate.

On the one hand, according to our analysis results, economic development led to an increase in fire occurrences, and the accumulation of people and goods associated with economic development increased the likelihood of fire accidents. This is not good for fire prevention and control. On the other hand, our results showed that economic development also has an inhibitory effect on fire occurrences as economic growth can increase fire safety investment and strengthen fire prevention and control, and improve fire safety at a management level. Therefore, economic development has a dual effect on the fire situation. There are different results for different economic development periods in different regions; this has been illustrated by a historical analysis undertaken by Liang and Ren [45]. The Chinese government first introduced the "Fire Statistic Management Rule" (implemented in 1990) in 1989, and this ruling was revised in 1996 (implemented in 1997); this explains the statistical standard change during this period and why the fire occurrence rate fluctuated sharply between 1990 and 1997, as shown in Figure 6.

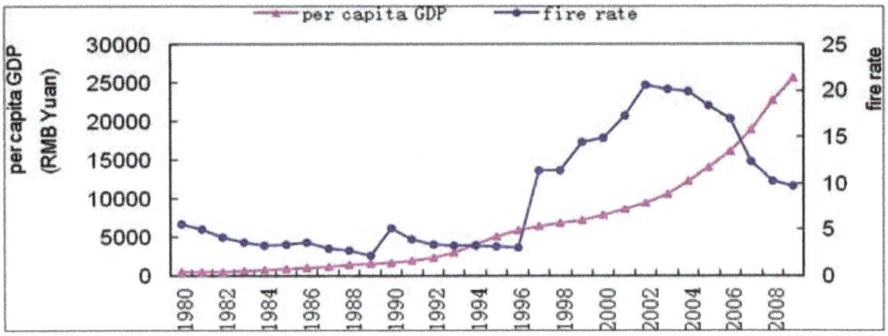

Figure 6. Per capita GDP and fire rate of China, 1980 to 2009.

(2) The influence of humidity change on fire situation change. The "National Climate Change Assessment Report I: the History and Future Trend of Chinese Climate Change" pointed out that China has the same pattern of climate change as other regions in the world, and, on average, the climate is warmer and drier than that in Northern hemisphere regions or other parts of the world [46]. Xu *et al.* proposed that there will be a visible warming/drying pattern in most Chinese regions in the future (2011 to 2080) [47]. Figure 4 also illustrates the significantly arid climate in 2000 and 2009.

The parameter of the climate factor (β_2) in the FSDM is negative; this means that a drier climate has a negative effect on the fire situation. Because combustible matter exists in the atmosphere, it exchanges energy and material and keeps in water equilibrium with the surrounding environment. When a combustible is burned, the water will first parch and then evaporate and decompose into combustible gas. Therefore, the drier the climate, the lower the water content, as well the lower the fire-burning energy, or latent heat of evaporation, the higher the fire risk. Considering only the effect of the humidity factor on fire, a 1% decrease in the annual average relative humidity can lead to a 0.99% increase in the national average fire occurrence rate.

Figure 4 shows the decreasing trend of the overall fire occurrence rate in China; the change in rate was the combined result of economic development and climate change. Although the warming/drying trend in China had a negative effect on the fire situation, economic growth reversed this trend and improved the overall fire situation in China after 2003. Given that the sensitivity (β_2 = 0.99) of the fire occurrence rate to the annual average relative humidity change was greater than that of the GDP (β_1 = 0.15), the Chinese government has made progress in facing the challenge of the negative effect of climate change on urban fires. As the climate becomes drier, the fire prevention and control challenge becomes much greater. Therefore, it is necessary to increase the investment in fire safety and prepare a contingency plan in order to reduce the fire occurrence rate.

4.3. The Spatio-Temporal Regulation of China's Urban Fires

(1) Fire assimilation effect. The fire assimilation effect refers to the process whereby people's attitude and behavior gradually moves toward the attitude and behavior of a reference population or people. It is an unconscious adjustment of individuals, who are imperceptibly influenced by the outside environment. In neighboring cities, people have similar living habits and the fire risk situation is broadly similar. There may be comparative and imitative behavior patterns that should be considered when managing and investing in fire safety. The fire situation in one region is affected by that in the

neighboring region. This is called the "fire assimilation effect". The spatial lag parameter (β_3) in the FSDM can be viewed as the fire assimilation effect; the larger the absolute value of β_3, the stronger the fire assimilation effect, and vice versa.

The spatial lag parameter (β_3) is positive and significant at a 99% confidence interval. This means that the fire situation has a significant assimilation effect. The fire situation in certain regions is not only affected by factors in these particular regions by self-factors, but also influenced by the fire situation in neighboring regions. An improvement or deterioration in the fire situation can lead to an improvement or deterioration in neighboring regions. The estimated value of the spatial lag parameter (β_3) is 0.57; this means that a 1% increase in the fire occurrence rate in neighboring regions will lead to a 0.57% increase in the occurrence rate for local region.

This result suggests that, owing to the fire assimilation effect, there needs to be improved cooperation between regions in order to reduce the rate of fire occurrences. Fire safety administration departments should take joint actions and communicate with each other regarding decisions about policies, planning, the provision of equipment, and the management and supervision of the firefighting itself. All such departments are encouraged to learn from and compete with each other in order to improve the level of fire safety. Furthermore, the department in charge can provide demonstrations on fire disaster prevention and control that encourage surrounding regions to improve their own levels of fire safety.

(2) Fire inertia effect. Plenty of economic magnitudes experience economic behavior inertia in socio-economic activities. Research shows that the fire system is linked to previous events in history and fire disasters have a structural continuity from the past to the present and from the present to the future [48]. This is called "fire inertia". The fire inertia effect refers to the fire situation trend of being consistent with the previous change trend, and there is a logical relationship between the fire situation at one time point and the fire situation at an earlier time point. The fire situation and its change experience inertia in two surrounding areas on account of similar living habits, fire safety management measures and socio-economic production behavior. Therefore, regions where fire accidents occur frequently in an earlier period will be most likely to suffer a fire accident in the future. Regions where the fire situation is improved in the early days will continue to improve in the future. The time lag parameter (β_4) in the FSDM represents the fire inertia effect. The larger the absolute value of β_4, the larger the fire inertia effect, and vice versa.

The time lag parameter (β_4) is positive and significant at a 99% confidence interval. This means that the fire inertia effect is significant based on fire disaster data in China. The estimated value of the time lag parameter (β_4) is 0.54, which indicates that a 1% decrease in the previous fire occurrence rate will lead to a 0.54% decrease in the current fire occurrence rate. The parameter of the fire inertia effect ($\beta_4 = 0.54$) and the assimilation effect ($\beta_3 = 0.57$) are roughly the same and reflect the similar positive effect of both the internal (fire inertia effect) and external (fire assimilation effect) factors on the fire situation in one region.

This result suggests that the fire inertia effect should be fully utilized. Specifically, for regions that have a fire deterioration trend, the relevant departments should increase the safety investment in large fire prevention and control and also strengthen the management of fire safety in order to eliminate the fire inertia effect; these actions would help to reverse the fire deterioration trend. In those areas where the fire situation is improving gradually, the relevant departments should continue their current fire safety management and continue to invest in fire safety activities. They, too, should use the fire inertia effect to further improve the fire situation. Nowadays, the fire situation is improving gradually in China's eastern areas. However, the middle and western regions still have terrible fire situations resulting from poor economic development and a low level of firefighting investment. Relevant departments should publicize their effective practical experiences to the middle and western regions and increase their firefighting investment in order to reduce the fire inertia effect.

(3) Fire caution effect. Previous cases of fire in one area will alarm surrounding areas, and the administrative department tends to prevent any similar incidents by engaging in a safety investigation

and examining fire risks. These measures decrease the fire risk; this is called the "fire caution effect". The spatial and time-united lag parameter (β_5) in the FSDM illustrates this effect; the larger the absolute value of β_5, the more significant the fire caution effect, and vice versa.

The spatial and time-united lag parameter (β_5) is negative and significant at a 99% confidence interval; this illustrates that fire accidents in China have a significant caution effect. Previous fire accidents in one region will alarm the surrounding area and reduce the fire occurrence rate in the surrounding area. A 1% increase in previous fire accidents in one area will cause a 0.20% decrease in fire accidents in the surrounding region. This result suggests that related departments can learn from fire accidents in the surrounding area, and can investigate and eliminate fire risk based on the fire caution effect.

There are differences and connections between the fire caution effect and the fire assimilation effect. The fire assimilation effect analyzes the spatial section at a specific time point and reflects the fire spatial autocorrelation; the fire inertia effect analyzes the time series of fire and demonstrates the time connection in some regions; the fire caution effect combines the assimilation effect and the inertia effect and analyzes the time series as well as the spatial section.

However, the fire caution effect is only a remedial strategy because the parameter of this effect ($\beta_5 = -0.20$) is obviously lower than that of the fire assimilation ($\beta_3 = 0.57$) and inertia effects ($\beta_4 = 0.54$). Therefore, strengthening the management of fire safety and regional cooperation are key points in improving the fire situation.

5. Conclusions and Discussion

Fire situation change results from a combination of economic development and climate change. The response sensitivity of fire situations to climate change is higher than the sensitivity to economic development. A 1% increase in GDP per capita will lead to a 0.15% decrease in the average fire occurrence rate in China; however, a 1% decrease in the annual average humidity will lead to a 0.99% increase in the average fire occurrence rate in China. As the future climate will be drier, this will present further challenges and put pressure on preventing and reducing fire accidents. Society must pay close attention to the effect of drier climates on fire situations and increase the safety investment in preventing conflagrations.

The urban fire situation in China experiences significant spatial effects, which may be referred to as "fire assimilation effects", "fire inertia effects" and "fire caution effects". With the fire assimilation effect, the improvement or deterioration of a fire situation in a surrounding area will bring about an improvement or deterioration of a fire situation in a local area. The fire inertia effect demonstrates that regions where fire accidents occur frequently previously are most likely to suffer from fire accidents in the future. Regions where the fire situation is improved early on will continue to improve in the future. With the fire caution effect, previous fires that occurred in adjoining areas will have a cautionary effect on the local area and reduce the fire rate in those areas. Fire safety administration departments should take full advantage of these effects and take active measures to improve fire situations.

This paper also has some shortcuts. Considering the data availability and the objective of microanalysis of fire change, this paper only included GDP per capita as the representative indicator of economic development. However, macroeconomic factors such as the economic system, industry structure, education level and income level may also have an influence on the frequency of urban fires. Further research should continue to analyze the influence of the aforementioned factors so that research-led policy may be adopted by government. Otherwise, the degree of spatial autocorrelation of variables depends on the size of spatial particles and zone design methods. However, there is no determined function relationship between the change of spatial units and the autocorrelation. So, how to choose the appropriate size of geographical unit particles and zone design are very important in the research of spatial autocorrelation [49]. It should be studied further.

Acknowledgments: This study was primarily funded by a grant from the State Key Program of National Natural Science Foundation of China (71433068). Support was also provided by the National Natural Science Foundation of China (41371177), and the National Natural Youth Science Foundation of China (41201168).

Author Contributions: Zhenbo Wang conceived and designed the research; Bo Xu performed the data collection; Zhenbo Wang and Xiaorui Zhang analyzed the data and wrote the paper; Bo Xu reviewed the paper and made several comments and suggestions for revision.

Conflicts of Interest: The authors declare no conflict of interest.

References

1. Turco, M.; Llasat, M.C.; von Hardenberg, J.; Provenzale, A. Impact of climate variability on summer fires in a Mediterranean environment (northeastern Iberian Peninsula). *Clim. Change* **2013**, *116*, 665–678. [CrossRef]
2. Moritz, M.A.; Morais, M.E.; Summerell, L.A.; Carlson, J.M.; Doyle, J. Wildfires, complexity, and highly optimized tolerance. *Proc. Natl. Acad. Sci. USA* **2005**, *102*, 17912–17917. [CrossRef] [PubMed]
3. Wang, X.L.; Thompson, D.K.; Marshall, G.A.; Tymstra, C.; Carr, R.; Flannigan, M.D. Increasing frequency of extreme fire weather in Canada with climate change. *Clim. Change* **2015**, *130*, 573–586. [CrossRef]
4. Bonan, G.B. Forests and climate change: Forcings, feedbacks, and the climate benefits of forests. *Science* **2008**, *320*, 1444–1449. [CrossRef] [PubMed]
5. Pyne, S.; Andrews, P.; Laven, R. *Introduction to Wildland Fire*; Wiley: New York, NY, USA, 1996.
6. Díaz, D.R.; Lloret, F.; Pons, X. Spatial patterns of fire occurrence in Catalonia. *Landsc. Ecol.* **2004**, *19*, 731–745.
7. Marlon, J.R.; Bartlein, P.J.; Carcaillet, C.; Gavin, D.G.; Harrison, S.P.; Higuera, P.E.; Joos, F.; Power, M.J.; Prentice, I.C. Climate and human influences on global biomass burning over the past two millennia. *Nat. Geosci.* **2008**, *1*, 697–702. [CrossRef]
8. Pausas, J.G.; Keeley, J.E. A burning story: The role of fire in the history of life. *Bioscience* **2009**, *59*, 593–601. [CrossRef]
9. Oliveras, I.; Gracia, M.; Moré, G.; Retana, J. Factors influencing the pattern of fire severities in a large wildfire under extreme meteorological conditions in the Mediterranean basin. *Int. J. Wildland Fire* **2009**, *18*, 755–764. [CrossRef]
10. Venevsky, S.; Thonicke, K.; Sitch, S.; Cramer, W. Simulating fire regimes in human-dominated ecosystems: Iberian Peninsula case study. *Glob. Chang. Biol.* **2002**, *8*, 984–998. [CrossRef]
11. Bowman, D.M.J.S.; Balch, J.K.; Artaxo, P.; Bond, W.J.; Carlson, J.M.; Cochrane, M.A.; D'Antonio, C.M.; DeFries, R.S.; Doyle, J.C.; Harrison, S.P. Fire in the earth system. *Science* **2009**, *324*, 481–484. [CrossRef] [PubMed]
12. Perez, V.G.; Marquez, L.M.A.; Cortes, O.A.; Salmeron, M.M. Spatial-temporal analysis of fire occurrence in Durango. *Madera Y Bosques* **2013**, *19*, 37–58.
13. Falk, D.A.; Heyerdahl, E.K.; Brown, P.M.; Farris, C.; Fule, P.Z.; Mckenzie, D.; Swetnam, T.W.; Taylor, A.H.; van Horne, M.L. Multi-scale controls of historical forest-fire regimes: New insights from fire-scar networks. *Front. Ecol. Environ.* **2011**, *9*, 446–454. [CrossRef]
14. Liu, Z.H.; Yang, J.; Chang, Y.; Weisberg, P.J.; He, H.S. Spatial patterns and drivers of fire occurrence and its future trend under climate change in a boreal forest of Northeast China. *Global. Chang Bio.* **2012**, *18*, 2041–2056. [CrossRef]
15. Wang, Z.; Fang, C.; Zhang, X. Spatial expansion and potential zoning of construction land use in the Yangtze River Delta. *J. Geogr. Sci.* **2015**, *25*, 1–14. [CrossRef]
16. Fang, C.; Wang, Z. Quantitative Diagnoses and Comprehensive Evaluations of the Rationality of Chinese Urban Development Patterns. *Sustainability* **2015**, *7*, 3859–3884. [CrossRef]
17. Syphard, A.D.; Radeloff, V.C.; Keeley, J.E.; Hawbaker, T.J.; Clayton, M.K.; Stewart, S.I.; Hammer, R.B. Human influence on California fire regimes. *Ecol. Appl.* **2007**, *17*, 1388–1402. [CrossRef] [PubMed]
18. Yang, J.; He, H.S.; Shifley, S.R. Spatial controls of occurrence and spread of wildfires in the Missouri Ozark highlands. *Ecol. Appl.* **2008**, *18*, 1212–1225. [CrossRef] [PubMed]
19. Duncanson, M.; Woodward, A.; Reid, P. Socioeconomic deprivation and fatal unintentional domestic fire incidents in New Zealand. *Fire Saf. J.* **2002**, *37*, 165–179. [CrossRef]
20. AFAC. *Accidental Fire Fatalities in Residential Structures: Who's at Risk?* Technical Report; Australian Fire Authorities Council: Canberra, Australia, 2005.

21. Smith, R.; Wright, M.; Solanki, A. *Analysis of Fire and Rescue Service Performance and Outcomes with Reference to Population Socio-Demographics—Fire Research Series 9/2008*; Department for Communities and Local Government: London, UK, 2008.
22. Yang, L.Z.; Jiang, D.B. The relationship between fire and socioeconomic factors in China. *Eng. Sci.* **2003**, *5*, 62–67.
23. Schwartz, N.B.; Uriarte, M.; Gutiérrez, V.V.H.; Baethgen, W.; DeFries, R.; Fernandes, K.; Pinedo, V.M.A. Climate, landowner residency, and land cover predict local scale fire activity in the Western Amazon. *Global Environ. Change* **2015**, *31*, 144–153. [CrossRef]
24. Westerling, A.L.; Turner, M.G.; Smithwick, E.A.H.; Romme, W.H.; Ryan, M.G. Continued warming could transform Greater Yellowstone fire regimes by mid-21st century. *Proc. Natl. Acad. Sci. USA* **2011**, *108*, 13165–13170. [CrossRef] [PubMed]
25. Scholze, M.W.; Knorr, N.W.; Arnell, N.W.; Prentice, I.C. A climate-change risk analysis for world ecosystems. *Proc. Natl. Acad. Sci. USA* **2006**, *103*, 13116–13120. [CrossRef] [PubMed]
26. Chrysoulakis, N.; Cartalis, C. Thermal detection of plumes produced by industrial accidents in urban areas based on the presence of the heat island. *Int. J. Rem. Sens.* **2002**, *23*, 2909–2916. [CrossRef]
27. Panao, M.J.N.O.; Goncalves, H.J.P.; Ferrao, P.M.C. Optimization of the urban building efficiency potential for midlatitude climates using a genetic algorithm approach. *Renew. Energ.* **2008**, *33*, 887–896. [CrossRef]
28. Corcoran, J.; Higgs, G.; Rohde, D. Investigating the association between weather conditions, calendar events and socio-economic patterns with trends in fire incidence: An Australian case study. *J. Geogr. Syst.* **2011**, *13*, 193–226. [CrossRef]
29. Gunther, P. Rural fire deaths: the role of climate and poverty. *Fire J.* **1982**, *76*, 34–39.
30. Sufianto, H.; Green, A.R. Urban fire situation in Indonesia. *Fire Tech.* **2011**, *2*, 357–387. [CrossRef]
31. Standardization Administration of the People's Republic of China. *Urban Fire-Danger Weather Ratings GB/T 20487-2006*; China Standards Publishing House: Beijing, China, 2006.
32. Langner, A.; Siegert, F. Spatiotemporal fire occurrence in Borneo over a period of 10 years. *Glob. Change Biol.* **2009**, *15*, 48–62. [CrossRef]
33. Frejaville, T.; Curt, T. Spatiotemporal patterns of changes in fire regime and climate: Defining the pyroclimates of south-eastern France (Mediterranean Basin). *Clim. Change* **2015**, *129*, 239–251. [CrossRef]
34. Zhang, J.Q.; Liu, X.P.; Tong, Z.J. The study of grassland fire disaster risk assessment and regionalization: A case study in the western Jilin Province. *Geogr. Res.* **2007**, *26*, 755–762.
35. Yan, F.; Ang, Y.J.; Wu, B. Spatial and temporal distributions of drought in Hebei Province over the past 50 years. *Geogr. Res.* **2010**, *29*, 423–430.
36. Jennings, C.R. Socioeconomic characteristics and their relationship to fire incidence: A review of the literature. *Fire Tech.* **1999**, *35*, 7–34. [CrossRef]
37. Chang, H.S. Study of the exploration of fire occurrence spatial characteristics and impact factors: a case study of Tainan City. In Proceedings of 14th International Conference on Urban Planning and Regional Development in the Information Society, Spain, 22–25 April 2009; Design Center Sitges: Sitges, Spain, 2009.
38. Tobler, W.A. A computer movie simulating urban growth in the Detroit region. *Econ. Geogr.* **1970**, *46*, 234–240. [CrossRef]
39. Houthakker, H.S. The Pareto Distribution and the Cobb–Douglas Production Function in Activity Analysis. *Rev. Econ. Stud.* **1955**, *23*, 27–31. [CrossRef]
40. Gao, T.M. *Econometric Methods and Modeling*; Tsinghua Publishing House: Beijing, China, 2009.
41. Zhang, Z.-Y.; Zhu, P.F. Improvement on asymptotic efficiency of QMLE for SDPD model. *J. Quant. Tech. Econ.* **2009**, *5*, 145–157.
42. Anselin, L.; Griffith, D.A. Do spatial effects really matter in regression analysis? *Pap. Reg. Sci. Assoc.* **1988**, *65*, 11–34. [CrossRef]
43. Legendre, P. Spatial autocorrelation: trouble or new paradigm? *Ecology* **1993**, *74*, 1659–1673. [CrossRef]
44. Osborne, P.E.; Foody, G.M.; Suarez-Seoane, S. Non-stationarity and local approaches to modelling the distributions of wildlife. *Divers. Distrib.* **2007**, *13*, 313–323. [CrossRef]
45. Liang, W.; Ren, B.P. Judgment of China's Economic Development Stages and the Analysis of the Characters at the Current Economic Development Stage. *J. Quant. Tech. Econ.* **2009**, *4*, 3–18.
46. Ding, Y.H.; Ren, G.Y.; Shi, G.Y. National assessment report of climate change (I): Climate change in China and its future trend. *Adv. Clim. Change Res.* **2006**, *2*, 3–8.

47. Xu, Y.L.; Huang, X.Y.; Zhang, Y. Statistical analyses of climate change scenarios over China in the 21st Century. *Adv. Clim. Change Res.* **2005**, *1*, 80–83.
48. Xu, B.; He, N.; Gong, P.; Li, Y. Spatiotemporal evolution characteristics of fire disasters in China: GIS-based statistic analysis. *J. Nat. Disasters* **2012**, *21*, 198–203.
49. Openshaw, S. *The Modifiable Areal Unit Problem*; Geo Books: Norwick, UK, 1983.

sustainability

MDPI

Article

Application of Environmental Change Efficiency to the Sustainability of Urban Development at the Neighborhood Level

Hsing-Fu Kuo [1,*] and Ko-Wan Tsou [2]

1 Department of Tourism, Leisure and Hospitality Management, National Chi Nan University, Nantou, No.1, University Rd., Puli., Nantou 54561, Taiwan

2 Department of Urban Planning, National Cheng Kung University, No.1, University Rd., East Dist., Tainan City 70101, Taiwan; uptkw@mail.ncku.edu.tw

* Author to whom correspondence should be addressed; luckiekuo@gmail.com; Tel.: +886-4-9291-0960 (ext. 4687)

Academic Editor: Tan Yigitcanlar

Received: 6 June 2015; Accepted: 31 July 2015; Published: 5 August 2015

Abstract: This study describes a new assessment model framework, termed the driving forces-pressure-state-impact-policy and pattern (DPSIP) model, for environmental change efficiency in urban land development, based on urban sustainable development and the theory of economic efficiency evaluation. A spatial and measurable efficiency value is defined for environmental changes in urban land development, which provides a comprehensive evaluation index for the efficiency of urban development and its environmental impact. This type of urban interior sustainability is considered new within the context of global environmental changes. We identify nine important indicators to evaluate the relative efficiency of 233 neighborhoods in Tainan, Taiwan. The results indicate that the average environmental change efficiency is 89.44%, which shows clear spatial differentiation. The key indicators affecting the efficiency score are area, population density, location, mixed land uses, the floor area ratio, and the impervious ratio. In the future, urban design can reduce environmental impacts and enhance efficiency values.

Keywords: urban land development; natural environmental impact; driving forces-pressure-state-impact-policy and pattern (DPSIP)

1. Introduction

Due to urbanization, the size of urban areas and urban populations have been increasing rapidly and continuously worldwide. According to the 2014 Revision of World Urbanization Prospects [1], the urban population is projected to increase by 2.46 billion, an increase from 3.88 billion (54%) in 2014 to 6.34 billion (66%) in 2050. This projected increase in the urban population will create various demands, and the resources necessary for daily life and production in cities are believed to be an important driving force that will contribute to changes in urban land use [2–4]. Additional agricultural and rural lands will be converted for urban development, resulting in an expanded urban landscape [5–7].

For many cities, changes in land use are required to meet the daily needs of residents and the demands of economic development. The natural environment may suffer the most direct and serious consequences of urbanization due to various interacting factors, such as the economy, society, and culture [8–10]. Although changes in land use will affect the environment, they will also provide income, employment, and other economic benefits to cities [11].

Based on the limited availability of land and natural resources and the irreversibility of the impacts on the natural environment, Whitford *et al.* (2001) and De Koeijer *et al.* (2003) [12,13] have argued that environmental impacts, or the excessive use of resources, can be reduced; the impact of global environmental change can be mitigated; and the efficiency of urban development can be improved by considering changes in urban land use from the perspective of sustainable development. Assessing the efficiency of economic and social development and their impact on the natural environment is an important issue for researchers and urban planners. Urban development efficiency has traditionally been considered an issue of economic efficiency based on economic theory [14]. Since the emergence of sustainable development, urban development efficiency has become more environmentally oriented. In 1990, Schaltegger and Sturm (1990) [15] introduced the concept of eco-efficiency, which represents the ratio of the increased value of development to the increased environmental impact. In 1992, the World Business Council for Sustainable Development (WBCSD) further defined this concept as the ratio of economic value to its associated impact on environmental indicators. Eco-efficiency emphasizes the integration of economic benefits and environmental impacts, and it thus attempts to effectively combine several components: business units on a micro-scale or individual level; urban or regional development planning and management on a larger spatial scale; and the macro-scale objective of sustainable development [16–20].

Although scholars have increasingly used eco-efficiency to assess the relationship between the economy and the environment, most studies focus on the observations of manufacturers [21–23] or the industrial sector at the regional or national level [16,18,19,24,25]. Little effort has been devoted to explicitly exploring and defining the eco-efficiency of urban land development. Therefore, the current knowledge of how to integrate and estimate the impacts of urban land development on the natural environment remains limited. A focus on the eco-efficiency of urban land development would lead to further assessments of the corresponding development efficiency of each spatial unit in cities under the threat of urbanization or urban sprawl. To address the research gap in relation to the environmental impacts of urban land use change, we proposed a new systematic model framework, termed the driving forces-pressure-state-impact-policy and pattern (DPSIP) model, to analyze and structure the relationship between environmental impacts and developmental effectiveness in the process of urban land development.

Additionally, we defined a measurable and detectable environmental change efficiency index based on urban sustainable development and the theory of efficiency evaluation. This efficiency index was used to analyze the influential factors of efficiency scores and to plan the strategies for efficiency improvement with the DPSIP model. Environmental change efficiency was defined as the proportional relationship between the input costs of environmental resources and the social and economic benefits of urban development within the context of global environmental changes.

We considered the following factors to validate the applicability of the method proposed here and to select the area for our empirical research: (a) urban development and its environmental impacts on the environment; (b) the diversification of land use patterns; and (c) the relationship between urban development strategies and improvements in efficiency. Based on these factors, Tainan City was selected as the area for this study. Finally, we employed data envelopment analysis (DEA) and geographically weighted regression (GWR) to design spatial development strategies to increase efficiency by comparing efficiency distributions and urban development patterns.

2. Materials and Methods

2.1. Empirical Research Area

Tainan City is located in the southern coastal plain of Taiwan (120°11′6.3″ E, 22°59′31.9″ N) (see Figure 1), covering an area of 175.65 km^2. It is the oldest and currently the fourth largest city in Taiwan, with a total population of 764,658 in 2007. The city is relatively flat, with no mountain ranges or high elevations. Its urban development process is influenced by both several hundred years of history and long-term natural factors as well as the interactions between the two. Tainan City is divided into 233 neighborhoods and six administrative districts: Annan, Anping, East, North, South, and West Central. Tainan City has diversified and complicated types of land use. The largest land area is used for agriculture (37.32%), followed by built-up areas (19.63%), miscellaneous purposes (14.88%), transportation (11.20%), water conservation (8.63%), and public facilities (3.62%), indicating that Tainan City still has a vast area of land available for development. Tainan City recently implemented numerous strategies related to urban development, environmental protection and healthy development, such as the Tainan Region Plan, a land flattening and reforestation program, a program for the transformation of vacant lots and houses and the Tainan Healthy City Plan. These initiatives provide a good opportunity to test whether the model proposed in this study can effectively assess the relationship between urban land development and the environment.

Four basic principles were developed to select the spatial analysis units based on the objectives of this study, the characteristics of study area, the quality of the data and the DEA theory, including the significance of a standard measurement for efficiency assessment and the practical applicability for urban planning, both of which reflect the development characteristics of the area for empirical research and meet the restrictions on the number of units for evaluation and decision making. A total of 233 neighborhoods in the smallest administrative region in Tainan City were considered the decision-making units (DMUs) of analysis. The relative locations are shown in Figure 2.

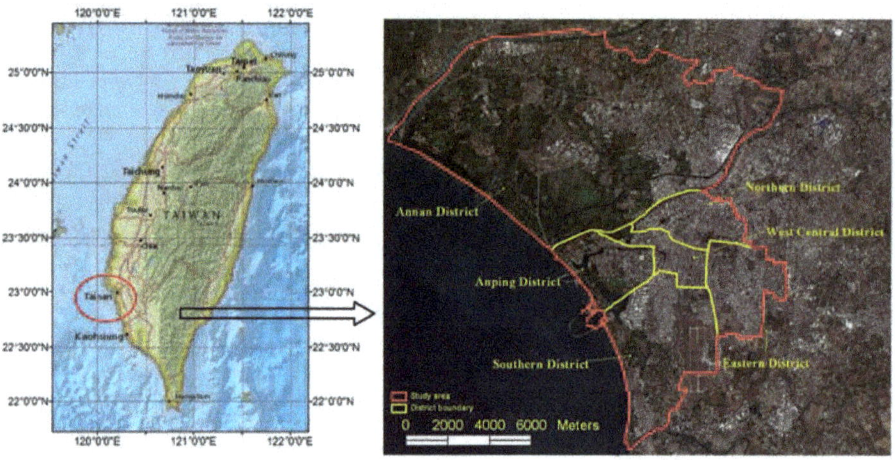

Figure 1. Location of the research area.

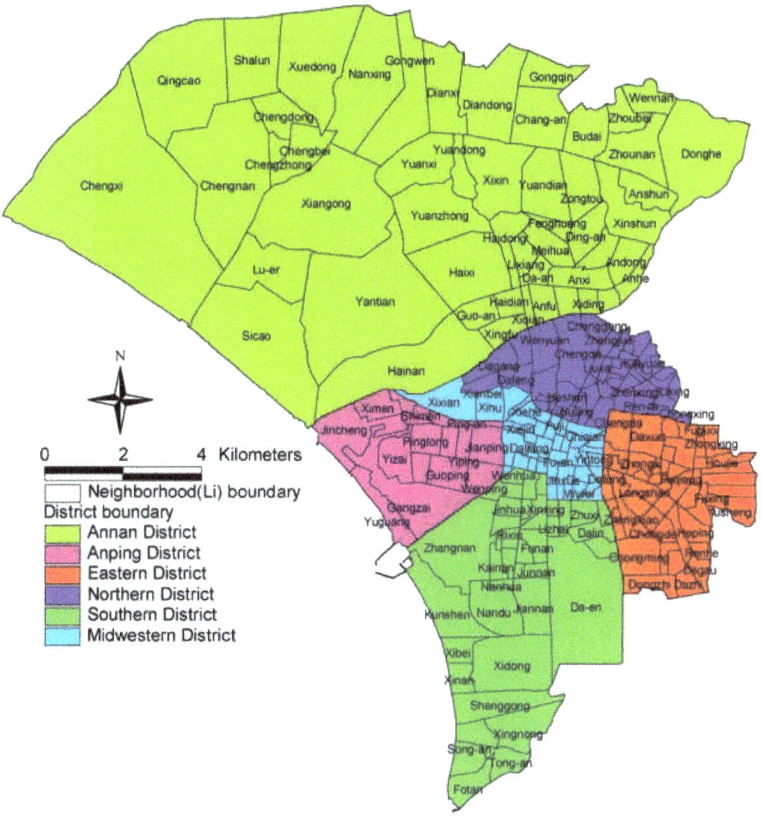

Figure 2. Relative locations of the decision-making units (DMUs).

2.2. The applicability of DPSIP in Assessing the Efficiency of Urban Land Development

Many recent studies have used the drivers-pressure-state-impact-response (DPSIR) model to solve environmental problems and diagnostically evaluate sustainable development [26–28].

DPSIR is a common theoretical model and interdisciplinary tool used to provide and communicate knowledge of the current state of the environment and causal factors regarding environmental issues. Partially due to its simplicity as a framework, DPSIR arguably cannot account for the dynamics of the system that it models or handle cause-consequence relationships. DPSIR has also shown itself to be incompatible with the multiple perspectives required by human interaction in global ecology; more-over, the framework can only assess one impact at a time, and it is difficult to apply to existing or future policies [29–31].

To address these limitations and propose a suitable method for assessing the complex relationships between urban land development, environmental change, and urban policy, we have enhanced and modified the original DPSIR framework and suggest a new assessment model framework: DPSIP (see Figure 3).

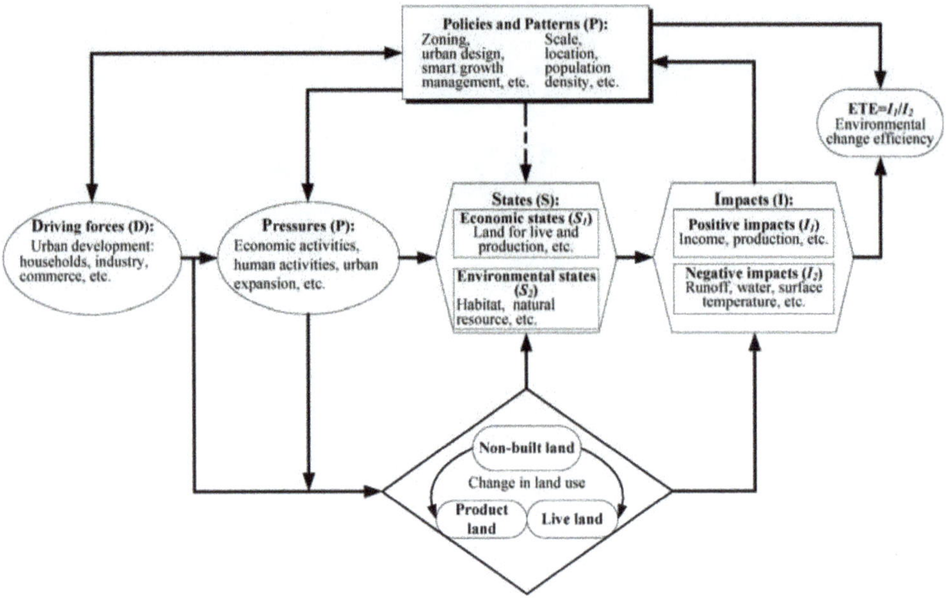

Figure 3. The driving forces-pressure-state-impact-policy and pattern (DPSIP) framework applied to environmental change efficiency.

The relationship between environmental impacts and developmental effectiveness in the process of urban land development was analyzed and structured using this assessment framework. The implications of the policies played a similar role and relationship to response (R) in the traditional DPSIR system. However, the influence of urban management policies on the entire system of urban development constituted the ultimate policy response for the entire system and, consequently, had an impact in the early stages through the driving force and pressure components.

In contrast to the input and output factors of spatial units for environmental change efficiency within the city, urban development policies and patterns are citywide and represent an external environmental variable. As factors affecting the efficiency of environmental change, urban development policies and patterns develop in the early stages and have high degrees of stability.

Overall, the basic concept of the DPSIP model is that, when driven by the economic and daily activities of humans, pressure is exerted on natural resources and the environment. Thus, policies and patterns change the current state of the environment and the quality of natural resources, resulting in internal and external impacts on the entire system of urban development. Urban planners can respond by adjusting the development strategies related to the environment and the economy to mitigate the pressure on the natural environment, thus achieving the goal of sustainability.

2.2.1. Driving Forces

Driving forces refer to the reasons for change in an urban land use system. Recently, many studies [32–35] have indicated that urban land use and land cover changes are important drivers of global environmental changes. They are the underlying causes of the changes in urban land use and in environmental conditions. Driving forces can be divided into natural driving forces and socio-economic driving forces [9].

2.2.2. Pressure

Eder and Narodoslawsky (1999) [36] have indicated that environmental pressure mainly results from human activities. Human and commercial activities interact with and influence the natural resources and ecological systems in the environment. The pressure placed on the environment is expressed as the current state of the competition among diverse social and economic driving forces in the quest for land resources [25,31,37].

2.2.3. State

The amount of land that a city's residents require for daily living and production is an important factor that generates a shift in land use, which is an inevitable component of urban development. In general, the change pattern in urban land development consists primarily of a shift from agricultural land, open space, or other non-built land to production or residential land [5–7,32]. These changes in land use patterns result in the alteration of a city's existing development stages or land characteristics. The same type of land use changes may affect a single parcel of land or an area containing adjacent parcels of land. To measure the characteristics of environmental change in urban development, the state is divided into two factors: S_1 of the economy and S_2 of the environment. The land use changes measured in S_1 of the economy include increases in commercial land, industrial land, residential land, transportation land, and land used for facilities or other related purposes. The land use changes measured in S_2 of the environment include changes in non-built lands that cause a decrease in biodiversity or the depletion of natural resources.

2.2.4. Impact

The characteristics of the individual land types are reflected in the state component of DPSIP. The traditional DPSIR framework is capable of analyzing only one impact at a time. In this study, the impact is considered to have two parts: the positive impact (I_1) and the negative impact (I_2) of urban development, both of which are inherent in land use changes. The positive impact refers to the actual impact and benefits to the economy and residents' quality of life, including population growth, increased income, and economic development [11,20,24,25,38,39]. Because of changes in land use patterns, the overall environmental impacts are characterized by accumulation and diffusion. These negative impacts involve ecological issues, including a reduction in habitat quality and a decrease in biodiversity due to declining habitats [40–42]. Furthermore, land use changes can result in an increase in the area of the impermeable layer of the earth's surface, affecting cities' hydrological systems and resulting in increased surface runoff [43–46] and surface temperature [47–49].

2.2.5. Policy and Pattern

Urban planners respond to the demand for urban development and the associated development issues with corresponding policies or guidance for urban development. The primary purpose of these policies is to respond to the driving forces and pressures of urban development.

In the theoretical framework proposed in this study, the urban policy and urban pattern components are the most external factors. Policies and patterns not only affect the initial driving forces of the entire urban development system but also generate direct and indirect effects on the pressure of development, the current state of the land, and the impacts of development, leading to further effects on the efficiency of environmental change.

The results of a literature review reveal that policies that may be relevant to the efficiency of environmental change in urban land development include traditional zoning systems for land use, urban design regulations, and practices adopted by smart growth management [50–54]. These urban development policies influence the current state of urban land use and socio-economic development, thereby affecting the efficiency of the eventual changes to the environment.

2.3. Definition of Environmental Change Efficiency

2.3.1. Definition of Efficiency

Environmental change efficiency involves balancing the natural environment with economic benefits. Efficient urban spatial units improve the management of the urban natural environment and create a competitive advantage by reducing the depletion of natural environmental resources and improving ecological protection. Environmental change efficiency has three characteristics based on the DPSIP model, the theory of efficiency and other studies [17,18,25,54,55]: a simultaneous improvement in the effectiveness of the economy and the natural environment; an increase in the value of urban development with a small environmental impact; and the promotion of urban economic growth with a reduction in the use of natural resources.

The definition of environmental change efficiency is expressed as follows:

$$\text{Environmental Change Efficiency (ECE)} = \frac{I_1(positive\ impacts)}{I_2(negative\ impacts)} = \frac{\sum_{j=1}^{n} O_j}{\sum_{i=1}^{m} I_i} \quad (1)$$

I_1 = the positive impacts resulting from the accumulation of current state changes in society and the economy, *i.e.*, the sum of outputs.
I_2 = the negative impacts resulting from the accumulation of current state changes in related natural environmental resources and the ecological environment, *i.e.*, the sum of inputs.

The numerator for overall outputs represents the change in urban land use patterns due to urban land development. The numerator represents the positive impact (I_1) on urban residents' quality of life and on the economy, which indicates an increase in socio-economic efficiency or the quality of services generated by urban development.

The denominator for overall inputs represents the change in natural environmental resources or changes in the depletion state and ecological environmental impact in the natural environmental system caused by a land use change due to urban development. The denominator represents the negative impact (I_2) on the natural environment and natural resources in the urban area at the time of development.

The efficiency scores in this study provide index values of a single composite type to reflect the state of sustainable development. In general, the efficiency index proposed in this study represents the composite performance results for sustainability in an area. With the ability to detect the factors that affect efficiency scores, we can clearly understand the factors that may improve the area's overall efficiency and thereby achieve the goal of sustainable development. If the value of a city's average environmental change efficiency can be increased annually and the standard deviation of the environmental change efficiency of a DMU within a city can be decreased annually, then the region can obtain a positive state of development within the context of global environmental change. Moreover, the impact generated by urban development and changes in the state of land use would also be positive.

2.3.2. Input and Output Indicators

When evaluating efficiency, the selection of representative indicators as inputs and outputs is both important and difficult [56]. To select the indicators, we began with a literature review [11,13,25,39,54,57] focusing on the indicators suitable for monitoring and measuring environmental impacts and urban sustainability at the neighborhood level; we selected indicators based on DPSIP and the primary aspects of sustainable development.

In this study, an indicator is defined as a measurable and detectable variable that reflects the causes and effects of land use change and that can be easily understood and used in both planning and decision making. We evaluated potential indicators using the following four criteria: reliability, usability, availability, and usefulness in decision making. An initial list of input and output indicators was developed based on earlier studies [11,13,24,25,39,57]. After the potential indicators had been evaluated, the most suitable indicators were chosen for demonstration in Tainan City. Finally, the final four input indicators and five output indicators were selected based on the availability of data and the practicality of use at the neighborhood level.

The inputs were primarily oriented toward natural environmental resources and the ecological environment, including factors such as surface temperature [57–59], surface runoff [2,13,43–46], habitat quality [13,40–42,60], and water consumption [2,44,61].

The outputs were chosen to reflect the benefits and goals of economic and social development according to the Tainan City Sustainable Development Action Plan proposed by the Tainan City Government as well as other studies [11,20,24,25,38,39]. The outputs included population, production, income, the service quality of infrastructure, and accessibility.

The data were primarily collected from the Landsat Thematic Mapper Satellite Image (2007); the Land Use Investigation of Taiwan (2007); the Taiwan Water Corporation (2007); the Department of Household Registration of the Ministry of the Interior (2007); the Industry, Commerce and Service Census (2006); the Financial Data Center of the Ministry of Finance (2007); and the Database of the Traffic Network Digital Map (2007) (see Table 1).

Table 1. Input and output indicators for the data envelopment analysis (DEA) efficiency analysis.

Indicators	Definition	Units	Mean	Std. Dev.	Range	Data Sources
Input						
Surface temperature	Global average surface temperature	°C	27.867	2.168	12.559	Landsat Thematic Mapper Satellite Image, 2007 [62]
Surface runoff	Total volume of overland flow draining off the land	m^3/s	12.475	20.743	195.824	Land Use Investigation of Taiwan, 2007 [63]
Habitat quality	Area-weighted mean patch fractal dimension value of habitats	None	1.420	0.098	0.485	Land Use Investigation of Taiwan, 2007 [63]
Water consumption	Total water use	m^3	381,268	268,421	1,629,889	Taiwan Water Corporation, 2007 [64]
Output						
Population	Total population	Persons	3262	1732	11,252	Department of Household Registration, Ministry of the Interior, 2007 [65]
Service quality of infrastructure	Density of infrastructure	m^2/ha	25.5668	13.1649	64.8379	Land Use Investigation of Taiwan, 2007 [63]
Production	Total production of industry and commerce	USD$	57,292	117,016	1,174,532	Industry, Commerce and Service Census, 2006 [66]
Income	Total household income	USD$	17,112	12,663	89,125	Financial Data Center, Ministry of Finance, 2007 [67]
Accessibility	Global integration value of streets	None	0.240	0.038	0.150	Database of the Traffic Network Digital Map, 2007 [68]

The common methods of efficiency assessment include ratio analysis, regression analysis, and DEA [69]. One characteristic of DEA is that no prior knowledge of the production function between the input and output attribute data is required; similarly, no relative weight needs to be set for the attribute data [54,60–72]. Therefore, DEA is useful for comprehensively assessing the indicators of different types and data patterns; it is widely used in economic science, agricultural economics, public economics, financial economics, and economic policy. Additionally, DEA is considered an appropriate analysis method for many studies related to eco-efficiency or environmental efficiency [21–24,54,73–77].

Considering the characteristics of complexity and development in urban areas and the advantages of DEA, we used the input-oriented CRS (constant returns to scale) model to obtain efficiency measures corresponding to the assumptions. The efficiency measures attained from CCR model are known as overall technical efficiency (OTE) scores.

2.4. Influential Factors in Environmental Change Efficiency

Several useful environmental assessment tools exist for promoting sustainable development, including strategic environmental assessment (SEA) [78,79], territorial impact assessment (TIA) [80,81] and spatial policy monitoring. These tools focus on the process by which environmental considerations are fully integrated into the preparation and adoption of policies, plans, and programs. The tools provide a means to consider the relationships between policies, plans, and environmental impacts. To assess the potential factors affecting environmental change efficiency, the efficiency factors were formed using urban characteristics based on the construction framework of DPSIP and the dimensions related to urban land development in SEA and TIA [82,83] and urban policies [52,84].

The policies and management strategies for urban land development have traditionally included zoning and urban design control. The concepts of growth management, smart growth, new urbanism, and sustainable development have been proposed in recent years [50,51,53,85]. Among these, smart growth and new urbanism are the most well-known and can effectively reduce the spread of urban areas and manage urban development. These concepts are currently believed to be the most important planning concepts and development policies.

We integrated urban characteristics and urban policy concepts, both of which affect the efficiency of urban lands and environments, and then cross-compared these factors. From this process, six quantitative indicators were identified, including land area, location, population density, the floor area ratio, the impervious ratio, and mixed land use [51–53,86–89].

Additionally, because the factors that were determined to affect efficiency had different spatial distributions, GWR model was used to analyze the efficiency factors.

GWR is a local multivariate regression function and a developed method for the analysis of spatial data in which the data samples are weighted based on their spatial proximity [90]. In contrast to the more traditional ordinary least-squares regression (OLS), GWR can solve the problem of spatial autocorrelation in residuals and can reduce the instability of space [90–94]. GWR produces a separate set of regression parameters for every observation across the study area. Thus, it avoids the assumed relationships between dependent and independent variables that are generated by traditional OLS models. Because the factors used in this study have different spatial distributions, GWR may be a useful analysis tool for assessing which of these factors influence efficiency.

3. Results and Discussion

3.1. Scores of Environmental Change Efficiency

Table 2 presents the results of the DEA for environmental change efficiency. The average efficiency score of the 233 neighborhoods was 89.44%, with a maximum efficiency score of 100% and a minimum score of 61.90%. A total of 48 neighborhoods had an efficiency score of 100%. Comparing the districts, the ratio of efficient DMUs to all DMUs was 28.95% in the Midwestern District, 27.66% in the Eastern District, 25.58% in the Northern District, 20.00% in the Anping District, 12.82% in the Southern District, and 84.03% in the Annan District. After comparing the means and standard deviations for each district, the Midwestern District emerged as the district with the highest average (95.96%) and the lowest standard deviation (3.63%), followed by the Anping District (92.42%, 7.35%) and the Northern District (91.79%, 8.29%). The Southern District and the Annan District fell below the overall average in Tainan City, with averages of 86.13% and 84.03%, respectively, which was caused by primarily agricultural land use types or the district's status as a recent growth area with a smaller population, less production and lower service quality of infrastructure.

Table 2. Efficiency results for 233 neighborhoods.

District	DMUs			Efficiency Score	
	Total	Efficiency	Efficiency Percentage (%)	Mean	Std. Dev.
Eastern	47	13	27.66	89.70	9.62
Southern	39	5	12.82	86.13	10.50
Midwestern	38	11	28.95	95.96	3.63
Northern	43	11	25.58	91.79	8.29
Anping	15	3	20.00	92.42	7.35
Annan	51	5	9.80	84.03	9.37
Tainan City	233	48	20.60	89.44	9.48

3.2. *Spatial Patterns and Slack Analysis of Environmental Change Efficiency*

The administrative boundaries of each district denote an actual geographical area of space. The spatial distribution of the DMU efficiency scores revealed an overall decreasing trend in the distribution of efficiency scores, moving outward from the city center and forming two spatial development axes, $\overleftrightarrow{AC}$ and $\overleftrightarrow{DE}$, and signifying relatively efficient urban development (see Figure 4). The $\overleftrightarrow{AC}$ efficiency axis ran from the southeast side of Tainan City, went through the downtown area, and extended to the northwest. The $\overleftrightarrow{DE}$ efficiency axis extended from northeast to southwest. Efficiency scores decreased in areas located farther from downtown. With the exception of a few efficient DMUs that represented dispersed development characteristics (e.g., Donghe and Anqing in the Annan District, Yuping in the Anping District, and Xingsheng and Ren-ai in the Northern District), the most efficient DMUs appeared in clustered or adjacent developments and were concentrated across the administrative districts, particularly in the areas adjacent to the Annan, Anping, and Northern Districts and in the DMUs in the Northern and Midwestern Districts.

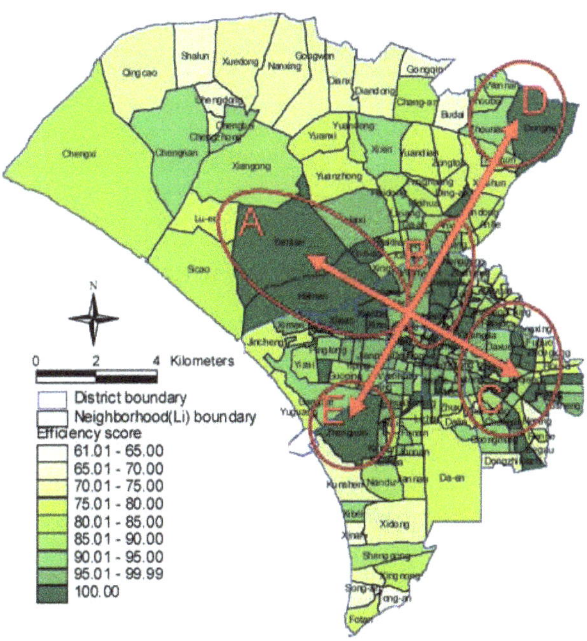

Figure 4. Spatial patterns of efficiency scores.

When the number of DMUs exhibiting differences in a certain input or output indicator increases, most of them must be able to improve their environmental change efficiency via an increase or decrease in this indicator. These indicators are particularly important for improving the overall efficiency of urban land development; they also serve as a reference for subsequent management and improvement strategies. Therefore, we calculated the number of DMUs with changeable differences with respect to nine input and output indicators to understand the importance of the direct input and output indicators that impact the efficiency of urban land development. The ratio of these DMUs to the total number of 233 DMUs was calculated and used as a reference point to determine the indicators for increased efficiency. When a significant number of DMUs revealed a difference in a certain input or output indicator, this indicator became the focus of efficiency improvements for the future; that is, in the future, a reduced value of these input indicators or an increased value of these output indicators will contribute to improved efficiency scores among the DMUs.

Based on the calculated difference between the variables, the indicators relevant to future efficiency improvements were identified as follows: an increase in the quality of public facilities and services (62.23%), an increase in income (57.51%), an increase in production (35.19%), a decrease in surface runoff (25.75%), a decrease in water consumption (18.88%), an increase in habitat quality (17.60%), an increase in the population of urban residents (16.31%), a decrease in the urban surface temperature (12.02%), and an increase in accessibility (8.58%).

3.3. Factors Affecting Environmental Change Efficiency

The adjusted R^2 value of the OLS regression was 0.37, whereas the adjusted R^2 value of the GWR was 0.53, indicating that the GWR can better explain the influential factors of environmental change efficiency.

The regression coefficients of the variables in the OLS and GWR models were organized, and the efficiency scores increased when the factor values increased. In contrast, a coefficient of less than 0 indicated that the efficiency scores decreased when this variable's values increased (see Table 3).

Table 3. Parameter summary for the OLS model and the GWR models.

Parameter	OLS	GWR							
		Mean	Minimum	Maximum	<0		>0		*p*-Value
					DMUs	(%)	DMUs	(%)	
Area	0.0107	0.0247	−0.0122	0.0877	22	9.44	211	90.56	0.05 **
Location	−0.0023	−0.0026	−0.0047	0.0000	233	100.00	0	0.00	0.03 **
Population density	0.0101	0.0275	0.0058	0.6246	0	0.00	233	100.00	0.00 ***
Floor area ratio	−0.0043	0.0061	−0.0657	0.2794	150	64.38	83	35.62	0.00 ***
Impervious ratio	−0.0532	−0.0353	−0.3485	0.2078	149	63.95	84	36.05	0.05 **
Mixed land use	−1.5029	−1.5202	−32.6921	10.9520	150	64.38	83	35.62	0.07 *

* Significance level $p < 0.1$; ** Significance level $p < 0.05$; *** Significance level $p < 0.001$.

Our results revealed that five of the six impact variables proposed had spatial differences. Regarding the variables' influence, the location and population density variables had the same effect in the OLS and GWR models. The effect of the location was negative, indicating that the efficiency scores decreased with increases in the distance from the city center. The result is in line with those of recent studies [83,95,96]. For the population density variable, the efficiency scores increased as the population density increased.

In this study, among the six variables influencing efficiency, with the exception of the two variables discussed above, four variables: area, the floor area ratio, the impervious ratio, and mixed land use had different effects based on the two regression analyses. In general, the GWR simultaneously revealed positive and negative effects in these four variables due to the development characteristics and spatial non-stationarity of the 233 neighborhoods. These differences must be considered when preparing a strategy to improve efficiency in different areas of the city. According to Table 3, the neighborhoods with positive values of influential factors could improve their environmental change efficiency by increasing their area, population density, floor area ratio and mixed land use. In addition, the neighborhoods with negative values of influential factors could improve their environmental change efficiency by decreasing their area, floor area ratio, impervious ratio and mixed land use.

3.4. Efficiency Improvement Strategy

As the focus of early development in Tainan City, some neighborhoods in the Midwestern District had characteristics of high population density, high building density, low plant coverage, and a significant impermeable layer. These characteristics were reflected in the input indicators of environmental resources; an assessment of the indicators of environmental change efficiency revealed relatively poor performance for surface runoff, surface temperature, and habitat quality. To improve efficiency throughout Tainan City, the current urban development strategies for the city, including urban renewal, local environmental change, reforestation, and landscaping in vacant lots, should be continued. The impact on the natural environment generated during the process of urban development can be reduced by increasing the green space and habitats for wildlife.

To improve efficiency through the sustainable development of urban areas and the environment, the foci of future improvements for Tainan City, in descending order, include reducing surface runoff, reducing water consumption, improving habitat quality, and reducing surface temperature. Urban planners can use phased and zoned development, retain existing natural open spaces in urban districts, and increase areas of green space with ecological functions in urban areas through a standardized review of urban design and the implementation of a comprehensive urban planning process. With improved input and output indicators, overall efficiency can be indirectly improved to achieve the long-term goal of reducing the impacts of global environmental change. When developing strategies to improve efficiency in areas that differ in terms of the quality of public facilities and services and in adjacent areas and spatially clustered areas, the development of public facilities should be prioritized.

In terms of the factors influencing efficiency, population density is significant with regard to compact urban development. The influence of different spatial units gradually increased from the junction of the Midwestern, Southern, and Anping Districts fanning out to the south, north, and east of the outlying areas, forming a nearly concentric circle. For the outermost neighborhoods in Tainan City, environmental change efficiency can be greatly improved by increasing the population density. The current land use types in these neighborhoods are primarily agricultural or recently developed with smaller populations and less production activity. We suggested that these neighborhoods be more compact by increasing their population densities. It is not only an important strategy for a compact city but can also increase its environmental efficiency.

A total of 149 neighborhoods in Tainan City (63.95%) could improve their environmental change efficiency by reducing their impervious area ratios. These areas are located primarily in the Midwestern District of the city center and in the nearby Northern and Anping Districts. According to the existing literature [47–49], a higher impervious area ratio leads to a greater surface temperature and surface runoff for the city. Therefore, future urban designs can effectively reduce their environmental impact and enhance their efficiency scores by focusing on the impervious area ratio.

4. Conclusions

Changes in a city's land use and the expansion of urban development are key drivers of global environmental change and are unavoidable aspects of economic and social development in many cities. Although urban development land use changes negatively impact the natural environment, they often produce important economic and social benefits for city residents.

In this study, we propose the DPSIP model framework to address the process, complexity and the relationship between environmental impacts and developmental effectiveness. This model can help researchers overcome the limitations associated with the traditional DPSIR, with which assessing two or more impacts at a time is difficult; DPSIR is also difficult to apply to existing or future policies [29–31]. For cities with rapid urbanization, which is similar to what is occurring in Tainan City, DPSIP could generate significant benefits for the entire urban development management system. The urban planners, urban managers, and policymakers could obtain useful information from relevant evaluation results and create policies to improve the efficiency of urban development and could mitigate the global environmental impact caused by urban land use changes according to the factors affecting environmental change efficiency.

Additionally, spatial and measurable efficiency scores were calculated for environmental changes in urban land development, providing a comprehensive evaluation index for the efficiency of urban development and environmental impacts. This index constitutes a new resource for achieving urban sustainability at the neighborhood level.

Finally, GWR was applied to the factors affecting efficiency and was successfully validated. The empirical results revealed that the GWR analysis had greater overall explanatory power than the traditional regression model; the GWR analysis also generated different coefficients based on the spatial difference of DMUs due to the effects of the various influential factors. This feature will help researchers understand the effects of differences across the different DMUs, which may contribute to the development of more specific efficiency strategies.

Acknowledgments: The authors would like to thank the Ministry of Science and Technology of the Republic of China, Taiwan, for financially supporting this research under contract no. MOST-98-2410-H-006-092-MY2.

Author Contributions: This study represents an effort from two authors. Hsing-Fu Kuo performed research, analyzed the data and wrote the paper; Ko-Wan Tsou made contribution to the research design. All authors read and approved the final manuscript.

Conflicts of Interest: The authors declare no conflict of interest.

References

1. United Nations. *World Urbanization Prospects: The 2014 Revision*; United Nations: New York, NY, USA, 2014.
2. Haase, D.; Nuissl, H. Does urban sprawl drive changes in the water balance and policy? The case of Leipzig (Germany) 1870–2003. *Landsc. Urban Plan.* **2007**, *80*, 1–13. [CrossRef]
3. He, C.Y.; Okada, N.; Zhang, Q.F.; Shi, P.J.; Li, J.G. Modelling dynamic urban expansion processes incorporating a potential model with cellular automata. *Landsc. Urban Plan.* **2008**, *86*, 79–91. [CrossRef]
4. Pauleit, S.; Duhme, F. Assessing the environmental performance of land cover types for urban planning. *Landsc. Urban Plan.* **2000**, *52*, 1–20. [CrossRef]
5. Carlson, T.N.; Arthur, S.T. The impact of land use—Land cover changes due to urbanization on surface microclimate and hydrology: A satellite perspective. *Glob. Planet. Change* **2000**, *25*, 49–65. [CrossRef]
6. Grimm, N.B.; Faeth, S.H.; Golubiewski, N.E.; Redman, C.L.; Wu, J.G.; Bai, X.M.; Briggs, J.M. Global change and the ecology of cities. *Science* **2008**, *319*, 756–760. [CrossRef] [PubMed]
7. Kalnay, E.; Cai, M. Impact of urbanization and land-use change on climate. *Nature* **2003**, *423*, 528–531. [CrossRef] [PubMed]
8. Hasse, J.E.; Lathrop, R.G. Land resource impact indicators of urban sprawl. *Appl. Geogr.* **2003**, *23*, 159–175. [CrossRef]
9. Lambin, E.F.; Meyfroidt, P. Land use transitions: Socio-ecological feedback *versus* socio-economic change. *Land Use Policy* **2010**, *27*, 108–118. [CrossRef]
10. Weng, Q.H. Modeling urban growth effects on surface runoff with the integration of remote sensing and GIS. *Environ. Manag.* **2001**, *28*, 737–748. [CrossRef]
11. Prato, T. Evaluating land use plans under uncertainty. *Land Use Policy* **2007**, *24*, 165–174. [CrossRef]
12. De Koeijer, T.J.; Wossinik, G.A.A.; Smit, A.B.; Janssens, S.R.M.; Renkema, J.A.; Struik, P.C. Assessment of the quality of farmers' environmental management and its effects on resource use efficiency: A Dutch case study. *Agric. Syst.* **2003**, *78*, 85–103. [CrossRef]
13. Whitford, V.; Ennos, A.R.; Handley, J.F. "City form and natural process"—Indicators for the ecological performance of urban areas and their application to Merseyside, UK. *Landsc. Urban Plan.* **2001**, *57*, 91–103. [CrossRef]

14. Ohls, J.C.; Pines, D. Discontinuous urban-development and economic efficiency. *Land Econ.* **1975**, *51*, 224–234. [CrossRef]
15. Schaltegger, S.; Sturm, A. Ökologische Rationalität Ansatzpunkte zur Ausgestaltung von ökologieorientierten Managementinstrumenten. *Die Unternehmung* **1990**, *4*, 273–290. (In German)
16. Cha, K.; Lim, S.; Hur, T. Eco-efficiency approach for global warming in the context of Kyoto Mechanism. *Ecol. Econ.* **2008**, *67*, 274–280. [CrossRef]
17. Huppes, G. Eco-efficiency: From focused technical tools to reflective sustainability analysis. *Ecol. Econ.* **2009**, *68*, 1572–1574. [CrossRef]
18. Li, D.Z.; Hui, E.C.M.; Leung, B.Y.P.; Li, Q.M.; Xu, X. A methodology for eco-efficiency evaluation of residential development at city level. *Build. Environ.* **2010**, *45*, 566–573. [CrossRef]
19. Seppala, J.; Melanen, M.; Maenpaa, I.; Koskela, S.; Tenhunen, J.; Hiltunen, M.R. How can the eco-efficiency of a region be measured and monitored? *J. Ind. Ecol.* **2005**, *9*, 117–130. [CrossRef]
20. Singh, R.K.; Murty, H.R.; Gupta, S.K.; Dikshit, A.K. An overview of sustainability assessment methodologies. *Ecol. Indic.* **2009**, *9*, 189–212. [CrossRef]
21. Hua, Z.S.; Bian, Y.W.; Liang, L. Eco-efficiency analysis of paper mills along the Huai River: An extended DEA approach. *Omega-Int. J. Manag. Sci.* **2007**, *35*, 578–587. [CrossRef]
22. Kuosmanen, T.; Kortelainen, M. Measuring eco-efficiency of production with data envelopment analysis. *J. Ind. Ecol.* **2005**, *9*, 59–72. [CrossRef]
23. Yang, W.; Jin, F.J.; Wang, C.J.; Lv, C. Industrial eco-efficiency and its spatial-temporal differentiation in China. *Front. Environ. Sci. Eng.* **2012**, *6*, 559–568. [CrossRef]
24. Sanjuan, N.; Ribal, J.; Clemente, G.; Fenollosa, M.L. Measuring and improving eco-efficiency using data envelopment analysis A case study of Mahon-Menorca cheese. *J. Ind. Ecol.* **2011**, *15*, 614–628. [CrossRef]
25. Yu, Y.D.; Chen, D.J.; Zhu, B.; Hu, S.Y. Eco-efficiency trends in China, 1978–2010: Decoupling environmental pressure from economic growth. *Ecol. Indic.* **2013**, *24*, 177–184. [CrossRef]
26. Holman, I.P.; Rounsevell, M.D.A.; Cojacaru, G.; Shackley, S.; McLachlan, C.; Audsley, E.; Berry, P.M.; Fontaine, C.; Harrison, P.A.; Henriques, C.; *et al.* The concepts and development of a participatory regional integrated assessment tool. *Clim. Change* **2008**, *90*, 5–30. [CrossRef]
27. Niemeijer, D.; de Groot, R.S. A conceptual framework for selecting environmental indicator sets. *Ecol. Indic.* **2008**, *8*, 14–25. [CrossRef]
28. Tscherning, K.; Helming, K.; Krippner, B.; Sieber, S.; Paloma, S.G.Y. Does research applying the DPSIR framework support decision making? *Land Use Policy* **2012**, *29*, 102–110. [CrossRef]
29. Bell, S. DPSIR = A Problem Structuring Method? An exploration from the "Imagine" approach. *Eur. J. Oper. Res.* **2012**, *222*, 350–360. [CrossRef]
30. Carr, E.R.; Wingard, P.M.; Yorty, S.C.; Thompson, M.C.; Jensen, N.K.; Roberson, J. Applying DPSIR to sustainable development. *Int. J. Sustain. Dev. World Ecol.* **2007**, *14*, 543–555. [CrossRef]
31. Svarstad, H.; Petersen, L.K.; Rothman, D.; Siepel, H.; Watzold, F. Discursive biases of the environmental research framework DPSIR. *Land Use Policy* **2008**, *25*, 116–125. [CrossRef]
32. Dalla-Nora, E.L.; de Aguiar, A.P.D.; Lapola, D.M.; Woltjer, G. Why have land use change models for the Amazon failed to capture the amount of deforestation over the last decade? *Land Use Policy* **2014**, *39*, 403–411. [CrossRef]
33. Grimmond, S. Urbanization and global environmental change: Local effects of urban warming. *Geogr. J.* **2007**, *173*, 83–88. [CrossRef]
34. Schaldach, R.; Alcamo, J.; Koch, J.; Kolking, C.; Lapola, D.M.; Schungel, J.; Priess, J.A. An integrated approach to modelling land-use change on continental and global scales. *Environ. Model. Softw.* **2011**, *26*, 1041–1051. [CrossRef]
35. Seto, K.C.; Satterthwaite, D. Interactions between urbanization and global environmental change. *Curr. Opin. Environ. Sustain.* **2010**, *2*, 127–128. [CrossRef]
36. Eder, P.; Narodoslawsky, M. What environmental pressures are a region's industries responsible for? A method of analysis with descriptive indices and input-output models. *Ecol. Econ.* **1999**, *29*, 359–374. [CrossRef]
37. Haase, D.; Totzer, T. Urban–Rural linkages-analysing, modelling, and understanding drivers, pressures, and impacts of land use changes along the rural-to-urban gradient. *Environ. Plan. B* **2012**, *39*, 194–197. [CrossRef]

38. Lee, Y.J.; Huang, C.M. Sustainability index for Taipei. *Environ. Impact Assess. Rev.* **2007**, *27*, 505–521. [CrossRef]
39. Zhao, C.J.; Fu, G.B.; Liu, X.M.; Fu, F. Urban planning indicators, morphology and climate indicators: A case study for a north-south transect of Beijing, China. *Build. Environ.* **2011**, *46*, 1174–1183. [CrossRef]
40. Kattwinkel, M.; Biedermann, R.; Kleyer, M. Temporary conservation for urban biodiversity. *Biol. Conserv.* **2011**, *144*, 2335–2343. [CrossRef]
41. Kohsaka, R. Developing biodiversity indicators for cities: Applying the DPSIR model to Nagoya and integrating social and ecological aspects. *Ecol. Res.* **2010**, *25*, 925–936. [CrossRef]
42. Sandstrom, U.G.; Angelstam, P.; Khakee, A. Urban comprehensive planning—Identifying barriers for the maintenance of functional habitat networks. *Landsc. Urban Plan.* **2006**, *75*, 43–57. [CrossRef]
43. Fox, D.M.; Witz, E.; Blanc, V.; Soulie, C.; Penalver-Navarro, M.; Dervieux, A. A case study of land cover change (1950–2003) and runoff in a Mediterranean catchment. *Appl. Geogr.* **2012**, *32*, 810–821. [CrossRef]
44. Lin, Y.P.; Hong, N.M.; Wu, P.J.; Wu, C.F.; Verburg, P.H. Impacts of land use change scenarios on hydrology and land use patterns in the Wu-Tu watershed in Northern Taiwan. *Landsc. Urban Plan.* **2007**, *80*, 111–126. [CrossRef]
45. Tang, J.; Wang, L.; Yao, Z. Analyses of urban landscape dynamics using multi-temporal satellite images: A comparison of two petroleum-oriented cities. *Landsc. Urban Plan.* **2008**, *87*, 269–278. [CrossRef]
46. Zhang, B.A.; Xie, G.D.; Zhang, C.Q.; Zhang, J. The economic benefits of rainwater-runoff reduction by urban green spaces: A case study in Beijing, China. *J. Environ. Manag.* **2012**, *100*, 65–71. [CrossRef] [PubMed]
47. Coseo, P.; Larsen, L. How factors of land use/land cover, building configuration, and adjacent heat sources and sinks explain urban heat islands in Chicago. *Landsc. Urban Plan.* **2014**, *125*, 117–129. [CrossRef]
48. Li, W.; Bai, Y.; Chen, Q.; He, K.; Ji, X.; Han, C. Discrepant impacts of land use and land cover on urban heat islands: A case study of Shanghai, China. *Ecol. Indic.* **2014**, *47*, 171–178. [CrossRef]
49. Miller, J.D.; Kim, H.; Kjeldsen, T.R.; Packman, J.; Grebby, S.; Dearden, R. Assessing the impact of urbanization on storm runoff in a peri-urban catchment using historical change in impervious cover. *J. Hydrol.* **2014**, *515*, 59–70. [CrossRef]
50. Cabrera, J.F.; Najarian, J.C. Can new urbanism create diverse communities? *J. Plan. Educ. Res.* **2013**, *33*, 427–441. [CrossRef]
51. Dierwechter, Y. The spaces that smart growth makes: Sustainability, segregation, and residential change across Greater Seattle. *Urban Geogr.* **2014**, *35*, 691–714. [CrossRef]
52. Miles, R.; Song, Y. "Good" neighborhoods in Portland, Oregon: Focus on both social and physical environments. *J. Urban Aff.* **2009**, *31*, 491–509. [CrossRef]
53. Wey, W.M.; Hsu, J. New Urbanism and Smart Growth: Toward achieving a smart National Taipei University District. *Habitat Int.* **2014**, *42*, 164–174. [CrossRef]
54. Yin, K.; Wang, R.S.; An, Q.X.; Yao, L.; Liang, J. Using eco-efficiency as an indicator for sustainable urban development: A case study of Chinese provincial capital cities. *Ecol. Indic.* **2014**, *36*, 665–671. [CrossRef]
55. Huppes, G.; Ishikawa, M. Why eco-efficiency? *J. Ind. Ecol.* **2005**, *9*, 2–5. [CrossRef]
56. Bosetti, V.; Locatelli, G. A data envelopment analysis approach to the assessment of natural parks' economic efficiency and sustainability. The case of Italian national parks. *Sustain. Dev.* **2006**, *14*, 277–286. [CrossRef]
57. Pauleit, S.; Ennos, R.; Golding, Y. Modeling the environmental impacts of urban land use and land cover change—A study in Merseyside, UK. *Landsc. Urban Plan.* **2005**, *71*, 295–310. [CrossRef]
58. Hoppe, P. Different aspects of assessing indoor and outdoor thermal comfort. *Energy Build.* **2002**, *34*, 661–665. [CrossRef]
59. Svensson, M.K.; Eliasson, I. Diurnal air temperatures in built-up areas in relation to urban planning. *Landsc. Urban Plan.* **2002**, *61*, 37–54. [CrossRef]
60. Hirzel, A.H.; Hausser, J.; Chessel, D.; Perrin, N. Ecological-niche factor analysis: How to compute habitat-suitability maps without absence data? *Ecology* **2002**, *83*, 2027–2036. [CrossRef]
61. Interlandi, S.J.; Crockett, C.S. Recent water quality trends in the Schuylkill River, Pennsylvania, USA: A preliminary assessment of the relative influences of climate, river discharge and suburban development. *Water Res.* **2003**, *37*, 1737–1748. [CrossRef]
62. United States Geological Survey. 2007 Landsat Thematic Mapper Satellite Image. Available online: http://glovis.usgs.gov/ (accessed on 10 January 2010).

63. National Land Surveying and Mapping Center. *2007 Land Use Investigation of Taiwan*; National Land Surveying and Mapping Center: Taipei, Taiwan, 2009.
64. Taiwan Water Corporation. *2007 Annual Consumption of Water for Villages*; Taiwan Water Corporation: Taipei, Taiwan, 2009.
65. Department of Household Registration of the Ministry of the Interior. *2007 Population for Villages*; Ministry of the Interior: Taipei, Taiwan, 2008.
66. Directorate General of Budget, Accounting and Statistics. *2006 Industry, Commerce and Service Census*; Directorate General of Budget, Accounting and Statistics: Taipei, Taiwan, 2008.
67. Financial Data Center of the Ministry of Finance. *2007 Family Income Statistics*; Ministry of Finance: Taipei, Taiwan, 2009.
68. Institute of Transportation of the Ministry of Transportation and Communications. *2007 Database of the Traffic Network Digital Map*; Ministry of Transportation and Communications: Taipei, Taiwan, 2009.
69. Thanassoulis, E. A comparison of regression-analysis and data envelopment analysis as alternative methods for performance assessments. *J. Oper. Res. Soc.* **1993**, *44*, 1129–1144. [CrossRef]
70. Bevilacqua, M.; Braglia, M. Environmental efficiency analysis for ENI oil refineries. *J. Clean. Prod.* **2002**, *10*, 85–92. [CrossRef]
71. Huang, J.H.; Yang, X.G.; Cheng, G.; Wang, S.Y. A comprehensive eco-efficiency model and dynamics of regional eco-efficiency in China. *J. Clean. Prod.* **2014**, *67*, 228–238. [CrossRef]
72. Lauwers, L. Justifying the incorporation of the materials balance principle into frontier-based eco-efficiency models. *Ecol. Econ.* **2009**, *68*, 1605–1614. [CrossRef]
73. Coelli, T. A multi-stage methodology for the solution of orientated DEA models. *Oper. Res. Lett.* **1998**, *23*, 143–149. [CrossRef]
74. De Koeijer, T.J.; Wossink, G.A.A.; Struik, P.C.; Renkema, J.A. Measuring agricultural sustainability in terms of efficiency: The case of Dutch sugar beet growers. *J. Environ. Manag.* **2002**, *66*, 9–17. [CrossRef]
75. Dyckhoff, H.; Allen, K. Measuring ecological efficiency with data envelopment analysis (DEA). *Eur. J. Oper. Res.* **2001**, *132*, 312–325. [CrossRef]
76. Ezcurra, R.; Iraizoz, B.; Rapun, M. Regional efficiency in the European Union. *Eur. Plan. Stud.* **2008**, *16*, 1121–1143. [CrossRef]
77. Hernandez-Sancho, F.; Picazo-Tadeo, A.; Reig-Martinez, E. Efficiency and environmental regulation—An application to Spanish wooden goods and furnishings industry. *Environ. Resour. Econ.* **2000**, *15*, 365–378. [CrossRef]
78. Victor, D.; Agamuthu, P. Strategic environmental assessment policy integration model for solid waste management in Malaysia. *Environ. Sci. Policy* **2013**, *33*, 233–245. [CrossRef]
79. White, L.; Noble, B.F. Strategic environmental assessment for sustainability: A review of a decade of academic research. *Environ. Impact Assess. Rev.* **2013**, *42*, 60–66. [CrossRef]
80. Camagni, R. Territorial Impact Assessment for European regions: A methodological proposal and an application to EU transport policy. *Eval. Program Plan.* **2009**, *32*, 342–350. [CrossRef] [PubMed]
81. Golobic, M.; Marot, N. Territorial impact assessment: Integrating territorial aspects in sectoral policies. *Eval. Progr. Plan.* **2011**, *34*, 163–173. [CrossRef] [PubMed]
82. Alonso, W. The economics of urban size. *Pap. Reg. Sci.* **1971**, *26*, 67–83. [CrossRef]
83. Prud'homme, R.; Lee, C.W. Size, sprawl, speed and the efficiency of cities. *Urban Stud.* **1999**, *36*, 1849–1858. [CrossRef]
84. Berke, P.; Godschalk, D.R.; Kaiser, E.J.; Rodriguez, D. *Urban Land Use Planning*; University of Illinois Press: Chicago, IL, USA, 2006.
85. Knaap, G.; Talen, E. New urbanism and smart growth: A few words from the academy. *Int. Reg. Sci. Rev.* **2005**, *28*, 107–118. [CrossRef]
86. Awuah, K.G.B.; Hammond, F.N.; Lamond, J.E.; Booth, C. Benefits of urban land use planning in Ghana. *Geoforum* **2014**, *51*, 37–46. [CrossRef]
87. Chou, T.L.; Chang, J.Y. Urban sprawl and the politics of land use planning in urban Taiwan. *Int. Dev. Plan. Rev.* **2008**, *30*, 67–92. [CrossRef]
88. Folmer, E.; Risselada, A. planning the neighbourhood economy: Land-use plans and the economic potential of urban residential neighbourhoods in the Netherlands. *Eur. Plan. Stud.* **2013**, *21*, 1873–1894. [CrossRef]
89. McLaughlin, J. Urban Land Use Planning. *Land Use Policy* **2009**, *26*, 511–511. [CrossRef]

90. Brunsdon, C.; Fotheringham, A.S.; Charlton, M.E. Geographically weighted regression: A method for exploring spatial nonstationarity. *Geogr. Anal.* **1996**, *28*, 281–298. [CrossRef]
91. Cardozo, O.D.; Garcia-Palomares, J.C.; Gutierrez, J. Application of geographically weighted regression to the direct forecasting of transit ridership at station-level. *Appl. Geogr.* **2012**, *34*, 548–558. [CrossRef]
92. Fotheringham, A.S.; Brunsdon, C. Local forms of spatial analysis. *Geogr. Anal.* **1999**, *31*, 340–358. [CrossRef]
93. Li, T.; Corcoran, J.; Pullar, D.; Robson, A.; Stimson, R. A geographically weighted regression method to spatially disaggregate regional employment forecasts for South East Queensland. *Appl. Spat. Anal. Policy* **2009**, *2*, 147–175. [CrossRef]
94. Sheehan, K.; Strager, M.; Welsh, S. Advantages of geographically weighted regression for modeling benthic substrate in two greater Yellowstone ecosystem streams. *Environ. Model. Assess.* **2013**, *18*, 209–219. [CrossRef]
95. Fang, C.L.; Guan, X.L.; Lu, S.S.; Zhou, M.; Deng, Y. Input-output efficiency of urban agglomerations in China: An application of data envelopment analysis (DEA). *Urban Stud.* **2013**, *50*, 2766–2790. [CrossRef]
96. Murphy, E. Urban spatial location advantage: The dual of the transportation problem and its implications for land-use and transport planning. *Transp. Res. Pt. A* **2012**, *46*, 91–101. [CrossRef]

sustainability

MDPI

Review

Ecologizing Our Cities: A Particular, Process-Function View of Southern California, from within Complexity

Ashwani Vasishth

Ramapo College of New Jersey, 505 Ramapo Valley Road, Mahwah, NJ 07430, USA; vasishth@ramapo.edu; Tel.: +1-201-684-6616; Fax: +1-201-684-7257

Academic Editor: Tan Yigitcanlar

Received: 8 July 2015; Accepted: 17 August 2015; Published: 25 August 2015

Abstract: Cities, as the quintessential socio-technological artifacts of human civilization, are seen to set us apart from nature. But an ecosystem view from nested scale-hierarchical process-function ecology shows us that cities are best seen as the emergent and nodal end points of interactive flows of matter, energy and information. From within such a view, a clear need emerges to ecologize our cities by better integrating them back with nature. Arguing from such an ecosystem approach to depicting reality, this paper proposes that tracing the processes and functions which constitute the morphology of the city leads us to articulate an urban ecology that incorporates heat island mitigations, urban forestry, and ecological landscape management (taken both as the introduction of native vegetation and the insertion of increased proportions of pervious paving), all considered within the framework of an integrative ecosystem approach to land use planning. More importantly, such an approach to urban ecology is useful because, as a mode of intervention, it rests on—indeed, requires—an acknowledgement in ecological planning of the often amorphous and usually only indirectly sensible atmospheric, biogeochemical and hydrological processes and functions.

Keywords: ecosystem approach; urban ecology; eco-cities; process-function ecology; heat island mitigation; urban forestry; green infrastructure; ecological landscape management

---

1. Introduction

It is clear that we are squarely in the Age of the Anthropocene—an epoch in which humans are the dominant geophysical force shaping planetary processes. As Steffen *et al.* [1] state, "Human activities have become so pervasive and profound that they rival the great forces of nature and are pushing the Earth into planetary terra incognita. The Earth is rapidly moving into a less biologically diverse, less forested, much warmer, and probably wetter and stormier state".

If this is true, then surely cities need to be at the center of this concern. Over half of humanity is already urban, and it is projected that by 2050, 80% of humanity will be living in cities. What this means is that if we care about reducing our impact upon the planet, we need to lighten the weight of our cities and make their ecological footprint smaller, on a per capita basis.

Grimm *et al.* [2] and Pickett *et al.* [3] collectively make the point that it is useful to distinguish between ecology *in* cities and ecology *of* cities. The first refers to the study of fragments of nature within the urban context. The second refers to the study of the urban context itself as an ecological system. While this is an important distinction, pointing as it does to two quite diverse ways in which the idea of urban ecology can be taken, the concern here is more with ways in which cities themselves can be ecologized—that is to say, rather than trying to increase fragments of nature within urban areas, we need to be concerned with the pervasive replacement of built, grey infrastructure by natural, green infrastructure.

One significant way of doing this is by allowing nature to do what nature does well, and by not going against the flow of natural processes and functions. In this particular light, it makes sense to transform existing cities into Eco-cities, retrofitting the built environment and make sure that all new urban development itself abides by the principles of Eco-city design.

Although there is not yet an accepted international standard specifying the development of Eco-cities, certain principles and practices can be distilled from the literature [4–7]. Key amongst these is the principle of working with nature, rather than trying to dominate and control it.

Register [4] (p. 182), in establishing the eco-cities meme, suggests that, "(t)o begin, we might try the Ecological Golden Rule: do unto others—including plants, animals, and the Earth herself—as you would have others do unto you. Dividing the golden rule into two, we might embrace the social ecological commandments taught to every pre-kindergarten child: be nice to others and clean up after yourself. Refining this a bit further, we could say that there are three major environmental prescriptions into which most others fit: Conserve, recycle, and preserve biodiversity".

Taking this deeper, we might look at the Hannover Principles devised by William McDonough [5] for EXPO 2000, the World Fair at Hannover, Germany. Along with the chapter by Head and Lam [6], titled "How Cities Can Enter the Ecological Age", in which they conclude that our cities might ecologize themselves most effectively by: (a) reducing their carbon footprint; (b) reducing their ecological footprint; (c) improving their Human Development Index scores; and (d) increasing biodiversity. As well as de Jong *et al.* [7], in which they parse out the diverse set of labels given to eco-cities with a frequency analysis.

As it turns out, there is a distinction worth making between the ecological footprint of cities, which is large, and the per capita ecological footprint of urban dwellers, which is small compared to that of non-urban dwellers. The ecological footprint of a city, typically, is orders of magnitude larger than the area of the city itself. But the ecological footprint of a typical urban dweller is a fraction of the footprint of rural or suburban dwellers. What this means is that there is something to be said for living in cities, but still, cities need help with reducing their total ecological footprint.

With this as background, and with a foremost concern for reducing our urban footprints—ecological, carbon and socio-cultural—the questions we are presented with are these: How do cities most press themselves upon nature? What is it that cities do, which most generate a load upon the planet?

Table 1. Resolving the pressures cities put on nature.

If the problem with cities is that . . .	Then the solution is for cities to . . .
They contain huge amounts of impervious surfaces (roofs, roads, driveways, pavement)	Entrain storm-water into the ground, using porous pavement and vegetation and trees and cisterns and roof gardens
They import huge quantities of energy (fossil fuel, electricity)	Increase energy efficiency and conservation Use distributed energy generation and renewable energy
They import huge quantities of nutrients (food)	Grow more food in cities
They import huge quantities of raw materials	Use less stuff (Sustainable Production and Consumption, SCP)
They export huge amounts of waste matter	Divert more solid waste from landfills (increase recycling, composting, and reuse of materials)
They are islands of heat (2–4 degrees C hotter than the countryside)	Heat Island Mitigation (cool roofs, green roofs, trees)
They consume nature (farmland, open space, parks, wetlands)	Create more nature within cities (urban farms, open space, parks, bioswales, wetlands)

It may be worthwhile to consider that, in one sense, there are three urban sectors that put the most load on nature: Buildings, Transportation, and Electricity [8] (p. 2). Here the emission of Greenhouse Gases (GHGs) is taken to be a proxy for "load on nature". In one sense, the carbon footprint of human activity is a useful way of characterizing natural pressures and stresses. But, clearly, this is only one piece of the global puzzle [9,10].

In the case of these three sectors—Buildings, Transportation, and Electricity—there are a few well-known ways of mitigating this load—green building and energy standards; mixed land-use development; transit-oriented development; distributed generation of electricity; and the societal promotion of renewable energy options.

Of course, the key to ecologizing cities is to let nature do what nature does well, and to do everything else in ways that put the least pressure on natural processes and functions. This last is central to the successful ecologization of our cities—that is to say, we must first learn to see our cities as physical entities emergent from the flows of matter, energy and information. Then we understand that the way to managing our cities is less by morphological manipulation and more by managing the flows of its constitutive processes and functions. Pruning and culling is necessary and useful, in its place. But Bonsai gives us a head-start to low-impact control over morphological growth. In that light, the sorts of solutions that rise to the surface then become the subject of our attention.

2. Setting a Glocal Context

There is a clear degree of urgency in this call for the ecologization of cities across the world. Over the next two decades, mid-sized cities everywhere are expected to see a marked surge in population, and the mega-cities of the world are likely to become even more unwieldy. Much of this growth is going to occur across the Asia-Pacific region, and most of it will be driven by rural-to-urban migration, rather than by internal population growth [11]. But we can be sure that cities everywhere are going to come under even more pressure to refresh, enhance and expand their infrastructure.

The question is, what form will this infrastructure expansion take? Crudely, there is a binary choice between conventional "grey" infrastructure and innovative "green", "blue" and ephemeral (digital) infrastructure. We can either try to build our way out of the problem of increasing urbanization, or we can, more intelligently, proceed to ecologize our way into a solution. Given the extent to which the cities of the world are already loading the planet, and given the massive ecological footprint of humanity, at large, it should be clear that further increasing our footprint is, quite simply, not an option. Then one root articulation of the situation is: how can we accommodate a growing urban population even as we shrink our global urban footprint.

We need, indeed, a way of *using* nature to facilitate the entirely foreseeable surge in urbanization, since, clearly, abusing nature is not giving us the sorts of outcomes we actually want. We need to grow our cities capacity to support a high quality of human life even as we quite dramatically shrink our urban footprint. And we need to do this very rapidly, and without hesitation. Here, fortunately, the two countervailing forces of Globalization and Localization can be made to work in our favor.

2.1. The Homogenizing Forces of Globalization

Even as vast populations across the world are urbanizing, just so are these populations—and their cities—westernizing. New Delhi begins to approach Taipei, which in its turn reaches out toward Tokyo. But, at the same time, though much more painfully, there is a south-to-north flow as well, with Curitiba landing memetic tentacles in Los Angeles, and thence to—could it be?—Singapore, as the idea of bus-rapid-transit (BRT) becomes increasingly contagious. Congestion pricing crystalizes first in Singapore, moving from there to Europe and the USA.

Anything we can do to nurture the free flow of "smart" ideas across the globalizing landscapes of the planet might well work in our favor. Without in any way seeking to suggest that organic foods are going to "save the day" for us, the fact of the matter is, "green" ideas are emerging far more rapidly in the surging countries of the south than they ever did in the entrenched cities of the north.

The cultural speed with which ideas spread is cause for hope. But at the same time the flow of modernization ideals from north to south is disturbing, undercutting as it does, local values and customs. Ideas *can* spread very rapidly. The question is, which ideas are deemed spreadable and which ideas do, indeed, diffuse.

2.2. The Hetrogenizing Forces of Localization

Ideas, like plants, tend to hybridize as they are transplanted out of their native context. And so they seek out novel contexts within which to flourish. Take, for example, the case of biogas digester technology. If we take the 1970s and the early 1980s to represent the first wave of appropriate technology—driven in no small part by E.F. Schumacher's 1973 opus [12], *Small is Beautiful: Economics As If People Mattered*—and if we take that first wave to have broken and dissipated itself, spent against the forces of an inexorably Westernizing world, then surely this decade is a time when the very same ideas of an appropriate technology can be revitalized.

In that first wave, and as one example, the Government of India expended significant resources in the development of technologies such as biogas plants and solar cookers. These technologies died at birth, by and large. Today, however, with localization more effectively beginning to balance out globalization, solar water heaters are endemic in at least some cities of the south. Similarly, new generations of biogas digesters are showing themselves to be entirely adaptable to the high-stakes game of contemporary urbanization.

2.3. The Integrative Forces of Glocalization

One of the reasons that this second wave of appropriate technology might stand a chance to help us innovate our way out of a dire predicament, is because the artifact of such technologies, though largely unchanged, is riding on the coat-tails of entirely different processes and functions. As we realize the strict limits of large, centralized, infrastructure solutions—power plants, sewage treatment plants, storm-water processing facilities, and other urban utilities—we become more aware of the power of decentralized, distributed solutions. Small-scale solar photovoltaic systems may offer tightly localized solutions to the highly global problem of energy poverty. Modular biogas digesters begin to solve the organic and sewage waste problems of an urbanization process that simply cannot keep pace with current growth rates.

In this spirit of innovation under "Glocalization", it may be entirely possible to shrink our burgeoning urban footprint even as we scramble to accommodate a surging urban population, provided we actually deploy "smart", amorphous, green and decentralized technologies that are chosen, strategically, for their ability to fill functional niches, in highly localized conditions but at a globalized scale.

3. Mitigation Measures

3.1. Managing Ecological Processes and Functions as a Way of Reintegrating Cities with Nature

Dark, heat-absorbing, impervious surfaces—roofs, roads, and parking lots—are one iconic hallmark of urbanization. Such surfaces, often unmitigated, have a range of significant and cumulative adverse effects on the ecological and biogeochemical processes and functions that underwrite our cities, and so shape our inhabited world. Conventional building practices result in increased ambient temperatures due to the proliferation of heat-absorbing surfaces, increased urban storm-water runoff, reduced groundwater recharge, disruptions of local landscape ecologies, fragmentation of natural habitats, increased air pollution, increased water pollution, increased biological and mechanical heat stress, and exacerbate as well the separation of humans from nature.

This diverse range of changes to processes and functions can partly be captured by the concepts of urban heat islands, urban forestry, and xeriscape. We can dramatically change how our cities work, and how they sit in nature, by paying conscious attention to these ecological phenomena in land use planning. Although these various effects, as well as the measures that can effectively mitigate them have been known for some time, the ways in which we choose to plan and build have just barely begun to take these factors into account—perhaps because we continue to treat the urban world as mechanical, rather than embracing its essentially organic basis.

Such dark, under-shaded surfaces absorb in-coming solar radiation and then re-radiate this heat into the lower atmosphere, raising localized temperatures, often by 2.5 to 5.0 degrees Celsius. This increase in ambient temperatures usually results in greater expenditures of energy for cooling the structures we inhabit, particularly in the mid- to low-latitudes and in the summer afternoons, when energy demand is often at its highest.

Mitigation measures—the use of lighter colored and heat reflecting surfaces for roofs and paving, as well as the increased planting of ecologically suitable species of trees and vegetation—are capable of reducing ambient temperatures by 2.0 to 4.0 degrees Celsius. This reduction is achieved partly by physically altering the heat-absorbing properties of surfaces, partly by increasing localized cooling due to evaporative transpiration from plant and soil systems, and partly by morphologically inserting shade into the urban landscape, thus reducing energy consumption in the summer and in the afternoons when energy demand is highest.

It should be mentioned here that there is a converse "winter penalty" that is incurred, in some cases, by the wide-spread application of these heat island mitigation measures, in that the cost of heating buildings in the winter would be increased somewhat. But Rosenfeld *et al.* [13] (p. 54) find that this is a small penalty, and the cumulative summer-time benefits of reducing air conditioning costs by far outweigh the winter penalty. Elsewhere, Rosenfeld *et al.* [14] (p. 57) note that this net energy saving applies as far north as New York City, explaining that, in all mid-latitude locations, winter sun is lower in the sky, and thus the ratio of sunlight striking the roof to the walls is also smaller. In addition, winter days are shorter, and so they suggest that the summer benefits of lighter colored roofs may substantially outweigh their winter penalty.

Taken together, these mitigation measures have a number of other quite substantial benefits as well. Tropospheric ozone formation is a temperature sensitive photochemical reaction, in which precursor gases—volatile organic compounds (VOCs) and oxides of nitrogen (NO_x)—react in the presence of sunlight to form smog. This reaction is temperature-sensitive. Thus, reductions in urban ambient temperatures carry the potential of reducing smog-formation, without physically reducing the volume of precursor gases exhausted into the lower atmosphere.

The extensive planting of ecologically suitable species of trees and shrubbery, besides increasing morphological shading and enhancing the locally cooling processes of evaporative transpiration in soil-plant systems, also greatly increase the surfaces available to capture ambient particulate matter (dust) generated by traffic and by urban activity, thus potentially benefiting respiratory health. In addition, the vegetation of the urban landscape increases the proportion of pervious to impervious surfaces, which, in turn, reduces storm-water run-off even as it increases ground-water recharge. A variety of habitat-enhancing ecological and community effects can also be ascribed to the increased native vegetation resulting from such measures. Not incidentally, these reductions in temperature also reduce the often considerable thermal stress on roofing and paving materials, measurably increasing their effective life span and reducing maintenance costs [15] (p. 1).

We have known about these processes and phenomena for some time, but the shape of how we plan and build has only just begun to take these factors into account in transformative ways. No doubt this lag in adoptive action is shaped most by disciplinary fragmentation in research, and by the professional segmentation of environmental planning into functional and siloed typologies such as land use planning, air quality planning, water quality planning, storm-water management, urban forestry, and so on. But it may as well be the case that conventional descriptions of the world are traditionally biased toward the morphological—in that, it is easier to mobilize action against pollution processes that are directly sensible to us, based on sight and smell, and harder to do so against pollution processes that can only be indirectly measured, using instruments and models, such as climate change.

3.2. *Albedo Modification, Vegetation and Urban Forestry as Heat Island Mitigation*

Impervious surfaces are a hallmark of urbanization. Vitousek [16] argues that land use and land cover change, taken together, are one of the three most significant global change processes that ecologists must take into account. From within an ecological perspective, roofs, roads and paving are perhaps the single most critical factor that set cities apart from the countryside [17–23]. The consequences of such a concentration of impervious surfaces, usually in the form of dark asphalt and roofing materials, extend to influencing the local climate and the local hydrology in varying degrees, depending upon the particulars of locational and ecological context. Taha [24] (p. 99) notes that "northern hemisphere urban areas annually have an average of 12% less solar radiation, 8% more clouds, 14% more rainfall, 10% more snowfall, and 15% more thunderstorms than their rural counterparts".

However, urban heat islands, like most ecological phenomena, are not a singularity. In general, urban areas are 2.5 to 5.0 degrees Celsius warmer than their surrounding countryside. Depending upon latitude, the surrounding ecology, and meso-scale climate, a heat island effect may show itself most either in the summer or in the winter, during the day or at night, and cause increases in heating, smog formation or rainfall. In the higher latitudes, urban heat islands may most markedly increase temperatures in the winter, thus reducing building heating costs. In lower latitudes, the effect may be most pronounced in the summer months, resulting in higher air-conditioning costs and increased smog formation. In coastal arid climates such as Los Angeles, the heat island effect may be most relevant in the afternoons, causing increased smog formation and energy consumption. Along the more humid Atlantic seaboard, heat islands may generate increased rainfall and thunderstorms [25–28]. While in desert locations, and with depressed topographies such as Phoenix, the effect may most show itself most at night, keeping the urban core hotter for hours after the sun has set [29,30] and thus increasing energy consumption long into the otherwise-cooler nights.

In Southern California, in the case of the urbanization in the region surrounding Los Angeles, the Mediterranean climate is influenced by its coastal location, juxtaposed with an inland desert ecology, and capped by a tropospheric inversion layer that tends to trap smog-forming precursor gases—namely, volatile organic compounds and oxides of nitrogen. In such a case, and broadly speaking, the urban heat island effect is most markedly manifest at about 2 p.m. in the afternoon, when increased ambient temperatures most severely affect peak demands for electricity, and when the temperature-sensitive photochemical smog-forming reactions most manifest their pollution effects.

Here, three sorts of strategies are available to mitigate the heat island effect. We can physically increase the albedo, or heat reflecting properties, of sunward oriented surfaces such as roofs, roads and paving, by using lighter colored or otherwise more heat-reflecting materials. We can increase the proportion of vegetation and shrubbery to hard landscapes, and promote the adoption of roof-top gardens or green roofs, thus increasing opportunities for the plant-soil based processes of evapo-transpirative cooling to find play. And we can use urban forestry programs to extensively plant strategically sited and ecologically suitable tree species throughout the urbanized area, increasing both evapo-transpiration and physical shading.

Rosenfeld *et al.* [13] describe a "cool communities" strategy for the inland urbanized Los Angeles area, in which they assess the energy conservation and tropospheric ozone (smog) air pollution reduction benefits of a two-pronged strategy that focuses on increasing the albedo of roofing and paving materials by an average of 0.30, and on the strategic planting of 11 million trees in the more densely inhabited parts of the region. Their analysis shows a 12% reduction in the number of days per year on which tropospheric ozone exceeds the National Ambient Air Quality Standards (NAAQS), and a 10% reduction in air-conditioning loads during peak early afternoon demand. They found that, at peak temperatures, around 2 p.m., an approximately 2.5 to 3.0 degrees Celsius reduction in ambient temperatures would be effected by their "cool communities" strategy.

Their research concludes that the proposed albedo modification component and the tree planting component of their "cool communities" strategy generate roughly equal amounts of ambient cooling in the lower atmosphere of the Los Angeles urbanized area. That is to say, if about one-third of the rooftops within the region, and if the paved surfaces concentrated within 25% of the inland urbanized area, were treated so as to increase the albedo of treated roofs by about 0.35 and the albedo of modified paving by about 0.25, this would generate an average increase in the albedo of sunward oriented surfaces in the order of about 0.30. And if, in addition, about 11 million ecologically suitable species of trees were to be planted strategically across the region, then about half the cooling in ambient temperatures would be attributable to each of these two strategies. "The cooling for 'albedo only' turns out to be equal to that of 'trees only,' and is additive" [13] (p. 53).

Estimating smog reduction benefits on the basis of the reduction in the number of days in the year that smog concentrations exceed the California ambient air quality standard of 90 parts per billion by volume (ppbv), their simulation shows that the combined benefits of the tree planting and albedo modification strategies result in a 12% reduction in the number of days in a year on which the air quality standards for tropospheric ozone are exceeded. "In apportioning how much of the benefits we calculated could be attributed to the three separate strategies (trees, roofs, and pavements), we found 50% of the temperature decrease (and thus 50% of the smog reduction) arises from tree planting. The remaining 50% was proportionally attributed to albedo changes resulting from light-colored roofs (0.35) and pavements (0.25), which translates to 29% of the benefits from light-colored roofs and 21% from light-colored pavements" [13] (p. 53–54).

Smog, or tropospheric ozone, is not a directly emitted pollutant, but rather is the product of a complex reaction involving two sets of precursor gases—oxides of Nitrogen (NO_X) and volatile organic compounds (VOC)—in the presence of sunlight. The photochemical reaction may be either NO_X-constrained or VOC-constrained, depending on the relative proportion of the gases present in the troposphere. In the case of Southern California, Rosenfeld *et al.* take the reaction to be NO_X-constrained. In assessing the smog-reduction benefits of their proposed heat island mitigation measures of shade tree planting and a change to lighter colored paving and roof surfaces, they consider two components in the reduction of NO_X gases—the direct reductions in NO_X emissions by power plants, due to reductions in peak-time electric power consumption, and the effective or "equivalent" reductions in NO_X, due to reductions in ambient temperatures.

In the base case for Southern California, they assume that 1225 t of NO_X and 1350 t of VOCs are present and available to the photochemical smog-formation reaction by the early afternoon peak reaction time. They find that the reductions in electricity consumption result in a small reduction in NO_X emissions by power plants, in the order of 6.35 t, or a direct reduction of 0.5% in NO_X. However, as they point out, "(r)educing smog by citywide cooling can be considered equivalent to reducing the formation of smog precursors at constant temperatures". Relying on research by Taha [31,32], Rosenfeld *et al.* conclude that the two strategies of shade trees and lighter colored or higher albedo surfaces, together result in a 10% reduction in smog. They conclude that this 10% reduction in smog is equivalent to a 25% reduction in precursor gases, with the tropospheric system behaving as though there had been a 317 t reduction in NO_X emissions within the air basin.

Albedo modification strategies, cool roofs and cool paving interventions that cumulatively increase regional albedo from 0.25 to 0.40, have been modeled to effectively reduce localized ambient temperatures by as much as 4.0 degrees Celsius in Southern California's mid-latitude climate [24] (p. 101). Taha concludes, "temperature decreases of this magnitude could reduce the electricity load from air conditioning by 10% and smog (ozone concentrations) by up to 20% during hot summer days". Elsewhere, Taha [32] (p. 1668) has found that the average albedo for sunward oriented land surfaces in Southern California is 0.14, and has concluded that the theoretical "maximum increase in albedo will probably never exceed 0.30", and that this should be established as the extreme upper bound for modeling purposes, while an albedo increase of 0.15 for sunward oriented surfaces is a reasonable moderate increase.

The results of Taha's simulation of such changes in albedo, for a clear and warm day in August, at 3 p.m., indicate that the urban core might see a decrease in temperature of about 1.5 to 2.0 degrees

Celsius in the case of moderate (0.15) increase in albedo, and up to 4.0 degrees Celsius in the case of an extreme (0.30) increase in albedo, with outlying areas showing a more modest decrease of about 1.0 and 2.0 degrees Celsius [32] (p. 1670). The estimated effect of such a temperature reduction on tropospheric ozone formation was considered to account for "(1) a decrease in some photochemical reaction rates; (2) a decrease in temperature-dependent biogenic hydrocarbon emissions; (3) a decrease in evaporative losses of organic compounds from mobile and stationary sources; and (4) a decreased need for cooling energy, generating capacity, and, thus, emissions from power plants" [32] (p. 1667).

3.3. *Changing the Albedo of Roofing and Paving Materials*

Vernacular architecture, in a cross-cultural context, is defined as the traditional, native, locally prevalent mode of building, using locally available materials and construction techniques, and based on a traditional and historically tested knowledge-base. Many "traditional", and hence by implication "primitive", modes of knowing may actually be more effective than modern-day beliefs and practices. Take, for instance, the traditional architectures of places that fall within desert climates. In most cases, structures in such places are regularly white-washed, including rooftops. For instance, "building owners in hot cities like Haifa and Tel Aviv are required to whitewash their roofs each spring, after the rains stop" [14] (p. 55). Modern day building practices are driven far more by the contemporary economics of air conditioning, which routinely fail to internalize many of the costs of not using such traditional building techniques.

An ecological approach to building would require attention to such knowledge processes. One key insight from process-function ecology is that direct human sensory perception is at best a limited means of "getting at" the processes and functions that actually shape our world. Conventional empiricism, being based on a reliance on our senses of sight, smell, hearing, taste and touch, has only limited value in an ecosystem approach. Processes and functions outside the scope of our senses drive many of the phenomena that matter most to us.

Albedo is one such phenomenon. In general, and very incompletely, the gradient from light to dark colors does approximate the gradient from high to low albedo—that is to say, from highly heat reflecting properties to highly heat absorbing properties. But a substantial part of the heating that occurs due to incoming solar radiation is in the near-infrared range of the spectrum, and so hidden from our direct sensory abilities. This explains why, for instance, "dark" terracotta roofing tiles may be measurably cooler than "white" asphalt-fiberglass shingles [14] (p. 57), and why old "white" shingles may be more heat reflective (by up to 10 degrees Celsius) than modern "white" shingles, which use one-sixth the thickness of white pigment than they did in 1960 [14] (p. 55).

What this means, of course, is that we are not strictly constrained to the aesthetic of "white", in our urban landscapes. The use of, for instance, titanium dioxide (TiO_2) as an additive to paints used to coat roof surfaces, allows us to apply a range of pastel shades which still have the high albedo properties in which we are most interested. Recent developments in building materials, particularly some very interesting contemporary research about the dirt-repelling properties of TiO_2-coated materials, for instance, raises interesting prospects for longer-lasting albedo-increasing effects in a variety of building materials [33,34]. Another facet of such an albedo-modification approach would focus on roads and pavements, where direct experiments show substantial heat reduction benefits as well.

3.4. *Tree Planting and Vegetation Change as Integrative Regional Environmental Interventions*

Landscape level land use change is one of the most significant ways in which we shape, and by which we can reshape, our lived environments. The displacement of native vegetative cover, first by small-scale agriculture, then by the more extensive irrigated agricultural systems that mark our recent industrializing history, resulted in a host of ecological changes upon the land.

Just as one example, Southern California saw a significant decreasing trend in ambient temperatures as large-scale agriculture and orchard cultivation took hold at the turn of the previous century, with yearly high temperatures dropping almost as low as 35 degrees Celsius by about 1930.

Then, urbanization became the ecologically dominant force in land cover and land use change, and the yearly high temperatures began a fairly steady increase, which has continued into the present [14] (p. 56).

The insertion of ecologically appropriate species of trees and vegetative cover into the urban fabric can be at least as powerful a transformation of the ecosystem processes and functions that support the city, as was their displacement by impervious surfaces. In the particular context of urban heat island mitigation, the most obvious way in which trees help is by physically interjecting shade into our built landscape, thus reducing the heat loads on the walls and immediate surroundings of our urban environment. Shade alone may provide a significant reduction in heat flux, reducing the amount of heat transferred through walls and roofs into the interior spaces by as much as 16 to 27 degrees Celsius, and thus directly reducing the amount of cooling work needed to be done by our air-conditioning systems.

3.4.1. Soil-Vegetation Evaporation and Transpiration as Cooling Processes

However, there is a subtler, though at least as effective, process of cooling that is a by-product of tree and plant growth. Vegetation draws up water from the soil below, through its root structures, and some of this water is released in the form of moisture by the foliage (transpiration) and by the soil itself (evaporation), so cooling the lower atmosphere. The soil-vegetation complex acts to enhance this natural process of evaporative transpiration, or evapo-transpiration. This process can be a major influence in micro-climate cooling, as walking under a broad, leafy tree on any hot, dry summer afternoon will directly demonstrate. Evapo-transpiration processes can generate estimated reductions in local ambient temperatures of 5.0 to 7.5 degrees Celsius, on a typical summer afternoon [35–37]. This cooling effect is more pronounced in dry, semi-arid climates such as Southern California.

3.4.2. Green Roofs for Heat Insulation and Storm-Water Retention

A different, but equally effective and promising strategy is the widespread introduction of what are coming to be called "green roofs", or roof-top gardens. As Oberndorfer *et al.* [38] (p. 823) point out, green roofs provide multiple ecosystem services in urban ecosystems, "including improved storm-water management, better regulation of building temperatures, reduced urban heat island effects, and increased urban wildlife habitat".

Both through extensive experimentation and through materials innovation, green roofs are now poised to significantly help restore nature and natural processes back into the built urban environment. Broadly speaking, there are two sorts of green roofs—extensive and intensive. Extensive green roofs are usually thin layers of vegetative growing media, typically six inches or less, spread over large expanses of roofing, with some suitably durable and hardy species of ground cover, such as one of the many varieties of sedum. "The challenge in designing extensive green roofs is to replicate many of the benefits of green open space, while keeping them light in weight and affordable. Thus, the new generation of green roofs relies on a marriage of the sciences of horticulture, waterproofing, and engineering" [39].

Green roofs have evolved, in recent years, from being thought of as an additional burden to be placed on roof structures to being seen now as an additional protective covering that helps shield the waterproofing membranes of conventional flat or very low slope roofs from heat stress. Experimental tests seem to indicate that well-designed and properly constructed extensive green roofs may help extend the life of the waterproofing membrane and of the roof structure itself, even as they insulate the enclosed spaces from the worst ravages of the summer sun [15].

As a heat island mitigation strategy, green roofs are different from albedo modification and urban forestry in that their primary functional action is to physically insulate the roof membrane. Certainly the albedo of such green roofs is likely to be higher than that of conventional (particularly normal asphalt) shingles. But, when compared to the albedo of most materials normally used for their heat-reflective properties (titanium-dioxide treated white shingles or some of the more contemporary membrane materials), the benefits are likely to be nominal. There is certainly an evapo-transpirative effect, but since it plays out in the rather narrow zone immediately above the ground cover, its heat-reducing actions, either locally or regionally, are again likely to be nominal at best.

However, extensive green roofs do have one additional advantage, in that they can be designed to deliver, at little increase in cost and performance, virtually any desired level of storm-water retention. A 50% reduction in runoff is almost the default setting, and additional gains are easily made. Designers across the world have worked very extensively with green roofs, and case studies are available across a very wide range of siting conditions and using different technologies, making comparative analysis possible. Most researchers who have worked with green roof technologies seem to be clear that these technologies, with some little care and attention in execution, are consistently reliable and do, indeed, deliver the range of benefits that theoretical calculations suggest.

3.4.3. Green Façades and Living Wall Systems for Heat Island Mitigation and Air Pollution Control

Extending the discussion of green roofs to the remaining skin of the building envelope, it is worth noting recent developments in our understanding of green facades and living wall systems [40,41]. Essentially, vertical panes of vegetation are used to envelope either exterior or interior walls of buildings. These provide multiple benefits—reducing energy consumption by improving the thermal performance of the building, mitigating the heat island effect, mitigating noise pollution, improving indoor air quality, improving health and well-being, and more generally, enhancing urban biodiversity [40] (p. 2).

There are a number of ways in which green façade and living wall systems can be implemented. Pre-planted panels can be attached structurally to the wall, with an integrated irrigation system. Alternatively, felt pockets with growing medium can be attached against a waterproof membrane, with nutrient-laced fluids being used to keep the system moist at all times. A third alternative involves the use of planter boxes and a system of trellis-work. Such systems can be used on both eternal walls and interior vertical surfaces. In the latter case, it is not uncommon to link the living wall with the air conditioning and circulation system of the building, to capture the air purification and humidification benefits of the vegetated system.

While the aesthetic, air quality and noise pollution mitigation benefits of such systems are quite clear, it is not at all obvious that—at a systems level—green façades and living wall systems are economically viable. But research has started to emerge that seeks to establish the comparative life cycle analysis and the cost-benefit analysis of such vertical vegetation systems [41,42].

3.4.4. Urban Forestry and Landscape Ecology in Air Pollution Mitigation

Heat island mitigation measures that include strategic and intensive tree planting can cumulatively reduce local ambient temperatures by between 2.0 to 4.0 degrees Celsius. As discussed earlier, this reduction in local temperatures can potentially reduce the formation of tropospheric ozone (smog) by up to 20%.

An additional and not insignificant benefit to urban ecology derives in the case of Southern California, from the implementation of tree planting ordinances for downtown surface parking lots and car dealerships. This is particularly salient in the case of Los Angeles County, where little effort is currently made to implement or enforce any such minimum tree cover measure, and acres of cars can be seen sitting baking in the sun all day. A 50% tree cover ordinance would go a long way to mitigating the range of adverse environmental impacts from these typically treeless expanses of impervious surfaces [43–47]. Not only are there measurable benefits to be realized from the reductions in evaporative emissions from such parked vehicles, but substantial storm-water and ground water benefits would accrue as well, both in terms of storm-water mitigation and in terms of ground water recharge. This is especially true if tree-planting ordinances are combined with land-cover management techniques such as the use of porous pavement and pervious concrete, implemented in appropriate ways [19,48–50].

These reductions in local ambient temperature have the additional benefit of decreasing the need for air conditioning during peak demand periods—that is to say, in the summer and in the mid-afternoon. This decrease reduces the region's need for cooling energy, particularly in the residential context, as Rosenfeld *et al.* [13] and Taha [32] point out, in turn reducing the demand for electricity generating capacity, and so indirectly reducing emissions from power plants. Of course, power plants supplying electricity to a particular region, such as Southern California, may or may not be located in that region. And nuclear power plants are also an exception to this case. But, in most instances, some air pollution benefits can be expected to accrue from this reduced demand for air conditioning energy. Beside toxic ozone-precursor emission reductions, a substantial abatement of greenhouse gas emissions can also be attributed to such heat island and urban forestry sorts of interventions.

An additional and related air pollution control benefit accruing directly from increased use of ecologically appropriate species of trees and vegetation is the capture and sequestration of carbon dioxide (CO_2), a significant greenhouse gas, through the natural process of photosynthesis. Rosenfeld *et al.* [13] (p. 57) suggest that urban trees may provide three times the CO_2 reduction benefits than the same trees planted in forests or in non-urban areas. This reduction occurs because, in urban environments and besides the direct sequestration of carbon into the biomass through photosynthesis (which might be in the order of about 5 kilograms of carbon), these urban trees may also reduce energy consumption for air conditioning if they are appropriately sited so as to provide direct shading to buildings, by as much as 15 kilograms each year. Nowak and Crane [51] (p. 387) estimate that urban trees, through a combination of direct carbon sequestration and carbon dioxide emission avoidance, may provide four times the GHG reduction benefits of the same tree planted in a forest stand. As such, projects that seek to implement tree planting as a net carbon sequestration strategy should consider prioritizing the planting of trees in urban environments, particularly in cases where these trees might directly and indirectly shade air conditioned buildings, as their return on investment will be much higher than if they were to fund similar projects in forest or rural areas.

Besides direct local shading and local cooling through evapo-transpiration, another local air quality benefit accrues from the ability of leafy trees to trap fine and ultra-fine particulate matter onto their leaf surfaces. The dense planting of otherwise low-biogenic emission tree species [52–54] downwind of dust pollution sources such as traffic corridors with high volumes of, for instance, truck traffic, would substantially reduce human and ecological exposures to toxic exhaust gases in strategically identified "hot spots", generating potentially substantial environmental health benefits.

3.5. Impervious Surface Management and Landscape Ecology for Storm-Water Retention and Groundwater Recharge

The extent to which our cities are marked by the spread of impervious surfaces is a powerful indicator of our ecological footprint, and of the weight of our tread upon the land. A variety of strategies are available to us to mitigate the ecological impacts of roofs, roads and paving. We can insert trees and vegetation into our urban landscapes far more copiously than is our current practice, taking care to choose species of trees, shrubs and ground cover vegetation that are well adapted to local ecosystem conditions. We can advocate strongly for the conversion of conventional urban and suburban lawns, which intensively use irrigation and chemicals, to xeriscape sorts of plants and vegetation. We can begin to popularize the use of green roofs, which will do quite well in Southern California with thoughtful design and appropriate selection of cover species, such as the wide variety of sedum, which are able to thrive with little on-going maintenance.

A variety of porous materials are also available for parking lots and for paving, which, when combined with rainwater harvesting technologies, can substantially increase groundwater recharge even as they dramatically reduce storm-water run-off [19,48–50].

4. Implications

4.1. Letting Nature Back into Our Cities, Using Ecological Processes, Functions and Landscape Management

Conventional building practices result in increased ambient temperatures due to the proliferation of heat-absorbing surfaces, reduced groundwater recharge and increased urban storm-water runoff. Constructing tree-less parking lots, placing non-native vegetation in ornamental gardens and synthetically maintained lawns, result in a patchwork appropriation of land uses, increased air and water pollution, more biological and material heat stress, and ultimately, the deepening separation of humans from nature.

Rather than using locally appropriate building materials and climatically adapted dwelling types, we choose instead to capitalize on what seem, in the short term and in some narrowly defined way, like the clear economic benefits of mass-production mass-culture. Of course, we must then compensate for the ecological consequences of such narrowly constructed choices through the increased use of air conditioning and heating, single-occupancy automotive transportation, and the ever-greater importation of water and electricity. And so it is, that by denying ecology, we come to live more heavily upon the land.

Fortunately, it need not be so. We can let nature back into our cities, using intelligence and trees and native vegetation to lighten our tread. These strategies from urban ecology can, together, provide many of the infrastructure benefits our contemporary society needs. Heat island mitigations, urban forestry, and impervious surface management can drastically reduce air and water pollution, significantly increase our natural water supply, substantially strengthen the connectivity of the rich and diverse habitats within which we dwell, and at the same time considerably mitigate that massive transfer of below-ground carbon into the atmosphere due to our civilization's reliance on fossil fuels.

Heat island mitigation measures use lighter colored and heat reflecting building and paving materials for sunward oriented surfaces, to reduce peak afternoon loads on our electricity supply infrastructure, and to substantially extend the life of the building materials themselves, by reducing heat stress. Urban forestry uses ecologically appropriate species of trees and shrubs strategically planted to shade our buildings, to cool the air through the entirely natural processes of evaporative transpiration, to capture dust particles upon their copious leaf surfaces, to capture and store rainwater, and penetrate the soils to increase groundwater recharge. Impervious surface management would use innovative and by now well-tested materials technologies to make our downtown parking lots more porous, while deploying drought-resistant xeriscape plants which naturally need less water to grow across our lawns and gardens. Together, and cumulatively, these green infrastructure measures would reduce our ecological footprint, and at the same time increase the effective carrying capacity of land.

The key elements to such an ecosystem approach to ecological planning require: that we give due consideration to the sometimes intangible processes and functions that drive occurrence in reality; that we conceptualize complex systems as being organized into nested levels; that we give attention to the power of multiple spatial, temporal and organizational scales to reveal different relevant aspects of reality; and that we properly select multiple depictive boundaries that simultaneously respect the ecological elements of structure, pattern and process [55,56].

4.2. Pulling It All Together: Humans as Components of Ecosystems

Integrating our cities and urban regions back into nature is an objective we should take seriously, both because it reduces adverse environmental impacts, thus reducing pollution treatment and remediation costs, and because such a strategy, if based on research-based knowledge derived from contemporary ecosystem ecology, landscape ecology, and urban ecology, would significantly reduce the ecological footprint of human habitation, thus effectively improving how we interact with planetary carrying capacity. Together, these potential benefits provide a sound and savvy science-based approach to contemporary regional sustainability planning.

Urbanizing habitat conservation planning, by percolating ecologically appropriate landscape elements back into the city, would move regions such as Southern California, with their high incidence of pressured, at-risk, threatened and endangered species, away from a reactive "crisis management" approach to a more sustainable, nurturing and proactive approach to integrating our cities back into nature.

The approach advocating for the adoption of albedo-modifying heat island mitigation measures, when combined with urban forestry, impervious surface management, and xeriscape sorts of ecologically appropriate vegetation, represents one example of innovative connections that wait to be made across conventionally disparate and insular sub-disciplines making up the planning structures of regional governance [55].

Taking a landscape ecology approach to regional land cover management in urbanizing areas would, in itself, strengthen habitat integrity, reduce pressures on nature conservation planning, reduce energy consumption, improve air quality, reduce storm-water runoff, reduce urban runoff pollution, enhance groundwater recharge, enhance community livability, allow the inculcation of a cultural connectivity with ecological processes and functions, and so, quite directly, with nature. Arguably, acting reflectively and self-consciously to take account of such usually ignored processes and functions would also strengthen the robustness of the way everyday planning and decision making happens at the community and city level.

And, at the very least, such an integrative approach to regional planning would foster synergistic support across conventional planning disciplines, with transportation planners, air quality planners, water quality and supply planners, urban foresters, land use planners, habitat conservation planners, community development planners, natural resource planners, energy planners, and so on, both providing support to, and receiving support from, one another.

A quite specific sort of ecosystem approach based on nested scale hierarchic or process-function ecosystem ecology does, in very pragmatic ways, provide us with the tools to create richly informative descriptions of otherwise complex spaces. The "dilemmas in a general theory of planning" posited by Rittel and Webber in their classic characterization of complex systems as "wicked problems", are indeed amenable to planning [57]. But only if we are astute enough to recognize their assertion that the tools from "tame" problem planning cannot be applied, in and of themselves, to complex systems. And then we need to see that there are indeed ways in which we can engage these complex systems in meaningful conversations that generate outcomes more desirable to the greater good and across levels of organization [55].

We need to grow from a merely reactive and mechanistic problem-solving approach that attempts to singularize issues to make them easier for us to wrap our heads around, toward a planning that embraces the adoption of an adaptive management-based ecosystem approach. This approach needs also to be grounded firmly in techniques for making rich depictions that allow us to get a better handle on complexity [56]. Respect for, and especially a deep appreciation of, complexity is necessary. But we have the means for constructive engagement as well.

5. Synthesis: Hallmarks of an Ecosystem Approach to Making Rich Depictions under Complexity

Attention to context and consequence are the hallmarks of an ecosystem approach to planning, under conditions of complexity. Nested scale-hierarchic process-function ecosystem ecology offers some very useful tools for generating pragmatic descriptions in environmental planning [56]. To summarize, and as prelude to demonstrating an application of such an ecosystem approach to regional urban planning in the particular case of Southern California, three key points need to be underscored:

(1). System connectivity counts—the ecological consequences of specific actions cut across spatial, temporal and functional scales, and must be traced across levels of organization.

(2). Processes and functions may matter more than morphological entities and events—occurring actuality may trump perceived reality in coming to know ecological consequences.
(3). Multiple boundaries and scales must be chosen deliberatively, and across levels of organization, using functionally appropriate and diverse spatial, temporal and organizational scales, to richly capture ecological context.

Put differently, the core elements of an ecosystem approach to decision making under conditions of complexity consist of the self-conscious and reflective use of a multi-perspective, multi-criteria, multi-scale, multi-boundary approach to making operational descriptions, using the levels-of-organization concept to structure the planning domain under consideration, and paying particular attention to the processes and functions that may or may not be directly evident to human sensory perception [56].

Planning, as the craft of societal deliberative decision making, has already begun the work of integrating the instrumental dimensions of the sustainability imperative. Two of the keys to the sustainability puzzle are embodied in: (a) the recognition of the need for a multi-factor, multi-criteria, multi-perspective approach to describing the contextual characteristics of decision phenomena; and (b) the need for an inter-temporal, inter-generational perspective in tracing consequence. The next step in the realization of a genuinely ecological approach—that is to say, taking on complexity fearlessly, while getting beyond the metaphoric imagery and the jargon of holism, the everything-is-connected-to-everything-else sorts of stuff—is an expansion of this view of sustainability in two directions.

First, we must broaden our ideas of system formation to accept ecological processes and functions as the real and proper "objects of concern" for regional environmental planning. This requires that we train ourselves in the sophisticated choosing of multiple functionally relevant boundaries. And also that we expand our ideas of scale beyond our current recognition of spatial and temporal dimensions, by integrating organizational and functional scales as well.

On the second front, and perhaps more urgently, we need to train ourselves in the trans-disciplinary application of knowledge, both across other disciplines and within our own. We must learn to speak in different tongues, becoming comfortable in diverse disciplinary cultures. Much of this work has already been done, as a survey of topics covered at almost any planning conference, or the interests of almost any cohort of planning students will demonstrate. We just need to get more systematic about this. And we need to reach across the divides that fragment our own discipline as well. Air quality planners should seek out habitat conservation planners, who should talk on a regular basis with water quality planners, who should be working closely with land use planners, who should be talking with the urban forestry folks, who should be working side-by-side with the community economic development planners, and so on.

Of course, none of this is really new. Calls to multi-disciplinary holism stretch back as far as recorded memory can see. What has changed is the ecological frame—evolutionary scale hierarchic ecosystem ecology—within which we can now situate a truly ecological planning practice. We have seen complexity in all its richness, and blinking doesn't make it disappear. We can move away and move back, and still see reasonable approximations of what we were looking at earlier. It is our conceptual imagination that has expanded—not displacing old ways of knowing, but incorporating them into an overarching ecosystem approach. And with the advent of new technologies in computer networking and the internet, we can begin to become savvy in decentralized and democratic information management, so as to lower the information costs of taking a truly adaptive, response-sensitive management-based approach to engaging the complex sorts of problems with which we most need to deal.

Living as we do in our four-dimensional world, the craft of planning practice requires us to broaden our scope so as to integrate across sub-disciplines. We need to extend our conception of how the world actually works so that we can give due consideration to processes and functions as the building blocks of nature. We need to extend our descriptions of phenomena both inward and outward, across levels of organization, so as to better take account of context.

And we need to reach across time-event horizons so as to better appreciate consequence. By accepting the premise—that reality is better described as exhibiting nested structures, which are shaped in their actuality by processes and functions, and requiring the use of multiple perspectives, boundaries and scales in their telling—we come to a place where we can begin the business of incorporating the constraints and principles articulated by Rittel and Webber [57]. The strategic and systematic breaching of the constructed, but now deeply entrenched, boundaries our technologies have allowed us to create between the "human" and the "natural", by integrating within and across levels of organization, and again by expanding our world-view to incorporate processes and functions as the stuff the world is actually made up of, allows us to both embrace the contextual richness illustrated by Holling and Goldberg [58] and to realize what some ecologists have already begun to see—that humans, properly, are indeed components of ecosystems [59,60]. Then we can get down to the business of getting humans back into nature, and thus of placing our cities back into their ecological context.

Acknowledgments: The author acknowledges the valuable comments and feedback received from the anonymous reviewers.

Conflicts of Interest: The author declares no conflict of interest.

References

1. Steffen, W.; Crutzen, P.J.; McNeill, J.R. The Anthropocene: Are humans now overwhelming the great forces of nature? *AMBIO* **2007**, *36*, 614–621. [CrossRef]
2. Grimm, N.B.; Grove, J.M.; Pickett, S.T.A.; Redman, C.L. Integrated approaches to long-term studies of urban ecological systems. *BioScience* **2000**, *50*, 571–584.
3. Pickett, S.T.A.; Cadenaso, M.L.; Grove, J.M.; Nilon, C.H.; Pouyat, R.V.; Zipperer, W.C.; Costanza, R. Urban ecological systems: Linking terrestrial ecological, physical, and socioeconomic components of metropolitan areas. *Annu. Rev. Ecol. Systemat.* **2001**, *32*, 127–157. [CrossRef]
4. Register, R. *Eco-Cities: Rebuilding Cities in Balance with Nature*; New Society Publishers: Gabriola Island, BC, Canada, 2006.
5. The Hannover Principles: Design for Sustainability. Available online: http://www.mcdonough.com/wp-content/uploads/2013/03/Hannover-Principles-1992.pdf (accessed on 18 August 2015).
6. Head, P.; Lam, D. How Cities Can Enter the Ecological Age. In *Ecocity Planning: Policies, Practices and Design*; Wong, T.C., Yuen, B., Eds.; Springer: Berlin, Germany, 2011.
7. De Jong, M.; Joss, S.; Schraven, S.; Zhan, C.; Weijnen, M. Sustainable-smarter-resilient-low carbon-eco-knowledge cities? Making sense of a multitude of concepts promoting sustainable urbanization. *J. Clean Prod.* **2015**, in press
8. USA EIA. *Emissions of Greenhouse Gases in the United Stated, 2009*; Energy Information Administration: Washington, DC, USA, 2011.
9. Rockstrom, J.; Steffen, W.; Noone, K.; Persson, A.; Chapin, F.S., III; Lambin, E.F.; Lenton, T.M.; Scheffer, M.; Folke, C.; Schellnhuber, H.J.; *et al.* A safe operating space for humanity. *Nature* **2009**, *461*, 472–475. [PubMed]
10. Steffen, W.; Richardson, K.; Rockstrom, J.; Cornell, S.E.; Fetzer, I.; Bennett, E.M.; Biggs, R.; Carpenter, S.R.; de Vries, W.; de Wit, C.A.; *et al.* Planetary boundaries: Guiding human development on a changing planet. *Science* **2015**. [CrossRef] [PubMed]
11. Urbanization trends in Asia and the Pacific. Available online: http://www.unescapsdd.org/files/documents/SPPS-Factsheet-urbanization-v5.pdf (accessed on 20 August 2015).
12. Schumacher, E.F. *Small is Beautiful: Economics as If People Mattered*; Harper Collins: New York, NY, USA, 1973.
13. Rosenfeld, A.H.; Akbari, H.; Romm, J.J.; Pomerantz, M. Cool communities: Strategies for heat island mitigation and smog reduction. *Energ. Build.* **1998**, *28*, 51–62. [CrossRef]
14. Rosenfeld, A.H.; Romm, J.R.; Akbari, H.; Lloyd, A.C. Painting the town white—And green. *Technol. Rev.* **1997**, *100*, 51–59.
15. Cool Roofs as an Energy Conservation Measure for Federal Buildings. Available online: https://escholarship.org/uc/item/9nh9n203#page-1 (accessed on 20 August 2015).

16. Vitousek, P.M. Beyond global warming: Ecology and global change–Macarthur award lecture. *Ecology* **1994**, *75*, 1861–1876. [CrossRef]
17. Arnold, C.L., Jr.; Gibbons, C.J. Impervious surface coverage: The emergence of a key environmental indicator. *J. Am. Plan. Assoc.* **1996**, *62*, 243–258. [CrossRef]
18. Pielke, R.A.; Marland, G.; Betts, R.A.; Chase, T.N.; Eastman, J.L.; Niles, J.O.; Niyogi, D.; Running, S.W. The influence of land-use change and landscape dynamics on the climate system: Relevance to climate-change policy beyond the radiative effect of greenhouse gases. *Philos. Trans: Ser. A* **2002**, *360*, 1705–1719. [CrossRef] [PubMed]
19. Brabec, E.; Schulte, S.; Richards, P.L. Impervious surfaces and water quality: A review of current literature and its implications for watershed planning. *J. Plan. Lit.* **2002**, *16*, 499–514. [CrossRef]
20. Berke, P.R.; MacDonald, J.; White, N.; Holmes, M.; Line, D.; Oury, K.; Ryznar, R. Greening development to protect watersheds. *J. Am. Plan. Assoc.* **2003**, *69*, 397–413. [CrossRef]
21. DeFries, R.; Asner, G.; Houghton, R. *Ecosystems and Land Use Change*; American Geophysical: Washington, DC, USA, 2004.
22. Stone, B., Jr. Paving over paradise: How land use regulations promote residential imperviousness. *Landsc. Urban Plan.* **2004**, *69*, 101–113. [CrossRef]
23. Jin, M.; Dickinson, R.E.; Zhang, D. The footprint of urban areas on global climate as characterized by Modis. *J. Clim.* **2005**, *18*, 1551–1565. [CrossRef]
24. Taha, H. Urban climates and heat islands: Albedo, evapo-transpiration, and anthropogenic heat. *Energ. Build.* **1997**, *25*, 51–62. [CrossRef]
25. Dixon, P.G.; Mote, T.L. Patterns and causes of Atlanta's urban heat island-initiated precipitation. *J. Appl. Meteorol. Clim.* **2003**, *42*, 1273–1284. [CrossRef]
26. Shepherd, J.M.; Pierce, H.F.; Negri, A.J. City-made rainfall patterns. *Bull. Am. Meteorol. Soc.* **2002**, *83*, 516–517.
27. Rozoff, C.M.; Cotton, W.R.; Adegoke, J.O. Simulation of St. Louis, Missouri, Land Use Impacts on Thunderstorms. *J. Appl. Meteorol. Clim.* **2003**, *42*, 716–738. [CrossRef]
28. Rozoff, C.M.; Cotton, W.R.; Adegoke, J.O. Thunderstorms and the urban heat island. *Bull. Am. Meteorol. Soc.* **2003**, *84*, 756–758.
29. Brazel, A.J.; Fernando, H.J.S.; Hunt, J.C.R.; Selover, N.; Hedquist, B.C.; Pardyjak, E. Evening transition observations in Phoenix, Arizona. *J. Appl. Meteorol. Clim.* **2005**, *44*, 99–112. [CrossRef]
30. Cooling an asphalt jungle: ASU works with utilities to ease urban heat buildup. Available online: http://caplter.asu.edu/docs/iis/TribuneArticle6_5_05.pdf (accessed on 7 June 2015).
31. Taha, H. Modeling impacts of increased urban vegetation on ozone air quality in the South Coast Air Basin. *Atmos. Environ.* **1996**, *30*, 3423–3430. [CrossRef]
32. Taha, H. Modeling the impacts of large-scale albedo changes on ozone air quality in the South Coast Air Basin. *Atmos. Environ.* **1997**, *31*, 1667–1676. [CrossRef]
33. Application of titanium dioxide photocatalysis to create self-cleaning building materials. Available online: http://www.protecsolutions.com.tr/uploads/0448e8a631771ff503d2f4633f1cdc96.pdf (accessed on 7 June 2015).
34. Frazer, L. Titanium dioxide: Environmental white knight? *Environ. Health Persp.* **2001**, *109*, A174–A177.
35. Meier, A.K. Strategic landscaping and air-conditioning savings: A literature review. *Energ. Build.* **1991**, *15*, 479–486. [CrossRef]
36. Sailor, D.J. Simulations of annual degree day impacts of urban vegetative augmentation. *Atmos. Environ.* **1998**, *32*, 43–52. [CrossRef]
37. Dimoudi, A.; Nikolopoulou, M. Vegetation in the urban environment: microclimatic analysis and benefits. *Energ. Build.* **2003**, *35*, 69–76. [CrossRef]
38. Oberndorfer, E.; Lundholm, J.; Rass, B.; Coffman, R.R.; Doshi, H.; Dunnett, N.; Gaffin, S.; Kohler, M.; Liu, K.K.Y.; Rowe, B. Green roofs as urban ecosystems: Ecological structures, functions, and services. *BioScience* **2007**, *57*, 823–833. [CrossRef]
39. Extensive vegetated roofs. Available online: https://www.wbdg.org/resources/greenroofs.php (accessed on 17 August 2015).
40. Loh, S. Living walls: A way to green the built environment. *BEDP* **2008**, *26*, 1–7.

41. Ottele, M.; Perini, K.; Fraaij, A.L.A.; Haas, E.M.; Raiteri, R. Comparative life cycle analysis for green facades and living wall systems. *Energ. Build.* **2011**, *43*, 3419–3429. [CrossRef]
42. Perini, K.; Rosasco, P. Cost-benefit analysis for green facades and living wall systems. *Build. Environ.* **2013**, *70*, 110–121. [CrossRef]
43. Actualizing microclimate and air quality benefits with parking lot tree shade ordinances. Available online: http://www.fs.fed.us/psw/programs/cufr/products/11/cufr_69.pdf (accessed on 7 June 2015).
44. Scott, K.I.; Simpson, J.R.; McPherson, E.G. Effects of tree cover on parking lot microclimate and vehicle emissions. *J. Arboric.* **1999**, *25*, 129–142.
45. Scott, K.I.; Simpson, J.R.; McPherson, E.G. Green parking lots: Can trees improve air quality? In Proceedings of the Best of the West Summit, San Francisco, CF, USA, 26–29 September 1998; pp. 86–87.
46. Where are all the "Cool" parking lots? Available online: http://www.fs.fed.us/psw/programs/cufr/products/newsletters/UF3.pdf (accessed on 7 June 2015).
47. Trees, parking and green law: Strategies for sustainability. Available online: http://www.naturewithin.info/Roadside/Trees_Parking_GreenLaw.pdf (accessed on 7 June 2015).
48. Cahill, T. A second look at porous pavement/underground recharge. *Watershed Prot. Tech.* **1994**, *1*, 76–78.
49. Booth, D.B.; Leavitt, J. Field evaluation of permeable pavement systems for improved storm water management. *J. Am. Plan. Assoc.* **1999**, *65*, 314–325. [CrossRef]
50. Understanding pervious concrete. Available online: http//www.perviouspavement.org/downloads/PerviousConcreteCSI_12-05.pdf (accessed on 7 June 2015).
51. Nowak, D.J.; Crane, D.E. Carbon storage and sequestration by urban trees in the USA. *Environ. Pollut.* **2002**, *116*, 381–389. [CrossRef]
52. Benjamin, M.T.; Winer, A.M. Estimating the ozone-forming potential of urban trees and shrubs. *Atmos. Environ.* **1998**, *32*, 53–68. [CrossRef]
53. Benjamin, M.T.; Sudol, M.; Bloch, L.; Winer, A.M. Low-emitting urban forests: A taxonomic methodology for assigning isoprene and monoterpene emission rates. *Atmos. Environ.* **1996**, *30*, 1437–1452. [CrossRef]
54. Beckett, K.P.; Freer-Smith, P.; Taylor, G. Effective tree species for local air-quality management. *J. Arboric.* **2000**, *26*, 12–19.
55. Vasishth, A. A scale-hierarchic ecosystem approach to integrative ecological planning. *Prog. Plan.* **2008**, *70*, 99–132. [CrossRef]
56. Vasishth, A. An ecosystem approach to adaptive decision making under complexity: Rich depictions, multiple perspectives and savvy stakeholders. *Int. J. Oper. Quant. Manag.* **2010**, *16*, 1–20.
57. Rittel, H.W.J.; Webber, M.M. Dilemmas in a general theory of planning. *Policy Sci.* **1973**, *4*, 155–169. [CrossRef]
58. Holling, C.S.; Goldberg, M.A. Ecology and planning. *J. Am. I. Plan.* **1971**, *37*, 221–230. [CrossRef]
59. McDonnell, M.J.; Pickett, S.T.A. *Humans as Components of Ecosystems: The Ecology of Subtle Human Effects and Populated Areas*; Springer-Verlag: New York, NY, USA, 1993.
60. Alberti, M.; Marzluff, J.M.; Shulenberger, E.; Bradley, G.; Ryan, C.; Zumbrunnen, C. Integrating humans into ecology: Opportunities and challenges for studying urban ecosystems. *BioScience* **2003**, *53*, 1169–1179. [CrossRef]

sustainability

MDPI

Case Report

Assessing Sustainability of Mixed Use Neighbourhoods through Residents' Travel Behaviour and Perception: The Case of Nagpur, India

Sarika Bahadure [†,*] and Rajashree Kotharkar [†]

Department of Architecture and Planning, Visvesvaraya National Institute of Technology (VNIT), Nagpur Pin-440010, India; rskotharkar@arc.vnit.ac.in

* Author to whom correspondence should be addressed; sarikabahadure@arc.vnit.ac.in; Tel.: +91-712-280-1290; Fax: +91-712-280-1376

† These authors contributed equally to this work.

Academic Editors: Tan Yigitcanlar and Md. Kamruzzaman

Received: 27 March 2015; Accepted: 17 August 2015; Published: 2 September 2015

Abstract: Mixed land-use development is the integration of different land-use functions like residential, commercial, recreational, and institutional in an urban sector or a neighbourhood. Integrating transport and land-use mix is one of the goals of planning policies around the world. Prior studies mention the benefits of mixed land-use development towards creating sustainable environment, but do not specify the proportion of the mix of compatible land uses. This study attempts to assess the sustainability of the neighbourhoods with mixed land-use in the context of the Nagpur city, India. Residents' travel behaviour in twelve neighbourhoods is studied by means of indicators namely trip lengths, mode of travel, vehicle ownership, and travel expenses. To investigate the users' insight, the study further examines residents' perception with the help of parameters such as safety, satisfaction, pollution, and mix. The sustainability indices are computed for both residents' travel behaviour and perception, for each neighbourhood. The study revealed that neighbourhoods with high and moderate land-use mix are sustainable with travel behaviour. Residents' perception sustainability index indicates neighbourhoods with moderate land-use mix are more sustainable than those with high and low land-use mixed neighbourhoods. This study advocates stakeholders' insight and the proportion of mix in land-use planning decisions.

Keywords: sustainability index; mixed land-use; neighbourhood; travel behaviour; perception

1. Introduction

Mixed land-use development is about co-existence of more than one type of land-use function like residential, commercial, recreational, educational, *etc.* in a specific area. It gives emphasis on higher intensity and diversity of land uses [1]. It is a key component of many development trends, including traditional neighbourhood development, transit oriented development, new urbanism, liveable communities, and smart growth principles [2]. These development trends act as a strong proponent to create a sustainable environment [3–7]. The land-use mix is associated with certain externalities both in terms of benefits (proximity, vibrancy, and social bonding) and cost (over loaded infrastructure and pollution), creating either a sustainable or unsustainable built environment. Prior studies mention that areas with mixed land-use are more sustainable than segregate areas [6]; however, the proportion of the mix of compatible land uses for sustainable development is not understood.

Traditionally, Indian cities had mixed land-use characteristics [8] with a high mix of commercial and residential uses. Over time, the residential land-use has been replaced by commercial activities due to economic market pressures [9,10]. Other areas in the city have a moderate mix in sub-centres or along a major road. It is obvious from the current state of Indian cities that land-use mix when left uncontrolled leads to the issues of environmental externalities related to mixed land-use. Thus, there is a need to investigate the range of the land-use mix that will enhance sustainability. To investigate this, the neighbourhood scale is appropriate as it provides both residential and non-residential functions [11] and connects communities and dwellings to the wider urban context.

There are various factors like the density, transport facility, travel behaviour, and socio-economic aspects which influence mixed land-use development. Amongst the various factors that affect mixed land-use, transport and land-use mix (as part of urban form or built environment) are strongly interrelated and integrating them is one of the goals of planning policies around the world [4,11–16]. Reid states, the measures of travel demand [trip frequency, trip length, vehicle-miles of travel (VMT), and mode choice] are the functions of socioeconomic and built environment [17]. Leck also supported these variables along with vehicle-hour travel and travel frequency influencing density and mix [18]. Trip generation, trip distribution, modal split, and route assignment also influence travel demand [19].

1.1. Mixed Land-Use and Transport

In a mixed land-use area, as various activities are in close proximity, they support sustainable forms of transport (walking, cycling or transit) [20]. Professional services are permitted in residential areas in few Master plans [21]. Different studies in the past reveal the advantages of mixed land-use supporting sustainable transport, health wellbeing, and reduction of air pollution. Mixing of uses near the employment destination is found to reduce travel demand [13]. Land-use mix reduces personal vehicle travel and significantly increases walking and cycling [16]. Increased walking has a direct connection to health wellbeing. The performance of the body mass index model with mixed land-use suggests that the presence of mixed land uses improves walk-ability [12]. Neighbourhoods with mixed uses facilitate the use of non-automobile travel modes and shorten car trips, thereby reducing vehicle greenhouse gas emissions [22,23].

1.2. Mixed Land-Use and Social Aspects

Social aspects also influence mixed land-use development. A mixed land-use environment creates social benefits favouring the residents [24]. A mix of diverse housing types based on income, functions, and forms provides residents with multiple choices when selecting their abode. A fine-grained mix of attractive destinations and good aesthetic qualities like tree cover in a neighbourhood supports walking for various activities and also for pleasure [25]. Dwellers in neighbourhoods with mixed land-use are more likely to use public spaces [26]. A neighbourhood with mixed land-use promotes a high level of activity on the street, keeping it safe. Children feel safe to play or cycle, elderly meet and talk, and strangers feel they are intruding [27]. Community cohesion, interaction, and sense of belonging are strong in a mixed land-use community. People perform diverse activities (live, shop, play, and work) in the same environs. The frequencies of their meeting, greeting, smiling, and talking increase. Neighbourhoods with mixed land-use provide convenient access to people and places and so residents and workers prefer such neighbourhoods for a balanced life [4].

An insight of the social aspects can be gained with the help of the stakeholders' perception. In a neighbourhood residents are the prime stakeholders. Residents' perception presents the inhabitants' views about the socio-physical services and other aspects of life in their local areas. An insight can be sought on various aspects of the neighbourhood such as diversity, vitality, attractiveness, health, safety, satisfaction, community cohesion, interaction, a sense of belonging, and quality of life. Residents' perception includes community participation, provides an insight on the current status, and reveals the focus area where improvement is needed. This tool is used to assess sustainability aspects. The residents' self-report tool is used to create, a neighbourhood sustainability framework for New Zealand. Behaviours, perceptions and experiences associated with their neighbourhoods are used to get information on the built environment (land-use mix and density) [11]. Residents' perception has been supported in prior studies to provide a robust understanding of various aspects of sustainability like mixed land-use, travel behaviour, environment, density, connectivity, social benefits, *etc.* [12,28–30].

1.3. Need Identification: Mixed Land-Use—Residents' Travel Behaviour and Perception

The interaction of mixed land-use with transport supports sustainability, but the proportion of mix has not been specified in prior studies. Hence, this study attempts to investigate the same with the help of residents' travel behaviour in the mixed land-use neighbourhoods. Travel behaviour is the study of how people use transport. It probes into the number, pattern, and purpose of trips, destinations, modes, route choices, *etc.* [31,32]. Besides these parameters age, gender, status, values, self-selection, attitude, and lifestyles also influence travel behaviour [28,33]. As residents' perception provides an insight of the neighbourhood, this study is further extended to investigate the residents' perception in the mixed land-use neighbourhoods.

The neighbourhoods are selected based on the high to low mix of commercial and residential land-use in the urban context of Nagpur, India. Residents' travel behaviour and perception in the neighbourhoods with mixed land-use is presented in terms of the sustainability indices.

Travel behaviour studies the residents' travel habits like commuting distance, mode of travel, vehicle ownership, and expenses on travel, while perception examines residents' opinion, feelings, safety, satisfaction, and mix indicators. Though the residents' travel behaviour and perception are less related, the study tries to compare the sustainability indices of these two domains. The purpose of this study is to explore the sustainability of varying land-use mixed neighbourhoods. It addresses the following research questions:

(1) Whether the inhabitants of mixed land-use neighbourhoods support mixing, and if they do, then how much mix is supported to achieve sustainability?
(2) Are the sustainability indices of residents' travel behaviour and perception similar or different?

The organization of the paper is as follows. The introduction presents the mixed land-use and its influencing factors. It further states the need to conduct the study in the neighbourhoods with mixed land-use (through residents' travel behaviour and perception indicators) to achieve sustainability. The second section describes the methodology (data, indicators, results, and observation) in three stages; (a) study area; (b) residents' travel behaviour; and (c) perception. The third section presents the sustainability index model (normalization, weights and aggregation) and discusses the findings. The last section concludes the study.

2. Methodology and Data Presentation

The methodology is divided into three stages (refer Figure 1). The first stage elucidates the study area selection criteria. Twelve neighbourhoods spread across the city, with varying residential and commercial mix were selected for the study. Entropy index is used to measure the land-use mix for these neighbourhoods.

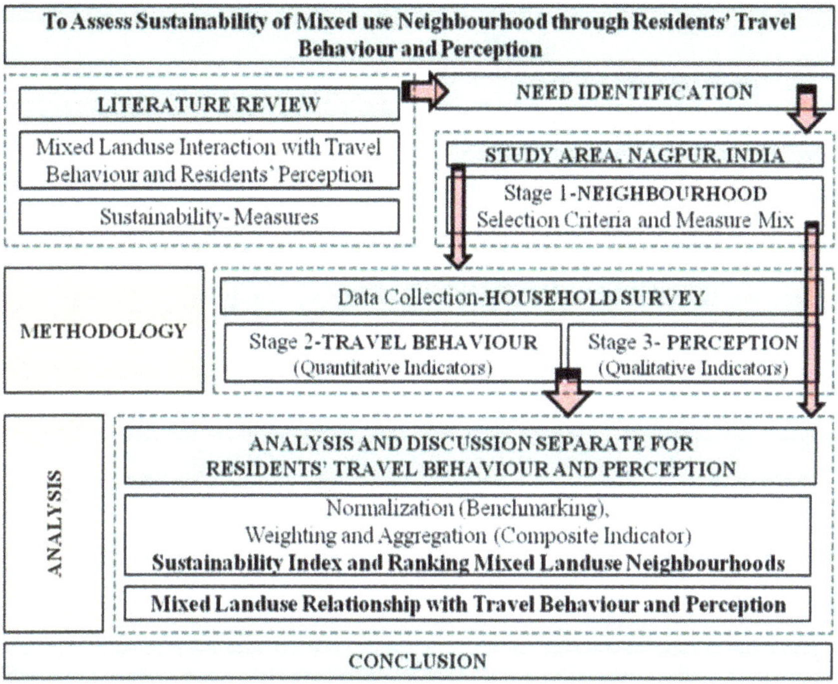

Figure 1. Methodology adopted for the study.

In the second and third stages, residents' travel behaviour and perception is studied. Household survey is conducted to understand various aspects like trip length to various destinations, travel mode, vehicle ownership, safety, satisfaction, pollution, social interaction, *etc.* of residents' travel behaviour and perception. Travel behaviour is studied with the help of quantitative indicators while perception is studied using qualitative indicators. Although a variety of indicators could have been used to operationalize sustainability index, the study is restricted to eleven travel behaviour indicators and twelve perception indicators.

Indicator scores are used to compute the sustainability index for each neighbourhood under travel behaviour and perception domains. At the end, the sustainability index score is correlated with the land-use mix measure (entropy index) to identify the relationship.

2.1. First Stage—Study Area: Neighbourhoods in Nagpur City, India

Nagpur is centrally located in the country and well connected to major cities in India by air, rail, and road. It acts as the key administrative, business and institutional centre for central India. The total area of Nagpur Municipal Corporation is 21,756 hectares. The Nagpur city has a population of 2,405,665 as per 2011 census with a gross density of 110 persons per hectare. Nagpur is ranked the thirteenth largest urban agglomeration in India [34].

There is an observed pattern of a varying land-use mix within the city. The core has a dense mixed land-use environment that has evolved over time. These areas have high land and rental value making it less feasible for pure residential use so mixed land-use exists. There are planned and semi-planned areas with a balance of residential and non-residential activities. Sprawled areas near the fringe are partly planned and partly unplanned plotted development with residential and no commercial use [27].

2.1.1. Selection Criteria for Neighbourhoods

There are different typologies of neighbourhoods within the Nagpur Municipal Corporation limit, each with a different land-use mix. The city-core has a dense mixed land-use environment evolved over time. In these areas more commercial land-use prevails due to the high land and rental value making it less feasible for pure residential use. The other areas of the city are planned and semi-planned with a moderate mix of residential and non-residential activities. The sprawled areas near the fringe are partly planned and partly un-planned plotted development with residential and no or little commercial use [27]. The development potential is bringing changes to the existing land-use pattern of the city. Whether this change is creating sustainable or unsustainable environ for the city needs to be explored. An attempt is made to study the sustainability status of neighbourhoods with varying mixed land-use.

Twelve neighbourhoods (NH1 to NH12) under four categories (L1 to L4) of varying residential and commercial mix were selected for the study (refer to Table 1 and Figure 2). Within a category the age of the included neighbourhoods and their socio-economic characteristics are similar.

Table 1. Twelve neighbourhoods under four categories of varying land-mix.

Category		Neighbourhoods	Population Density	Population Density Including Floating Population	Age of Neighbourhood (Years)	Spatial Distribution	Mix Characteristics
Level-1 (L1)	NH1	Buldi—CBD and centrally located	230.29	394.62	150	core area, compact, organic, medium rise high density	high mix of commercial and residential land-use
	NH2	Itwari—known for wholesale market	466.46	596.56	150		
	NH3	Mahal—known for heritage and retail market	323.62	442.48	150		
Level-2 (L2)	NH4	Sadar—in North West Nagpur	75.65	102.77	100	partly old core compact, organic, dense and partly new planned	moderate mix of commercial with residential land-use
	NH5	Dharampeth—in West Nagpur	158.04	193.76	75	moderately dense, low to medium rise and planned	
	NH6	Buddha Nagar—in North Nagpur	234.49	250.20	75		
Level-3 (L3)	NH7	Nagsen Nagar, Jaripatka—along Nara Road and Ring Road in North Nagpur	257.70	259.13	40	moderately dense, low to medium rise and planned	mix of commercial only along the major road with no influx of commercial in the residential areas
	NH8	Trimurti Nagar—along Ring Road in South West Nagpur	176.16	177.05	30		
	NH9	Friends Colony—along Katol Road in North West Nagpur	142.04	142.79	25	partly moderately dense and planned and partly sprawled low rise	

Table 1. *Cont.*

Category		Neighbourhoods	Population Density	Population Density Including Floating Population	Age of Neighbourhood (Years)	Spatial Distribution	Mix Characteristics
Level-4 (L4)	NH10	Civil Lines—Residential and administrative area planned and designed during British period in West Nagpur	39.32	41.95	90	planned and sprawled low rise	no or very less ingress of commercial activity and has only residential land use
	NH11	Nara—Sprawl growth in North suburb Nagpur	32.64	32.73	5–10	partly planned and partly unplanned sprawled low rise	
	NH12	Manish Nagar—Sprawl growth in South suburb Nagpur	32.64	32.69	5–10		

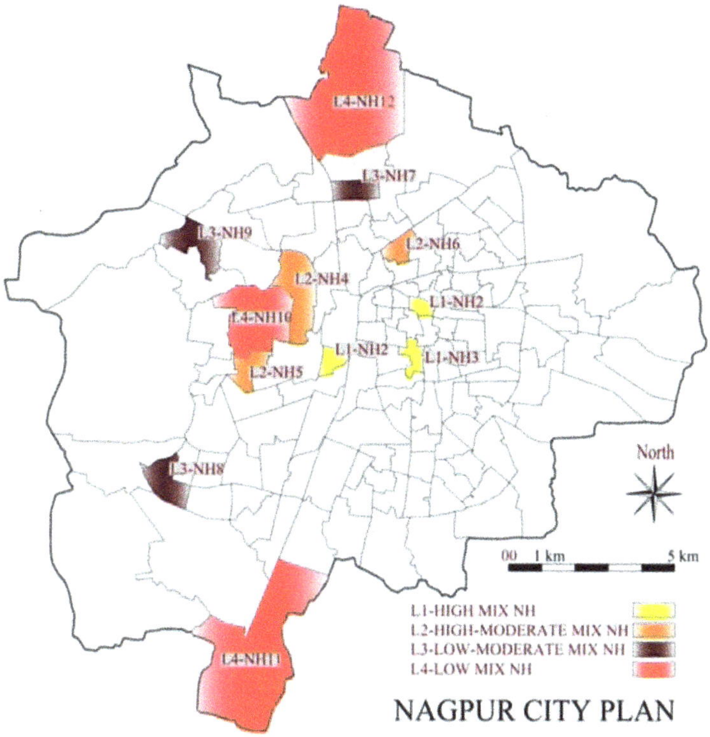

Figure 2. Spatial location of the selected neighbourhoods in Nagpur City.

2.1.2. Sample Selection

The procedure for selecting the sample size is important since the study depends on data surveyed. Simple random sampling across the twelve neighbourhoods with uniform stratification was carried out. The initial sample size n_1 is calculated by Equation (1a). The size of population is adjusted using Equation (1b), but this has an effect on the small and moderate size populations. In this study n_1/N is negligible, so assumed $n_2 = n_1$. The design effect is the ratio of the sampling variance of an estimator under a given design to the sampling variance of an estimator under the simple random sampling of the same sample. The design effect is accommodated by Equation (1c). Finally, the response rate is adjusted to determine the final sample size by Equation (1d) [35].

$$n_1 = \frac{z^2 P(1-P)}{e^2} \tag{1a}$$

$$n_2 = n_1 \frac{N}{N + n_1} \tag{1b}$$

$$n_3 = d_{eff} \times n_2 \tag{1c}$$

$$n = \frac{n_3}{r} \tag{1d}$$

The assumptions and constants used for the above equations are: the total population of twelve neighbourhoods (N) which is 216,829; margin of error (e) is taken to be 6%; confidence level is 95% so (z) is 1.96; population variance (P) is 50%; design effect (d_{eff}) for simple random sampling is 1 and response rate (r) is 75%.

Using the above technique n_1, n_2, and n_3 is 268 and the final sample n is 357. Thus, total samples considered for the study is 360. Carrying out uniform stratification, each study area is allotted 30 samples.

2.1.3. Measuring Land-Use Mix

Measuring land-use mix is critical to mix land-use studies and there have been numerous methods of measuring like entropy index, dissimilarity index, distance to walkable destination (amenities and transit stops), and the number of amenities available within a certain distance [12]. Entropy index quantifies randomness, segregation, diversity or compressibility in the data. Land-use mix exhibits a pattern of combination and segregation of different land uses. So, entropy index is the most widely accepted and commonly used index by researchers for representing the land-use mix within geographic area [12,13,15,36,37]. Crevero [13] derived the entropy Equation (2a) as:

$$Entropy\ Index = (-1) \times \sum_{j} \frac{P_j \times \ln(P_j)}{\ln(J)} \tag{2a}$$

where, P_j is the proportion of developed land in the Jth land-use type. Frank described the evenness of the distribution of built square footage among seven land-use categories as single family, multi-family, retail and service, office, entertainment, institutional and industrial and later for a three-category mix of residential, commercial, and office [36,37].

Entropy index varies between 0 and 1, wherein 0 indicates single use (homogenous) and 1 maximum land-use mix (heterogeneous).

The existing land-use distribution for each neighbourhood is plotted on the base map through a field survey. High resolution is used for the base map which is at plot or property level. The entropy index is computed for commercial and residential land-use by Equation (2b).

$$Landuse\ Mix\ Entropy\ Index\ (EI) = (-1) \times \frac{\left[\left(\frac{b1}{a}\right) \times \ln\left(\frac{b1}{a}\right) + \left(\frac{b2}{a}\right) \times \ln\left(\frac{b2}{a}\right)\right]}{\ln(n)} \tag{2b}$$

where, a is the total area in square meter of two land uses, b1 is the commercial land-use area in square meters, b2 is the residential land-use area in square meters, and n is 2 (total number of land uses in the mix). Table 2 shows the entropy index of the neighbourhoods. The neighbourhoods are placed in the descending order of entropy index values in each category.

Table 2. Entropy index of the selected twelve neighbourhoods.

Category Levels	L1			L2			L3			L4		
Neighbourhoods	NH1	NH2	NH3	NH4	NH5	NH6	NH7	NH8	NH9	NH10	NH11	NH12
Entropy Index	1.00	0.95	0.78	0.78	0.77	0.71	0.35	0.30	0.17	0.21	0.08	0.12

2.2. Second Stage—Residents' Travel Behaviour: Indicators, Results, and Observations

Travel behaviour explores the accessibility, mode, vehicle ownership, and expenses related to travel. The study on residents' travel behaviour is based on measurable indicators (refer Table 3). Each indicator score is computed from the household survey and the average score is taken for analysis.

Table 3. Travel behaviour (quantitative indicators).

Indicator Set	Indicator Description
Accessibility to land-use destination (Trip length of travel in kilometres)	1. Trip length for Work 2. Trip length for Shop 3. Trip length for Vegetable market 4. Trip length for School 5. Trip length for College 6. Trip length to Health facility 7. Trip length for Restaurant/Theatre 8. Trip length for Garden
Travel Mode to work	9. Percentage of Sustainable Mode (Walk + Cycle + Transit)
Vehicle Ownership	10. Two-wheelers and Four-wheelers {Converted to Passenger Car Unit (PCU) for uniformity}
Travel Expenses	11. Percentage of Income spend on travelling

2.2.1. Accessibility to Land-Use Destinations

The trip length is the actual network distance travelled by the residents from their own residence to various destinations. The trip length to various destinations is computed from the survey report. The selected destinations are workplaces, shops, educational institutions, health amenities, and recreational areas.

The trip lengths to different destinations vary across the neighbourhood categories (refer Figure 3). The proximity to the amenities is directly proportional to the number of trips. In NH1, due to commercial dominance, only those people having their workplaces in the same neighbourhood reside there, while others have migrated to different parts of the city leaving their residential space for commercial use. It is observed that the high land-use mix in neighbourhoods results in more residents having their workplaces within 25 meters (assumed as work from home). This indicates the proximity to the workplace in high land-use mix areas. In the higher income group (HIG), children travel more than 5 km in all neighbourhoods to reach school. Residents from low land-use mix areas (NH11 and NH12) move outside the neighbourhood to access almost all amenities and travel around 9.5, 5.5 and 4.5 km on an average for work, school, and health facilities respectively.

2.2.2. Mode of Travel

Mode of travel to the destination includes two-wheeler, four-wheeler, walking or traveling by cycle, para-transit (motorized cycle rickshaw), and bus. High walkability is noted in areas with high land-use mix (refer Figure 4). The mode share of public transport does not show a relationship with the land-use mix which indicates that the land-use mix does not promote or discourage the use of public mode. Instead, residents prefer to use a personal vehicle rather than public transport. In L4 category, due to greater commuting distance to work, residents do not prefer to walk or cycle.

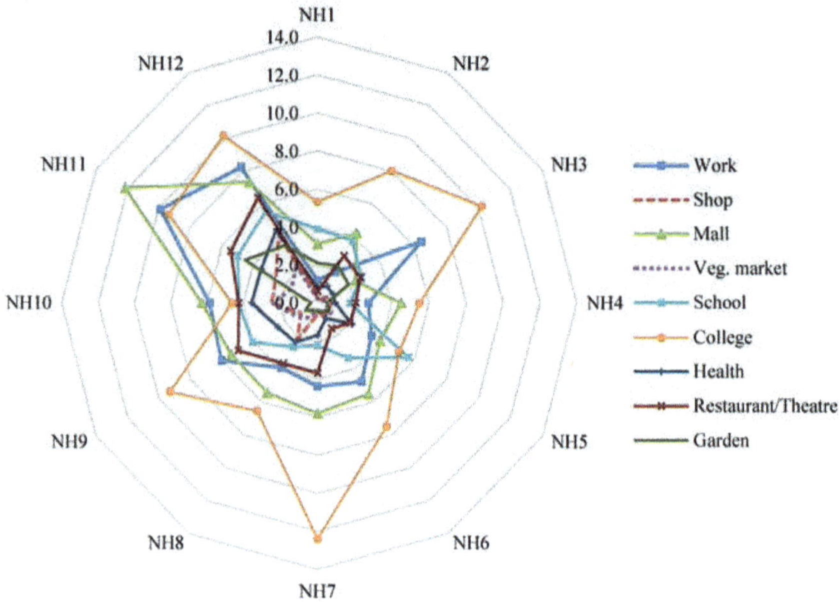

Figure 3. Access to various land-use destinations.

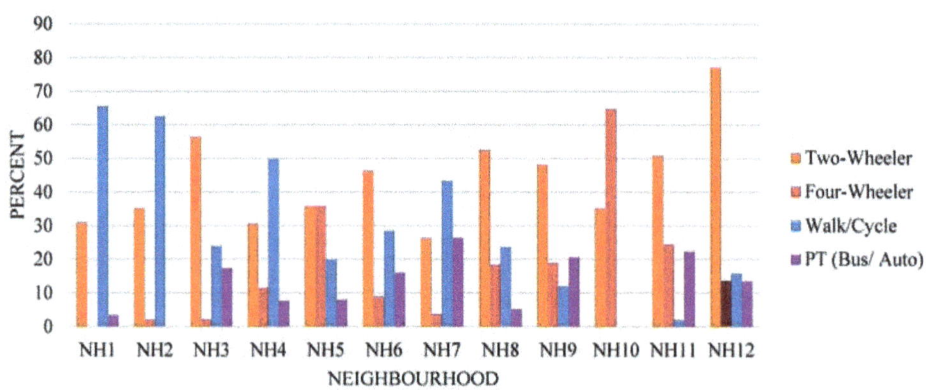

Figure 4. Travel mode percentage for work trip.

2.2.3. Vehicle Ownership

Vehicle ownership is recorded in terms of the number of cycles, two-wheelers, and four-wheelers. Vehicle ownership per household is calculated in terms of percentage (refer Figure 5). Nagpur is a two-wheeler city, where two-wheelers contribute to 55 percent of all mode of transport share and 65 percent of non-motorized mode share [38]. It is observed that irrespective of the income group and neighbourhood type, two-wheeler ownership is high and constant. Vehicle ownership is observed to be linked with the income levels as cycle ownership is high in neighbourhoods with low income residents (NH7) whereas, car ownership is higher in neighbourhoods with high income residents (NH5 and NH10). However, many residents in high mix (L1) neighbourhoods reported that, they can afford a car but are discouraged to own one due to the unavailability of parking space and congestion.

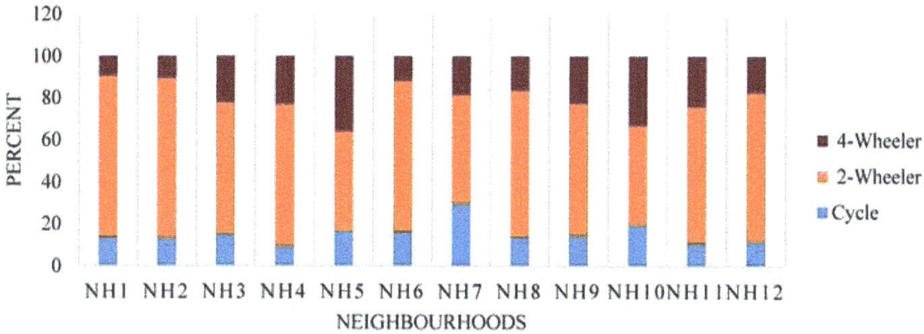

Figure 5. Vehicle ownership in percentage.

2.2.4. Travel Expenses

The average household income of residents from NH4, NH5, and NH10 is high, and that of residents from NH7 is low; the income of inhabitants from other neighbourhoods is in a similar range. Total travel expenses per month are calculated as a sum of spending for different destination trips and the maintenance of the vehicles (refer to Figure 6). It is observed that residents in neighbourhoods with high land-use mix spend less and those living in neighbourhoods with low land-use mix spend more on travel.

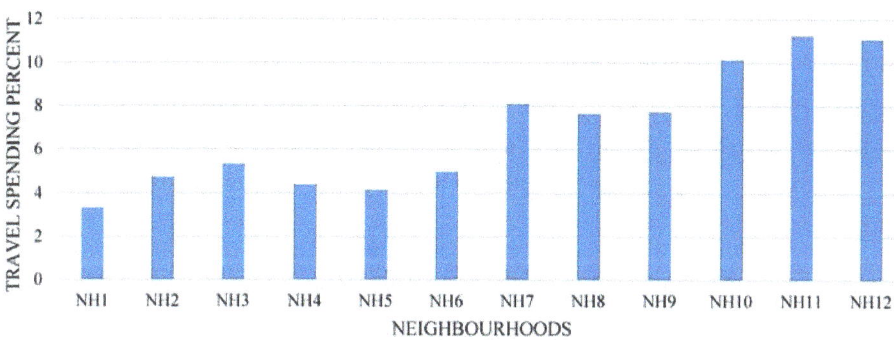

Figure 6. Percentage of household income on travel expenses per month.

2.3. *Third Stage—Residents' Perception: Indicators, Results and Observations*

The study of perception tries to understand residents' perception about the neighbourhood in the context of safety, satisfaction, pollution, interaction, and mix. This insight is obtained through a household survey. The questionnaire is formulated on the five-point Likert scale (LS) (scored as 1-strongly satisfied; 2-satisfied; 3-neither; 4-dissatisfied; and 5-strongly dissatisfied) and "yes/no" (Y/N) answers nominal scale as mentioned in Table 4.

Table 4. Residents' perception (qualitative indicators).

Indicator Set	Indicator Description
Safety and Satisfaction	1. On street Safety (Y/N) 2. Satisfied with amenities in NH (Y/N) 3. Wish to shift from current NH (Y/N) 4. Part of Social Group (Y/N) 5. Neighbourhood Safety (LS) 6. Neighbourhood Quality (LS) 7. Sense of Belonging (LS)
Pollution	8. Feel Traffic Problem (Y/N) 9. Feel Noise Pollution(Y/N) 10. Feel Air Pollution(Y/N)
Favour Mix	11. Favour mix of residential and commercial land-use (LS) 12. Favour mix LIG, MIG and HIG (LS)

2.3.1. Safety and Satisfaction

The road and neighbourhood safety is studied. Road safety is studied through accidents and neighbourhood safety in terms of natural surveillance, citizen's safety, and safety from theft. Satisfaction is studied in terms of neighbourhood quality, availability of amenities, a sense of belonging, residents' wish to migrate to another neighbourhood, and their participation in a social group. Figure 7a indicates the residents' aspiration towards safety and satisfaction measured as the percentage of "yes choices". The high score on "yes choices" for street safety, satisfaction about amenities, and participation indicates that residents feel safe and are satisfied, however a high score on migration indicates dissatisfaction. Figure 7b indicates the cumulative count of the Likert scale for thirty households in each neighbourhood. The higher score on the Likert scale indicates higher neighbourhood safety, quality and sense of belonging.

The amount of satisfaction felt by the residents towards their neighbourhood is studied. Quality amenities within the neighbourhood satisfy the residents. In many neighbourhoods some amenities are highly accessible, while others are less accessible. Dissatisfaction with the quality of the neighbourhood leads to an urge to migrate to a better neighbourhood. The sense of belonging is highly related to the social factor rather than the physical factor.

The overall satisfaction level is above average even when there is inadequate infrastructure. Areas with high land-use mix (L1) are safer in terms of natural surveillance and theft, but not in road safety. They suffer from poor quality of life due to high congestion and, lack of infrastructure and open spaces. Due to the environmental externalities and the pressure to accommodate space for commercial use, L1 residents prefer migration. L1 and L2 residents have a strong sense of belonging and social bonding. They also interact and take active part in social activities and know more people in their neighbourhood. Residents in moderate mix (L2, L3, and NH10) areas perceive their neighbourhoods as safe and have a higher level of satisfaction. Low land-use mix (NH11 and NH12) residents feel unsafe and dissatisfied due to sprawl, poor road infrastructure, lack of street lights, and little natural surveillance. They show the lowest sense of belonging.

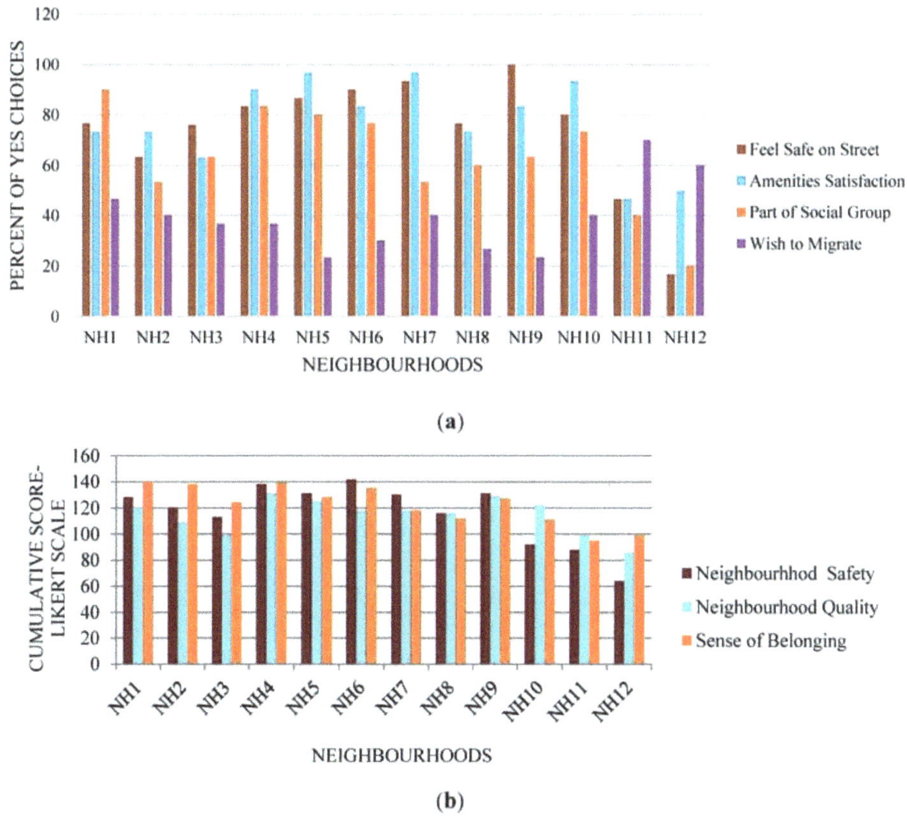

Figure 7. (**a**) Safety and satisfaction—percentage; (**b**) safety and satisfaction—cumulative scores.

2.3.2. Pollution and Traffic Problem

Figure 8 shows the percentage of "yes choices" for pollution and traffic problems. High pollution and the problem of traffic congestion are observed in L1 neighbourhoods. However, some residents from the highly noisy L1 area mentioned that they are used to the high noise level and feel uncomfortable in complete silence. L3 neighbourhoods are perceived as less polluted neighbourhoods. NH10 does not have pollution issues due to high greenery and low density. Some pollution in NH11 and NH12 is seen due to continuous construction work. Neighbourhoods with a high land-use mix see pollution and traffic congestion as externalities.

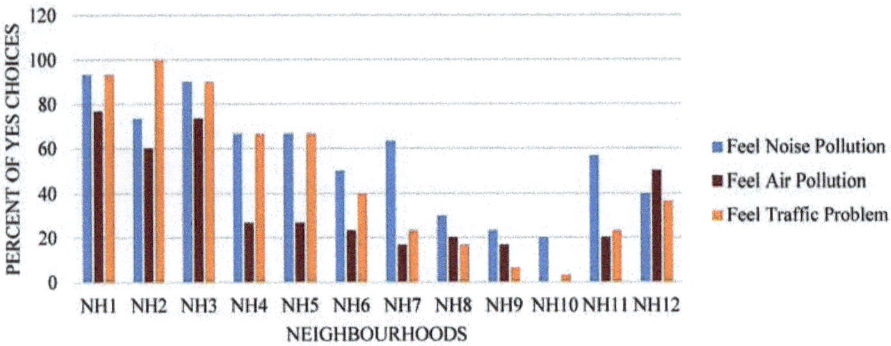

Figure 8. Pollution and traffic related problems.

2.3.3. Favour Mix

This section studies residents' opinion on the mixing of land uses and income. Dwellers favoured a mix of land uses. They mentioned that the commercial area should be on the main road and the residential area should be in the inner part of the neighbourhood by which they would benefit from the proximity to daily needs without having to suffer the externalities associated with commercial use. Most residents expressed their inclination towards a mix of Middle Income Group (MIG) or HIG but not Lower Income Group (LIG) (refer Figure 9). They also stated they are more comfortable with a similar income group.

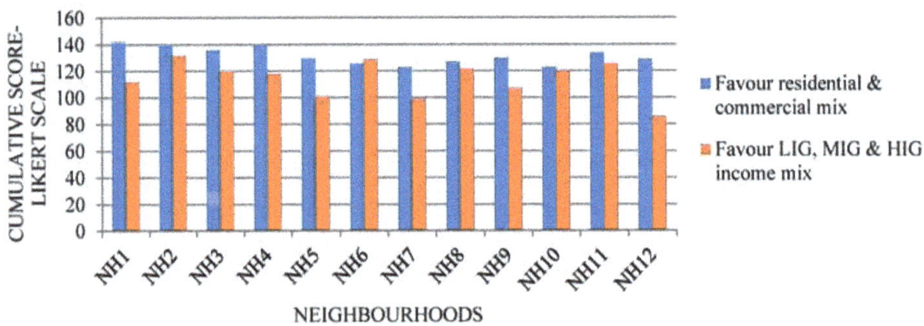

Figure 9. Residents favouring mix—cumulative scores.

3. Sustainability Index Based on Composite Indicators

Sustainability index is calculated using composite indicators which are assigned performance based benchmarking and weights for each neighbourhood.

3.1. Normalization and Benchmarking

Normalization is used to convert different indicator units into a common and comparable scale. The various normalization methods are: (a) ranking based on the indicators' performance over time; (b) standardization or z score; (c) rescaling or min-max method; (d) categorical scale [39].

This study uses normalization on a five-point scale as shown in the Table 5. The colours in the table act as a visual aid for representing the levels of sustainability. To achieve sustainability, the goal for the neighbourhood is to attain a five score.

Table 5. Sustainability benchmarks scores based on normalization.

Five-Point Scale	Sustainability Benchmark (Colours represents levels)
1	Low (extremely unsustainable situation)
2	Medium-Low (not suitable but not as bad as in previous case)
3	Medium (discrete level of sustainability)
4	Medium-High (satisfactory level of sustainability)
5	High (target level of sustainability)

Benchmarking establishes a baseline measure for the performance of individual indicators and acts as a yardstick to measure and compare effectiveness. It helps in the appraisal of performance and comparative analysis of neighbourhoods and allows for improvement on current performance.

3.1.1. Benchmarking for Travel Behaviour

The benchmarking for travel behaviour is based on the literature review, expert opinions, and the comparative scale. A lower distance to the destination results in high sustainability. Studies mention that the walking distance to the destination should be less than a quarter of a mile (~0.4 km) [40,41]. Preferred trip length for various purpose of travel varies. The benchmark for a shopping trip length starts from a range of 0.0–0.4 km with an increment of 0.4 km as shops must be accessible by foot. For school, health, and recreational facilities, benchmarking range starts at 0.0–0.8 km as these can be accessible by foot or cycle. Work places are scattered all over the city and residents move out of the neighbourhood for work, thus the benchmarking range starts at 0.0–1.5 km. Institutes for higher education are usually not within or near the neighbourhoods but are spread at different locations in the city and suburbs, thus college going students commute long distances. So, the benchmarking range for higher education starts at 0.0–1.5 km.

Walk, cycle, and transit are the sustainable modes of travel whereas two-wheelers and four-wheelers are non-sustainable modes. The percentage mode share is used for benchmarking.

To compute vehicle ownership two-wheeler and four-wheeler ownership is converted to Passenger Carriage Unit (PCU) for uniformity. The final PCU (vehicle ownership) is computed by the following Equation (3).

$$Passanger\ Carriage\ Unit\ (PCU) = 0.5 \times \left(\frac{Avg\ 2W \times 4.8}{Avg\ HH}\right) + 1 \times \left(\frac{Avg\ 4W \times 4.8}{Avg\ HH}\right) \quad (3)$$

where, 0.5 and 1.0 is the conversion factor for two-wheeler and four-wheeler respectively [42], *Avg 2W* and *Avg 4W* is the two-wheeler and four-wheeler ownership per neighbourhood respectively, *Avg HH* is the average household size in each neighbourhood which varies from 5.27 in NH1 to 3.37 in NH9 and 4.8 is the average national household size [43]. For benchmarking of vehicle ownership, a lower PCU score specifies higher sustainability.

Household spending on travel is calculated in the percentage of household income. Expenses more than 3% are considered for the benchmark value. The higher the percentage of travel expenses, the lower is the sustainability (refer to Table 6).

3.1.2. Benchmarking for Perception

The benchmarking for perception is based on the expert opinions and comparative scale. The "yes/no" nominal scale questions are converted into the percentage of the households who answered yes. For street safety, satisfaction, and social participation, higher sustainability means a higher percentage of "yes" and *vice-versa*. The residents' "urge to migrate" is recorded in the reverse scale, thus its higher percentage means lower sustainability. Similarly, if perception for pollution and traffic problems is high then the neighbourhood has low sustainability. Neighbourhood safety, quality, sense of belonging, and favouring mix are recorded on a five-point Likert scale. Scores are converted to a 1 to 5 sustainability scale (refer Table 7).

3.2. Weights and Aggregation

Assigning weights to indicators is critical because weighting reflects the relative importance given to each indicator. For travel behaviour indicators, weights are assigned based on researcher's insight, peers opinion, and residents' view. The share of trip length in indicators is 70%, as it has eight sub-indicators under it. Other indicators (travel mode, vehicle ownership and travel expenses) have a share of 10% each.

The equal weighting technique is adopted for perception data. This is adopted because diversified indicators are present, and it is difficult to judge which is better than the other. There are twelve indicators and each has one twelfth of the weight as shown in Table 5.

Aggregation groups all the indicator scores into a composite index score. To determine the composite indicator score, linear aggregation of the weighted normalized indicator is derived by the Equation no (4).

$$Composite\ Indicator\ Score = \sum_{i=1}^{n} w_i I_i \tag{4}$$

where, i = 1,2,3, ...n, and n is 11 for travel behaviour and 12 for perception indicator; w is the weight of each indicator as mentioned in Tables 5 and 6; and I correspond to the normalized indicator score between 1 to 5.

Table 6. Benchmarking and weights for travel behaviour indicators.

Indicator Description	Sustainability Benchmarking Range						Weights	
	Low	Medium-Low	Medium	Medium-High	High	A	Fraction of A	Weights
	1	2	3	4	5	%		
1. Trip length for Work	6.01–more	4.51–6.00	3.01–4.50	1.51–3.00	0–1.50	70	(2/8) × A	0.175
2. Trip length for Shop	1.61–more	1.21–1.60	0.81–1.20	0.41–0.80	0.00–0.40		(1/8) × A	0.088
3. Trip length for Vegetable market	1.61–more	1.21–1.60	0.81–1.20	0.41–0.80	0.00–0.40		(0.75/8) × A	0.066
4. Trip length for School	3.21–more	2.41–3.20	1.61–2.40	0.81–1.61	0.00–0.80		(1/8) × A	0.088
5. Trip length for College	6.01–more	4.51–6.00	3.01–4.50	1.51–3.00	0–1.50		(0.75/8) × A	0.066
6. Trip length to Health facility	3.21–more	2.41–3.20	1.61–2.40	0.81–1.61	0.00–0.80		(1/8) × A	0.088
7. Trip length for Restaurant/Theatre	3.21–more	2.41–3.20	1.61–2.40	0.81–1.61	0.00–0.80		(0.75/8) × A	0.066
8. Trip length for Garden	1.61–more	1.21–1.60	0.81–1.20	0.41–0.80	0.00–0.40		(0.75/8) × A	0.066
9. Percentage of Sustainable mode (Walk + Cycle + Transit)	0.00–20.00	20.01–40.00	41.01–60.00	60.01–80.00	80.01–100	30	(1/3) × A	0.1
10. Two-Wheelers and Four-Wheelers Converted in Passenger Car Unit (PCU)	4.01–more	3.01–4.00	2.01–3.00	1.01–2.00	0.00–1.00		(1/3) × A	0.1
11. Percentage of Income spent on Travelling	12.01–more	9.01–12.00	6.01–9.00	3.01–6.00	0.00–3.00		(1/3) × A	0.1

Table 7. Benchmarking and weights for perception indicators.

Indicator Description	Units	Sustainability Normalization Range					Equal Weights
		Low	Medium-Low	Medium	Medium-High	High	
		1	2	3	4	5	
1. On Street Safety	Percentage of household answered Yes	59–less	60–69	70–79	80–99	90–100	1/12 = 0.083 each
2. Satisfied with Amenities in NH		59–less	60–69	70–79	80–99	90–100	
3. Part of Social Group		59–less	60–69	70–79	80–99	90–100	
4. Wish to Shift from Current NH		61–more	46–60	31–45	16–30	0–15	
5. Neighbourhood Safety	5 point Likert scale	1	2	3	4	5	
6. Neighbourhood Quality		1	2	3	4	5	
7. Sense of Belonging		1	2	3	4	5	
8. Feel Traffic Problem	Percentage of household answered Yes	86–100	71–85	56–70	41–55	40–less	
9. Feel Noise Pollution		86–100	71–85	56–70	41–55	40–less	
10. Feel Air Pollution		86–100	71–85	56–70	41–55	40–less	
11. Favour Mix of Residential and Commercial use	5 point Likert scale	1	2	3	4	5	
12. Favour Mix of LIG, MIG and HIG		1	2	3	4	5	

3.3. Discussion

The composite indicator, sustainability index, sustainability level, and ranking of the neighbourhoods for both residents' travel behaviour and perception is presented in Table 8. The composite indicators' score is in the range of 0 to 5. It is converted to 0 to 1 scale to compute the sustainability index. The sustainability index of travel behaviour is in the range of 0.27 (for NH11) to 0.78 (for NH1) and perception is in the range of 0.47 (for NH12) to 0.83 (for NH6 and NH9).

Table 8. Sustainability index for travel behaviour and perception.

Category Levels	Selected Neighbourhoods		Travel Behaviour				Residents Perception			
			Composite Indicator Score	Sustainability Index	Comparative Sustainability level	Ranking	Composite Indicators Score	Sustainability Index	Comparative Sustainability level	Ranking
L1	Buldi	NH1	3.90	0.78	4	1	3.25	0.65	3	8
	Itvari	NH2	3.27	0.65	3	3	3.11	0.62	3	9
	Mahal	NH3	2.65	0.53	3	7	3.02	0.60	3	10
L2	Sadar	NH4	3.50	0.70	4	2	3.97	0.79	4	4
	Dharampeth	NH5	2.89	0.58	3	6	3.92	0.78	4	6
	Buddha Nagar	NH6	3.10	0.62	3	5	4.17	0.83	4	1
L3	Nagsen Nagar	NH7	3.20	0.64	3	4	3.84	0.77	4	7
	Trimurti Nagar	NH8	2.41	0.48	2	8	3.97	0.79	4	4
	Friends Colony	NH9	1.84	0.37	2	9	4.17	0.83	4	1
L4	Civil Lines	NH10	1.61	0.32	2	10	4.11	0.82	4	3
	Manish Nagar	NH11	1.37	0.27	1	12	2.51	0.50	3	11
	Nara	NH12	1.40	0.28	1	11	2.34	0.47	2	12

Note: Refer Table 5 for color coding.

The resultant sustainability scores of the neighbourhoods are grouped in five comparative sustainability levels. The final ranking is decided based on the sustainability index score. The neighbourhood with a high index is sustainable and ranked first in the list. Neighbourhoods with a high land-use mix are ranked at the top in travel behaviour, but ranked eighth to tenth in a perception survey. Neighbourhoods with moderate land-use mix are ranked between the third and ninth position for travel behaviour and first to seventh for perception. L4 neighbourhoods having a low land-use mix are ranked last in both cases except NH10 which is placed at third position in perception due to green ambiance and superior infrastructure.

3.3.1. Residents' Travel Behaviour

The travel behaviour study revealed that areas with a high mix of land-use (with entropy index between 0.71 and 1.00) have less trip length, less vehicle ownership, and less travel expenses. Thus, they are more sustainable with a score of 3 and 4 out of 5. However, the neighbourhoods with less land-use mix are scaled as extremely unsustainable (score 1) to unsustainable (score 2) as they need to travel long distances for all amenities and thus, have high vehicle ownership and more travel expenses (refer to Figure 10). Neighbourhoods with moderate land-use mixed are placed at the satisfactory level of sustainability. The findings in this study confirm the results of prior studies on land-use and travel relationship.

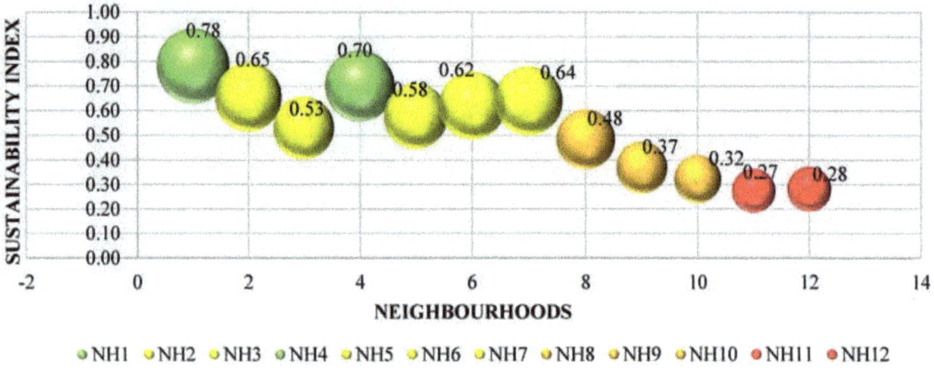

Figure 10. Sustainability index for travel behaviour.

3.3.2. Residents' Perception

Prior studies mention the social benefits of the mix land-use [6], but the study presented here reveals that neighbourhoods with moderate land-use mix are better liveable spaces than neighbourhoods with high land-use mix. This may be due to the fact that most prior studies have been in the context of developed countries where the land-use mix characteristics are very different from developing countries. Moreover, peoples' adaptability to various impacts of land-use mix also varies. It is observed that overall residents express a level of satisfaction even with inadequate facilities and infrastructure. Thus, none of the neighbourhood is extremely-sustainable (score 5) or extremely-unsustainable (score 1) and only one is unsustainable (score 2). Neighbourhoods with moderate land-use mix are sustainable, areas with high and low land-use mix have average sustainability (score 3) (refer to Figure 11).

The sustainability index of residents' travel behaviour with perception shows low correlation (0.30). The findings suggest that, even though neighbourhoods with high land-use mixed are sustainable in travel domain, the residents do not prefer to be part of these neighbourhoods; nor do they want to be part of neighbourhoods with low land-use mix (NH11 and NH12). They prefer neighbourhoods with moderate land-use mix. This advocates that residents wish to stay in neighbourhoods with moderate land-use mix and not in high or low land-use mix.

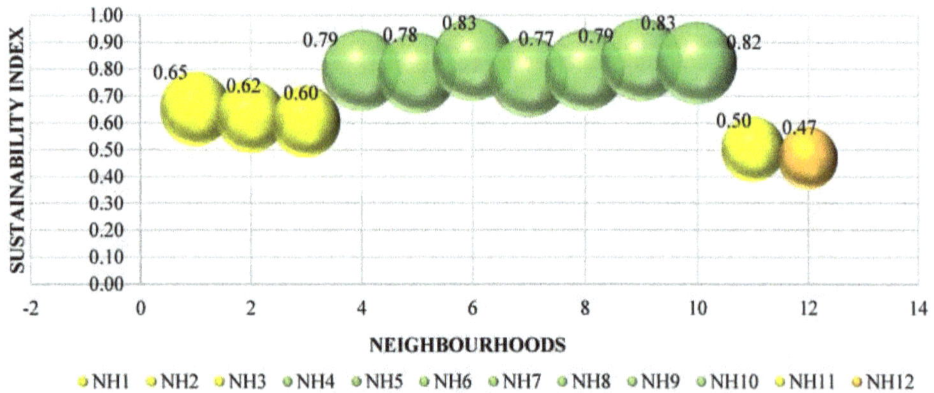

Figure 11. Sustainability index for residents' perception.

3.3.3. Relationship of Mixed Land-Use Measure with Residents' Travel Behaviour and Perception

Figure 12 shows the relationship of land-use mix measure (entropy index) with resident's travel behaviour and perception. The sustainability index score of resident's perception is higher than that of travel behaviour. Low entropy index score (*i.e.*, low mix land-use) is related to low sustainability. Moderate to high entropy index score (*i.e.*, moderate to high mix land-use) is related to moderate to high sustainability.

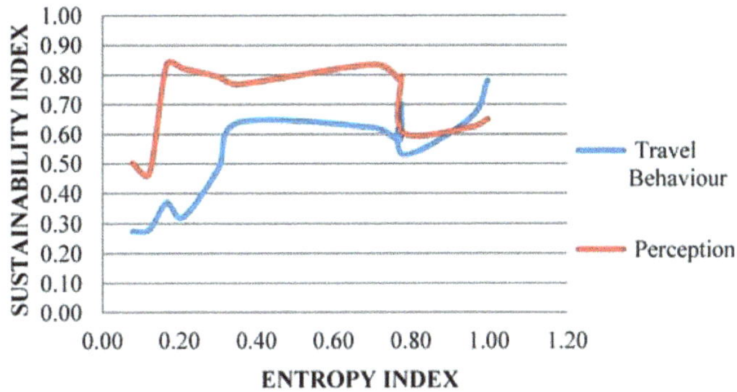

Figure 12. Scatter plot showing relationship of entropy index and sustainability index.

The correlation of the land-use mix (entropy index) with travel behaviour is 0.88 indicating both are highly correlated. Travel behaviour is sustainable when the land-use mix is high and *vice versa*.

The correlation of the land-use mix entropy index with perception is 0.10 indicating poor correlation. It signifies that land-use mix does not directly encourage perception, however, adequate amenities, good infrastructure, and less pollution contribute to an affirmative opinion. In study areas, neighbourhoods with high and low land-use mix lacked infrastructure, leading to dissatisfaction. Thus the poor correlation is perceived.

4. Conclusions

There is a strong advocacy to bring different land uses together for achieving sustainability. Indian cities have a legacy of mixed land-use character. There is an evident mix of land uses in the present context which hints towards a similar trend in the future as well. Here the concern is the proportion of the mix of different land uses for creating sustainable built environment.

Residential and commercial land-use mix is the most compatible mix type. However, the proportion of the mix of these two compatible land uses (commercial and residential) which leads to sustainability has not been explored in prior studies. The change in the land-use mix affects the inhabitants' travel behaviour, their safety and their satisfaction. The current study tries to identify sustainability indices for neighbourhoods with different land-use mix. By surveying twelve neighbourhoods (NH1 to NH12) categorized under four levels (L1 to L4) with different land-use mix, the study has attempted to comprehend how the neighbourhood performs with the change in land-use mix proportion. Mixed land-use is measured as entropy index of commercial and residential land-use.

Perception is carried out to encompass participatory approach, behaviour and experiences with the neighbourhood. The relationship between mixed land-use and travel behaviour has been well established in prior studies and is confirmed in the current study. Mixed land-use measure is highly correlated with the travel behaviour sustainability index, but is weakly correlated to the residents' perception sustainability index.

It can be concluded that the neighbourhoods with high land-use mix (L1 category with entropy nearing 1) are sustainable when it comes to travel behaviour but residents are not satisfied here due to the externalities associated with pollution, congestion, and lack of infrastructure.

The neighbourhoods with low land-use mix (L4 category with entropy nearing 0) are extremely unsustainable due to greater travel distance, poor accessibility of amenities, inadequate infrastructure, as well as safety and satisfaction issues. NH10 even though being a low and segregated land-use mixed neighbourhood, is more sustainable as it is planned and maintained with a good physical infrastructure and is the greenest part of the city.

Moderate land-use mix neighbourhoods (L2 and L3 category with entropy in the range of 0.77 to 0.33) are sustainable both in terms of residents' travel behaviour and perception, and are preferred by the residents. The citizens prefer to be a part of a green, aesthetically appealing, with adequate infrastructure, mixed land-use neighbourhood.

The outcome of the study suggests that high land-use mixed areas though sustainable in the travel domain are not likewise sustainable in the perception domain, due to dissatisfaction amongst the residents here. Low land-use mix and sprawled areas are unsustainable both for the travel and the perception domain. Moderate mix neighbourhoods lead to sustainable development.

Acknowledgments: The authors would like to acknowledge support provided by Visvesvaraya National Institute of Technology, Nagpur, India for publishing this work. The authors would also like to acknowledge support provided by residents during the household surveys conducted in various parts of Nagpur city.

Author Contributions: Sarika Bahadure, first author, took the initiative and conceived the study, shortlisted the indicators, collected data through household surveys, and carried out the statistical analysis. Rajashree Kotharkar, the second author, provided a methodology, framework and revisions for study, and structure of the paper.

Conflicts of Interest: The authors declare no conflict of interest.

References

1. Jill, G. Mixed use in theory and practice: Canadian experience with implementing a planning principle. *J. Am. Plann. Assoc.* **2002**, *68*, 71–84.
2. Miller, N.A.; Miller, J. Defining Mixed-Use Development. Available online: http://designforhealth.net/wp-content/uploads/2012/12/DPmixed_use.pdf (accessed on 12 March 2014).
3. Graeme, E.; Foord, J. The Generation of Diversity: Mixed-Use and Urban Sustainability. In *Urban Sustainability through Environmental Design: Approaches to Time-People-Place Responsive Urban Spaces*; Porta, S., Romice, O., Greaves, M., Thwaites, K., Eds.; Routledge: New York, NY, USA, 2007.
4. Tombari, E.A. *Smart Growth, Smart Choices Series: Mixed-Use Development*; AICP Land Development Series; National Association of Home Builders (NAHB): Washington, DC, USA, 2005.
5. Jones, P.; Roberts, M.; Morris, L. *Rediscovering Mixed-Use Streets, the Contribution of Local High Streets to Sustainable Communities*; The Policy Press: Bristol, UK, 2007.
6. Saville-Smith, K.; Dwyer, M.; Warren, J. Valuing Sustainable Neighbourhoods Final. s.l.: Report NH3112/2 for Beacon Pathway Limited. Available online: http://www.beaconpathway.co.nz/images/uploads/Final_Report_NH3112(2)_Valuing_neighbourhoods.pdf (accessed on 17 August 2015).
7. Jane, J. *The Death and Life of Great American Cities*; Vintage: New York, NY, USA, 1961.
8. Hidalgo, D.; Pai, M.; Carrigan, A.; Bhatt, A.; Owen, B. *National Investment in Urban Transport Towards People's Cities through Land Use and Transport Integration*; Shakti Sustainable Energy Foundation: New Delhi, India, 2012.
9. Lucas, R.E.; Rossi-Hansberg, E. On the Internal Structure of Cities. *Econometrica* **2002**, *70*, 1445–1476. [CrossRef]
10. Sita, K.; Phadke, V.; Guha, S. *The Declining City-Core of an Indian Metropolis: A Case Study of Bombay*; Concept Publishing Company: New Delhi, India, 1988.
11. Saville-Smith, K.; Lietz, K.; Bijoux, D.; Howell, M. *Neighbourhood Sustainability Framework: Prototype. s.l.: NH101 Prepared for Beacon Pathway Limited*; Beacon Pathway Limited and the Foundation for Research, Science and Technology: Auckland, New Zealand, 2005.

12. Brown, B.B.; Yamada, I.; Smith, K.R.; Zick, C.D.; Jones, L.K.; Fana, J.X. Mixed land use and walkability: Variations in land use measures and relationships with BMI, overweight, and obesity. *Health Place* **2009**, *15*, 1130–1141. [CrossRef] [PubMed]
13. Cervero, R. Land Use Mixing and Suburban Mobility. *Transp. Q.* **1988**, *42*, 429–446.
14. Stead, D.; Marshall, S. The Relationships between Urban Form and Travel Patterns. An International Review and Evaluation. *Eur. J. Transp. Infrastruct. Res.* **2001**, *1*, 113–141.
15. Dur, F.; Yigitcanlar, T. Assessing land-use and transport integration via a spatial composite indexing model. *Int. J. Environ. Sci. Technol.* **2015**, *12*, 803–816. [CrossRef]
16. Ewing, R.; Cervero, R. Travel and the Built Environment, A Meta-Analysis. *J. Am. Plan. Assoc.* **2010**, *76*, 265–294. [CrossRef]
17. Ewing, R.; Cervero, R. Travel and the Built Environment: A Synthesis. Transportation Research Record. *Transp. Res. Rec.* **2001**, *1780*, 87–114. [CrossRef]
18. Leck, E. The Impact of Urban Form on Travel Behavior: A Meta-Analysis. *Berkeley Plan. J.* **2006**, *19*, 37–58.
19. Boyce, D.E.; Williams, H.C.W.L. *Forecasting Urban Travel: Past, Present and Future*; Edward Elgar Publishing: Glos, UK, 2015.
20. Healthy Spaces and Places. Available online: http://www.healthyplaces.org.au/site/ (accessed on 12 August 2015).
21. Chandigarh Master Plan 2031, Chandigarh Administration. Available online: http://chandigarh.gov.in/cmp_2031.htm (accessed on 13 January 2015).
22. Spears, S.; Boarnet, M.G. *Draft Policy Brief on the Impacts of Landuse Mix Based on a Review of the Empirical Literature*; California Environmental Protection Agency Air Resources Board: Sacramento, CA, USA, 2014.
23. Frank, L.D.; Greenwald, M.J.; Kavage, S.; Devlin, A. *An Assessment of Urban Form and Pedestrian and Transit Improvements as an Integrated GHG Reduction Strategy*; Washington State Department of Transportation: Washington, DC, USA, 2011.
24. Sonia, H. The Mixed-Use Trend: Planning Attitudes and Practices in Northeast Ohio. *J. Archit. Plan. Res.* **2007**, *24*, 224–244.
25. Handy, S. Urban Form and Pedestrian Choices: Study of Austin Neighborhoods. *Transp. Res. Rec.* **1996**, *1552*, 135–144. [CrossRef]
26. Saville-Smith, K. The Value of Neighborhood Intensification: The Interface between Urban Dynamics & Housing Change Dwelling, Neighborhood Design and Affordability, New Zealand. In Proceedings of the Crossing into the 2nd Decade of the 3rd Millennium 22nd International Housing Research Conference, Istanbul, Turkey, 4–7 July 2010.
27. Bahadure, S.P.; Kotharkar, R.S. Social Sustainability and Mixed Landuse, Case Study of Neighbourhoods in Nagpur, India. *Bonfring Int. J. Ind. Eng. Manag. Sci.* **2012**, *4*, 76–83.
28. OECD. Household Behaviour and the Environment Reviewing the Evidence. Organisation For Economic Co-Operation and Development: France, 2008. Available online: http://www.oecd.org/environment/consumption-innovation/42183878.pdf (accessed on 26 October 2014).
29. Lesliea, E.; Saelensb, B.; Frankc, L.; Owena, N.; Baumand, A.; Coffeee, N.; Hugo, G. Residents' perceptions of walkability attributes in objectively different neighbourhoods: A pilot study. *Health Place* **2005**, *11*, 227–236. [CrossRef] [PubMed]
30. Chow, A.S.Y. Urban Design, Transport Sustainability and Residents' Perceived Sustainability: A Case Study of Transit-oriented Development in Hong Kong. *J. Comp. Asian Dev.* **2014**, *13*, 73–104. [CrossRef]
31. Travel Behavior. Available online: http://en.wikipedia.org/wiki/Travel_behavior.WikimediaFoundation,Inc (accessed on 20 May 2015).
32. Kotharkar, R.S.; Bahadure, S.P. Mixed Landuse and Sustainable Urban Development, A Case Study of Nagpur. In Proceedings of the PLEA 2012—28th Conference, Opportunities, Limits & Needs Towards an Environmentally Responsible Architecture, Lima, Perú, 7–9 November 2012.
33. Cao, X.; Handy, S.L.; Mokhtarian, P.L. The influences of the Built Environment and Residential Self-Selection on Pedestrian Behaviour: Evidence from Austin, TX. *Transportation* **2006**, *33*, 1–20. [CrossRef]
34. GoI. Provisional Population Totals, Census of India 2011, Urban Agglomerations/Cities Having Population 1 Million and above, 2011. Available online: http://censusindia.gov.in/2011-prov-results/paper2/data_files/india2/Million_Plus_UAs_Cities_2011.pdf (accessed on 13 December 2014).

35. Statistics Canada. *Survey Methods and Practices*; Minister Responsible for Statistics Canada Minister of Industry: Ottawa, ON, Canada, 2010; Catalogue no. 12-587-X. Available online: http://www.statcan.gc.ca/ (accessed on 15 January 2015).
36. Frank, L.D.; Pivo, G. Impacts of Mixed Use and Density on Utilization of Three Modes of Travel. *Transp. Res. Rec.* **1994**, *1466*, 44–52.
37. Frank, L.D.; Schmid, T.L.; Sallis, J.F.; Chapman, J.; Saelens, B.E. Linking objectively measured physical activity with objectively measured urban form: Findings from SMARTRAQ. *Am. J. Prev. Med.* **2005**, *2*, 117–125. [CrossRef] [PubMed]
38. Urban Mass Transit Company Limited and Nagpur Improvement Trust. *Comprehensive Mobility Plan for Nagpur Draft Final Report*; Urban Mass Transit Company Limited and Nagpur Improvement Trust: Nagpur, India, 2014.
39. Nardo, M.; Saisana, M.; Saltelli, A.; Tarantola, S. Workpackage 5, Input to Handbook of Good Practices for Composite Indicators' Development, Deliverable 5.2. Available online: https://www.uni-trier.de/fileadmin/fb4/projekte/SurveyStatisticsNet/KEI-WP5-D5.2.pdf (accessed on 15 November 2014).
40. Duncan, D.T.; Aldstadt, J.; Whalen, J.; Melly, S.J.; Gortmaker, S.L. Validation of Walk Score for Estimating Neighborhood Walkability: An Analysis of Four US Metropolitan Areas. *Int. J. Environ. Res. Public Health* **2011**, *8*, 4160–4179. [CrossRef] [PubMed]
41. Ewing, R.; Hodder, R. *Best Development Practices: A Primer for Smart Growth*; Smart Growth Network: Washington, DC, USA, 1998.
42. Congress, Indian Roads. *Guidelines for Capacity of Urban Roads in Plain Areas, IRC-106*; IRC Code of Practice: New Delhi, India, 1990.
43. MHFW, GoI. *National Family Health Survey (NFHS-3) 2005–2006 India*; International Institute for Population Sciences Deonar: Mumbai, India, 2007.

sustainability

MDPI

Article

Visualization of a City Sustainability Index (CSI): Towards Transdisciplinary Approaches Involving Multiple Stakeholders

Koichiro Mori [1,*], Toyonobu Fujii [2,†], Tsuguta Yamashita [3,†], Yutaka Mimura [4,†], Yuta Uchiyama [5,†] and Kengo Hayashi [4,†]

1. International Center, Shiga University, 1-1-1 Banba, Hikone, Shiga 522-8522, Japan
2. Oceanlab, School of Biological Sciences, University of Aberdeen, Main Street, Newburgh, Aberdeenshire AB41 6AA, UK; t.fujii@abdn.ac.uk
3. Department of Sociology, Kyoto University Graduate School of Letters, Yoshida Honmachi, Sakyo-ku, Kyoto 606-8501, Japan; yamashita.tsuguta.24a@st.kyoto-u.ac.jp
4. Research Institute for Humanity and Nature, 457-4 Motoyama, Kamigamo, Kita-ku, Kyoto 603-8047, Japan; mimura@chikyu.ac.jp (Y.M.); kensuke@chikyu.ac.jp (K.H.)
5. Institute of Human and Social Sciences, Kanazawa University, Kakuma, Kanazawa, Ishikawa 920-1192, Japan; y-uchiyama@staff.kanazawa-u.ac.jp

* Author to whom correspondence should be addressed; ko-mori@biwako.shiga-u.ac.jp; Tel.: +81-749-27-1109.

† These authors contributed equally to this work.

Academic Editors: Tan Yigitcanlar and Md. Kamruzzaman

Received: 31 March 2015; Accepted: 27 August 2015; Published: 10 September 2015

Abstract: We have developed a visualized 3-D model of a City Sustainability Index (CSI) based on our original concept of city sustainability in which a sustainable city is defined as one that maximizes socio-economic benefits while meeting constraint conditions of the environment and socio-economic equity on a permanent basis. The CSI is based on constraint and maximization indicators. Constraint indicators assess whether a city meets the necessary minimum conditions for city sustainability. Maximization indicators measure the benefits that a city generates in socio-economic aspects. When used in the policy-making process, the choice of constraint indicators should be implemented using a top-down approach. In contrast, a bottom-up approach is more suitable for defining maximization indicators because this technique involves multiple stakeholders (in a transdisciplinary approach). Using different materials of various colors, shapes, sizes, we designed and constructed the visualized physical model of the CSI to help people evaluate and compare the performance of different cities in terms of sustainability. The visualized model of the CSI can convey complicated information in a simple and straightforward manner to diverse stakeholders so that the sustainability analysis can be understood intuitively by ordinary citizens as well as experts. Thus, the CSI model helps stakeholders to develop critical thinking about city sustainability and enables policymakers to make informed decisions for sustainability through a transdisciplinary approach.

Keywords: indicator; stakeholder; megacity; co-design; co-production; bottom-up; anthropogenic impact; environmental threshold

1. Introduction

1.1. Background

Cities are among the most important entities to be controlled for achieving a sustainable future of human well-being on earth. Urban population and its impact have been steadily increasing. According to the United Nations Department of Economic and Social Affairs (UNDESA) [1], 3.6 billion people of the 7.0 billion world population live in urban areas, and this urban population is projected to increase to 6.3 billion in 2050 (while the world population will be 9.3 billion). Cities negatively impact local and global environments directly and indirectly through resource consumption and trade. For example, cities in the world account for between 71% and 76% of global carbon dioxide (CO_2) emissions [2], but are considered engines of economic development that foster socio-economic prosperity [3]. However, prosperity is also accompanied by an expansion of urban inequality [4]. Thus, striking a balance among environmental, economic and social needs is critical to securing sustainable human well-being. For this reason, there is an urgent need for a new and reliable system for assessing city sustainability that is capable of providing relevant and requisite information for policy making [5]. Furthermore, during the decision-making process, discourse would be desirable among all the relevant stakeholders so that information on city sustainability can be shared, and that feasible and effective policies can be contrived and implemented.

In response to such demands, we have developed an objective system called the "City Sustainability Index" (CSI), which provides a scientific basis for evaluating and comparing cities along the three dimensions of sustainable development (*i.e.*, environmental, economic and social considerations). In addition, we have created a visualized physical model of the CSI to deliver multi-dimensional information in a simple and straightforward manner so that the contents can be readily understood by diverse audiences. The purposes of this paper are to: (1) provide a brief explanation of the concept of city sustainability and its assessment methods; (2) introduce a visualized model of CSI with the results of sustainability assessment applied to 18 world megacities; and (3) discuss several applications and development of a visualized CSI model.

1.2. Sustainability and the Transdisciplinary Approach

There have been increasing demands to construct indicators that can assess urban sustainability, and many lists of urban sustainability indicators have been provided [6–9]. For example, Shen *et al.* examined and compared nine different lists of urban sustainability indicators used for nine regions/cities, and derived a primary list of urban sustainability indicators on a comparative basis [10]. These early studies have focused on making lists of environmental, economic, social and governance indicators to cover as many aspects of cities as possible. However, methods to integrate these different indicators have not been adequately developed [5]. Furthermore, most urban sustainability indicators have focused on the sustainability of the target cities themselves, not the sustainability of cities across the world; such indicators include the Green City Index [11,12], and City Development Index (or Global Urban Indicators) [13]. In addition, some research has considered the issues of environmental sustainability only at local scale, not on a global scale; these studies include that of Bettencourt and West, who quantitatively analyzed and discussed the impact of economic growth in cities on society and the local environment [14]. Therefore, a more comprehensive set of indicators is needed that can assess the impact exerted by cities upon the sustainability of the global environment.

The main reason for developing sustainability indicators is to provide tools for policy making, information sharing and community improvement [15]. The frameworks adopted for the existing sustainability indicators can be divided into two paradigms, namely "bottom-up" and "top-down" approaches [16]. The top-down approach is based on the knowledge of experts and professionals, with the intention of providing an objective and macroscopic viewpoint, which in turn enables comparative analysis among cities in different geographical locations. However, the interpretation of the concept of sustainability often is different among professionals [17]. Moreover, since the decision-making process is done only by a small number of experts and policymakers, the top-down approach tends to preclude active participation by members of the public, who will actually be most affected by the decision. In other words, the top-down approach fails to take into consideration the views and opinions that only local people can possess. As a result, decisions made through such a process might well be biased in favor of policymakers [18]. Therefore, the top-down approach is not based on a democratic process to represent the collective opinion of a community.

As the limitations and defects inherent in top-down approaches have been revealed, the merits of bottom-up approaches gradually have been examined. Bottom-up approaches rely on the participation of local communities in the decision-making process, and encourage individual members of a local community to select relevant indicators; therefore, a bottom-up approach reflects local needs and issues [19,20]. According to Fraser *et al.* [21], the implementation of a bottom-up approach has two main advantages. Firstly, by incorporating the knowledge of local community members who are most familiar with the local situation, this approach can increase the meaningfulness of indicators in the society. The bottom-up approach also reflects diverse opinions of the community, making the process of indicator selection more democratic than that in the top-down approach. Secondly, the process of community participation itself contributes to the empowerment and education of the community [22]. There even appear to be synergistic interactions between these two aspects (*i.e.*, inclusiveness and empowerment), since the purpose of establishing the sustainability indicators is the improvement of local and global environment, an aspect of which is the vitality and functioning of the local community itself.

However, the bottom-up approach also has disadvantages. Compared to the top-down approach, engaging the participation of a community in the selection of indicators requires much more time and resources, especially if there are conflicts of opinions within the community. The bottom-up approach also can result in selecting too many indicators within a particular city that are not comparable among cities.

Fortunately, top-down and bottom-up approaches are not mutually exclusive. In fact, the introduction of a bottom-up approach does not deny the necessity for quantitative and aggregated methods that are used often in top-down approaches. Comparative analysis with other locations or at wider geographical scales provides more comprehensive understanding of the whole, as well as the local situations, than can a narrowly focused (*i.e.*, strictly local) analysis. Integration of bottom-up and top-down approaches could compensate for the shortcomings in each, and has therefore been considered as an effective strategy for developing sustainability indices.

The approach that involves not only professionals of different disciplines, but also various stakeholders, is called a "transdisciplinary approach" [17,23,24]. To conduct a transdisciplinary approach, top-down and bottom-up approaches must be integrated. However, synthesizing these two contrasting approaches is not an easy process. There is always a possibility that conflicts between different stakeholders will occur, especially between citizens and policymakers. Since it is the policymakers who normally have the overriding power to make and to implement policies, these conflicts might result in disregard of bottom-up processes. On the other hand, it also can be expected that in some cases, the choices made by a community will be based solely on self-benefits, contradicting the needs for sustainable development at bigger scales. Therefore, a transdisciplinary approach must endeavor "to provide a balance between community and higher level actors" [25]. Although technical criteria can be measured, the perception of a citizen cannot be measured. Therefore, the method of assessing the level of convergence between different stakeholders, especially between professionals and citizens, can be helpful in building a consensus. Such a method has been developed by, for example, Battaglia *et al.* [24,26].

Fraser *et al.* also observed that one of the biggest challenges in the integration of top-down and bottom-up approaches is to identify the extent to which the public should be engaged in the decision-making process and the scales at which indicators are to be perceived as relevant by the public [21]. Because different stakeholders and indicators operate at different scales that range from the local community and administrative boundary to international and transboundary areas, identifying the ideal scale for indicators is not an easy task. When sustainability issues are addressed on a global scale, they are not likely to be perceived as relevant by local people or be reflected in their surrounding environment at local levels. Therefore, global environmental issues must be considered by the professionals from a scientific perspective, *i.e.*, through a top-down approach. In addition, local community members often focus merely on issues relating to the current generation, and not on the potential linkages between current and future generations. Because the notion of sustainability assumes intergenerational equality, the challenge to establish a relevant timescale is also an obstacle for the bottom-up approach. In the CSI, these challenges are handled by combining two different types of indicators, namely, "constraint" and "maximization" indicators. We assume that constraint and maximization indicators are suitable for top-down and bottom-up approaches, respectively, and a detailed account of the two types of indicators will be presented in Section 2.2.

1.3. *Visualization and the Transdisciplinary Approach*

A critical aspect in integrating of the two approaches (*i.e.*, bottom-up and top-down) is to present the process of data collection and interpretation, as well as the final output, clearly to the community in such a way that all non-experts can easily understand them and thereby readily participate in the whole process [27]. Visualization is an effective method by which to convey scientific discoveries to non-professional audiences [28–32]. Because visual data can be understood instinctively, visualization does not require high literacy about the scientific information, especially numerical and mathematical literacy [33]. Visualization can also provide an interactive interface and facilitate a participatory approach [34]. Methods of visualization have been elaborated widely, especially since the establishment of a special conference and journal dedicated to visualization in the 1980s [35].

Methods of data visualization frequently have been used to present scientific research on global environmental issues. Because studies on the global environment inevitably involve various pieces of information from different disciplines and their complex integration processes, visualization plays a critical part in making research results readily understood by an audience. Existing applications include the visualization of atmospheric data [36], CO_2 emissions [37] and simulation of water supply and demand [38]. "Tangible Earth" [39] and "Science on a Sphere®" (SOS) [40] developed a digital terrestrial globe that allows users to acquire interactive and educational experience about the global environment. Visualization also provides a platform that enhances viewers' spatial cognition, especially when the visualization deals with urban data [41–43]. Density distribution is a particularly important concept for understanding cities [44].

2. City Sustainability

2.1. *Concept of City Sustainability*

The term "sustainability" has become increasingly important, particularly in connection with global environmental issues, and what it means conceptually also has been continually evolving [45]. Sustainability is, therefore, not a general term that has a clear and fixed definition or application although the notion of sustainable development as presented in the Brundtland Report is famous. A review of the notion of sustainability is beyond the scope of this paper, but the interested reader can be referred to numerous pieces of academic literature [5,45–49]. Hereafter, we would like to focus on the notion of "city" sustainability.

Many urban sustainability indicators focus only on whether the city under evaluation is sustainable within its boundaries. Importantly, the concept of city sustainability in this paper (*i.e.*, CSI) is different from others in this respect. Herein, we use the notion of city sustainability based on Mori and Yamashita [50] because this interpretation provides a clear framework that describes how environmental, economic and social states should be related in terms of city sustainability. Accordingly, city sustainability denotes maximization of the total economic and social net benefits that a city produces, without exceeding environmental limits and while staying within acceptable limits of socio-economic inequity. In regard to environmental limits, when the limitations are on a global scale a set of threshold values is assigned based on the published study, "Planetary boundaries" [51]. This concept of city sustainability recognizes the global environmental limitations of the "leakage effects" that a city has beyond its urban boundaries. Therefore, city sustainability assumes that whether a city under evaluation continues to be in a healthy condition in terms of local environmental aspects is of no consequence; in contrast, whether the global environment is sustainable while the current socio-economic activities of the city are maintained is significantly important.

Constraint conditions should also be applied in the context of intergenerational equity. That is, for a specified future time period, the extent of socio-economic activities in a city should not exceed the environmental limits. Even if a city does not currently exceed a given threshold of environmental limits, the negative impacts may accumulate and exceed the threshold at some point in the future. The accumulated total environmental burden may have a serious negative impact on the environment that future generations should enjoy. If the accumulated burden into the future exceeds the environmental threshold within the given time scale, its current state should be considered to be unsustainable. Hence, a sustainable city is defined as a city that maximizes economic and social net benefits (degree of satisfaction) while meeting constraint conditions of the environment and socio-economic equity in both opportunities and distribution into the indefinite future.

However, it is insufficient to merely satisfy the constraint conditions. A city exists for the pursuit of economic and social prosperity based on agglomeration effects, and this point should not be viewed lightly. So long as a city fulfills conditions of limits in regard to the environment and equity, economic and social benefit must be increased to a maximum capacity. When maximization has not occurred in this city, there is still room to promote pursuit of benefit because the city has not reached an optimal condition.

2.2. Constraint and Maximization Indicators

Based on the foregoing concept of city sustainability, the CSI is composed of two types of indicators: constraint indicators and maximization indicators. Constraint indicators are used to judge whether a city meets the minimum necessary conditions to be sustainable, based on relevant criteria and thresholds in terms of environmental limitations and socio-economic equity. The thresholds related to environmental limitations should be provided by scientific research, considering leakage effects on the global environment beyond the boundaries of cities. Constraint indicators for environmental limits that are considered to be appropriate for the CSI include, among others, the annual amount of greenhouse gas emissions; water footprint; the atmospheric concentrations of PM_{10} (particulate matter 10 micrometers in size or less), nitrogen oxides (NO_x), sulfur oxides (SO_x) and mercury (Hg); and the amount of direct and indirect consumption of forest resources. Appropriate indicators for socio-economic equity include the Gini coefficient of household income, poverty ratio, and the population ratio of access to safe drinking water, among others.

Maximization indicators measure the benefits that cities generate in economic and social aspects. As cities create more benefits, prosperity increases. Concerning the maximization indicators, city performance can be evaluated in a succinct way: the higher the benefits are, the better the performance of the city is. Then, if the increase in the benefits was free from exceeding any environmental and socio-economic constraints, the city could simply continue to pursue economic growth and social amenities. Maximization indicators that are considered to be appropriate for the CSI include, among others, indicators of economic outputs, such as gross domestic product (GDP) per capita; agglomeration costs, such as traffic congestion and housing costs; indicators of social amenities, such as the extent of public transportation; the number of hospitals per a unit of area; the number of physicians per population; and the number of universities per population.

3. Visualized Model of CSI

3.1. Application of the CSI to Megacities

"Large" cities are often said to be eco-efficient; but megacities (cities with a population of at least 10 million) have large negative impacts on the global environment due to the total environmental burden of their large population. We have created a prototype CSI for 18 megacities using five constraint indicators and seven maximization indicators; these were selected based on data availability and comparability across the megacities.

The five constraint indicators are: the amount of emissions of CO_2; the atmospheric concentration of Hg; the atmospheric concentration of PM_{10}; water footprint based on blue water, which includes surface and groundwater and is measured by the consumptive use of the run-off flow [52]; and the Gini coefficient of household income. Grid data on a global scale were used to determine the amounts of CO_2 emissions and the atmospheric concentration of Hg. The World Health Organization (WHO) database of PM_{10} in cities was used to establish the atmospheric concentration of PM_{10}. Although data for the water footprint of countries were available, the data for specific cities were not. Thus, because there was a strong correlation between water footprint and GDP at national scales, we estimated the water footprint of respective megacities from the national data by using both population and GDP data sets for each megacity.

The seven maximization indicators are: the quantity of solid waste generated; ratio of GDP per capita; congestion cost; green rate (the ratio of green areas to urban areas); suicide rate; university density; and the number of physicians. Grid data on a global scale were used for calculating the ratio of GDP per capita, congestion cost (population-weighted average distance to a city center as a proxy) and green rate. Data defining waste generation, university coverage and suicide rate were available from the municipalities. For the university indicator, university rankings were considered in addition to the number of universities. The number of physicians is assumed to be a significant indicator with respect to social security; national data were used as a proxy for this indicator due to the lack of data at smaller scales. The sources of data are provided in Table 1.

The thresholds were derived from published research [51], and were used for judging whether a city meets the necessary minimum conditions for city sustainability. For example, the concept of planetary boundaries [51] has provided some threshold values for a few global environmental indicators including water footprint (global freshwater use). The sources of data and thresholds are provided in Table 1.

The prototype City Sustainability Index consisting of the 12 indicators was applied to the 18 megacities; however, the actual urban areas of the megacities were not clearly defined beforehand. In fact, various definitions and methods exist for delineating urban boundaries of cities, including administrative boundaries [53], functional boundaries [54,55], and morphological boundaries [56,57]. In this paper, we defined the spatial extent of urban boundaries based on population density data.

Although pair-wise comparisons between maximization indicators would be possible, presentation of all possible combinations of indicators for the 18 megacities would cause information overload. Therefore, information had to be distilled to facilitate comparisons among cities. For this purpose, we standardized maximization indicators on a relative scale using a z-value, where the "worst" city was assigned a ranking of 0 and the "best" city was given a ranking of 100. "Best" and "worst" were dependent on the context of the indicator; for some indicators (e.g., suicide rate) "best" was the smallest value, but for others (e.g., number of physicians) the largest value indicated the "best" performance.

The method that we employed to establish and standardize scores for the maximization indicators was the same as the method used to produce composite indicators consisting of multiple individual indicators [58]. Constraint indicators did not require scaling or standardization because the necessary minimum conditions for city sustainability (*i.e.*, threshold values) were derived from the literature; these thresholds were treated as unbiased standards in this study.

Table 1. Indicators used in a prototype City Sustainability Index (CSI).

Category		Indicator	Unit	Data Source	Threshold
Environmental	1	Water footprint based on blue water	$m^3/(\text{person year})$	The Water Footprint Network. Mekonnen, M.M. and Hoekstra, A.Y. [59]	4000 $km^3/(\text{person}\cdot\text{year})$ (from Planetary Boundaries)
	2	Amount of emissions of CO_2	$t/(\text{person year})$	EDGAR	Judge whether the state of the average temperature in 2100 will be 2 degrees on the Celsius scale higher than that in 1850. (from IPCC)
	3	Quantity of solid waste generated	$kg/(\text{person year})$	Karak *et al.* [60]. UN-HABITAT [61]. Japanese Ministry of the Environment [62]	No threshold. This is a maximization Indicator. Negative contributor.
	4	Atmospheric concentration of Hg	$\mu g/m^3$	AMAP/UNEP	1 $\mu g/m^3$ (annual average) (from WHO)
	5	Atmospheric concentration of PM_{10}	$\mu g/m^3$	WHO	70 $\mu g/m^3$ (Annual mean) (from WHO)
Economic	6	Ratio of GDP per capita (GDP per capita in the city/GDP per capita in the country in which the city is located)	dimensionless	The World Bank	No threshold. This is a maximization Indicator. Positive contributor.
	7	Gini coefficient of household income	dimensionless	UN HABITAT	0.4 (from UN-Habitat)
	8	Congestion cost (population-weighted average distance to a city center)	m	ORNL	No threshold. This is a maximization Indicator. Negative contributor.
Social	9	Green rate (Ratio of green areas to urban areas)	%	GLCF	No threshold. This is a maximization Indicator. Positive contributor.
	10	Suicide rate	per 100,000 people	Various sources	No threshold. This is a maximization Indicator. Negative contributor.
	11	University density	per one million people, considering the ranking of universities	Various sources	No threshold. This is a maximization Indicator. Positive contributor.
	12	Number of physicians	Per 1000 people	WHO	No threshold. This is a maximization Indicator. Positive contributor.

Figure 1 shows the assessment of the 18 megacities using the prototype CSI. The downward-sloping trend line shows the total of the standardized numerical values of the maximization indicators, or in other words, the size of total economic and social benefits. These numerical values do not have any units, and thus no scale on the vertical axis is shown. In Figure 1, the megacities are arranged (from right to left) in an increasing order of the total maximization indicator value. The bar graphs depict the determinations as to whether the necessary conditions for city sustainability are met for the five constraint indicators. Shaded sections indicate that the city is not sustainable in terms of that particular constraint indicator. To our surprise, no megacity satisfied the necessary minimum condition for city sustainability.

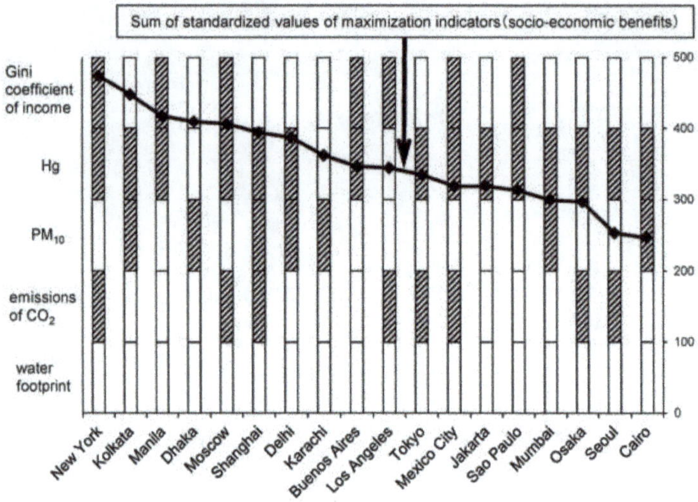

Figure 1. Results of city sustainability assessments of 18 world megacities using the City Sustainability Index. Bars represent determinations of sustainability for five constraint indicators; shaded segments indicate failure to meet threshold values for a particular indicator.

3.2. *Visualizing CSI*

Transdisciplinary research is a challenge of sustainability science that is formulated by Future Earth, a major international research platform coordinating new interdisciplinary approaches to investigate effective transformations to a sustainable world. The integration of scientific disciplines and social priority in terms of sustainability is of paramount importance. In this light, the CSI study described in Section 3.1 also aimed to involve multiple stakeholders effectively through the processes of co-design, co-production and co-dissemination (Figure 2). In the co-design process, scientifically trained academicians and social stakeholders jointly frame a definition of the required knowledge and propose a research definition to establish a commonly shared understanding of the research goals, to identify the relevant research disciplines, and to agree on the roles of different groups of stakeholders [63]. In the co-production process, all the stakeholders including academicians, policymakers, and the other public participants are required to continuously exchange their respective knowledge so as to ensure the societal relevance of the research [63]. The co-dissemination process includes publication of the acquired knowledge, conversion of it into usable and understandable information, and open discussions on the evaluation, application and relevance of the results, particularly among conflicting stakeholders in real society [63].

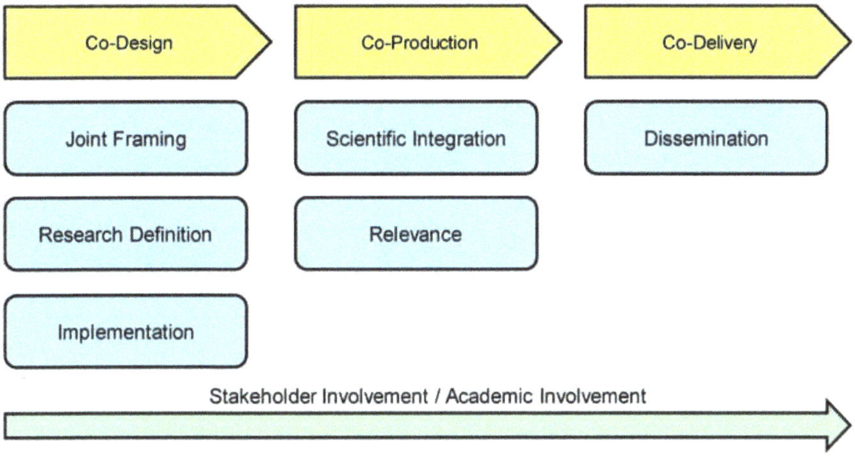

Figure 2. Framework for transdisciplinary co-creation of the knowledge. Source: Mauser *et al.* [63].

The importance of co-design, co-production and co-dissemination to relevant stakeholders has been increasingly recognized in the field of sustainability research because actions of stakeholders play a significant role in providing real solutions to the problems of sustainability. Academic researchers are required to share the scientific knowledge obtained from academic research with plural stakeholders in real society and link it to a specific social movement towards sustainable society. In this respect, we have created a visualized model of the CSI with which all stakeholders, including non-researchers, can intuitively understand the extent to which an evaluated city is sustainable (Figure 3). The visualized model of an evaluated city consists of four components: conditions of constraint indicators; distribution of population density; conditions of social maximization indicators and conditions of economic maximization indicators (Figure 3).

The top section of the visualization model in Figure 3 displays five globes, which represent the five constraint indicators used in the prototype CSI described in Section 3.1; by their color, the globes indicate whether the necessary minimum conditions for city sustainability have been satisfied. A blue globe, well-known imagery that depicts a healthy planet, indicates that the city is sustainable in terms of a given constraint indicator. A red globe is a warning sign; it implies that a city is unsustainable in its relationship with the global environment if the constraint indicator is an environmental indicator. The number of red globes indicates to what extent a city does not meet the necessary minimum conditions for city sustainability. Thus, an audience can easily see if a city is sustainable, or the extent to which it is not, simply by looking at the number of colored globes.

The section immediately below the globes shows the distribution of population density within the city being evaluated. Population density is denoted by the height of bar, and each bar represents a 5 km × 5 km physical area within the city. The higher the bar is, the higher the population density is. Normally, the spatial differences in population within a city are difficult to visualize. However, from the visualized model an audience can readily understand which parts of a city are more densely populated than others. Moreover, by comparing the spatial patterns of population densities across several cities, an audience can also judge whether a given city has a tendency to be "sprawled" or whether the city tends to have its population concentrated in a particular area, e.g., the city center.

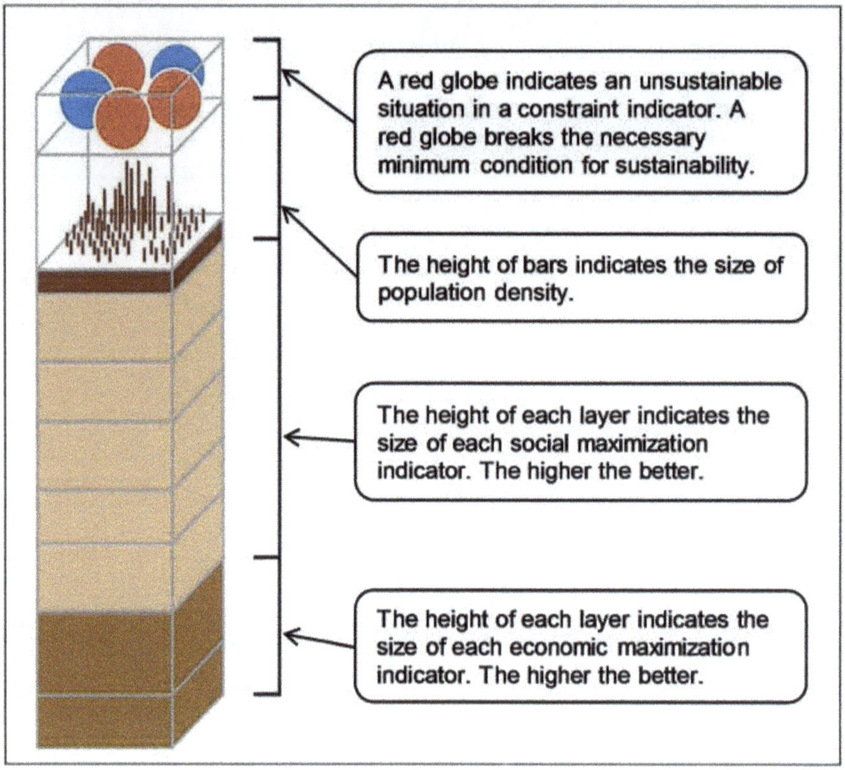

Figure 3. Structure of a visualized CSI model for a city.

The bottom two parts of the visualized CSI model contain "blocks", each of which corresponds to a particular economic or social maximization indicator. Each type of indicator (economic or social) is depicted in a different color, and the height of each block implies the size of benefit measured by the indicator. Thus, higher blocks indicate larger benefits than smaller blocks, and the cumulative height of all blocks shows the size of the total benefits that a city produces. Thus, by comparing the heights of blocks in the models for several cities, an audience can readily understand which city generates the largest benefits in total.

3.3. *Usage of the Visualized Model*

The visualized CSI model allows a user to assess and compare different cities easily based on the concept of city sustainability described in Sections 2.1 and 2.2. For example, Figure 4 shows visualized models for 12 megacities. A user can simply check whether there are any red globes to see how many cities meet the necessary minimum conditions for city sustainability. A user also can compare the physical heights of the models for the various cities to see which city has the highest overall socio-economic benefit (*i.e.*, the tallest model in Figure 4). Lastly, a user can readily identify the most sustainable city by selecting models of cities without red globes, and then picking out the city that is the highest among them. In short, the city for which the visualized CSI model has the greatest height and no red globe is the most sustainable.

Figure 4. Samples of visualized City Sustainability Index (CSI) models for 12 megacities.

Figure 5 illustrates how two cities can be compared to decide which is more sustainable. The top section of the model shows that City A is not sustainable (*i.e.*, it has red globes) and City B is sustainable (*i.e.*, it has no red globes). In this case, City B should always be assessed as being better than City A, even though the height of the model for City A is greater than that for City B. The heights of the two models indicate that City A generates greater benefits than City B in maximization indicators; however, this advantage is gained at the expense of the global environment. Neither of the models for City B and City C contain red globes, thus it is possible to compare the two cities in a straightforward way. The higher the model is, the more sustainable the city is. Because the model for City C is taller than that for City B, City C produces more socio-economic benefits than City B does, and does so in a sustainable way.

However, it is not necessarily possible to find the most sustainable city. If all the cities were to have at least one red globe (Figure 4), none of them would be considered to be sustainable because they do not satisfy the necessary minimum conditions for city sustainability. Under such circumstances, it is meaningless to compare the cities in search of the best city because all are unsustainable. Nevertheless, comparisons of both the model height and the number of red globes among cities can be useful in identifying the way to make such cities sustainable.

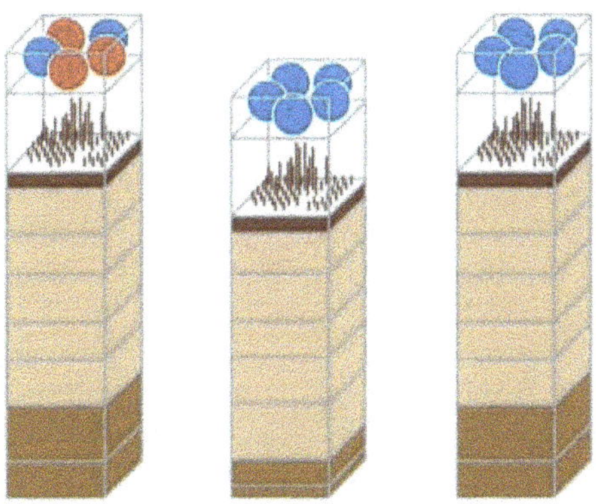

Figure 5. Comparison among cities.

On the other hand, the user in search of the most sustainable city might need to compare each maximization indicator among cities in addition to the total benefits even when all the cities under evaluation meet the necessary minimum conditions for city sustainability (e.g., City B and City C in Figure 5). Such a comparison is necessary because aggregation of individual benefits into a cumulative measure (*i.e.*, total benefits) loses detailed information about each indicator. Consequently, the sizes of benefits of individual indicators among cities may be vastly different, yet produce identical total benefits. Unless one city is better than another in every comparable indicator, the comparison of performance in each indicator is still critical. To accomplish such a comparison, the user can flexibly select the blocks of maximization indicators, and can choose a subset of blocks for comparing the height among cities.

It is also possible to make a comparative analysis of cities that are similar in terms of the distribution of population density. Such a comparison may reveal that some indicators exhibit larger differences between the cities than do any other pairs, thereby providing useful insight into how relevant policies and management actions may be focused on improving the conditions pertinent to those indicators. The feasibility and potentiality of extending such policies or actions to an "inferior" city might be high if similar cities were compared and analyzed.

4. Discussion

4.1. Bottom-up and Top-down Approaches in CSI

We believe that the visualized CSI model introduced in this paper provides a useful tool that allows a wide range of stakeholders to participate in the decision-making process on subjects that influence city sustainability. However, when applying a transdisciplinary approach to complex decision-making, arguments often eventuate in deliberations regarding whether to take a top-down approach or a bottom-up approach [15,21,25]. In the CSI, top-down and bottom-up approaches apply to constraint and maximization indicators, respectively. Because each approach has its own advantages and disadvantages, a CSI study must strive for the appropriate balance between the two approaches; this balance may be society-specific. Because we constructed the visualized CSI model to harmonize with the concept of a transdisciplinary approach, our focus has been placed more on the use of bottom-up approaches in maximization indicators than on the use of top-down approaches. The disadvantage of a bottom-up approach, however, is that there is a high probability that stakeholders will by themselves select the indicators or make judgements about them based purely on their lay opinions, which often lack both expert knowledge and scientific grounding. The solution derived from bottom-up decision-making processes could ignore long-term views, potential impacts of leakage effects and negative externality, and hence fail to take the well-being of future generations into consideration. Although a bottom-up approach seems to be possible for determining constraint indicators in CSI, one must remember that these indicators must be selected, and thresholds set, in a top-down approach based on a long-term view, global perspective, and scientific knowledge. If a bottom-up approach is incorporated into defining constraint indicators in a CSI study, important indicators on a global scale may be omitted, depending on the preferences of local stakeholders. If this happens, the purpose of promoting the sustainability of a city may be lost.

Maximization indicators measure the extent of the benefits a city generates in economic and social aspects, and these benefits are assessed after the conditions of the constraint indicators are met. In evaluating maximization indicators, there is a strong probability that the size of the benefit may be influenced unduly by the values and preferences of local stakeholders, such as residents of that city. If a city meets the necessary minimum conditions for city sustainability, there should be no objection to the city's maximization of economic and social benefits in any form. In other words, in a CSI study we are able to appreciate diversity in the values and preferences of stakeholders as long as the necessary minimum conditions for city sustainability are satisfied. Therefore, stakeholders may be allowed to freely select and manage the metrics for the maximization indicators by adopting a bottom-up approach. In this respect, we need to create a model of CSI in which weightings can be put flexibly on maximization indicators based on certain values expressed by stakeholders; this feature is not available in the current version of the CSI prototype.

4.2. Visualized CSI Model for Education

When city sustainability is assessed within a global perspective, a wide range of people from different communities and regions across the world must be engaged to raise awareness of the possible links between their respective societies' lifestyles and global environmental issues. One possibility for accomplishing such engagement would be to effectively disseminate scientific knowledge to a wide range of stakeholders through educational activities [25,64,65].

The visualized model of the CSI can convey complicated information in a simple and straightforward manner to diverse audiences so that the content of complex sustainability analyses can be readily understood by people who are not experts. The appeal and success of visualization in communicating about similarly complex issues has been demonstrated in the applications of Geographic Information Systems (GIS) [42,43] and the NOAA Science on a Sphere® (SOS) program [40]. In view of these successful approaches, the visualized CSI model can provide an ideal teaching resource and/or learning tool to raise awareness and explore real-world issues in a way that helps learners develop critical thinking about the society and environment to which they belong. The teaching and learning activities accomplished through the CSI can provide opportunities for learners to: (1) explore ideas and issues on the environmental effect each city has on the planet; (2) consider the choices citizens make and the consequences of those choices on the environment; (3) identify important local issues and link them to national and international issues of sustainable development; (4) compare and contrast the state of the environmental, economic and social systems that shape ways of living in the cities across the world; and (5) analyze the relationship between human activities and the environment on a global scale. The incorporation of CSI in education may therefore make an effective contribution to raising awareness, changing public behavior and eventually providing a basis for tackling the challenges of sustainable development faced by cities in the long term.

4.3. Visualized CSI Model for Policy-Making

Indicators have been playing an increasingly important role in providing vital information on subjects such as sustainable development to allow for informed decisions to be made. As a result, the focus is now shifting towards a process in which the development of urban sustainability indicators is integrated into policy institutions and decision-making processes of city planners [66]. Likewise, sustainability indicators have steadily gained acceptance as reliable tools to gauge the extent to which a community is moving towards sustainability [67]. In this regard, the CSI allows for the effective communication of visualized and quantitative information amongst diverse stakeholders, and the output is therefore expected to provide an ideal basis for informed decisions on policy making.

A workshop for discussions and information-sharing may be introduced for the purpose of transforming a city into a sustainable one [68,69]. Certainly, in the local policy-making process a workshop could be organized that involves a variety of stakeholders, and in which the visualized CSI model can be used to transform multi-dimensional information into a readily digestible form. The output of such a workshop could provide the respective stakeholders with opportunities to: (1) identify both positive and negative aspects of the city in relation to sustainable development; (2) evaluate possible links between alternative decisions that stakeholders make and the consequences of such decisions for the environment; and (3) consider the best options for decisions affecting the environment at local, national and international scales.

Alternatively, it may be more desirable to establish a consortium that involves cities across the world and in which the visualized CSI model is used optimally in the policy-making process. The output of the CSI provides the respective participating parties with the opportunity to compare the state of the environmental, economic and social systems between cities and evaluate alternatives in decisions affecting issues relating to city sustainability. The participants can then exchange ideas and discuss issues about sustainable development with reference to the city assessed to exhibit the best sustainability performance based on output of the visualized CSI model. This approach will allow poorly performing cities (with respect to sustainability) to develop critical thinking about the lifestyles of their societies. Importantly, cities with different cultural backgrounds must discuss if it is feasible for any social best practices and/or policies to be transferred among these cities. The visualized CSI model thus encourages many parties to participate in the development of the consortium and thereby facilitates multi-stakeholder processes and collaborations, as well as social learning, simultaneously.

Currently the visualized CSI model represents information about the sustainability of cities based on both constraint and maximization indicators. However, an ability to identify and indicate any cause-effect relationships among those indicators would be a desirable addition to the model, as this feature would allow policymakers to specify what controls (*i.e.*, policies and actions) should be implemented to achieve the best possible outcome for sustainable development. In addition, the types of indicators considered to be appropriate for a CSI study are not limited to the constraint and maximization indicators discussed in this paper; we are currently exploring additional unique aspects of city characteristics that are difficult to be assessed in terms of city sustainability (e.g., local climate, demographic structure, and composition of buildings). These additional indicators may be of more relevance than, say, the mean travel distance to a city center, in understanding the links between features of cities and their sustainability performance. Overall, the establishment of a consortium will strengthen the ties between cities, and the further development of the CSI will help the participants in such a consortium to better understand both differences and commonalities across cities in terms of city sustainability. This will in turn help the respective societies identify potential practices from their consortium colleagues to be incorporated in their own policy-making process, as well as to identify unique practices that may only be suitable for a particular society.

4.4. Limitations and Future Research

The visualized CSI model provides both a foundation for the transdisciplinary approach and a framework for measuring and comparing the performance of cities in the context of sustainability. However, the return of meaningful information for use in policy making is highly dependent on the way in which the indices that comprise CSI are incorporated in the actual participatory and/or decision-making processes. In this regard, there are mainly three issues to be addressed with respect to limitations and future prospects of the CSI methodology.

First of all, the visualized CSI model provides a simplified means to help a range of stakeholders comprehend and compare the states of different cities in terms of city sustainability. However, stakeholders have yet to be involved in the process of selecting indicators and deciding how the individual indicators are weighted and presented within the model. In essence, the transdisciplinary approach encourages interactive behavior through the good use of co-design and co-production in participant-led decision-making processes. In this regard, the CSI currently aims to provide an interactive platform for policy making through which stakeholders can decide which indicators to be considered, how they are weighted and how they are presented; once these decisions are made, a CSI study can be performed and stakeholders can discuss the results to come up with the best solution for the society. However, Lockton also points out that the goal of the design process is to modify or redesign the assessment system to influence users' behavior towards a particular 'target behavior' [70]. From this point of view, the CSI study may require stakeholders to be involved even at the designing stage of the model development so that the resulting outcome can influence the participants and then encourage their further engagement in support of a participatory approach.

Secondly, the current CSI model can only present a transient description of the state of a city at a fixed point in time; therefore, the model is not amenable to visually interactive manipulation. In the CSI, the maximization indicators are normalized and then aggregated so that their heights can be visually compared. However, if following analysis, one indicator is deemed to be more important than another, revised weightings must be assigned to the indicators and the indicators re-aggregated. The current CSI model does not accommodate this type of real-time demand ("what-if" scenarios) and this inability may hamper advancement in the use of the visualized CSI model in transdisciplinary programs. Fortunately, integrated technologies are available that provide multimodal display of interaction with information in real time [71]; these tools must be used to develop a computer-based, visually interactive CSI model that allows easy manipulation of weightings among indicators and thereby provides instant alternative graphical representations of different scenarios. In future developments of the CSI model, cause-effect relationships among the sustainability indicators must be identified and included. This feature will allow stakeholders to use a computer-based visualized CSI model to explore various decision choices for city sustainability through the use of co-design and co-production approaches. Furthermore, the identification and inclusion of the unique aspects of city characteristics into the CSI model, as discussed earlier, will add another dimension in the exploration of the cause-effect relationships among the existing indicators, the newly identified characteristics of cities and the performance of the urban sustainable development.

Finally, the effective use of the visualized CSI model is strongly dependent on the availability of a large quantity of data. In addition, the boundaries of cities in this study are defined based on population density (*i.e.*, urban settlement area), which do not necessarily match the politically defined city boundaries. For this reason, the collection and processing of the required data sets can be highly labor-intensive as well as prohibitively expensive. However, public participation through co-design and co-production processes, aided by a consortium of major cities, has gradually gained popularity as an essential component in the management of urban development, because such a participatory approach can help close resource gaps [72]. To help reduce data-gathering costs, public participants can be recruited and trained to conduct surveys for collecting data [67,73]. The involvement of citizens in data collection could provide the CSI with better and larger data sets. If this enhanced supply of data could be continually processed and updated in a refined visualized CSI model, further public engagement would be encouraged, as well as interactions between stakeholders, which may eventually generate a positive feedback in the context of city sustainability (Figure 6). The importance of continually revising and refining the visualized CSI model is crucial to its meeting the challenges of these critical issues in the context of city sustainability.

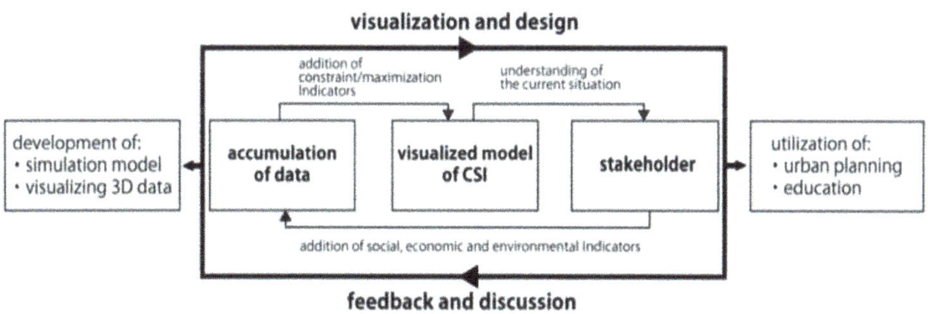

Figure 6. Schematic diagram of a feedback loop facilitated by use of the City Sustainability Index model.

5. Conclusions

We have developed a 3-D model of a City Sustainability Index (CSI), based on our original concept of city sustainability. A sustainable city is one that maximizes the total socio-economic net benefits within constraint conditions of environmental limits and socio-economic equity. In accordance with this concept, the system of the CSI comprised two types of indicators: maximization indicators and constraint indicators. The former measure the benefits that a city generates in socio-economic aspects. The latter judges whether a city meets the necessary minimum conditions for city sustainability.

The 3-D model of the CSI shows the results of the analysis in a user-friendly succinct way. To provide this visualization, the height of a model for a city represents the results for the maximization indicators, whereas blue or red globes displayed at the top of the model indicate how many city sustainability constraint conditions are satisfied.

When pursuing city sustainability in practice, the choice of constraint indicators should be made using a top-down approach on the basis of scientific knowledge. On the other hand, maximization indicators should be selected and weighted by implementing a bottom-up approach because this method respects the diversity of values that multiple stakeholders hold, which may be specific to a particular city. The flexibility in the choice and the prioritization of maximization indicators could be the key to successfully involving relevant stakeholders in transdisciplinary research towards city sustainability.

Crucially, for the involvement of appropriate stakeholders and the implementation of relevant policies towards sustainability in cities, complicated information must be conveyed to diverse stakeholders in a simple and straightforward manner so that the current and future situations on sustainability can be readily understood by all the key parties in the process, including non-experts. The visualized model of the CSI can contribute to this conveyance, and also potentially play a significant role in sustainability education.

Finally, it is worth noting here that transdisciplinary co-creation processes consisting of co-design, co-production and co-dissemination have been urgently required for practically solving global sustainability issues according to Future Earth (www.futureearth.org). However, these terms are relatively new and hence they have still remained equivocal even conceptually. As a result, it is not yet clear as to what kinds of practical activities in the transdisciplinary processes need to be conducted. This line of research should therefore be further carried out. This paper suggests that the visualized model of CSI can contribute to transdisciplinary co-creation processes for city sustainability.

Acknowledgments: The authors would like to thank the Research Institute for Humanity and Nature (RIHN), Shin Muramatsu (the leader of the project on Megacities and the Global Environment) and the other project members. This research has also been supported by the Future Earth program in the Future Earth program in the Research Institute of Science and Technology for Society (RISTEX) and the Japan Science and Technology Agency (JST).

Author Contributions: The corresponding author is a leader of this research, but all the authors contributed equally to this work.

Conflicts of Interest: The authors declare no conflict of interest.

References

1. United Nations Department of Economic and Social Affairs (UNDESA). *World Urbanization Prospects, the 2011 Revision*; United Nations: New York, NY, USA, 2012.
2. Seto, K.C.; Dhakal, S.; Bigio, A.; Blanco, H.; Delgado, G.C.; Dewar, D.; Huang, L.; Inaba, A.; Kansal, A.; Lwasa, S.; *et al.* Human Settlements, Infrastructure and Spatial Planning. In *Climate Change 2014: Mitigation of Climate Change. Contribution of Working Group III to the Fifth Assessment Report of the Intergovernmental Panel on Climate Change*; Edenhofer, O., Pichs-Madruga, R., Sokona, Y., Farahani, E., Kadner, S., Seyboth, K., Adler, A., Baum, I., Brunner, S., Eickemeier, P., *et al.*, Eds.; Cambridge University Press: Cambridge, UK, 2014; Chapter 12; pp. 923–1000.
3. United Nations Human Settlements Programme (UN-Habitat). *State of the World's Cities 2012/2013: Prosperity of Cities*; Routledge: New York, NY, USA, 2013.
4. United Nations Human Settlements Programme (UN-Habitat). *State of the World's Cities 2010/2011: Bridging the Urban Divide*; Routledge: New York, NY, USA, 2010.
5. Mori, K.; Christodoulou, A. Review of Sustainability Indices and Indicators: Towards a New City Sustainability Index (CSI). *Environ. Impact Assess. Rev.* **2012**, *32*, 94–106. [CrossRef]
6. Haghshenas, H.; Vaziri, M. Urban sustainable transportation indicators for global comparison. *Ecol. Indicators* **2012**, *15*, 115–121. [CrossRef]

7. Wang, Y.; Lam, K.-C.; Harder, M.K.; Ma, W.-C.; Yu, Q. Developing an Indicator System to Foster Sustainability in Strategic Planning in China: A Case Study of Pudong New Area, Shanghai. *Ecol. Indic.* **2013**, *29*, 376–389. [CrossRef]
8. López-Ruiz, V.-R.; Alfaro-Navarro, J.-L.; Nevado-Peña, D. Knowledge-city Index Construction: An Intellectual Capital Perspective. *Expert Syst. Appl.* **2014**, *41*, 5560–5572. [CrossRef]
9. Pires, S.M.; Fidélis, T.; Ramos, T.B. Measuring and comparing local sustainable development through common indicators: Constraints and achievements in practice. *Cities* **2014**, *39*, 1–9. [CrossRef]
10. Shen, L.-Y.; Ochoa, J.J.; Shah, M.N.; Zhang, X. The Application of Urban Sustainability Indicators—A Comparison between Various Practices. *Habitat Int.* **2011**, *35*, 17–29. [CrossRef]
11. Siemens. *The Green City Index: A Summary of the Green City Index Research Series*; Siemens: Munich, Germany, 2012. Available online: http//www.siemens.com/entry/cc/en/greencityindex.htm (accessed on 19 May 2014).
12. Meijering, J.V.; Kern, K.; Tobi, H. Identifying the Methodological Characteristics of European Green City Rankings. *Ecol. Indic.* **2014**, *43*, 132–142. [CrossRef]
13. United Nations Human Settlements Programme (UN-Habitat). *Global Urban Indicators Database, Version 2*; United Nations Publications: New York, NY, USA, 2001.
14. Bettencourt, L.; West, G. A Unified theory of urban living. *Nature* **2010**, *467*, 912–913. [CrossRef] [PubMed]
15. Singh, R.K.; Murty, H.R.; Gupta, S.K.; Dikshit, A.K. An overview of sustainability assessment methodologies. *Ecol. Indic.* **2009**, *9*, 189–212. [CrossRef]
16. Bell, S.; Morse, S. Breaking through the glass ceiling: Who really cares about sustainability indicators? *Local Environ.* **2001**, *6*, 291–309. [CrossRef]
17. Bond, A.J.; Viegas, C.V.; Coelho, C.C.D.S.R.; Selig, P.M. Informal Knowledge Processes: The Underpinning for Sustainability Outcomes in EIA? *J. Clean. Prod.* **2010**, *18*, 6–13. [CrossRef]
18. Morse, S. Putting the pieces back together again: An illustration of the problem of interpreting development indicators using an African case study. *Appl. Geogr.* **2004**, *24*, 1–22. [CrossRef]
19. Chambers, R. Participatory rural appraisal PRA: Analysis of experience. *World Dev.* **1994**, *22*, 1253–1268. [CrossRef]
20. Chambers, R. Participatory rural appraisal PRA: Challenges, potentials and paradigm. *World Dev.* **1994**, *22*, 1437–1454. [CrossRef]
21. Fraser, E.D.G.; Dougill, A.J.; Mabee, W.E.; Reed, M.; McAlpine, P. Bottom up and top down: Analysis of participatory processes for sustainability indicator identification as a pathway to community empowerment and sustainable environmental management. *J. Environ. Manag.* **2006**, *78*, 114–127. [CrossRef] [PubMed]
22. Pretty, J.N. Participatory learning for sustainable agriculture. *World Dev.* **1995**, *23*, 1247–1263. [CrossRef]
23. Scholz, R.W.; Lang, D.; Wiek, A.; Walter, A. Transdisciplinarity Case Studies as Means of Sustainability Learning. Historical Framework and Theory. *Int. J. Sustain. High. Educ.* **2006**, *7*, 226–251.
24. Maiello, A.; Battaglia, M.; Daddi, T.; Frey, M. Urban sustainability and knowledge: Theoretical heterogeneity and need of a transdisciplinary framework. A tale of four towns. *Futures* **2011**, *43*, 1164–1174. [CrossRef]
25. Reed, M.S.; Fraser, E.D.G.; Dougill, A.J. An adaptive learning process for developing and applying sustainability indicators with local communities. *Ecol. Econ.* **2006**, *59*, 406–418. [CrossRef]
26. Battaglia, M.; Meloni, E.; Cautillo, A. Technical Assessment and Public Perception of Environmental Issues: The Case of the Municipality of Pisa. *Local Environ.* **2014**, *19*, 786–802. [CrossRef]
27. Herweg, K.; Steiner, K.; Slaats, J. *Sustainable Land Management—Guidelines for Impact Monitoring*; Centre for Development and Environment: Bern, Switzerland, 1998.
28. McCormick, B.H.; DeFanti, T.A.; Brown, M.D. *Visualization in Scientific Computing*; ACM SIGGRAPH: New York, NY, USA, 1987; p. 3.
29. Kreuseler, M.; Schumann, H. A Flexible approach for visual data mining. *IEEE Trans. Visual. Comput. Graph.* **2002**, *8*, 39–51. [CrossRef]
30. Keim, D.A. Information visualization and visual data mining. *IEEE Trans. Visual. Comput. Graph.* **2002**, *8*, 1–8. [CrossRef]
31. Wu, L.; Hsu, P. An interactive and flexible information visualization method. *Inf. Sci.* **2013**, *221*, 306–315. [CrossRef]
32. Ware, C. *Information Visualization: Perception for Design*, 2nd ed.; Morgan Kaufmann: Waltham, MA, USA, 2004.

33. Ebert, D.S. Extending Visualization to Perceptualization: The Importance of Perception in Effective Communication of Information. In *The Visualization Handbook*; Hansen, C.D., Johnson, C.R., Eds.; Academic Press: Burlington, ON, USA, 2004; pp. 771–780.
34. Card, S.K.; Mackinlay, J.D.; Shneiderman, B. *Readings in Information Visualization: Using Vision to Think*; Academic Press: Waltham, MA, USA, 1999.
35. Hornbæk, K. The Notion of Overview in Information Visualization. *Int. J. Hum. Comput. Stud.* **2011**, *69*, 509–525. [CrossRef]
36. Liang, J.; Gong, J.; Li, W.; Ibrahim, A.N. Visualizing 3D atmospheric data with spherical volume texture on virtual globes. *Comput. Geosci.* **2014**, *68*, 81–91. [CrossRef]
37. Nieman, A. Concrete *vs* Abstract Visualization. In *Making Visible the Invisible: Art, Design and Science in Data Visualization*; Mohl, M., Ed.; University of Huddersfield: Huddersfield, UK, 2012; pp. 49–56.
38. White, D.D.; Wutich, A.; Larson, K.L.; Gober, P.; Lant, T.; Senneville, C. Credibility, Salience, and legitimacy of boundary objects: Water managers' assessment of a simulation model in an immersive decision theater. *Sci. Public Policy* **2010**, *37*, 219–232. [CrossRef]
39. Earth Literacy Program. Tangible Earth Project. Since 2008. Available online: http://tangible-earth.com/ (accessed on 10 March 2015).
40. Goldman, K.H.; Kessler, C.; Danter, E. *Science on a Sphere: Cross-Site Summative Evaluation*; Institute for Learning Innovation: Edgewater, FL, USA, 2010.
41. Acevedo, W.; Masuoka, P. Time-series animation techniques for visualizing urban growth. *Comput. Geosci.* **1997**, *23*, 423–435. [CrossRef]
42. Ribarsky, W. Virtual Geographic Information Systems. In *The Visualization Handbook*; Hansen, C.D., Johnson, C.R., Eds.; Elsevier: Burilngton, ON, USA, 2005; pp. 449–477.
43. Ribarsky, W.; Wasilewski, T.; Faust, N. From Urban Terrain Models to Visible Cities. *IEEE Comput. Graph. Appl.* **2002**, *22*, 10–15. [CrossRef]
44. Burdett, R.; Çavuşoğlu, Ö.; Verdis, S. *City Transformations*; LSE Cities, The London School of Economics and Political Science: London, UK, 2013. Available online: http://files.lsecities.net/files/2013/10/city-transformations-newspaper_en.pdf (accessed on 15 March 2015).
45. Handoh, I.; Hidaka, T. On the timescales of sustainability and futurability. *Futures* **2010**, *42*, 743–748. [CrossRef]
46. Dresner, S. *The Principles of Sustainability*, 2nd ed.; Earthscan: London, UK, 2008.
47. Pezzey, J.C.V.; Toman, M.A. *The Economics of Sustainability: A Review of Journal Articles. Discussion Paper 02–03*; Resources for the Future: Washington, DC, USA, 2002.
48. Fischer, J.; Manning, A.D.; Steffen, W.; Rose, D.B.; Daniell, K.; Felton, A.; Garnett, S.; Gilna, B.; Heinsohn, R.; Lindenmayer, D.B.; *et al.* Mind the Sustainability Gap. *TRENDS Ecol. Evol.* **2007**, *22*, 621–624. [CrossRef] [PubMed]
49. Baumgärtner, S.; Quaas, M. What is sustainability economics? *Ecol. Econ.* **2010**, *69*, 445–450. [CrossRef]
50. Mori, K.; Yamashita, T. Methodological framework of sustainability assessment in City Sustainability Index (CSI): A concept of constraint and maximisation indicators. *Habitat Int.* **2015**, *45*, 10–14. [CrossRef]
51. Rockstrom, J.; Steffen, W.; Noone, K.; Persson, A.; Chapin, F.S., III; Lambin, E.; Lenton, T.M.; Scheffer, M.; Folke, C.; Schellnhuber, H.; *et al.* Planetary boundaries: Exploring the safe operating space for humanity. *Ecol. Soc.* **2009**, *14*. Article 32.
52. Hoekstra, A.Y.; Chapagain, A.K.; Aldaya, M.M.; Mekonnen, M.M. *The Water Footprint Assessment Manual: Setting the Global Standard*; Earthscan: London, UK, 2011.
53. Aguilar, A.G.; Ward, P.M.; Smith, C.B. Globalization, regional development, and mega-city expansion in Latin America: Analyzing Mexico City's periurban hinterland. *Cities* **2003**, *20*, 3–21. [CrossRef]
54. Douglass, M. Mega-urban regions and world city formation: Globalisation, the economic crisis and urban policy issues in pacific Asia. *Urban Stud.* **2000**, *37*, 2315–2335. [CrossRef]
55. Hidle, K.; Farsund, A.A.; Lysgard, H.K. Urban rural flows and the meaning of borders functional and symbolic integration in Norwegian city-regions. *Eur. Urban Reg. Stud.* **2009**, *16*, 409–421. [CrossRef]
56. Benediktsson, J.A.; Pesaresi, M.; Amason, K. Classification and Feature Extraction for Remote Sensing Images from Urban Areas Based on Morphological Transformations. *IEEE Trans. Geosci. Remote Sens.* **2003**, *41*, 1940–1949. [CrossRef]

57. Rashed, T.; Weeks, J.R.; Roberts, D.; Rogan, J.; Powell, R. Measuring the Physical Composition of Urban Morphology Using Multiple Endmember Spectral Mixture Models. *Photogramm. Eng. Remote Sens.* **2003**, *69*, 1011–1020. [CrossRef]
58. Organisation for Economic Co-operation and Development (OECD). *Handbook on Constructing Composite Indicators—Methodology and User Guide*; OECD: Paris, France, 2008.
59. Mekonnen, M.M.; Hoekstra, A.Y. *National Water Footprint Accounts: The Green, Blue and Grey Water Footprint of Production and Consumption*; Value of Water Research Report Series No. 50; UNESCO-IHE: Delft, The Netherlands, 2011.
60. Karak, T.; Bhagat, R.M.; Bhattacharyya, P. Municipal Solid Waste Generation, Composition, and Management: The World Scenario. *Crit. Rev. Env. Sci. Technol.* **2012**, *42*, 1509–1630. [CrossRef]
61. UN-HABITAT. *Collection of Municipal Solid Waste—Key Issues for Decision-Makers in Developing Countries*; UNON, Publishing Services Section: Nairobi, Kenya, 2011.
62. Japanese Ministry of Environment. Waste Disposal in Japan. 2011. Available online: http://www.env.go.jp/recycle/waste_tech/ippan/h21/ (accessed on 2 September 2015). (In Japanese)
63. Mauser, W.; Klepper, G.; Rice, M.; Schmalzbauer, B.S.; Hackmann, H.; Leemans, R.; Moore, H. Transdisciplinary Global Change Research: The Co-creation of Knowledge for Sustainability. *Curr. Opin. Environ. Sustain.* **2013**, *5*, 420–431. [CrossRef]
64. Brewer, G.; Gajendran, T.; Landorf, C.; Williams, A. Educating for Urban Sustainability: A Transdisciplinary Approach. *Proc. Inst. Civil Eng. Eng. Sustain.* **2008**, *161*, 185–193. [CrossRef]
65. Reed, M.S. Stakeholder Participation for Environmental Management: A Literature Review. *Biol. Conserv.* **2008**, *141*, 2417–2431. [CrossRef]
66. Pintér, L.; Swanson, D.A.; Barr, J. *Use of Indicators in Policy Analysis: Annotated Training Module Prepared for the World Bank Institute*; IISD: Winnipeg, MB, Canada, 2004; Available online: http://www.iisd.org/pdf/2006/measure_use_indicators.pdf (accessed on 25 March 2015).
67. Kohsaka, R. Developing biodiversity indicators for cities: Applying the DPSIR Model to Nagoya and integrating social and ecological aspects. *Ecol. Res.* **2010**, *25*, 925–936. [CrossRef]
68. Nevens, F.; Frantzeskaki, N.; Gorissen, L.; Loorbach, D. Urban transition labs: Co-creating transformative action for sustainable cities. *J. Clean. Prod.* **2013**, *50*, 111–122. [CrossRef]
69. Sellberg, M.M.; Wilkinson, C.; Peterson, G.D. Resilience assessment: A useful approach to navigate urban sustainability challenges. *Ecol. Soc.* **2015**. [CrossRef]
70. Lockton, D.; Harrison, D.; Stanton, N.A. The design with intent method: A design tool for influencing user behavior. *Appl. Ergon.* **2010**, *41*, 382–392. [CrossRef] [PubMed]
71. Loftin, R.B.; Chen, J.X.; Rosenblum, L. Visualization Using Virtual Reality. In *The Visualization Handbook*; Hansen, C.D., Johnson, C.R., Eds.; Elsevier: Burlington, VT, USA, 2004; pp. 479–489.
72. Evans, B.; Joas, M.; Sundback, S.; Theobald, K. *Governing Sustainable Cities*; Earthscan: London, UK, 2005.
73. Elmqvist, T.; Alfsen, C.; Colding, J. Urban Systems. In *Ecosystems, Volume [5] of Encyclopedia of Ecology*; Jørgensen, S.E., Fath, B.D., Eds.; Elsevier: Oxford, UK, 2008; pp. 3665–3672.

sustainability

MDPI

Article

Moving towards Sustainability: Road Grades and On-Road Emissions of Heavy-Duty Vehicles—A Case Study

Wendan Zhang [1,2], Jian Lu [3], Ping Xu [1,2] and Yi Zhang [1,2,*]

[1] State Key Laboratory of Ocean Engineering, Shanghai Jiao Tong University, Shanghai 200240, China; fruits@sjtu.edu.cn (W.Z.); shanon@sjtu.edu.cn (P.X.)

[2] Transportation Research Center, Shanghai Jiao Tong University, Shanghai 200240, China

[3] Key Laboratory of Road and Traffic Engineering of Ministry of Education, Tongji University, Shanghai 200092, China; jianjohnlu@sjtu.edu.cn

[*] Author to whom correspondence should be addressed; darrenzhy@sjtu.edu.cn; Tel./Fax: +86-21-3420-7275.

Academic Editor: Tan Yigitcanlar

Received: 13 June 2015; Accepted: 7 September 2015; Published: 15 September 2015

Abstract: On-road vehicle emissions are one of the major sources of transport emissions. As a key design factor, road grades (or road slopes) have significant effects on on-road vehicle emissions, particularly on Heavy-Duty Vehicles (HDVs). However, the research into the relationship between road grades and on-road vehicle emissions is very rare in China. Taking a road network in Taiyuan, China, as a study area, this paper explored the influences of road grades on carbon monoxide (CO), hydrocarbon (HC), and nitrogen oxide (NO_x) emissions of HDVs. Combining emissions data collected by Portable Emission Measurement System (PEMS) with Vehicle Specific Power (VSP), we developed an emission rate model of HDVs. Then, we integrated it with the traffic simulation model VISSIM to attain the emissions of HDVs on nine scenarios differentiated by road grades. The results showed that the three emissions are found to be highly correlated to road grades, among which the CO emissions are most sensitive to the change of road grades and the HC emissions least. Compared to the emissions at 0% grade, the emissions at 4% grade will be boosted from 39.0% to 60.6%. The CO and NO_x emissions increase with the road grades in all nine scenarios, while the variations of HC emissions in different scenarios were complicated. The findings of this research will provide insights for policy-makers, scholars, and practitioners into strategies for improving road design to reduce traffic emissions and develop sustainable transportation in China.

Keywords: road grades; heavy-duty vehicles; emissions; sustainable transportation; China

1. Introduction

The combustion of fossil fuels by vehicles leads to road traffic emissions, which are one of the most significant sources of primary air pollutants. Air pollution is a major risk to health and to the environment, in developed and developing countries alike. Most sources of outdoor air pollution are well beyond the control of individuals and demand action by cities, as well as national and international policymakers in sectors like transport and energy waste management [1]. Road transport often appears as the single most important source of urban pollutant emissions in source apportionment studies, and road transport is likely to remain a large contributor to air pollution in the coming decades [2–4].

Motor vehicle emissions are a primary concern because of their potential detriment to local air quality and global atmospheric conditions. Although significant progress has been made to limit the pollutant emissions from individual vehicles, emissions from conventional gasoline-powered motor vehicles remain a concern in the United States (U.S.) and account for the majority of petroleum consumption [5]. In Europe, with the successful implementation of catalytic converters in gasoline vehicles, the transport sector, which used to be the highest emitter of Carbon Monoxide (CO), has reduced its CO emissions significantly (61% from 2003 to 2012). The transport sector is also the largest contributor to Nitrogen Oxides (NO_x) emissions, accounting for 46% of total European Union (EU)-28 emissions in 2012 [6]. The precise information of on-road vehicle emissions is crucial for the evaluation of the contribution of road traffic in general. For this reason, major efforts are being made for the reduction of polluting emissions from road transport all over the world [7].

Humanity has entered a new era of sustainability challenges, in striving to prevent our society and future generations from tipping into disastrous states, sustainable development has remained one of the primary policy goals in the large majority of countries around the world. Even the transport infrastructure sector has come to realize that development must not come at the expense of environmental and social objectives [8–10]. Over the past several decades, with rapid urbanization, many cities across the world, especially in developed countries, have experienced explosive trends in road construction and motorization. Although this has greatly increased people's mobility in daily life, it has also led to a series of problems, such as exhausting the finite fuel reserves and raising vehicle emissions. Therefore, the sustainability of cities is threatened by excessive automobile dependency [11].

How do road design and traffic management factors (e.g., road grades, lane width, and speed limit) influence on-road vehicle emissions? Currently, studies focusing on the relationships between significant factors and on-road vehicle emissions remain generally weak, providing insufficient support for strategies and guidelines from the perspective of improving road design and management. Therefore, it is imperative for policy-makers, scholars, and practitioners to investigate the factors significantly associated with on-road vehicle emissions to draw up strategies and guidelines to achieve the goal of reducing vehicle emissions and developing sustainable transportation. With data collected in Taiyuan, China, the aim of this paper is to explore the relationship between road grades and CO, Hydrocarbons (HC) and NOx emissions of Heavy-Duty Vehicles (HDVs) in, and provide guidance for the design of road grades that facilitates the reduction of on-road vehicle emissions and the development of sustainable transportation.

2. Literature Review

2.1. Road Grades and On-Road Vehicle Emissions

Over the past several decades, the volume of literature on the relationship between various factors and on-road vehicle emissions in the Western context has exploded, explaining why and how those factors might influence on-road vehicle emissions from diversified perspectives [12–20]. Among those studies, an increasing number of studies have explored the influences of road design factors, e.g., template (width, full bench/side cast), curve widening, and grade (or slope), with an attempt to reduce emissions through the improvement of road design methodology [15,17]. In Table 1, we listed some studies on the changes of fuel consumption and emissions (Carbon Dioxide (CO_2), CO, HC, and NO_x) with the change of road grades. The information in Table 1 indicated that:

(a) Three out of six papers considered uphill only and the other three analyzed both uphill and downhill.

(b) Generally, positive road grade yields higher emissions and negative road grade results in lower emissions.

(c) Among the three papers analyzing both uphill and downhill, Boriboonsomsin and Barth [15], and Frey *et al.* [21] concluded that the fuel change at locations with negative grade is higher than the fuel change at locations with positive grade. However, Boroujeni and Frey had the opposite conclusion [22].

It must be noted that these literatures deduced their conclusions under the condition of many restrictions, such as the test sites, categories of vehicles and fuels, driving behaviors, *etc.* Taking vehicles as an example, the first and third papers in Table 1 used light-duty vehicle while the sixth used passenger cars and trucks. Therefore, each paper in Table 1 has its own limited conditions, and they may not represent the situation at other conditions. In addition to studies in Table 1, some other papers also discussed the relationship of road grades on engine performance or vehicle emissions. Wyatt *et al.* found it was not correct to assume that the increased emissions on uphill slopes will be offset by the decreased emissions on paired downhill sections [23]. On the contrary, Prati *et al.*, indicated that the presence of both positive and negative grades along the same route has the effect of balancing these opposite influences on engine performance [24]. Alam *et al.*, observed that at lower grades, the change rates of emissions were higher and as the grade increases the change rates decrease [25]. Considering the brakes, Kern *et al.*, through the use of a model, found that vehicle emissions rose sharply once hills were sufficiently steep to warrant the use of brakes [26]. In the study of Silva *et al.*, the average HC and CO emissions for the entire segments slightly decreases for scenarios involving grades between 1% and 3% and then increase relatively to the baseline [13].

Table 1. Studies on the changes of fuel consumption and emissions with the change of road grades.

NO.	References	Grade Range (%)	Change Rate of Emission (%)				Change Rate of Fuel Consumption (%)	Mile-Per-Gallon Fuel Economy
			CO	HC	NO_x	CO_2		
1	Silva *et al.* [13]	from 0% to + 8%	4–73	3–47	24–380	2–64		
2	Zhang and Frey [27]	≥5%	60–140	60–110	180–450	40–90		
3	Boroujeni and Frey [22]	negative actual grade (1102 *)	1	−1	−7		−12	
		positive actual grade (1111 *)	38	22	42		14	
		all segments(2213 *)	10	4	10		1	
4	Boriboonsomsin and Barth [15]	6% followed by −6%					15–20	
		positive						from −2 to −1.5 times
		negative						2 times
5	Frey *et al.* [21]	positive					16–22	
		negative					from −22 to −24	
		total					from −5 to 0	
		from 0% to + 5%					40–100	
6	Cicero-Fernândez *et al.* [17]	from 0% to + 7%	3.0g/mile/ + 1%grade	0.04 g/mile/ + 1% grade				
		4.5% (vehicle with four passengers)	10.2 g/mile	0.07 g/mile				
		4.5% and 6.7% CO (air conditioning)	31.9 g/mile	0.07 g/mile				

* denotes the number of the segments with slope.

2.2. Review of Previously Research Methods

The most-frequently used method of "road design–on-road vehicle emissions" related studies is constituted by the collection of emissions data, development of an emission rate model, integration of emission rate model and traffic simulation models, and statistical analysis [12,14].

2.2.1. Collection of Emissions Data

On-board Portable Emission Measurement System (PEMS) is widely used in on-road vehicle emissions studies [13]. PEMS is able to collect the data of second-by-second emissions and speed variation of the vehicle under real-world conditions at any location traveled by the vehicle [17]. The relationship between second-by-second speed variation and on-road vehicle emissions can be used to build emission rate model. It is certain that this kind of emission rate model is more reliable than other microscopic emission models, because measurements are obtained during real-world driving, eliminating the concern about non-representativeness that is often an issue with dynamometer testing, and at any location, eliminating the site restrictions inherent in remote sensing [28].

2.2.2. Development of Emission Rate Model

Since the 1970s, different emission rate model were developed to estimate transport emissions with the purpose of focusing control strategies and transportation planning on those that are most effective, predicting how different strategies will effect local emissions, and measuring progress in reducing emissions over time. The Mobile Source Emissions Factor (MOBILE) model, first developed in the late 1970s in the U.S., is an emission rate model for predicting gram per mile emissions of CO_2, CO, HC, and NO_x, particulate matter (PM), and toxics from cars, trucks, and motorcycles under various conditions [29]. MOBILE has been replaced by Motor Vehicle Emission Simulator (MOVES) as U.S. Environmental Protection Agency's (EPA) official model for estimating emissions from cars, trucks and motorcycles. MOVES is able to estimate emissions for mobile sources covering a broad range of pollutants and allows multiple scale analysis [30]. Another emission rate model, Comprehensive Modal Emissions Model (CMEM), was initially developed in the late 1990s with sponsorship from the National Cooperative Highway Research Program (NCHRP) and EPA to fulfill the need for microscopic emissions modeling. CMEM is microscopic in the sense that it predicts second-by-second tailpipe emissions and fuel consumption based on different modal operations from the in-use vehicle fleet [31]. Outside the U.S., Davis *et al.*, developed the International Vehicle Emissions (IVE) model to make estimates of local air pollutants (criteria pollutants), greenhouse gas emissions, and toxic pollutants from motor vehicles [32].

Previous studies found out that the instantaneous vehicle emissions were highly correlated to the power output, *i.e.*, Vehicle Specific Power (VSP, kw/ton) [33]. It can be defined as the engine power output per unit vehicle total mass and takes into account aerodynamic drag, tire rolling resistance and road grade. VSP is a variable that relates the second-by-second vehicle emissions with the vehicle's instantaneous driving modes, which can be calculated from the instantaneous speed, acceleration and road grade [18]. Recently, VSP has been well accepted and widely applied in studies on emission modeling and prediction [34]. In China, the VSP-based emission rate model has been studied in the past several years. Qi *et al.* developed microscopic emissions models that predict vehicle fuel consumption and emissions by using instantaneous speed and acceleration and deceleration as explanatory variables [35].

2.2.3. Integration of Emission Rate Model and Traffic Simulation Models

PTV VISSIM is a microscopic multi-modal traffic flow simulation software package developed by PTV Planung Transport Verkehr AG in Karlsruhe, Germany. The name is derived from "Verkehr In Städten—SIMulationsmodell" (German for "Traffic in cities—simulation model"). VISSIM was first developed in 1992 and the latest version is PTV VISSIM 7 which was released in 2014. Song *et al.*, studied the applicability of traffic simulation model VISSIM in vehicle emission estimations based on VSP distribution [36]. Bing *et al.*, and Zhang *et al.*, investigated the relationship of intersection lane configuration and speed limit to vehicle emissions with VSP-based emission rate model [12,14].

Microscopic traffic simulation models, e.g., VISSIM, Paramics, were employed together with emission rate model for more complicated analysis. With the integration of VISSIM and CMEM, Noland *et al.*, evaluated the impact of the changes of road capacity on vehicle emissions [19]. Stevanovic *et al.* linked VISSIM and CMEM to optimize signal timings and minimize fuel consumption and CO_2 emissions [20]. By integrating VISSIM with MOVES, Lv *et al.* investigated the effects of traffic signal coordination on emissions and compared it with their effects on operational performance measures of delay and stops [37]. Paramics, another commonly used microscopic traffic simulation model, has also been employed in analyzing on-road vehicle emissions. Using Paramics, Chamberlin *et al.*, compared the emissions estimates from MOVES to those generated by CMEM [16]. Bartin *et al.* integrated MOBILE6.2 in Paramics and presented a microscopic simulation-based estimation of the spatiotemporal change in air pollution levels as a result of Electronic Toll Collection (ETC) deployment on a large-scale traffic network [38]. By integrating Paramics with CMEM, Boriboonsomsin *et al.*, estimated and compared vehicle emissions contributed from two types of High Occupancy Vehicle (HOV) lane configurations [15].

2.3. Studies on On-Road Emissions of HDVs in China

Although some scholars have preliminary established VSP based emission rate model with data collected in China, they rarely focused on on-road emissions of HDVs. HDVs, as diesel vehicles, are responsible for nearly 20 percent of CO_2 emissions from transportation in China, and at the same time, produce high levels of CO, HC, NO_x, and PM. These emissions are not only related to global warming and environmental disruption, but are also associated with bronchitis, asthma, and other lung diseases, and are responsible for millions of premature deaths worldwide [39]. Meanwhile, as a key geometric design factor, road grades are found to have significant effects on on-road vehicle emissions, particularly on HDVs [15,40,41]. However, research into the relationship between road grades and emissions of HDVs is very rare, especially in China.

In this paper, we present research results from a case study aiming at unraveling the influences of road grades on CO, HC and NO_x emissions of HDVs, with data collected from Taiyuan Metropolitan Area, China. The reason we focus on CO, HC and NO_x emission is because they account for nearly 90% of total on-road vehicle emissions in China [42]. In accordance with national standard, the HDVs in this study are diesel vehicles with a weight of over 3500 kg. Firstly, we establish an emission rate model of HDVs under VSP-bins with the data collected from a PEMS, OBEAS-3000 (developed by Xiamen Tongchuang Detection Technology Co., Ltd, Xiamen, China), in Taiyuan. Secondly, with the integration of emission rate model of HDVs and VISSIM, we calculate the CO, HC, and NO_x emissions of HDVs in 9 scenarios differentiated from road grades (from 0% to 4%). Then, we statistically analyze how road grades are associated with the CO, HC, and NO_x emissions and the total equivalent emissions of HDVs. Finally, based on the research findings, we discuss strategies oriented towards reducing emissions of HDVs. The findings of this study will provide insights for policy-makers, scholars, and practitioners into the improvement of road grade design with an attempt to reduce traffic emissions and develop sustainable transportation in China.

3. Data and Method

3.1. Study Area

Since the implementation of reform and opening up policy in 1978, China has experienced a major demographic transition of rapid and intense urbanization and motorization [43–45]. From 1978–2014, the level of urbanization in China tripled from 17.9%–54.77%, and the level of motorization boomed at 20% annually [46,47]. This situation has contributed to China's increasingly severe transport and environmental problems [48]. Among those problems, transport energy consumption and emissions have become issues of public concern. By the end of 2014, in China's total emissions, the transport emissions have covered over 25% [49], among which on-road vehicle emissions covered over 85% of all transport emissions [8]. In the Copenhagen Accord of the 2009 United Nations Climate Change Conference, the Chinese government has committed to reducing the emissions per unit Gross Domestic Product (GDP) by 40%–45% by 2020 as compared to 2005 [50]. In response to that goal, the Ministry of transport of the People's Republic of China has also committed to reduce on-road vehicle emissions by 11% in 2015 as compared to in 2005.

We chose a road network in Taiyuan Metropolitan Area, China, as the study area. Taiyuan is the capital and largest city of North China's Shanxi province, with a total population of 429.89 million inhabitants on 6959 km^2 (by the end of 2014) [51]. Mountainous regions and hills make up 80.1% of the whole Shanxi province, leading to bigger road grades than plain areas [51]. Shanxi Province is a leading producer of coal in China, with more coal companies than any other province and an annual production exceeding 300 million metric tons [51]. HDVs are major players in Shanxi's transport sector for coal transportation. However, they also produce vast emissions, raising increasing concerns about environmental disruption and public health. Therefore, examining the influences of road grades on HDVs in Shanxi would facilitate emissions reduction through the improvement in road design.

The road network we chose in Taiyuan included five roads, Wucheng Road (1.4 km), Renmin Road (1.4 km), Kangning Street (1.5 km), Changsheng Street (1.3 km), and Tongda Street (1.3 km) (Figure 1). This road network is located at south Taiyuan, a newly urbanized area where there used to be villages and farmlands. The road network was chosen because of its representativeness in Shanxi Province, featuring high ratios of HDVs and big road grades. The HDV ratios of traffic volume on the five roads, especially on Wucheng Road and Renmin Road, are as high as 20%. That is partly due to the fact they are major arterials connecting the urban core of Taiyuan to the suburban and rural areas. The road grades of the chosen network range from 0%–4%, which is in accordance with the range of lowest and highest grades of ascent road segments regulated in the national road design manual.

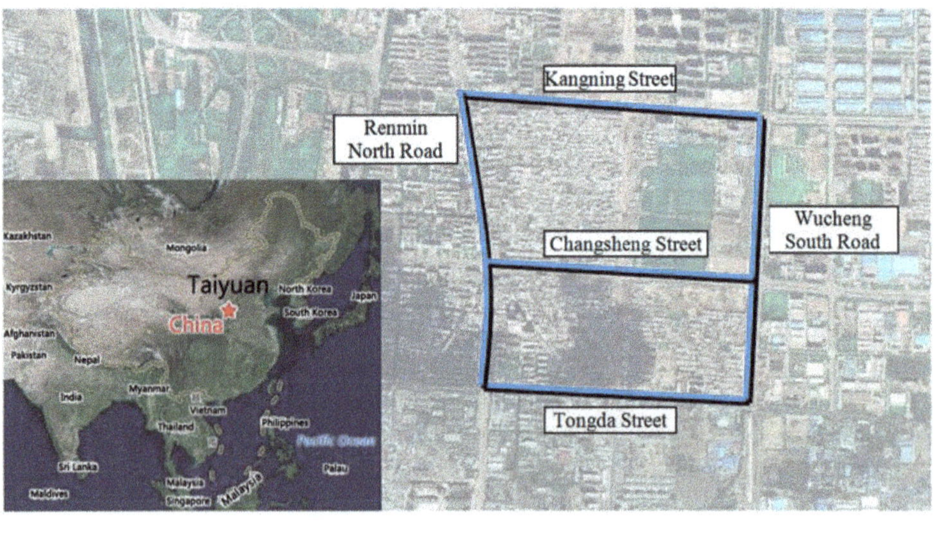

Figure 1. Study area (Courtesy of Baidu).

3.2. Development of Emission Rate Model Based on On-Board PEMS and VSP

3.2.1. Data Collection

In this study, we collected three types of data: traffic volume of HDVs and light-duty vehicles, road grades design, and on-road HDV emissions. The traffic volume data were counted on the morning and afternoon peak hours during 7:30–9:30 a.m. and 4:30–6:30 p.m. on 12 November 2013. The data of road grades design were collected from Transport Department of Shanxi Province. The on-road vehicle emissions data were collected by an on-board PEMS, OBEAS-3000, on the same day of traffic volume count.

On-board PEMS is able to collect second-by-second speed and emissions variation of the vehicle under real-world conditions. OBEAS-3000 is an integrated and high precision PEMS developed by the Xiamen Tong Chuang Detection Technology Co, Ltd. and can be used in petrol and diesel engines in light and heavy-duty vehicles [12]. OBEAS-3000 is comprised of five-gas analyzer, engine diagnostic scanner, on-board computer and Global Positioning System (GPS) (Figure 2). The five-gas analyzer E-BOX PC is imported with original packaging from SIEMENS, Germany, which is used to measure the volume percentage of CO, CO_2, HC, NO_x, and O_2 in the vehicle exhaust. Simultaneously, the engine scanner is connected to the On-Board Diagnostics (OBD) link of the vehicle, where engine and vehicle data are downloaded during vehicle operation. Engine data is used to calculate the volume flow rate of CO, CO_2, HC, NO_x, and O_2 in the vehicle exhaust, which is combined with the volume percentage to output the instantaneous mass flow rate (mg/s). The laptop is used to record the different parameters, such as temperature, humidity, and information regarding each test vehicle. The GPS is used to improve the precision of speed and acceleration. In addition, events during trips are also recorded with the GPS and laptop computer, including the time at which the vehicle crosses the centerline of key intersections and the path the vehicle travels.

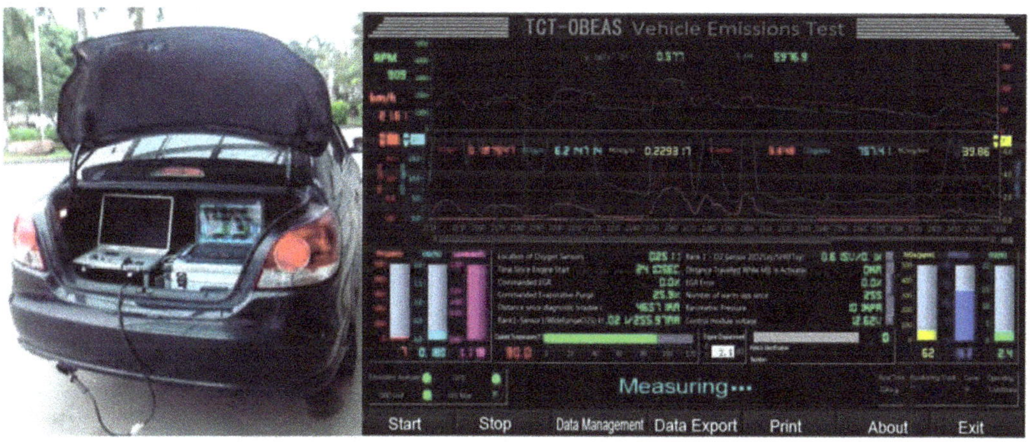

Figure 2. The application and user interface of OBEAS-3000.

According to the vehicle type and vehicle weight data from Transport Department of Shanxi Province, the HDVs weighing over 10 ton occupied nearly 70% of the HDV fleet in the chosen road network. Therefore, one heavy-duty vehicle produced by Chinese manufacturer First Automotive Works (FAW) (Table 2) was selected to collect on-road emissions data by OBEAS-3000. The data included second-by-second CO, HC and NO_x emissions, and the speed and acceleration. The durations of each test were about 30 min and 10 tests were conducted. The average test distance is about 15 km. More than 16,000 groups of valid records were collected from the vehicles.

Table 2. Basic information of the testing vehicles.

Parameters	HDV
Manufacturer	FAW * Car Co., Ltd
Brand	Jie Fang
Year of Manufacture	2012
Total Mass (kg)	15790
Engine Displacement (L)	6.6
Fuel Type	diesel
Emission Standard	Euro IV with EGR * + FEUP *
Mileage	10000

* FAW stands for First Automotive Works, EGR stands for Exhaust Gas Recirculation, and FEUP stands for FAW Electronic Unit Pump.

3.2.2. Definition of VSP & VSP-Bin

Recent studies have shown that the instantaneous vehicle emissions were highly correlated to VSP [34]. Therefore, the variable based on VSP, rather than instantaneous speed or average speed, is the core explanatory variable for analyzing energy consumption and emission of vehicles. In this study, we employed VSP to develop an emission rate model. VSP is a widely accepted explanatory variable, generally defined as the power required per unit mass of the vehicle, also known as the ratio of the power of the vehicle. According to existing literature, the formula of VSP for HDVs in China is expressed as follows [52]:

$$VSP_{HDV} = v \times (a + 9.807 \times \sin\theta + 0.186333) + \frac{3.702456}{m} \quad (1)$$

where v is the vehicle speed (m/s); a is the acceleration (m/s^2); θ is the angle of road grade; m is the vehicle weight (ton).

In order to better capture the emission characteristics under the concept of VSP, VSP values were usually distributed into bins, and the vehicle emissions are averaged under the same VSP-bin [53].

3.2.3. Emission Rate Model Based on On-Board PEMS and VSP

With collected data, we were able to calculate the VSP values and group the second-by-second emissions data into different VSP-bins. The principle of VSP-bin division is to keep the balance of each bin in emission rates and emission contributions [12]. After an analysis of emission data, it was found that 97.6% of all data fell into the VSP range of −25 to 25 (kw/t). Furthermore, we made sure that the numbers of emission data under each VSP-bin were no less than 20. The specific frequency distribution of VSP and proportion of different VSP range are illustrated in Figure 3 and Table 3. In Table 4, we listed the resulting emission rates of HDVs under each VSP-bin.

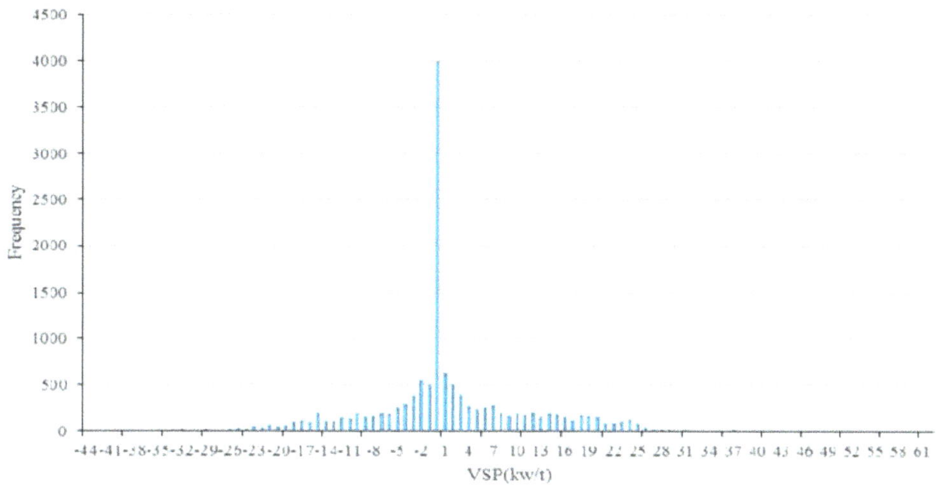

Figure 3. Frequency distribution of VSP.

Table 3. Proportion of different VSP range.

VSP Range	Proportion (%)
[−5, 5]	57.80
[−10, 10]	72.08
[−15, 15]	83.74
[−20, 20]	92.26
[−25, 25]	97.59
[−30, 30]	98.65
[−35, 35]	99.24
[−40, 40]	99.57

Table 4. Emission rates of HDV under VSP-bin (based on experimental values measured on one vehicle).

VSP-bins	Range	CO (mg/s)	HC (mg/s)	NO_x (mg/s)	Frequency
1	(−∞, −25)	1.3250	1.7616	7.1723	100
2	[−25, −20]	0.9457	2.6399	7.6521	230
3	[−20, −14]	1.4794	3.2946	8.7245	617
4	[−14, −10]	1.3027	3.3108	12.0206	498
5	[−10, −5]	3.6174	4.7910	6.4120	888
6	[−5, −2]	5.1957	6.9809	11.4792	935
7	[−2, −1]	11.4613	2.2593	33.8444	553
8	[−1, 0]	7.9797	1.0422	21.4202	498
9	[−1, 0]	9.9653	0.9430	36.9548	3993
10	[1, 2]	15.1145	3.0147	60.0636	631
11	[2, 5]	23.8573	2.9065	62.1429	1168
12	[5, 10]	37.2366	4.6050	93.3285	1144
13	[10, 14]	41.1333	3.1037	66.9862	738
14	[14, 20]	24.9549	3.5288	81.3605	989
15	[20, 25]	12.5323	2.7615	94.7466	574
16	[25, +∞)	16.6543	3.1497	106.1606	322

3.3. Integration of Emission Rate Model and VISSIM

The simulation model used in this study is VISSIM (version 4.3) as developed by PTV Planung Transport Verkehr AG in Karlsruhe, Germany. VISSIM is a stochastic, microscopic, time-step and behavior-based simulation model [54], integrating a psychophysical driver behavior model developed by Weidmann and a traffic model that is discrete and stochastic using driver-vehicle-units as single entities [55]. The parameters such as instantaneous speed, acceleration and vehicle weight, which are input for VSP calculation, could be generated from VISSIM.

3.3.1. Calibration and Validation of Microscopic Traffic Simulation Model, VISSIM

Microscopic traffic simulation models have been widely used because simulation is user-friendly, safer, faster, and less expensive than field implementation and testing. In practice, simulation model based analyses have often been conducted under default parameter values or best guessed values. At times, simulation model outputs could result in unrealistic estimates if the simulation model is not properly calibrated and validated. Thus, calibration and validation of simulation models are crucial steps in assessing their value in transportation operations, planning, and policy [56].

Microscopic simulation models contain a lot of independent parameters with default values for each variable. However, the users can input a range of values for the parameters. Model calibration is defined as the process by which the individual components of the simulation model are adjusted or tuned so the model can accurately represent field measured or observed traffic conditions, and model validation tests the accuracy of the model by comparing traffic flow data generated by the model with that collected from the field [57].

In this paper, we referred to the calibration and validation procedure developed by Park *et al.* [1] and Sun *et al.* [58,59] whose case study is VISSIM and the test-bed network is a little similar to our research. Specifically, the procedure has 10 steps: (a) measure of effectiveness selection (include target value and calibration target); (b) data collection; (c) calibration parameter identification; (d) initial evaluation; (e) experimental design; (f) run simulation; (g) surface function development; (h) candidate parameter set generations; (i) evaluation and (j) validation through new data collection.

The implementation of the procedure includes:

(a) Two measures of performance were selected for the calibration and validation procedure, which were (1) eastbound left-lane travel times on Tongda Road; and (2) the maximum queue length between the intersections of Renmin Road and Changsheng Road. These performance measures were chosen because of their ease of collection from the field and from VISSIM output files. The target value of average field eastbound left-lane travel time was 155.74 s, and the field maximum queue length was 43.2 m. The calibration target is the field data is in the 95% confidence interval of the simulation data.

(b) Field data of the two performance measures was extracted from the video surveillance system of Shanxi Provincial Highway Bureau.

(c) Calibration parameters here included number of observed preceding vehicles, minimum headway, average standstill distance and waiting time before diffusion (as shown in Table 5). We chose the acceptable ranges of values by Sun *et al.*, for this research.

(d) In the initial evaluation part, simulation data from VISSIM with default parameters were collected from 50 random seeded runs. With the limitation of the time and space of field measurement, the field data may be lower or higher than the true mean. So it is better to compare the field data with the distribution of the 50 simulation runs instead of average simulation data. The analysis results with Statistic Package for Social Science (SPSS) were shown in Figure 4.

Table 5. Default and acceptable range of candidate parameters value.

Calibration Parameters	Default Value	Acceptable Ranges of Values by Park *et al.*	Acceptable Ranges of Values by Sun *et al.*
number of observed preceding vehicles	2	1–4	1–4
minimum headway (m)	0.5	0.5–7	0.5–4
average standstill distance (m)	2	1–3	1–3
waiting time before diffusion (s)	60	20–40	45–120

From the one-sample test of SPSS, we could find that:

(a) $p = 0.128 > 0.05$, which means that there is no obvious difference between the average field travel time and the average simulation travel time ($p > 0.05$).

(b) 95% Confidence Interval of the Difference is (153.7280, 154.8712), and the average field travel time is in the 95% confidence interval.

From the one-sample test of SPSS, we could find that:

(a) $p = 0.224 > 0.05$, which means that there is no obvious difference between the average field maximum queue length and the average simulation maximum queue length ($p > 0.05$).

(b) 95% Confidence Interval of the Difference is (42.8802, 43.2766), and the average field maximum queue length is in the 95% confidence interval.

Figure 4 above indicated that both average field travel time and field maximum queue length were in the given confidence interval, therefore, calibration was not necessary for this research, nor the next several steps. This also means VISSIM with default parameters could be used in the following simulation.

This study used several days of data collection and two measures of performance. It is recommended to collect more field data and to consider variability of field data if possible. In future research, we will determine whether the process is applicable to other networks or simply to this specific one.

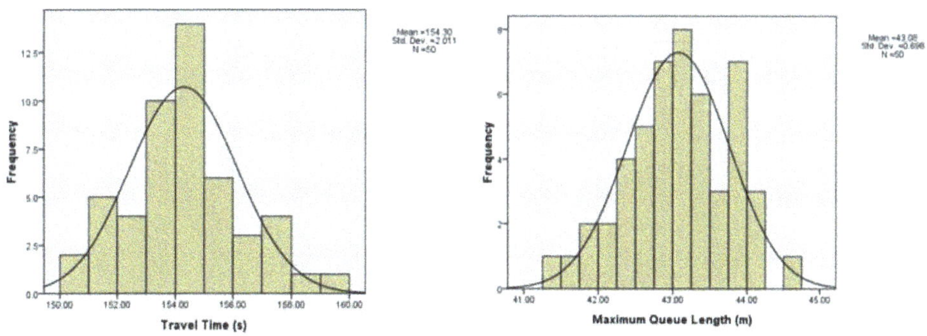

Figure 4. Eastbound left-Lane travel time distributions (**left**) and maximum queue length distributions (**right**) of default parameters.

3.3.2. Simulation Process

Process flow diagram of the simulation of highway segments is shown in Figure 5. The simulation in this paper was divided into three steps.

- Step 1: Determine the simulation parameters, such as the length and width of road segment, the road grades, traffic flow characteristics, vehicle velocity distribution, simulation time and the start time of data statistics, *etc.* (Table 6).

- Step 2: Build simulation scenarios: Add and edit road links, define the traffic distribution and the composition of traffic flow, input traffic flow and set the simulation parameters. We set nine research scenarios (from G1 to G9) with road grades changing from 0%–4% at an increment of 0.5% (Figure 6).
- Step 3: Run simulation and generate data: Run the simulation 10 times across all nine scenarios and output the data that were needed in calculating VSP, such as acceleration, speed, and road grades. Other related selected parameters are lane number, link number, simulation time, speed, start time, total path distance, total time in network, vehicle number, vehicle type, vehicle type name and weight (Figure 7).

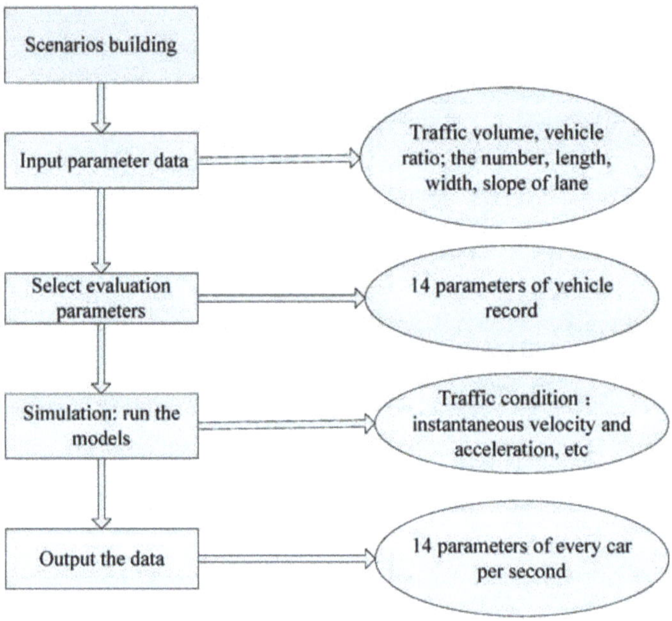

Figure 5. Process flow diagram of the simulation of highway segments.

Table 6. Simulated operation parameters based on the real condition of study area.

Characteristic Indexes	Parameters
The length of simulation road segment (m)	800 *
The lane width (m)	3.5
The traffic flow per lane (veh/h)	300
Simulation time (s)	300
Time to start recording data (s)	30
Distribution of vehicle velocity (km/h)	30–60
Range of road grades (%)	0–4
The number of lanes (ascent)	2
The proportion of HDV (%)	20

* denotes the length of the road link was set based on China's national standard, Design Specification for Highway Alignment (JTGD20-006) and a previous paper of Pei *et al.*, in which the ideal grades and the maximum unlimited grade lengths were stated [60].

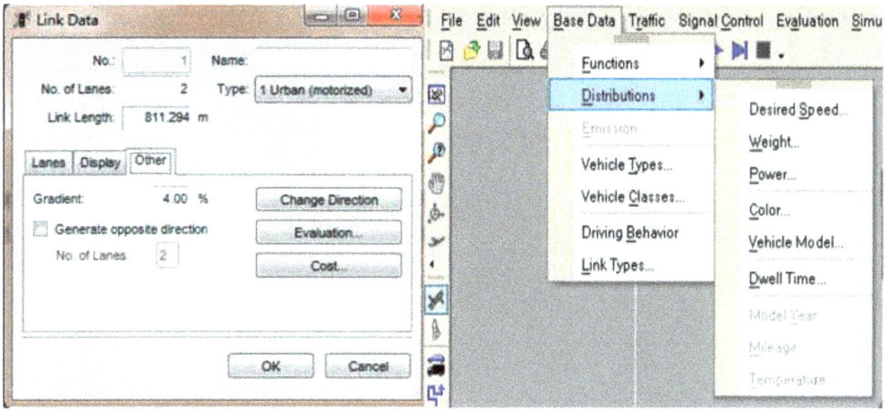

Figure 6. Set of link data and base data in scenario building.

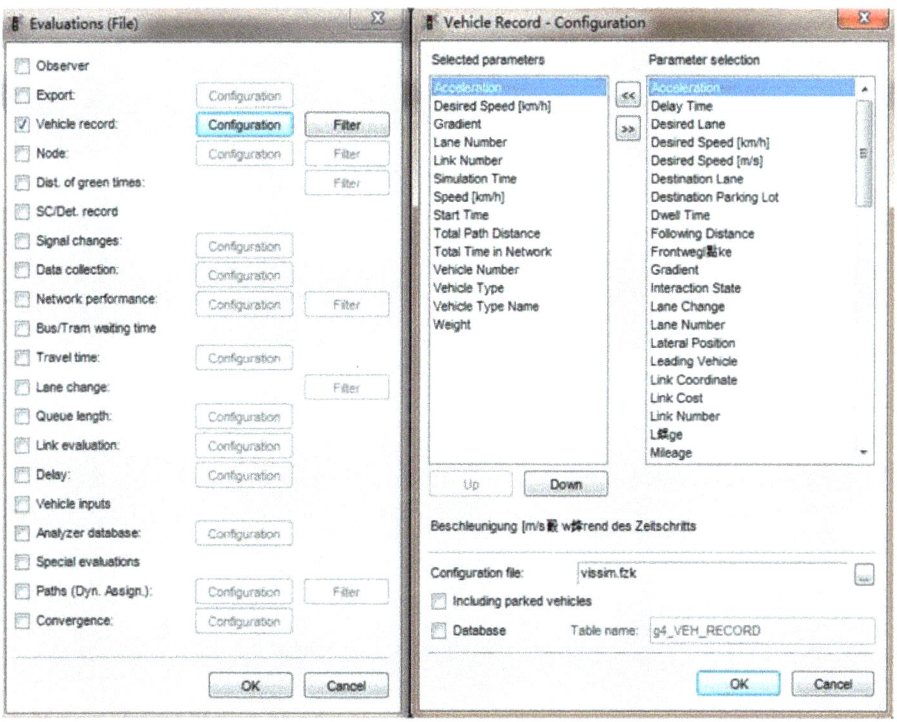

Figure 7. Set of output parameters.

4. Results

4.1. The CO, HC, and NO_x Emissions across Different Scenarios

Nearly 7800 groups of valid records were collected from VISSIM for each scenario in one simulation, adding up to over 700,000 groups of valid records for all scenarios in total. We input the results of VISSIM into formula (1) to calculate the second-by-second VSP of each HDV and the corresponding VSP-bin. Then, we attain the CO, HC, or NO_X emissions of an HDV by adding up second-by-second emission in coordination with VSP-bin. Finally, we obtain the CO, HC, or NO_X emissions of the road segment by adding up the emissions from all the 120 HDVs (=300veh/lane × 2 lanes × 20%) in the simulation. For each scenario, we averaged the results from 10 repeated simulations to increase the reliability of the results. The emission rate and total emissions of the three pollutants were listed in Tables 7 and 8.

Table 7. Emission rate of CO, HC, and NO_X under different simulation scenarios.

No.	G (%)	CO (mg/s × veh)	HC (mg/s × veh)	NO_X (mg/s × veh)
G1	0.0	23.037	3.032	64.774
G2	0.5	25.395	3.503	74.171
G3	1.0	26.763	3.567	74.882
G4	1.5	28.496	4.087	76.866
G5	2.0	30.694	3.975	78.117
G6	2.5	32.950	3.969	78.314
G7	3.0	35.085	4.320	84.083
G8	3.5	36.347	4.278	87.018
G9	4.0	36.493	4.249	88.520

Table 8. Total emissions of CO, HC, and NO_X under different simulation scenarios.

No.	G (%)	CO (g)	HC (g)	NO_X (g)
G1	0.0	11.6790	1.5308	32.8692
G2	0.5	12.9485	1.7639	37.9185
G3	1.0	13.6412	1.8267	38.5050
G4	1.5	14.4942	2.0907	39.4503
G5	2.0	15.5938	2.0346	39.9430
G6	2.5	16.7920	2.0290	40.0033
G7	3.0	17.9477	2.2002	43.0880
G8	3.5	18.6450	2.2095	44.7864
G9	4.0	18.7627	2.2074	45.6840

4.2. *Change Rate Analysis*

In this section, we presented the change of CO, HC and NO_X emissions when road grades change. Specifically, we employed change rate of other scenarios and baseline scenario, which indicated the variations of CO, HC and NO_X emissions between scenarios 2–9 and the baseline scenario 1. We also used change rate of adjacent intervals to illustrate the variations of emissions between different scenarios which indicated the variations between a scenario and the adjacent scenario.

4.2.1. CO Emissions

The CO emissions grew with the increase of road grades, reflected by the increasing change rate of other scenarios and baseline scenario and positive change rates of adjacent intervals. The change rate of adjacent intervals was biggest (0.1087) when the grade increased from 0%–0.5%, followed by the grades increasing from 2%–2.5% (0.0768) and from 1.5%–2% (0.0759) (Figure 8 & Table 9). On the contrary, when the grade increased from 3.5%–4%, the rate of change was the smallest (0.0063).

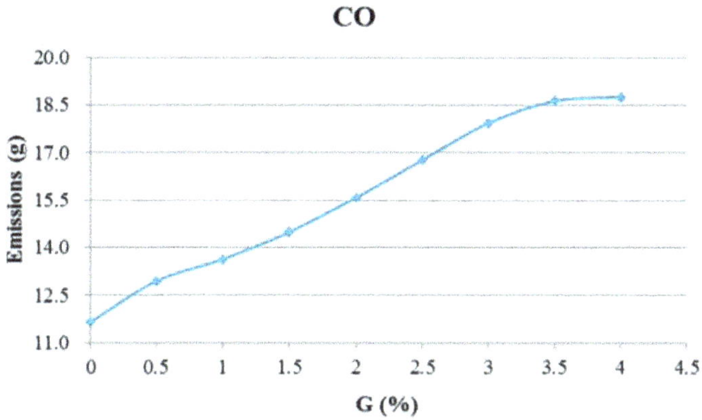

Figure 8. CO emissions of HDV in different scenarios.

Table 9. Change rate of other scenarios and baseline scenario and change rate of adjacent intervals of CO emissions in different scenarios.

G (%)	CO (g)	Change Rate of Other Scenarios and Baseline Scenario	Change Rate of Adjacent Intervals
0.0	11.6790	-	-
0.5	12.9485	0.1087	0.1087
1.0	13.6412	0.1680	0.0535
1.5	14.4942	0.2410	0.0625
2.0	15.5938	0.3352	0.0759
2.5	16.7920	0.4378	0.0768
3.0	17.9477	0.5367	0.0688
3.5	18.6450	0.5965	0.0389
4.0	18.7627	0.6065	0.0063

4.2.2. HC Emissions

Generally, with the growth of grades, the overall HC emissions were increasing, as indicated by 2.2074 g of HC emissions at 4% grade and 1.5308 g at 0% grade. However, the variations of HC emissions in different scenarios were complicated, reflected by both positive and negative change rates' adjacent intervals. For example, the change rate was the biggest (0.1523) when the grade increased from 0%–0.5%, followed by the change rate of 0.1445 from 1%–1.5% (Figure 9 & Table 10). On the contrary, when the grade changed from 1.5%–2%, the change rate was negative (−0.0268), so were the change rates from 2%–2.5% (−0.0027) and from 3.5%–4% (−0.0010).

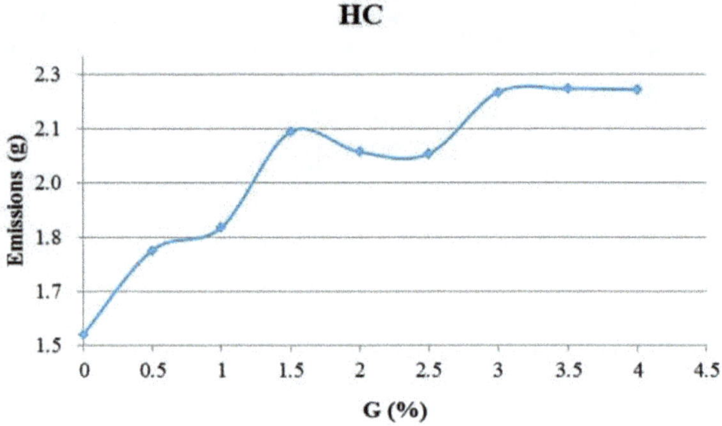

Figure 9. HC emissions of HDV in different scenarios.

Table 10. Change rate of other scenarios and baseline scenario and change rate of adjacent intervals of HC emissions in different scenarios.

G (%)	HC (g)	Change Rate of Other Scenarios and Baseline Scenario	Change Rate of Adjacent Interval
0.0	1.5308	-	-
0.5	1.7639	0.1523	0.1523
1.0	1.8267	0.1933	0.0356
1.5	2.0907	0.3658	0.1445
2.0	2.0346	0.3291	−0.0268
2.5	2.0290	0.3254	−0.0027
3.0	2.2002	0.4373	0.0844
3.5	2.2095	0.4434	0.0043
4.0	2.2074	0.4420	−0.0010

4.2.3. NO_x Emissions

Similar to the CO emissions, the NO_x emissions grew with the increasing of road grades. The change rate of adjacent intervals was biggest (0.1536) when the grade increased from 0%–0.5%, followed by the change rate of grade increasing from 2.5%–3% (0.0771) (Figure 10 & Table 11). When the grade increased from 1.5%–2%, the change rate was the smallest (0.0015).

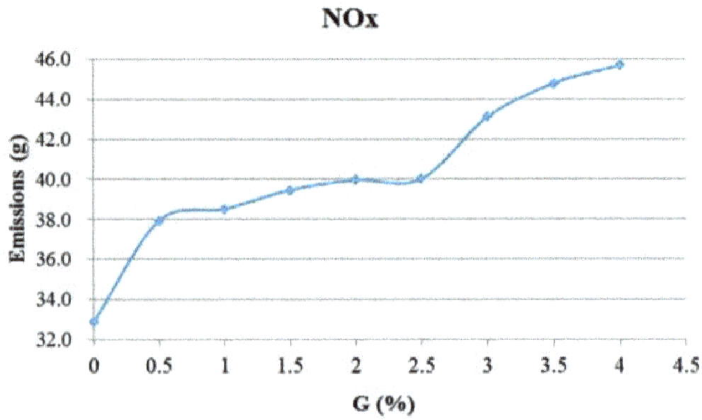

Figure 10. NO_x emissions of HDV in different scenarios.

Table 11. Change rate of other scenarios and baseline scenario and change rate of adjacent intervals of NO_x emissions in different scenarios.

G (%)	NO_x (g)	Change Rate of Other Scenarios and Baseline Scenario	Change Rate of Adjacent Intervals
0.0	32.8692	-	-
0.5	37.9185	0.1536	0.1536
1.0	38.5050	0.1715	0.0155
1.5	39.4503	0.2002	0.0246
2.0	39.9430	0.2152	0.0125
2.5	40.0033	0.2170	0.0015
3.0	43.0880	0.3109	0.0771
3.5	44.7864	0.3626	0.0394
4.0	45.6840	0.3899	0.0200

4.3. Regression Analysis

To explore the relationship between road grades and three traffic emissions, we applied correlation analysis and polynomial regression in SPSS. The results of the correlation analysis showed that the R^2 of CO, HC, and NO_x were 0.9855, 0.8433, and 0.9099, respectively. That demonstrated fine goodness of fit of correlations of the three traffic emissions with road grades. The results of polynomial regression demonstrated better goodness of fit than correlation regression, reflected by bigger R^2 (Table 12). Generally, the CO emissions showed the greatest goodness of fit while the HC emissions the least in both correlation analysis and polynomial regression.

Table 12. Polynomial regression correlation analysis between grades and three emissions.

Three Emissions	R^2 (for Quadratic)	R^2 (for Cubic)
CO	0.9891	0.9947
HC	0.9326	0.9422
NO_x	0.9116	0.9485

4.4. Analysis of Total Equivalent Emissions

The CO, HC and NO_x emissions have different degrees of negative impacts on public health and environment (4). In 2012, the Ministry of Environmental Protection of China published a national standard, *Method for Estimation of Environmental Impact Index of Vehicles*, to regulate the evaluation of the impacts of different traffic emissions [61]. This standard involves the Weight Value (WV) that demonstrates the weighted scores of different vehicle emissions considering their impact factors on public health and environment. In specific, the WVs of CO, HC and NO_x emissions are 5 points, 10 points, and 15 points, respectively [61], showing that the negative impact of NO_x is equivalent to two times of HC's and three times of CO's. Employing the Vehicle Environmental Impact Index (VEI) and WV, the calculation of the total equivalent emissions of the HDV in the study is expressed as follows:

$$Total\ Equivalent\ Emission = CO + 2 \times HC + 3 \times NO_X \tag{2}$$

We illustrated the total equivalent emissions in different scenarios (Figure 11). The overall trend of the total equivalent emissions is similar with that of the NO_x emissions. The reason is because among the three emissions, the NO_x emissions have the biggest exhaust and WV. Specifically, the rate of change of adjacent intervals (0.1490) was the biggest when the grade increased from 0%–0.5%, followed by the rate change from 2.5%–3% (Table 13). Meanwhile, when the grade increased from 2%–2.5%, the rate of change rate (0.0098) was the smallest.

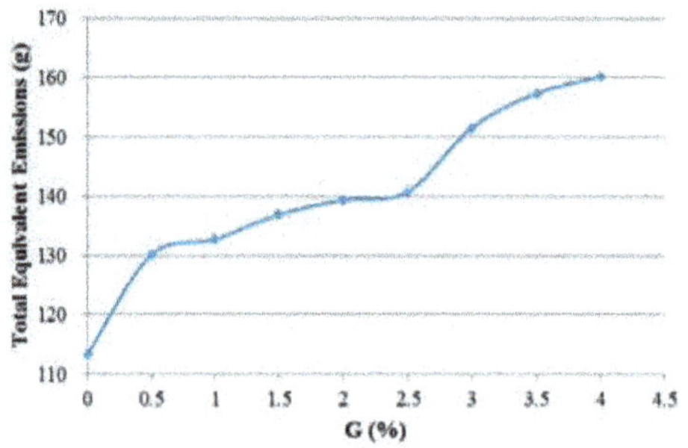

Figure 11. Total equivalent emissions of CO, HC and NO_x in different scenarios.

Table 13. Change rate of other scenarios and baseline scenario and change rate of adjacent intervals of total equivalent emissions in different scenarios.

G (%)	Total Equivalent Emissions (g)	Change Rate of Other Scenarios and Baseline Scenario	Change Rate of Adjacent Intervals
0.0	113.3483	-	-
0.5	130.2317	0.1490	0.1490
1.0	132.8095	0.1717	0.0198
1.5	137.0266	0.2089	0.0318
2.0	139.4919	0.2306	0.0180
2.5	140.8598	0.2427	0.0098
3.0	151.6122	0.3376	0.0763
3.5	157.4233	0.3888	0.0383
4.0	160.2294	0.4136	0.0178

5. Discussion and Policy Implications

Considering the challenges in moving towards sustainability, this paper, taking the case of Taiyuan as an example, provided the guidance for the design of road grades that facilitates the reduction of on-road vehicle emissions and the development of sustainable transportation. This paper presented findings from a study aiming at unraveling the relationship between road grades and on-road CO, HC, and NO_x emissions of heavy-duty vehicles in China. Taking a road network in Taiyuan, China, as a study area, this paper studied the relationship between road grades and CO, HC, and NO_x emissions of HDVs in China for the first time. Using emissions data collected by PEMS and VSP, we first developed an emission rate model of HDVs in Taiyuan Metropolitan Area, China. This is one of the contributions of this research, which will facilitate future studies on the reduction of on-road vehicle emissions in China. We will conduct further research to refine this model by using data from more types of vehicles.

Then, we integrated the emission rate model and simulation model VISSIM to attain three traffic emissions of HDVs on nine scenarios differentiated by road grades. The results of change rate analysis and regression analysis demonstrated how road grades are associated with on-road emissions of HDVs. To reveal the different impacts of CO, HC, and NO_x emissions on public health and environment, we then analyzed of the influences of road grades on total equivalent emissions based on the Chinese national standard.

For this research, the chosen vehicle provided by the Shanxi Provincial Highway Bureau represents the majority of HDVs operating in Taiyuan Metropolitan Area, China, and thus the results in this paper may not generally be valid on other roads and areas. However, using vehicles with different characteristics, such as emission control technology, fuel specifications and operating conditions, researchers can reproduce our method to obtain specific emission models for their research purpose. In addition, this paper focuses on the total emissions in the preset scenarios than the emissions in relation to the vehicles characteristics, especially the engines, so the emission results presented in the paper expressed in grams may be more appropriate than to evaluate results which are normalized in respect to the engine work, *i.e.*, emissions in g/kWh.

The relationship of road grades on CO, HC, or NO_x emissions shows some similar characteristics. Generally, road grades are highly correlated to CO, HC, or NO_x emissions, as indicated by the correlation analysis and polynomial regression. Looking at the variations between two adjacent grades, for CO and HC emissions, the smallest variation of emissions appears when the grade increases from 3.5%–4.0%. Meanwhile, the NO_x emissions experience the smallest change when the grade rises from 2.0%–2.5%. The reason for this situation remains unclear and requires further study.

With respect to the CO emissions, the rates of change of adjacent intervals are similar (5.35%–7.68%) between the road grades of 0.5% and 3.0%. When the grade is over 3.0%, the rates of change of adjacent are considerably smaller (0.63%–3.89%). In terms of the HC emissions, apart from the substantial increase when the grade changes from 0%–0.5%, they also drastically increase by 14.45% and 8.44% when the grade changes from 1.0%–1.5% and from 2.5%–3.0%. In other scenarios, the rates of change of adjacent are small and some are even negative. The NO_x emissions will increase significantly by 15.36% and 7.71% when the grade changes from 0%–0.5% and 2.5%–3.0%. Otherwise, the variations of NO_x emissions are remarkably smaller (0.15%–3.94%).

The CO and NO_x emissions increase with the road grades at all nine scenarios, demonstrating the positive effects of road grades. However, the variations of HC emissions in different scenarios were complicated. When the grade changed from 1.5%–2.5% and from 3.5%–4%, the emissions even drop, albeit to varying degrees. This may be due to the fact that HC is exhausted through incomplete combustion. Previous studies indicated that a low speed leads to a low combustion temperature, which is closely related to more emissions due to incomplete combustion, especially the HC emissions [62].

One of the main purposes of this study is to illuminate strategies and interventions towards public health and environmental protection. Therefore, it is necessary to unravel the total impacts of different emissions in consideration their varying degrees of impacts on public health and the environment. To this end, we calculated the total equivalent emissions with the weight values of different emissions from the Chinese national standard *Method for Estimation of Environmental Impact Index of Vehicles*. As the NO_x emissions have the biggest exhaust and weight value, the road grades are found to have similar influences on the total equivalent emissions as on the NO_x emissions.

Currently, with the urbanization and motorization surge in China, the urban land use is developing with high density of population, employment and activities, and the on-road traffic volume is booming. For arterials connecting urban core and outskirts, the ratio of HDVs among the traffic volume is considerably high, exhausting a large amount of emissions. Hence, it is a rigorous challenge to reduce the worsening impact of on-road vehicle emissions on public health and the environment. For road networks with a high ratio of HDVs, the road grades, if applicable, are suggested to be at 0% in road design. Otherwise, the road grades should be as small as possible due to the high sensitivity of on-road emissions of HDVs to road grades.

For policy-makers and practitioners, it is important to draw up policies and strategies oriented towards reducing total equivalent emissions in road design and traffic management, so as to alleviate environmental disruption, promote public health, and develop sustainable transportation. Specifically, the total equivalent emissions are always the smallest when the road grade is 0% and biggest when the road grade is 4%. Therefore, it is recommended that we design the road grade at 0% if the landform conditions allow. The total equivalent emissions witnessed the most significant increase when the grade changed from 0%–0.5% and from 2.5%–3.0%. Therefore, if the landform conditions do not permit designing a level road, it is suggested that the design of road grades changing from 0%–0.5% or from 2.5%–3.0% be avoided.

6. Strengths and Limitations

This study had a number of strengths and limitations. In terms of the strengths, the study focused on the on-road emissions of heavy-duty vehicles, which are one of the largest sources of transport emissions in China. Secondly, the study investigated the relationship between road grades and CO, HC, and NO_x emissions of HDV in China, which has been rarely studied in existing literature. Thirdly, this study for the first time established an emission rate model for HDVs in China based on PEMS and VSP. In spite of the strengths above, this paper also had some limitations. The first one was the restriction of study area and the scope of test times and vehicles. In terms of the road grades alone, a study on the distribution of the real road grades, which includes uphill road, downhill road and the combined uphill-and-downhill roads, may be needed for building the next simulation scenarios. For this reason, the results might not necessarily apply to other road networks in China with characteristics that were different from the study area. Secondly, this study focused on three major kinds of traffic emissions. Future studies may expand to other emissions including CO_2 and particle pollution (PM).

7. Conclusions

On-road vehicle emissions are one of the major sources of transport emissions. As a key geometric design factor, road grade significantly affects on-road vehicle emissions, particularly for heavy-duty vehicles (HDVs). In this paper, we presented research results from a study aiming at unraveling the influences of road grades on CO, HC and NO_x emissions of HDVs, with data collected from Taiyuan Metropolitan Area, China. Firstly, in the section on on-road emission testing, we developed an emission rate model of HDVs under VSP-bins with the data collected from a PEMS, OBEAS-3000, in Taiyuan. The establishment of this model contributes to facilitating future studies on the reduction of on-road vehicle emissions in China. Secondly, in the section on simulation, with the integration of an emission rate model of HDVs and VISSIM, we calculated the CO, HC, and NO_x emissions of HDVs in nine simulation scenarios differentiated according to road grades (from 0%–4%). Then, we statistically analyzed how road grades are associated with the CO, HC, and NO_x emissions and the total equivalent emissions of HDVs. Finally, based on the research findings, we discussed strategies oriented towards reducing emissions of HDVs.

The relationship of road grades with the CO, HC, or NO_x of HDVs shows some similar characteristics. The three emissions are found to be highly correlated to road grades, among which the CO emissions are most sensitive to the change of road grades and the HC emissions least. All three emissions experience the biggest change, ranging from 10.9% (CO) to 15.4% (HC), when the road grade rises from 0%–0.5%. HDVs exhaust least CO, HC, or NO_x emissions when the road grade is 0% and most at 4%. Compared to the emissions at 0%, the emissions at 4% grade will be substantially boosted from 39.0%–60.6%, among which the CO emissions see the biggest rise and HC the smallest. The CO and NO_x emissions increase with the road grades in all nine scenarios, demonstrating the positive effects of road grades. However, the variations of HC emissions in different scenarios were complicated. This may be due to the fact that HC is exhausted through incomplete combustion. Looking at the total equivalent emissions, road grades are found to have similar influences on them as on the NO_x emissions.

The findings of this study will facilitate our understanding on the relationship between road grades and on-road vehicle emissions of HDVs. The findings will also provide insights for policy-makers, scholars and practitioners in China into the improvement of road grade design with an attempt to reduce traffic emissions, alleviate environmental disruption, promote public health, and develop sustainable transportation, with the ultimate aim of moving towards sustainability in the future.

Acknowledgments: The authors acknowledge the financial support of Project 2012-1-15 by Transport Department of Shanxi Province, Project 12N11 by Transport Department of Jiangsu Province, Project 12&ZD203 and 14BSH032 by National Social Science Foundation of China, and Project 51308336 and 51408065 by National Natural Science Foundation of China.

Author Contributions: In this paper, Wendan Zhang designed the research programs, participated in the data collection, analyzed the results, and completed the writing of the introduction, results and discussion; Jian Lu committed to the design of the survey and data collection, and the writing of corresponding parts; Ping Xu carried out the data process and completed the writing work of corresponding parts; and Yi Zhang conceived and developed the research ideas and led the research in general.

Conflicts of Interest: The authors declare no conflict of interest.

Abbreviation List

HDVs	Heavy-Duty Vehicles
CO	Carbon Monoxide
HC	Hydrocarbons
NOx	Nitrogen Oxides
CO2	Carbon Dioxide
PEMS	Portable Emission Measurement System
VSP	Vehicle Specific Power
U.S.	United States
EU	European Union
MOBILE	Mobile Source Emissions Factor Model
MOVES	Motor Vehicle Emission Simulator
EPA	Environmental Protection Agency
CMEM	Comprehensive Modal Emissions Model
NCHRP	National Cooperative Highway Research Program
IVE	International Vehicle Emissions
VISSIM	Verkehr In Städten—SIMulationsmodell
ETC	Electronic Toll Collection
HOV	High Occupancy Vehicle
GDP	Gross Domestic Product
GPS	Global Positioning System
OBD	On-Board Diagnostics
FAW	First Automotive Works
EGR	Exhaust Gas Recirculation
FEUP	FAW Electronic Unit Pump
SPSS	Statistic Package for Social Science
WV	Weight Value
VEI	Vehicle Environmental Impact Index

References

1. World Health Organization. *Ambient (Outdoor) Air Quality and Health*; Fact sheet N. 313; World Health Organization: Geneva, Switzerland, 2014.
2. Maykut, N.N.; Lewtas, J.; Kim, E.; Larson, T.V. Source apportionment of PM 2.5 at an urban IMPROVE site in Seattle, Washington. *Environ. Sci. Technol.* **2003**, *37*, 5135–5142. [CrossRef] [PubMed]
3. Querol, X.; Viana, M.; Alastuey, A.; Amato, F.; Moreno, T.; Castillo, S.; Pey, J.; de la Rosa, J.; de La Campa, A.S.; Artíñano, B. Source origin of trace elements in PM from regional background, urban and industrial sites of Spain. *Atmos. Environ.* **2007**, *41*, 7219–7231. [CrossRef]
4. Franco, V.; Kousoulidou, M.; Muntean, M.; Ntziachristos, L.; Hausberger, S.; Dilara, P. Road vehicle emission factors development: A review. *Atmos. Environ.* **2013**, *70*, 84–97. [CrossRef]
5. U.S. Environmental Protection Agency. *Development of Emission Rates for Light-Duty Vehicles in the Motor Vehicle Emissions Simulator (MOVES2010)*; U.S. Environmental Protection Agency: Washington, DC, USA, 2011.
6. Guerreiro, C.; de Leeuw, F.; Foltescu, V.; Horálek, J. *Air Quality in Europe—2014 Report*; European Environment Agency: Copenhagen, Denmark, 2014.
7. EC European Commission. WHITE PAPER roadmap to a single European transport area towards a competitive and resource efficient transport system. Available online: http://ec.europa.eu/transport/themes/strategies/doc/2011_white_paper/white_paper_com%282011%29_144_en.pdf (accessed on 5 January 2015).

8. Fang, K.; Heijungs, R.; Duan, Z.; de Snoo, G.R. The Environmental Sustainability of Nations: Benchmarking the Carbon, Water and Land Footprints against Allocated Planetary Boundaries. *Sustainability* **2015**, *7*, 11285–11305. [CrossRef]
9. Disley, Y.P. Sustainable development goals for people and planet. *Nature* **2013**, *495*, 305–307.
10. Yang, J.; Yuan, M.; Yigitcanlar, T.; Newman, P.; Schultmann, F. Managing Knowledge to Promote Sustainability in Australian Transport Infrastructure Projects. *Sustainability* **2015**, *7*, 8132–8150. [CrossRef]
11. Qin, F.; Zhang, X. Designing an Optimal Subsidy Scheme to Reduce Emissions for a Competitive Urban Transport Market. *Sustainability* **2015**, *7*, 11933–11948. [CrossRef]
12. Zhang, C.; Jiang, Y.; Zhang, Y.; Bing, X.; Lu, J. An optimal speed limit investigation on highway for emission reductions. *Adv. Transp. Stud.* **2014**, *33*, 33–46.
13. Silva, C.; Farias, T.; Frey, H.C.; Rouphail, N.M. Evaluation of numerical models for simulation of real-world hot-stabilized fuel consumption and emissions of gasoline light-duty vehicles. *Transp. Res. Part D Transp. Environ.* **2006**, *11*, 377–385. [CrossRef]
14. Bing, X.; Jiang, Y.; Zhang, C.; Zhang, Y.; Lu, J. Effects of intersection lane configuration on traffic emissions. *Adv. Transp. Stud.* **2014**, *32*, 23–36.
15. Boriboonsomsin, K.; Barth, M. Impacts of road grade on fuel consumption and carbon dioxide emissions evidenced by use of advanced navigation systems. *Transp. Res. Rec. J. Transp. Res. Board* **2009**, *2139*, 21–30. [CrossRef]
16. Chamberlin, R.; Swanson, B.; Talbot, E.; Dumont, J.; Pesci, S. Analysis of MOVES and CMEM for evaluating the emissions impact of an intersection control change. In Proceedings of the Transportation Research Board 90th Annual Meeting, Washington, DC, USA, 23–27 January 2011.
17. Cicero-Fernández, P.; Long, J.R.; Winer, A.M. Effects of grades and other loads on on-road emissions of hydrocarbons and carbon monoxide. *J. Air Waste Manag. Assoc.* **1997**, *47*, 898–904. [CrossRef] [PubMed]
18. Hao, Y.; Yu, L.; Song, G.; Xu, Y.; Wang, H. Analysis of Driving Behavior and Emission Characteristics for Diesel Transit Buses Using PEMS'Measurements. In Proceedings of the Transportation Research Board 89th Annual Meeting, Washington, DC, USA, 10–14 January 2010.
19. Noland, R.B.; Quddus, M.A. Flow improvements and vehicle emissions: effects of trip generation and emission control technology. *Transp. Res. Part D* **2006**, *11*, 1–14. [CrossRef]
20. Stevanovic, A.; Stevanovic, J.; Zhang, K.; Batterman, S. Optimizing traffic control to reduce fuel consumption and vehicular emissions: Integrated approach with VISSIM, CMEM, and VISGAOST. *Transp. Res. Rec. J. Transp. Res. Board* **2009**, *2128*, 105–113. [CrossRef]
21. Frey, H.C.; Zhang, K.; Rouphail, N.M. Fuel use and emissions comparisons for alternative routes, time of day, road grade, and vehicles based on in-use measurements. *Environ. Sci. Technol.* **2008**, *42*, 2483–2489. [CrossRef] [PubMed]
22. Boroujeni, B.Y.; Frey, H.C. Road grade quantification based on global positioning system data obtained from real-world vehicle fuel use and emissions measurements. *Atmos. Environ.* **2014**, *85*, 179–186. [CrossRef]
23. Wyatt, D.W.; Li, H.; Tate, J.E. The impact of road grade on carbon dioxide (CO_2) emission of a passenger vehicle in real-world driving. *Transp. Res. Part D* **2014**, *32*, 160–170. [CrossRef]
24. Prati, M.V.; Costagliola, M.A.; Tommasino, C.; Della Ragione, L.; Meccariello, G. Road Grade Influence on the Exhaust Emissions of a Scooter Fuelled with Bioethanol/Gasoline Blends. *Transp. Res. Procedia* **2014**, *3*, 790–799. [CrossRef]
25. Alam, A.; Hatzopoulou, M. Investigating the isolated and combined effects of congestion, roadway grade, passenger load, and alternative fuels on transit bus emissions. *Transp. Res. Part D Transp. Environ.* **2014**, *29*, 12–21. [CrossRef]
26. Kern, J.; Clark, N.; Nine, R. Factoring terrain effects into vehicle emissions modeling and inventory. In Proceedings of the 10th Annual Coordinating Research Council On-Road Vehicle Emissions Workshop, San Diego, CA, USA, 27–29 March 2000.
27. Zhang, K.; Frey, H.C. Road grade estimation for on-road vehicle emissions modeling using light detection and ranging data. *J. Air Waste Manag. Assoc.* **2006**, *56*, 777–788. [CrossRef] [PubMed]
28. Frey, H.C.; Unal, A.; Rouphail, N.; Colyar, J. Use of on-board tailpipe emissions measurements for development of mobile source emission factors. In Proceedings of the US Environmental Protection Agency Emission Inventory Conference, Atlanta, GA, USA, 15–18 April 2002; pp. 1–13.

29. Yerramalla, A. *Vehicular Emissions Models Using MOBILE6. 2 and Field Data*; The University of Texas at Arlington: Austin, TX, USA, 2007.
30. Koupal, J.; Michaels, H.; Cumberworth, M.; Bailey, C.; Brzezinski, D. EPA's plan for MOVES: A comprehensive mobile source emissions model. In Proceedings of the 12th CRC On-Road Vehicle Emissions Workshop, San Diego, CA, USA, 15–17 April 2002; Coordinating Research Council: Alpharetta, GA, USA, 2002; pp. 15–17.
31. Rakha, H.; Ahn, K.; Trani, A. Comparison of MOBILE5a, MOBILE6, VT-MICRO, and CMEM models for estimating hot-stabilized light-duty gasoline vehicle emissions. *Can. J. Civ. Eng.* **2003**, *30*, 1010–1021. [CrossRef]
32. Nicole, D.; Lents, J.; Osses, M.; Nikkila, N.; Barth, M. Development and application of an international vehicle emissions model. *Transp. Res. Rec. J. Transp. Res. Board* **2005**, *1939*, 155–165.
33. Jimenez-Palacios, J.L. Understanding and Quantifying Motor Vehicle Emissions with Vehicle Specific Power and TILDAS Remote Sensing. Ph.D. Thesis, Massachusetts Institute of Technology, Cambridge, MA, USA, 1998.
34. Koupal, J.; Cumberworth, M.; Michaels, H.; Beardsley, M.; Brzezinski, D. *Draft Design and Implementation Plan for EPA's Multi-Scale Motor Vehicle and Equipment Emission System (MOVES)*; US Environmental Protection Agency: Washington, DC, USA, 2002.
35. Qi, Y.G.; Teng, H.H.; Yu, L. Microscale emission models incorporating acceleration and deceleration. *J. Transp. Eng.* **2004**, *130*, 348–359. [CrossRef]
36. Song, G.; Yu, L.; Zhang, Y. Applicability of traffic microsimulation models in vehicle emissions estimates: Case study of VISSIM. *Transp. Res. Rec. J. Transp. Res. Board* **2012**, *2270*, 132–141. [CrossRef]
37. Lv, J.; Zhang, Y. Effect of signal coordination on traffic emission. *Transp. Res. Part D* **2012**, *17*, 149–153. [CrossRef]
38. Bartin, B.; Mudigonda, S.; Ozbay, K. Impact of electronic toll collection on air pollution levels: Estimation using microscopic simulation model of large-scale transportation network. *Transp. Res. Rec. J. Transp. Res. Board* **2007**, *2011*. [CrossRef]
39. McCormick, R.L. The impact of biodiesel on pollutant emissions and public health. *Inhal. Toxicol.* **2007**, *19*, 1033–1039. [CrossRef] [PubMed]
40. Aashto, A. *Policy on Geometric Design of hIGHWAYS and Streets*; American Association of State Highway and Transportation Officials: Washington, DC, USA, 2001.
41. Ko, M.; Lord, D.; Zietsman, J. Environmentally Conscious Highway Design for Vertical Grades. *Transp. Res. Rec. J. Transp. Res. Board* **2013**, *2341*, 53–65. [CrossRef]
42. Ministry of Environmental Protection of the People's Republic of China; General Administration of Quality Supervision Inspection and Quarantine of the People's Republic of China. *Limits and Measurement Method for Exhaust Pollutants from Compression Ignition and Gas Fuelled Positive Ignition Engines of Vehicles (III, IV, V)*; China Environmental Publishing: Beijing, China, 2005.
43. Zhang, Y.; Yang, X.; Liu, Q.; Li, C. Who Will Use Pre-Trip Traveler Information and How Will They Respond? Insights from Zhongshan Metropolitan Area, China. *Sustainability* **2015**, *7*, 5857–5874. [CrossRef]
44. Zhang, Y.; Yang, X.; Li, Y.; Liu, Q.; Li, C. Household, Personal and Environmental Correlates of Rural Elderly's Cycling Activity: Evidence from Zhongshan Metropolitan Area, China. *Sustainability* **2014**, *6*, 3599–3614. [CrossRef]
45. Wu, W.; Li, P.K.; Zhang, Y. Modelling and Simulation of Vehicle Speed Guidance in Connected Vehicle Environment. *Int. J. Simul. Model.* **2015**, *14*, 145–157. [CrossRef]
46. Zhang, Y.; Li, Y.; Liu, Q.; Li, C. The Built Environment and Walking Activity of the Elderly: An Empirical Analysis in the Zhongshan Metropolitan Area, China. *Sustainability* **2014**, *6*, 1076–1092. [CrossRef]
47. Zhang, Y.; Li, Y.; Yang, X.; Liu, Q.; Li, C. Built Environment and Household Electric Bike Ownership. *Transp. Res. Rec. J. Transp. Res. Board* **2013**, *2387*, 102–111. [CrossRef]
48. Zhang, Y.; Wu, W.; Li, Y.; Liu, Q.; Li, C. Does the Built Environment Make a Difference? An Investigation of Household Vehicle Use in Zhongshan Metropolitan Area, China. *Sustainability* **2014**, *6*, 4910–4930. [CrossRef]
49. Fulton, L.; Cazzola, P.; Cuenot, F.; Kojima, K.; Onoda, T.; Staub, J.; Taylor, M. *Transport, Energy and CO2: Moving toward Sustainability*; International Energy Agency: Paris, France, 2009.
50. McKibbin, W.J.; Morris, A.C.; Wilcoxen, P.J. Comparing climate commitments: A model-based analysis of the Copenhagen Accord. *Clim. Chang. Econ.* **2011**, *2*, 79–103. [CrossRef]

51. National Bureau of Statistics of the People's Republic of China. *China Statistical Yearbook 2014*; China Statistics Press: Beijing, China, 2014.
52. Wang, H.; Yu, L.; Hao, H.; Song, G.; Guo, Y. Calculation of Vehicle Specification Power for Heavy-duty Vehicles. *Saf. Environ. Eng.* **2011**, *18*, 124–128.
53. Xu, Y.; Yu, L.; Song, G. VSP-Bin division for light-duty vehicles oriented to carbon dioxide. *Acta Sci. Circumst.* **2010**, *30*, 1358–1365.
54. Fellendorf, M.; Vortisch, P. Validation of the microscopic traffic flow model VISSIM in different real-world situations. In Proceedings of the Transportation Research Board 80th Annual Meeting, Washington, DC, USA, 7–11 January 2001.
55. Yadlapati, S.; Park, B. Development and Testing of Variable Speed Limit Logics at Work Zones Using Simulation. Master's Thesis, University of Virginia, Charlottesville, VA, USA, 2004.
56. Park, B.; Schneeberger, J. Microscopic Simulation Model Calibration and validation: case study of VISSIM simulation model for a coordinated actuated Signal system. *Transp. Res. Rec. J. Transp. Res. Board* **2003**, *1856*, 185–192. [CrossRef]
57. Milam, R.T.; Choa, F. Recommended guidelines for the calibration and validation of traffic simulation models. In Proceedings of the Eighth TRB Conference on the Application of Transportation Planning Methods, Corpus Christi, TX, USA, 22–26 April 2002.
58. Jian, S.; Xiao-guang, Y. Research into microscopic traffic simulation model systematic parameter calibration: A case study of VISSIM. *Comput. Commun.* **2004**, *22*, 3–6.
59. Sun, J.; Yang, X.-G.; Liu, H.-D. Study on Microscopic traffic simulation model systematic parameter calibration. *J. Syst. Simul.* **2007**, *1*, 48–50.
60. Pei, Y.-L.; Xing, E.-H. Grade and its length limitation analysis of highway. *J. Harbin Inst. Technol.* **2005**, *37*, 629–632.
61. Ministry of Environmental Protection of the People's Republic of China. *Method for Estimation of Environmental Impact Index of the Light-Duty Vehicle*; Ministry of Environmental Protection of the People's Republic of China: Beijing, China, 2010.
62. Gao, M.; Huang, W.; Wan, X. Study on the emission characteristics of LNG buses based on PEMS. *Environ. Eng.* **2014**, *11*, 151–154.

sustainability

MDPI

Article

Sustainable Water Infrastructure Asset Management: A Gap Analysis of Customer and Service Provider Perspectives

Sangjong Han [1,2], Hwankook Hwang [2], Seonghoon Kim [1], Gyu Seok Baek [1,3] and Joonhong Park [1,*]

1 Department of Civil and Environmental Engineering, Yonsei University, 50 Yonsei-Ro, Seodaemun-Gu, Seoul 03722, Korea; sjhan@kict.re.kr (S.H.); ebtl_sh@yonsei.ac.kr (S.K.); gsbaek@hotmail.com (G.S.B.)

2 Environmental and Plant Engineering Institute, Korea Institute of Civil Engineering and Building Technology, 283, Goyangdae-Ro, Ilsanseo-Gu, Goyang, Gyeonggi-Do 10223, Korea; hkhwang@kict.re.kr

3 Ministry of Environment, Sejong Government Complex Building 6, Doum 6-Ro 11, Sejong 30103, Korea

* Author to whom correspondence should be addressed; parkj@yonsei.ac.kr; Tel.: +82-2-2123-5798; Fax: +82-2-364-5300.

Academic Editors: Tan Yigitcanlar and Md. Kamruzzaman

Received: 5 June 2015; Accepted: 24 September 2015; Published: 29 September 2015

Abstract: The ultimate goal of urban water infrastructure asset management may be sustainable water supply with satisfaction for customers. In this work, we attempted to evaluate the gaps between the perspectives of customers and service providers in Korea's water infrastructure asset management. To evaluate the customers' perspective, a hierarchical questionnaire survey was conducted to estimate the weights of influence for six customer values and their attributes on Korean water utility management. To evaluate the service providers' perspective, an AHP (Analytic Hierarchy Process) analysis was performed to estimate the weights of influence for the customer values and their PIs (performance indicators). The gap analysis results show that customers place higher value on customer service satisfaction (emotion and information) than do the service providers (managers), whereas the managers place more value on affordability than do the customers. The findings from this work imply that improving customer service is effective in satisfying the desirable water LOS (level of service) for customers. Recommendations have also been provided for administrators and engineers to develop integrated decision-making systems that can reflect customer needs regarding the improvement of their water infrastructure asset management. The findings from this work may be helpful for the Korean government and water supply utilities in improving the sustainability of their water infrastructure asset management.

Keywords: water asset management; Gap analysis; LOS (level of service); PIs (performance indicators); customer value

1. Introduction

Sustainability may be one of the key goals of water infrastructure asset management. To make water infrastructure asset management sustainable, it is necessary not only to reduce maintenance costs but also to improve user satisfaction. Asset management in the water sector is an emerging concept that has been adapted from many successful applications in the management of other infrastructure assets, such as the transportation sector and power plants [1]. Many studies regarding water and wastewater asset management have focused on cost-effective and preventive management using risk-based methods [2–4], cost-benefit methods [5,6], and GIS-based asset data integrations [7]. In the past, some research into the methodology of evaluating the levels of service (LOSs) from the customer's perspective has been performed [8–10].

According to the overall asset management process suggested by the New Zealand NAMS (National Asset Management Steering) group, the starting point of strategic planning for asset management is to determine the LOSs that are currently provided [11,12]. Defining the current LOS provisions and setting performance targets play an important role in planning asset management [12]. Schulting and Alegre [1] defined the asset management process as follows: "in addition to the key parameters 'performance-risk-costs' which represent the internal process of asset management in an organization, there are external factors/drivers including customers and regulatory bodies which will influence and set the boundaries for the overall balance between these parameters".

This definition suggests that a comprehensive approach is required to evaluate and set appropriate LOS provisions for water asset management. Understanding the needs of external stakeholders is an area that has not been extensively studied. Customer satisfaction evaluations have been mainly investigated in the marketing sector. However, in the water sector, current customer satisfaction analyses have been limited to a few areas of water quality [13,14] and the cost of water service [15,16], without a comprehensive assessment of its weights of influence in the context of other customer values.

Gap analysis is a useful tool that can help companies or utilities identify gaps between their current status and future goals [17]. In water asset management, gap analysis has been used to identify gaps among different water utilities and to set targets for managing water infrastructure through benchmarking [18–20]. Although several previous studies have addressed the issues regarding gaps between the perspectives of customers and service providers in other sectors (e.g., marketing, airports, and banks) [21–23], no such analysis has been reported in the water asset management field.

In our previous study, Han *et al.* [24] developed a new method to quantitatively evaluate customer needs in the Korean water and wastewater service utility sector (independent of the service provider's perspective) using six customer values: (1) sustainability, (2) affordability, (3) quality (functional performance), (4) health and safety, (5) reliability and responsiveness, and (6) customer service. The developed methodology for quantitatively assessing customer values allowed us to analyze the gaps between the perspectives of water service customers and providers in Korean water and wastewater infrastructure asset management. To compare the LOS evaluations from the customer's perspective to those from the provider's perspective, it is necessary to select and develop PIs to measure the provider's values that are equivalent to the six customer's values considered in the LOS evaluation of the customer's perspective [24] Although PIs [25–36] and methodologies [37–40] to measure water infrastructure asset management from the provider's perspective already exist, it is necessary to further investigate which PIs are suitable for comparison with LOS evaluations from the customer's perspective. Based on a literature review, the research objectives of this work were (i) to select and develop customer-based key performance indicators (KPIs) to evaluate water infrastructure asset management in Korea and (ii) to compare the LOS evaluations between the perspectives of customers and providers.

2. Research Approach and Methodology

The research methodology for comparing the LOS evaluations between the perspectives of customers and service providers is briefly shown in Figure 1. A wellbeing approach [12] has been adopted to analyze the LOS evaluations based on the aspects of environmental, economic, and social wellbeing, which are matched with the six customer values suggested in our previous study [24]. Accessibility to an unrestricted water supply in the community is not included in the six customer values according to our previous customer survey because the weight of influence of accessibility had the lowest customer value. The customer value definitions are obtained from the New Zealand asset management steering group [12]. The "environmental sustainability" value is defined as the quantity and quality of water resources that should be provided for present and future generations. The "affordability" value is defined as the water service level at the lowest possible cost. The "quality" value indicates the structural and functional stability and quality of the pipe system and water supply pressure. The "health/safety" value is defined as the availability of drinking water and the perspective on the risk to customer health and safety. The "reliability/responsiveness" value represents the reliable and responsive provisions of water services without inconvenience or interruption. The "customer service" value is defined as the availability of polite and transparent provisions regarding water services. To evaluate the six values from a provider's perspective, a triple bottom line (TBL) approach [38,39] was used in this study.

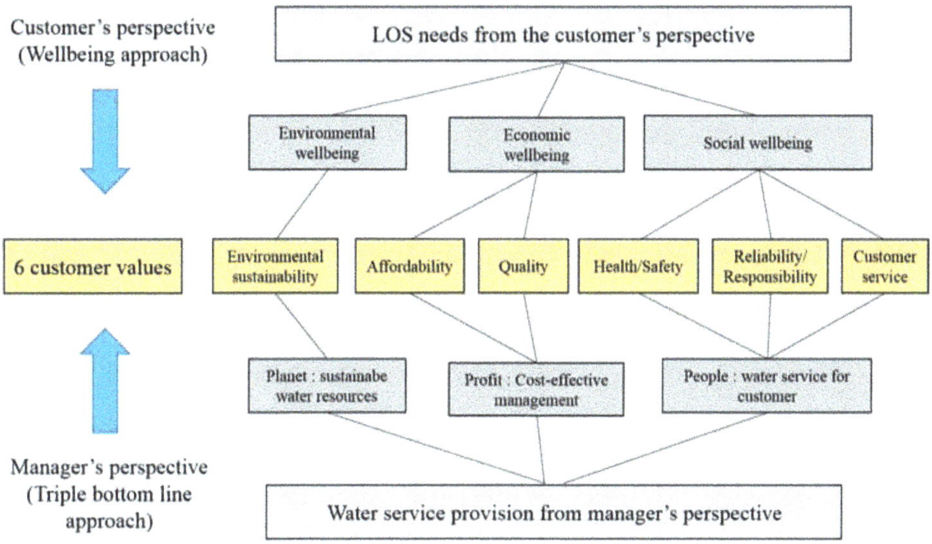

Figure 1. A schematic of the methodology used to compare the LOS evaluations between the perspectives of customers and providers (managers).

The LOS evaluations from the two perspectives were compared using the SERVPERF model [41]. Cronin and Taylor [41] proposed a performance-based methodology for measuring service quality (SERVPERF). They noted that the SERVPERF model was developed to resolve the difficulty in measuring the SERVQUAL scale based on the gap theory of Parasuraman *et al.* [21,42]. Compared with the SERQUAL Equation (1), the weighted SERVPERF equation proposed by Cronin and Taylor is expressed as Equation (2). They suggested that service quality is sufficient to measure only the performance that is perceived as having a direct relation to customer satisfaction [41].

$$Service\ Quality = (Performance - Expectations) \tag{1}$$

$$Service\ Quality = Importance\ (Performance) \tag{2}$$

To evaluate the customer values from the customer's perspective, the levels of customer value attributes (LCVAs) were defined, and their weights of influence were measured using hierarchical customer opinions based on surveys [24]. To evaluate the customer values from the service provider's perspective, KPIs were selected and developed as described below; their weights of influence were measured using the Analytic Hierarchy Process (AHP) to represent the opinion of water service experts. The weights of influence of the LCVAs and KPIs were compared to identify the gaps between the perspectives of the customers and providers regarding water infrastructure asset management. The methodology framework for the gap analysis conducted in this work is summarized in Figure 2.

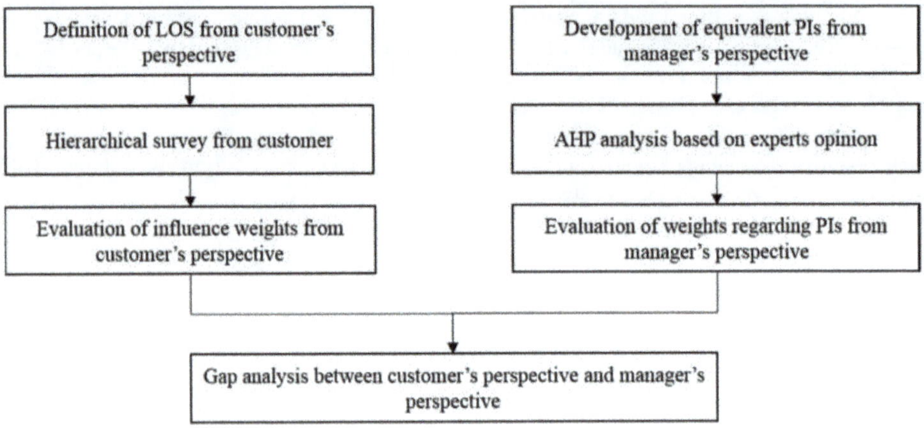

Figure 2. Scheme of the gap analysis between the perspectives of customers and managers on water infrastructure asset management.

3. LOS Evaluation from the Customer's Perspective

A new hierarchical survey method was developed in our previous study [24] to assess the response of customers regarding the LOSs for water utilities in Korea. Because it is difficult for general customers to understand the technical definitions of the customer values and their attributes and meanings in the LOS evaluations, it was necessary to stratify the survey questionnaire using simple to more complex questions. Questions regarding the LCVAs (levels of customer value attributes) were addressed first, followed by questions regarding the LOCVs (levels of customer values) and the LOWs (levels of wellbeing) (Figure 3). The answering system for the questionnaire was based on a seven-point Likert scale, *i.e.*, (1) "strongly disagree" (0 points), (2) "disagree" (16.7 points), (3) "disagree somewhat" (33.3 points), (4) "neutral" (50 points), (5) "agree somewhat" (66.7 points), (6) "agree" (83.3 points), and (7) "strongly agree" (100 points). An electronic survey was conducted via the Internet for 800 tap water customers from seven cities in Korea. When the customers were asked about water being "safe to drink", the concept of "safe to drink" included attributes regarding taste, color, and odor along with attributes regarding toxic chemicals and microbial hazards.

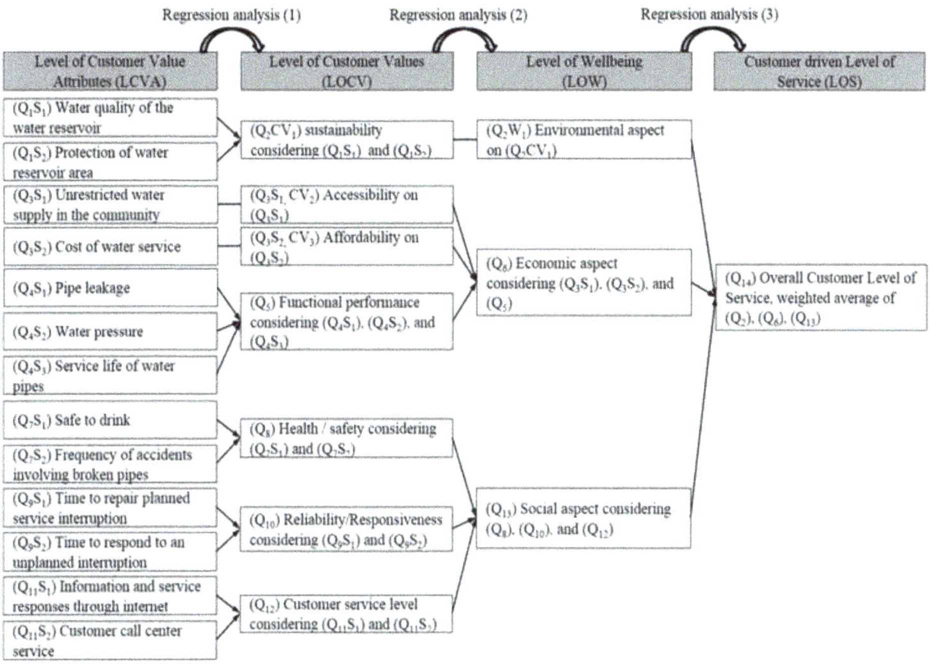

Figure 3. The hierarchical structure of the questionnaire used in the LOS evaluation from the customer's perspective [24].

The weights of influence between two adjacent levels were analyzed using multi-variant regression [43]. A standardized coefficient for the multi-variant statistical analysis (β) was used as the statistical indicator for the weight of influence. Based on the determined β values, the relative weight CVs (customer values) and CVAs (customer value attributes) were calculated (Table 1). For the LOCA, the relative weight of the 'customer service' factor was the highest (0.312) among the six CAs. For the LCVA, the relative weights of "information on Internet" and "customer call center service" (0.162 and 0.150, respectively) were higher than the other CVAs used in the analysis. Considering the fact that the "information on Internet" and "customer call center service" CVA factors are categorized into the "customer service" category for the LOCA, "customer service" was identified as the most influential CV factor in the LOS evaluation from the customer's perspective.

Table 1. The weights of influence (β) and relative weights of CVs and CVAs determined from the customer's perspective.

Factors	LOW (A)	LOCV (B)	LCVA (C)	Relative Weights of CVs ($= \frac{A \times B}{\sum A \times B}$)	Relative Weights of CVAs ($= \frac{A \times B \times C}{\sum A \times B \times C}$)
		β			
Level 1					
Environmental	0.131				
Economic	0.274				
Social	0.596				
Level 2-Environmental					
Sustainability	0.131	1		0.131	
Level 2-Economic					
Affordability	0.274	0.189		0.052	
Quality		0.810		0.222	
Level 2-Social					
Health/Safety	0.596	0.249		0.148	
Reliability/Responsiveness		0.227		0.135	
Customer service		0.524		0.312	
Level 3-*Sustainability*					
Water quality of water reservoir	0.131	1	0.449		0.059
Protection of water reservoir area			0.551		0.072
Level 3-*Affordability*					
Cost of water service	0.274	0.189	1		0.052
Level 3-*Quality(Physical)*					
Water losses	0.274	0.810	0.143		0.032
Water pressure			0.326		0.072
Useful life of water mains			0.531		0.118
Level 3-*Health/Safety*					
Safe to drink	0.596	0.249	0.754		0.112
Probability of pipe accidents			0.246		0.037
Level 3-*Reliability/Responsiveness*					
Time to respond to planned service interruption	0.596	0.227	0.297		0.040
Time to respond to unplanned interruption			0.703		0.095
Level 3-*Customer service*					
Information on internet	0.596	0.524	0.519		0.162
Customer call center service			0.481		0.150

4. LOS Evaluation from the Manager's Perspective

The procedure for evaluating the water supply utility LOSs from the water utility manager's perspective is shown in Figure 4.

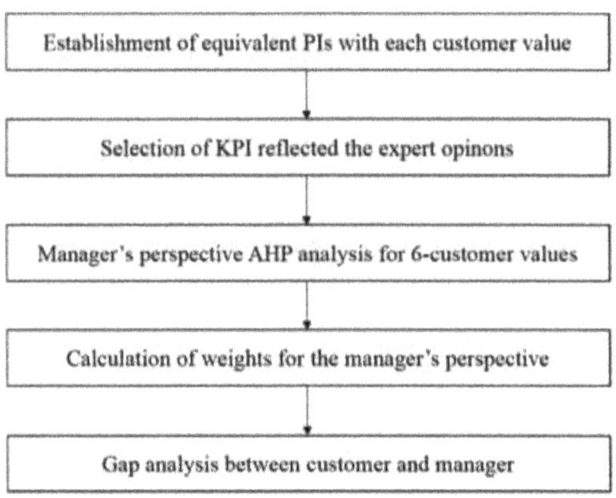

Figure 4. The evaluation process for the manager's perspective.

The AHP decomposes the decision problem into a hierarchy of sub-problems. The decision maker then evaluates the relative importance of its various elements via pairwise comparisons. The AHP converts those evaluations to numerical values (weights or priorities), which are used to calculate a score for each alternative [44]. Importance scales for pairwise comparison are shown in Table 2.

Table 2. Intensity of importance for pairwise comparison [44].

Linguistic terms (the comparison of A to B)	Score
Absolutely strong (AS)	9
Between AS and VS	8
Very strong (VS)	7
Between VS and FS	6
Fairly strong (FS)	5
Between FS and SS	4
Slightly strong (SS)	3
Between SS and E	2
Equal (E)	1
Between E and SW	1/2
Slightly weak (SW)	1/3
Between SW and FW	1/4
Fairly weak (FW)	1/5
Between FW and VW	1/6
Very weak (VW)	1/7
Between VW and AW	1/8
Absolutely weak (AW)	1/9

The hierarchy of the AHP analysis is presented in Figure 5. The advice from eight professional experts in the Korean water sector and information from selectively used literature from the IWA PI system and CARE-W rehab PIs were used in this study [26]. Level 1 is considered the strategic aspect and is equivalent to the LOW in the LOS evaluation from the customer's perspective. Levels 2 and 3 address tactical and operational aspects, respectively, and are equivalent to the LOCV and LCVA in the water infrastructure LOS evaluation from the customer's perspective. At the strategic level (Level 1), the three TBL goals (*i.e.*, planet, profit, and people) were selected as the major factors to be considered in the LOS evaluation from the water service provider's perspective. At the tactical level (Level 2), six factors that are equivalent to the six CVs were chosen for comparison with the weight of influence results from the analysis of the customer's perspective. At the operational level (Level 3), the KPIs for the planet ("environmental sustainability") and profit ("affordability" and "quality [physical]") goals were selected among the previously established PIs from the IWA PI system and the CARE-W rehab PIs [26]. Due to the unavailability of PI information, KPIs for the Level 1 objective for the customers' goals (the Level 2 objectives of "health/safety", "reliability/responsiveness", and "customer service") were newly developed in this work based on expert opinions.

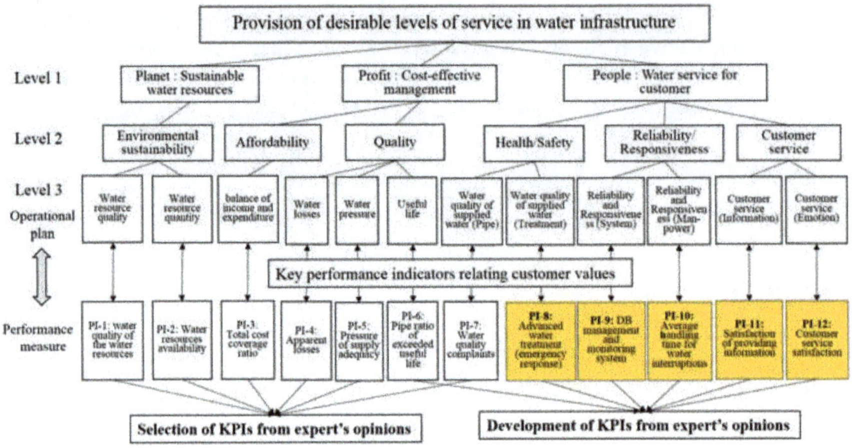

Figure 5. Hierarchy for LOS evaluation from the service provider's perspective and the KPIs for the six CVs.

Table 3. An example of a five-grade system for evaluating the PIs. (The example is for advanced water treatment for emergency management [PI-8]).

Grade	Description
1	Ultra-advanced purification technology application (membrane filtration), 100% automated WTP, expanded facilities for source water management, 100% compliance rate for the water treatment plant safety inspection
2	Advanced purification technology application, more than 80% automated WTP, 100% compliance rate for the water treatment plant safety inspection
3	More than 60% automated WTP, more than 80% compliance rate for the water treatment plant safety inspection
4	More than 40% automated WTP, more than 60% compliance rate for the water treatment plant safety inspection
5	Less than 40% automated WTP, less than 60% compliance rate for water treatment plant safety inspection

The new PIs were intended to be evaluated through rating scales. A five-grade system was developed to evaluate the new PIs based on the collected expert opinions. Advanced water treatment with emergency response capacity (PI-8) includes factors such as advanced water purification capacity, capacity for automatic facility operation, expanded capacity facilities for source water treatment, and safety of the water treatment plant (Table 3). The capacity of the database management and monitoring system (PI-9) refers to the use of a GIS-based management system, an adequate water distribution block system, a flow monitoring system, and a water metering automation system. With regard to the average handling time for water interruptions (PI-10), the results of the expert opinions indicated that Grade 1 is within 2.5 h, Grade 2 is within 5.6 h, and Grades 3, 4, and 5 are within 10.8, 24, and 40.7 h, respectively. According to a previous study [24], 5 h per year (46.6% of respondents) and 6–10 h per year (24.9% of respondents) were given as the accepted durations of water service disruptions for handling various incidents, such as freezing or pipe damage. The following satisfaction factors for providing information (PI-11) are considered: detailed water service information that contains the asset management goals for water utilities, water quality monitoring data in the water pipe network, the request process for all water service types, financial statements, statistics of water services, announcements regarding planned rehabilitation, database management for frequent complaints, provision of water charge calculation for each household, and bidding information. Customer service satisfaction (PI-12) includes items such as staff training for customer complaints (including complaint database management), the provision of tap water quality testing, and automatic meter reading that eschews visits to customers' homes to inspect water usage.

To assess the weights of influence for the water LOSs from the manager's perspective, a KPI-based AHP analysis was conducted [44]. This survey included eight experts: one professor, three researchers, and four public water utility managers. The collective working experience of the selected experts included more than 15 years in the water sector. As shown in Table 4, all PI weights of influence were derived using the AHP analysis results. Pairwise comparisons were used to evaluate the relative weights of the six CVs and their corresponding PIs. At the tactical level, the relative weights of the "affordability" and "quality" factors (0.317 and 0.204) were higher than those of the other CVs. Unlike the customer's perspective (Table 1), the "customer service" factor was considered the least influential factor from the provider's perspective. At the operational level PIs, the relative weight of the "total water cost coverage ratio" factor (0.317) was the highest among the PIs tested in the AHP analysis, whereas the customer service PIs (the "satisfaction with customer service on information" and "satisfaction with customer service on emotion" factors) were considered the least influential PIs from the provider's perspective.

Table 4. The weights of influence (β) and relative weights of the CVs and PIs determined from the AHP analysis of the service provider's perspective.

Factor	Strategic Level (A)	Tactical Level (B)	Operational Level (C)	Weights of CVs ($=\frac{A \times B}{\sum A \times B}$)	Weights of PIs ($=\frac{A \times B \times C}{\sum A \times B \times C}$)
Level 1					
Planet	0.114				
Profit	0.522				
People	0.365				
Level 2-Planet					
Sustainability	0.114	1		0.114	
Level 2-Profit					
Affordability	0.522	0.608		0.317	
Quality		0.392		0.204	
Level 2-People					
Health/Safety	0.365	0.453		0.165	
Reliability/Responsiveness		0.348		0.127	
Customer service		0.198		0.072	
Level 3-*Sustainability*					
Water resource quality	0.114	1	0.438		0.050
Water resource quantity			0.562		0.064
Level 3-*Affordability*					
Total cost coverage ratio	0.522	0.608	1		0.317
Level 3-*Quality(Pipe)*					
Apparent losses	0.522	0.392	0.239		0.049
Pressure of supply adequacy			0.485		0.099
Pipe ratio of exceeded useful life			0.276		0.056
Level 3-*Health/Safety*					
Water quality of supplied water from Pipe	0.365	0.453	0.633		0.105
Water quality of supplied water from treatment plant			0.367		0.061
Level 3-Reliability/Responsiveness					
System	0.365	0.348	0.597		0.076
Manpower			0.403		0.051
Level 3-*Customer service*					
Satisfaction with customer service on information	0.365	0.198	0.503		0.036
Satisfaction with customer service on emotion			0.497		0.036

5. The Gap Analysis Results between the Perspectives of Customers and Managers

The relative weights of the six CVs were compared between the perspectives of customers and managers (Figure 6). From the customer's perspective, the "customer service" CV factor was identified as the most influential. Moreover, from the provider's perspective, the "affordability" CV factor was found to be the most influential. Other CV factors (e.g., "sustainability", "quality", "health and safety", and "reliability and responsiveness") were insignificantly different between the perspectives of customers and managers. The results suggest the possible existence of significant gaps between customers' needs and managers' provision, and "customer service" and "affordability of cost" were identified as the customer values that had gaps between customers' needs and water utility provision.

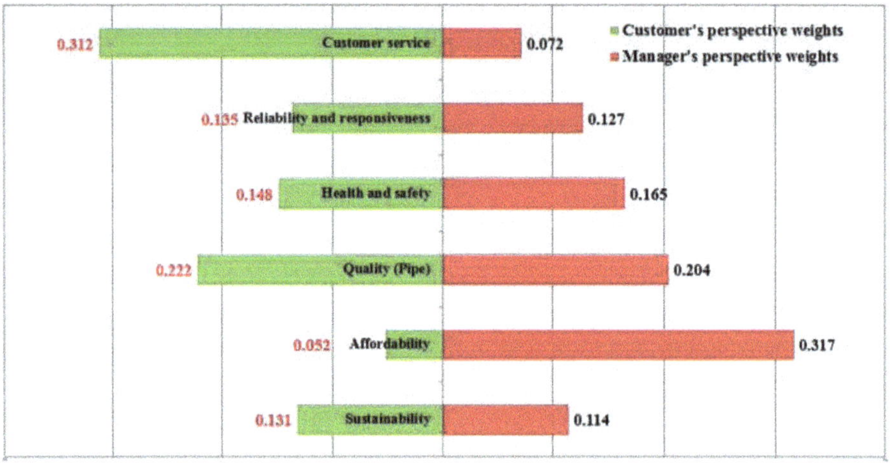

Figure 6. The gap analysis results between the perspectives of customers and managers on six CVs.

The gap analysis at Level 3 identified the CVAs and PIs with significant gaps between customers and service providers (Figure 7). The customers may place the highest values on satisfaction of customer service on emotion (relative influence weight = 0.150) and information (relative influence weight = 0.162), whereas the customer service attributes are given the lowest priority by water utility managers (corresponding relative weights = 0.036). Further, affordability was identified as the lowest priority by customers (relative influence weight = 0.052) but as the highest priority by managers (relative influence weight = 0.317). The results provide implications regarding what tactical and operational actions may be needed to improve customer satisfaction.

To improve customer service, information provision and the emotional aspect of contact with customers may have to be emphasized further in water utility management. Water service information includes announcements regarding water asset management planning, water quality monitoring data, and service. In other developed countries, such water service information is publicly available and reachable [45]; however, information on (i) water asset management planning and its evaluation outcomes and (ii) water quality monitoring data is often closed to the public or is unreachable in Korea. This lower public availability and reachability may cause citizens in many Korean cities to feel that the quality and price of tap water is not as good or reasonable as they expect them to be, even though the water quality and water infrastructure assets are scientifically monitored and externally evaluated by civilian experts with standards as high as those of other developed countries. There are some unique differences in water utility management in Korea compared with other countries. Most municipalities in Korea directly operate and manage water supply utilities, whereas water utilities are operated and managed mainly by private companies in European countries [46]. In addition, the managers of

Korean water utilities often have non-engineering education backgrounds, and change jobs frequently in local government; these factors hinder the development of long-term credibility from citizens.

The findings from this work suggest that the difference in the perception of reasonably cost-effective and safe water service between Korean customers (citizens) and water service providers may have resulted from a lack of such information being provided to the customers. To address this issue, announcements of asset management planning and subsequent evaluation summaries, posts about water quality monitoring data on publicly available websites or social network systems (SNS), and periodic customer surveys can be effective tools for communication with customers. In addition, improving customer service is an effective way to increase customer satisfaction. Hunter Water in Australia reported that their customer and commercial development teams succeeded in achieving 94% customer satisfaction during the fiscal year 2009–2010 as a result of consistent commitment to customer service excellence [47]. A sufficiently large service staff and customer call service professionals with technical knowledge are required to address customers' questions and problems regarding the quality and price of water service. Customers' opinions on customer service must be regularly surveyed and evaluated, and the feedback should be reflected in improvements to customer service in terms of information provision as well as emotional aspects.

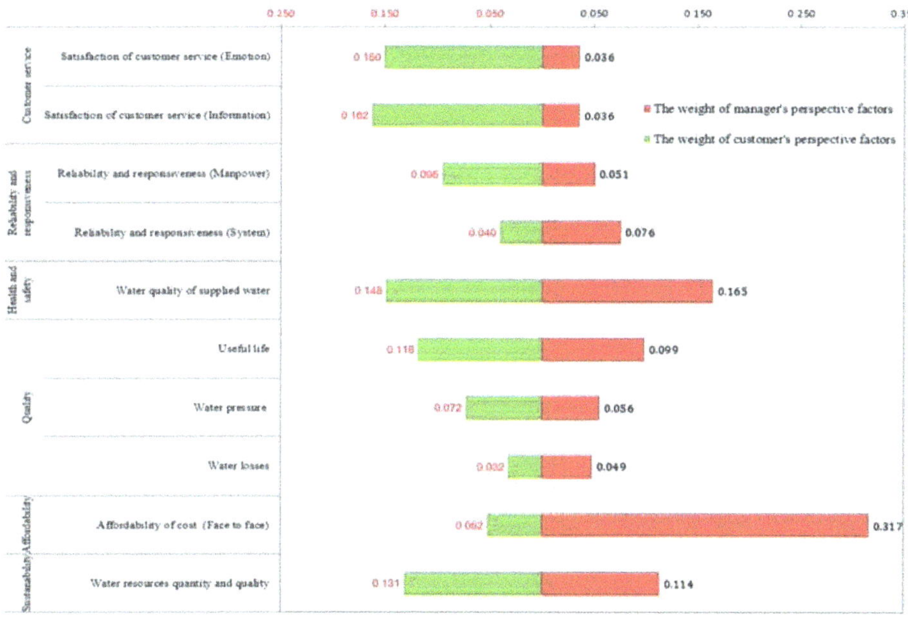

Figure 7. The gap analysis results between the perspectives of customers and managers on the water LOS.

6. Conclusions

Local administrators are interested in selecting the highest priority for a management area through a risk-based approach or by allocating additional funds for sustainable water management. However, in light of the results of this study, these actions may not represent the best approach for asset management. This study identified the gaps between customer needs and service provisions. The gap analysis results show that utility managers do not consider customer satisfaction, use information, and service-related emotions. This reflects the current problem of water infrastructure asset management in Korea. At the same time, the gap analysis results suggest that customer service improvement is an effective way to fill the gaps between the perspectives of customers and providers on water service.

To improve customer service, steps must be taken to increase the flow of information, particularly regarding water quality monitoring data, and reinforcement of staff training and education are recommended based on the results of gap analysis of the weights given to CVAs and PIs between the perspectives of customers and managers. The findings from this work may be useful in assisting local government administrators in making better decisions on water infrastructure asset management by integrating the perspectives of service providers and customers, and in turn may be helpful in improving the sustainability of water infrastructure asset management.

Acknowledgments: This research was supported by the Eco-Innovation Research Fund (414-111-002) in Korea, the National Research Foundation of Korea (NRF), and a grant funded by the Korea government (MSIP) (No. NRF-2013R1A2A2A03016475). We thank the eight experts who participated in the surveys and discussion for the AHP analysis and PI development.

Author Contributions: Sangjong Han and Hwankook Hwang developed the concept and conducted the surveys. Sangjong Han was the main writer of the manuscript. Seonghoon Kim analyzed the survey data. Gyu Seok Back developed the KPIs. Joonhong Park, as Sangjong Han's PhD dissertation advisor, edited the manuscript.

Conflicts of Interest: The authors declare no conflict of interest.

References

1. Schulting, F.; Alegre, H. Report: Global developments of strategic asset management. In Strategic Asset Management of Water Supply and Wastewater Infrastructure: Invited Papers from the IWA Leading Edge Conference on Strategic Asset Management (LESAM); Proceedings of the IWA Leading Edge Conference on Strategic Asset Management (LESAM), Lisbon, Portugal, 17–19 October 2007; IWA Publishing: London, UK, 2009; pp. 13–29.
2. Kleiner, Y.; Sadiq, R.; Rajani, B. Modeling failure risk in buried pipes using fuzzy markov deterioration process. In Proceeding of the ASCE International Conference on Pipeline Engineering and Construction, San Diego, CA, USA, 1–4 August 2004.
3. Rogers, P.D.; Grigg, N.S. Failure assessment modeling to prioritize water pipe renewal: Two case studies. *J. Infrastruct. Syst.* **2009**, *15*, 162–171. [CrossRef]
4. Baah, K.; Dubey, B.; Harvey, R.; McBean, E. A risk-based approach to sanitary sewer pipe asset management. *Sci. Total Environ.* **2015**, *505*, 1011–1017. [CrossRef] [PubMed]
5. Giustolisi, O.; Laucelli, D.; Savic, D.A. Development of rehabilitation plans for water mains replacement considering risk and cost-benefit assessment. *Civil Eng. Environ. Syst.* **2006**, *23*, 175–190. [CrossRef]
6. Gaivoronski, A.; Sechi, G.M.; Zuddas, P. Cost/risk balanced management of scarce resources using stochastic programming. *Eur. J. Oper. Res.* **2012**, *216*, 214–224. [CrossRef]
7. Halfawy, M.R.; Dridi, L.; Baker, S. Integrated decision support system for optimal renewal planning of sewer networks. *J. Comput. Civil Eng.* **2008**, *22*, 360–372. [CrossRef]
8. KWR (Kiwa Water Research). *Consumer Satisfaction, Preference and Acceptance Regarding Drinking Water Services: An Overview of Literature Findings and Assessment Methods*; Kiwa Water Research: Nieuwegein, The Netherlands, 2008.
9. Ojo, V.O. Customer satisfaction: A framework for assessing the service quality of urban water service providers in Abuja, Nigeria. Ph.D. Thesis, Loughborough University, Loughborough, UK, 2011.
10. Hanson, J.J.; Murrill, S.D. *South Tahoe Public Utility District 2012 Customer Satisfaction and Perceptions Survey Report of Results*; Meta Research, Inc.: Sequim, WA, US, 2013.
11. NAMS Group. *International Infrastructure Management Manual. NZ National Asset Management Steering Group*; Association of Local Government Engineers NZ Inc (INGENIUM): Thames, New Zealand, 2006.
12. NAMS Group. *Developing Levels of Service and Performance Measures. NZ National Asset Management Steering Group*; Association of Local Government Engineers NZ Inc (INGENIUM): Thames, New Zealand, 2007.
13. Levallois, P.; Grondin, J.; Gingras, S. Evaluation of consumer attitudes on taste and tap water alternatives in québec. *Water Sci. Technol.* **1999**, *40*, 135–139. [CrossRef]
14. Lou, J.C.; Lee, W.L.; Han, J.Y. Influence of alkalinity, hardness and dissolved solids on drinking water taste: A case study of consumer satisfaction. *J. Environ. Manag.* **2007**, *82*, 1–12. [CrossRef] [PubMed]

15. Kayaga, S.; Calvert, J.; Sansom, K. Paying for water services: Effects of household characteristics. *Util. Policy* **2003**, *11*, 123–132. [CrossRef]
16. Wahid, N.A.; Hooi, C.K. Factors determining household consumer's willingness to pay for water consumption in Malaysia. *Asian Soc. Sci.* **2015**, *11*. [CrossRef]
17. Wikipedia. Gap Analysis. Available online: https://en.wikipedia.org/wiki/Gap_analysis (accessed on 23 September 2015).
18. Mehan, G.T. *The Clean Water and Drinking Water Infrastructure Gap Analysis*; EPA-816-R-02-020; US Environmental Protection Agency Office of Water: Washington, DC, USA, 2002. Available online: http://www.epa.gov/safewater/gapreport.pdf (accessed on 23 September 2015).
19. Lafferty, A.K.; Lauer, W.C. *Benchmarking Performance Indicators for Water and Wastewater Utilities: Survey Data and Analyses Report*; American Water Works Association: Denver, CO, USA, 2005; Available online: http://www.amazon.ca/Benchmarking-Performance-Indicators-Wastewater-Utilities/dp/158321366X (accessed on 23 September 2015).
20. Graf, W. *Assessing Utility Practices with the Strategic Asset Management Gap Analysis Tool (SAM Gap)*; SAM2C06; Water Environment Research Foundation: Alexandria, CO, USA, 2010; Available online: http://simple.werf.org/UploadFiles/SAM2C06 (accessed on 23 September 2015).
21. Parasuraman, A.; Zeithaml, V.A.; Berry, L.L. A conceptual model of service quality and its implications for future research. *J. Mark.* **1985**, *49*, 41–50. [CrossRef]
22. Tsai, W.H.; Hsu, W.; Chou, W.C. A gap analysis model for improving airport service quality. *Total Qual. Manag. Bus. Excell.* **2011**, *22*, 1025–1040. [CrossRef]
23. Anand, S.V.; Selvaraj, M. Evaluation of service quality and its impact on customer satisfaction in indian banking sector—A comparative study using SERVPERF. *Life Sci. J. Acta Zhengzhou Univ. Overseas Ed.* **2013**, *10*, 3267–3274.
24. Han, S.; Chae, M.; Hwang, H.; Choung, Y. Evaluation of customer-driven level of service for water infrastructure asset management. *J. Manag. Eng.* **2014**, *31*. Article 04014067. [CrossRef]
25. Matos, R.; Cardoso, A.; Ashley, R.; Duarte, P.; Molinari, A.; Schulz, A. *Performance Indicators for Wastewater Services*; IWA Publishing: London, UK, 2003; Volume 1.
26. Alegre, H. *Performance Indicators for Water Supply Services*; IWA Publishing: London, UK, 2006.
27. Palme, U.; Tillman, A.-M. Sustainable development indicators: How are they used in Swedish water utilities? *J. Clean. Prod.* **2008**, *16*, 1346–1357. [CrossRef]
28. Sklar, D.C. Performance measurement and reporting, Asset Management for Water and Wastewater Utilities. In Proceedings of the Water Environment Federation "Hot Topics" Workshop, New Orleans, LA, USA, 18–19 June 2008.
29. Mutikanga, H.; Sharma, S.; Vairavamoorthy, K.; Cabrera, E., Jr. Using performance indicators as a water loss management tool in developing countries. *J. Water Suppl. Res. Technol. Aqua* **2010**, *59*, 471–481. [CrossRef]
30. OFWAT (Office of Water Services). Key performance indicators-guidance. Available online: http://www.ofwat.gov.uk/regulating/compliance/gud_pro1203kpi.pdf (assessed on September 8 2015).
31. Shinde, V.R.; Hirayama, N.; Mugita, A.; Itoh, S. Revising the existing performance indicator system for small water supply utilities in Japan. *Urban Water J.* **2013**, *10*, 377–393. [CrossRef]
32. Haider, H.; Sadiq, R.; Tesfamariam, S. Performance indicators for small-and medium-sized water supply systems: A review. *Environ. Rev.* **2013**, *22*, 1–40. [CrossRef]
33. Singh, M.; Mittal, A.K.; Upadhyay, V. Efficient water utilities: Use of performance indicator system and data envelopment analysis. *Water Sci. Technol. Water Supply* **2014**, *14*, 787–794. [CrossRef]
34. Haider, H.; Sadiq, R.; Tesfamariam, S. Selecting performance indicators for small and medium sized water utilities: Multi-criteria analysis using ELECTRE method. *Urban Water J.* **2015**, *12*, 305–327. [CrossRef]
35. Haider, H.; Sadiq, R.; Tesfamariam, S. Multilevel performance management framework for small to medium sized water utilities in Canada. *Can. J. Civil Eng.* **2015**. [CrossRef]
36. Haider, H.; Sadiq, R.; Tesfamariam, S. Inter-utility performance benchmarking model for small-to-medium-sized water utilities: Aggregated performance indices. *ASCE J. Water Resour. Plan. Manag.* **2015**. [CrossRef]
37. Slaper, T.F.; Hall, T.J. The triple bottom line: What is it and how does it work. *Indiana Bus. Rev.* **2011**, *86*, 4–8.
38. Koo, D.-H.; Ariaratnam, S.T.; Kavazanjian, E., Jr. Development of a sustainability assessment model for underground infrastructure projects. *Canadian J. Civil Eng.* **2009**, *36*, 765–776. [CrossRef]

39. Liner, B.; de Monsabert, S. Balancing the triple bottom line in water supply planning for utilities. *J. Water Resour. Plan. Manag. ASCE* **2011**, *137*, 335–342. [CrossRef]
40. Kang, D.; Lansey, K. Dual water distribution network design under triple-bottom-line objectives. *J. Water Resour. Plan. Manag. ASCE* **2012**, *138*, 162–175. [CrossRef]
41. Cronin, J.J., Jr.; Taylor, S.A. Measuring service quality: A reexamination and extension. *J. Mark.* **1992**, *56*, 55–68. Available online: http://www.researchgate.net/publication/225083621_Measuring_Service_Quality_-_A_Reexamination_And_Extension (accessed on 23 September 2015).
42. Zeithaml, V.A.; Berry, L.L.; Parasuraman, A. Communication and control processes in the delivery of service quality. *J. Mark.* **1988**, *52*, 35–48. Available online: http://areas.kenan-flagler.unc.edu/Marketing/FacultyStaff/zeithaml/Selected%20Publications/Communication%20and%20Control%20Processes%20in%20the%20Delivery%20of%20Service%20Quality.pdf (accessed on 23 September 2015).
43. Fox, J. *Applied Regression Analysis, Linear Models, and Related Methods*; Sage Publications Inc.: Thousand Oaks, CA, USA, 1997.
44. Saaty, T.L. *Multicriteria Decision Making: The Analytic Hierarchy Process*; AHP series; RWS Publications: Pittsburgh, PA, USA, 1990; Volume 1.
45. Orange County Sanitation District (OCSD). Orange County Sanitation District Asset Management Plan 2006. Available online: http://www.ocsd.com/Home/ShowDocument?id=11138 (assessed on 31 July 2015).
46. OECD. *frastructure to 2030: Mapping Policy for Electricity, Water and Transport*; OECD: Paris, France, 2007.
47. Hunter Water. Hunter Water 2009–2010 Annual Report. Available online: http://www.hunterwater.com.au/Resources/Documents/Annual-Reports---Past-Reports/Annual-Report-09-10.pdf (assessed on 31 July 2015).

sustainability

MDPI

Article

The Influence of Low-Frequency Noise Pollution on the Quality of Life and Place in Sustainable Cities: A Case Study from Northern Portugal

Juliana Araújo Alves [1], Lígia Torres Silva [2,*] and Paula Cristina C. Remoaldo [1]

1 Lab2PT—Landscape, Heritage and Territory Laboratory, University of Minho, Campus de Azurém, 4800-058 Guimarães, Portugal; jalves.geografia@gmail.com (J.A.A.); premoaldo@geografia.uminho.pt (P.C.C.R)

2 CTAC—Centre for Territory, Environment and Construction, Department of Civil Engineering, University of Minho, Campus Gualtar, 4710-057 Braga, Portugal

* Author to whom correspondence should be addressed; lsilva@civil.uminho.pt; Tel.: +351-253510200.

Academic Editor: Tan Yigitcanlar

Received: 8 June 2015; Accepted: 23 September 2015; Published: 19 October 2015

Abstract: Discussing urban planning requires rethinking sustainability in cities and building healthy environments. Historically, some aspects of advancing the urban way of life have not been considered important in city planning. This is particularly the case where technological advances have led to conflicting land use, as with the installation of power poles and building electrical substations near residential areas. This research aims to discuss and rethink sustainability in cities, focusing on the environmental impact of low-frequency noise and electromagnetic radiation on human health. It presents data from a case study in an urban space in northern Portugal, and focuses on four guiding questions: Can power poles and power lines cause noise? Do power poles and power lines cause discomfort? Do power poles and power lines cause discomfort due to noise? Can power poles and power lines affect human health? To answer these questions, we undertook research between 2014 and 2015 that was comprised of two approaches. The first approach consisted of evaluating the noise of nine points divided into two groups "near the source" (e.g., up to 50 m from power poles) and "away from the source" (e.g., more than 250 m away from the source). In the second approach, noise levels were measured for 72 h in houses located up to 20 m from the source. The groups consist of residents living within the distance range specified for each group. The measurement values were compared with the proposed criteria for assessing low-frequency noise using the DEFRA Guidance (University of Salford). In the first approach, the noise caused discomfort, regardless of the group. In the second approach, the noise had fluctuating characteristics, which led us to conclude that the noise caused discomfort.

Keywords: noise pollution; low-frequency noise; DEFRA; human well-being; sustainability; power poles

1. Introduction

The current accelerated urbanization process in Europe has been accompanied by a number of environmental and social problems arising from consumption patterns and lifestyle, such as greenhouse gas emissions, waste and wastewater and environmental noise. These problems have significant impacts on the environment, public health and people's quality of life. Noise pollution is a current public health problem associated with modern life and urbanization. This urban problem continues to increase in extent, frequency and severity as a result of urbanization, population growth and technological development.

Although the concept of sustainability has been in existence and use for nearly 30 years, even today it is still easier to define the concept than to apply it in practice.

Environmental noise can be regarded as one of the agents of deterioration in people's quality of life in an urban environment [1]. In Europe, noise achieves a significant level and is considered an environmental problem of major proportions and great impact. Accordingly, the European Noise Directive was established in 2002. The European Network on Noise and Health (ENNAH) is a Network funded by the European Union that gathers expert groups on noise and health in Europe. The network brings together some 33 European research centers located in 16 countries to support public policies to cope with noise and health, specifically, traffic noise [2]. The World Health Organization (WHO) estimates that Disability-Adjusted Life Years (DALYs) lost due to environmental noise are "6000 years for ischemic heart disease, 45,000 years for cognitive impairment of children, 903,000 years for sleep disturbance, 21,000 years for tinnitus and 587,000 years for annoyance" [3]. In addition, noise exposure is increasing in Europe.

Scientific evidence has proved that noise affects human health. On the other hand, it can be observed that large multinational organizations are increasingly concerned about this type of urban pollution. On the other hand, attention given to this topic focuses particularly on sources such as urban traffic, occupational exposure and wind turbines. Yet, little or no attention has been given to low-frequency noise originating from power poles.

After considering the literature review, it can be stated that, in general, the effects of low-frequency noise are not as well researched as other noise sources. Some authorities, such as the WHO, recognize the importance of environmental noise [3]. Nevertheless, the assessments adopted for low-frequency components and their effects on human health have received less focus than those of high frequencies in the academic literature, even though low frequencies are considered more annoying for humans [4–8]. Moreover, the United Kingdom has been progressing towards tackling the issues surrounding Low-Frequency Noise (LFN).

Noise and its impacts on health must be considered as quality of life indicators in sustainable cities. Studies on low-frequency noise emitted by power poles and its influence on human health are scarce, even if noise exposure has harmful effects and is a risk factor for human health. In Portugal, studies concerning low-frequency noise have focused on sources such as traffic, wind turbines and occupational exposure, specifically research carried out by the Center for Human Performance, Alverca/Portugal, on vibroacoustic disease [9,10]. This research is motivated by the absence of studies regarding low-frequency sources, such as power poles and their effects on human health. This leads us to pose the following questions: (1) Do power poles and power lines cause noise? (2) Do power poles and power lines cause discomfort? (3) Do power poles and power lines cause discomfort due to noise? (4) Do power poles and power lines cause an impact on human health?

The main objective of this paper is to analyze the exposure of the population to low-frequency noise from power poles in residential areas, especially in the village of Serzedelo (Municipality of Guimarães, northwestern Portugal), and its impacts on human health.

To accomplish this objective, two groups were considered: the "near the source" (e.g., distance to 400 kV power poles less than 50 m) group and the "away from the source" (e.g., distance to 400 kV power poles more than 250 m) group. In the second approach, a more in-depth analysis was carried out. which consisted of taking measurements for 72 h. All the measurement values were compared with the values of the criterion curve proposed by the Department of Environment, Food and Rural Affairs [11] to evaluate whether the measured values cause annoying conditions.

This study followed the sequence of an exploratory one, using surveys undertaken in 2010 in the village of Serzedelo using the same two groups: the "near the source" group (118 individuals were interviewed) and the "away from the source" group (55 individuals were interviewed) [12]. Although this study did not focus on the influence of low-frequency noise pollution, respondents only in the "near the source" group stated that they experienced discomfort from the noise.

This paper is organized as follows: Section 2 presents a literature review of low-frequency noise and human well-being; the material and methods are discussed in Section 3; the results and analysis are presented in Section 4; in Section 5, the evaluation of the perception of noise discomfort is presented. Finally, Section 6 presents the discussion and conclusions are drawn in Section 7.

2. Low-Frequency Noise and Human Well-Being

After the 1970s, several studies began highlighting the discomfort caused by environmental noise [3,8], especially traffic noise [1,13–15]. The most cited effects on human health refer to emotional changes, namely agitation, distraction, disappointment, stress, hypertension [7,16–18] and the association of low-frequency noise with cognitive impairments [19], the development of cardiovascular diseases [20,21], disturbances in sleep and heart rate [22–24] and hypertension [25,26].

In the field of occupational medicine, several studies claim that low-frequency noise is an agent that interferes with the performance of work tasks [25,27–29] and that low-frequency noise can affect mental and physical health.

Exposure to noise has harmful effects and constitutes a risk factor for human health. Some authors have treated these effects under the name as vibroacoustic disease [10], vibroacoustic pathology or vibronoise pathology [30], *i.e.*, systemic pathology encompassing the entire body, characterized by abnormal proliferation of extracellular matrices caused by excessive and prolonged exposure to low-frequency noise (LFN) [10,31].

In Portugal, the first study of low-frequency noise dates back to 1979. The study focused on Portuguese Air Force health workers in the General Aeronautical Material Workshops (GAMW) (Table 1) [9,10]. The study led to a definition of three clinical stages of vibroacoustic disease: Stage I—Mild (1–4 years), characterized by slight mood swings, indigestion and heartburn, oropharyngeal infections and bronchitis; Stage II—Moderate (4–10 years), characterized by chest pain, mood swings, back pain, fatigue, skin infections, inflammation of the gastric surface, pain and blood during urination, conjunctivitis and allergic processes; and Stage III—Severe (>10 years) characterized by psychiatric disorders, hemorrhages of nasal, digestive and conjunctive mucosa, varicose veins and hemorrhoids, duodenal ulcers, spastic colitis, decreased visual acuity, headaches, severe joint pain, intense muscular pain and neurological disturbances [9].

Table 1. Low-frequency noise and its effects on human health from different sources.

Authors	Object	Year	Main Results	Sample
Donald Laird (Psychological Laboratory, Colgate University)	Physiological effects of noise on typists	1928	Increase in energy expenditure when subjected to noisy environments, when the noise is smoothed on site the average increase in calorie spending is lower.	4 experienced typists.
E. Dart (Ford Occupational Physician—Detroit, USA)	Description of symptoms observed in aircraft technicians.	1946	Report of pain in the hands, swelling, tenosynovitis and increased vascular tone.	112 aircraft technicians.
G. I. Rumancev	Description of the symptoms of a group of workers in a concrete factory in the Soviet Union exposed to noise.	1961	Report of pain in the hands, swelling, tenosynovitis and increased vascular tone.	*
Alexander Cohen (National Institute for Occupational Safety and Health, USA)	Description of clinical complaints of boiler plant workers before and after the implementation of a hearing protection programme.	1976	Report of pain in the hands, swelling, tenosynovitis and increased vascular tone.	400 boiler plant workers.

Table 1. *Cont.*

Authors	Object	Year	Main Results	Sample
Yiming Zhao *et al.* (Research Center of Clinical Epidemiology; Research Center of Occupational Medicine, Beijing—China)	Description of the effects of industrial noise on the prevalence of hypertension in a group of 1101 female workers in a textile mill in Beijing, in 1985.	1991	Report that noise is a significant determinant of the prevalence of hypertension.	1101 female workers.
N. V. Grechkovskaia and I. A. Parpalei	Mention the vibronoise pathology in workers of the aircraft industry in Kiev, Ukraine.	1997	Report of disturbances such as psychovegetative syndrome, vegetative-vascular-trophic syndrome, cerebral anguidystonic syndrome as initial disturbances of vibronoise pathology.	103 assemblers-fitters
V. D. Balunov, A. F. Barsukov, V. G. Artamonova	Research involving the evaluation of the health condition of building industry workers engaged into ferro-concrete production in Saint Petersburg, Russia, submitted to infrasound, noise and general vibration.	1998	The health status was considered under the combined action of infrasound, noise and vibration generally presented greater morbidity for this group.	62 molders
Nuno A. A. Castelo Branco (Center of Human Performance, Alverca, Portugal)	Research report initiated in 1979 about the systemic changes caused by the continuous exposure to noise in the General Aeronautical Material Workshops in Portugal.	1999	Prolonged exposure to low-frequency sounds, below 500 Hz, presented clinical concern, especially with the high incidence of cases of sudden epilepsy and fury. 20 years of research have led to the definition of a new disease: vibroacoustic disease.	Data collected for 20 years
Kerstin Persson Waye (Department of Environmental Medicine, Göteborg University, Sweden)	Description the effect of ventilation noise was further examined along with traffic noise, in two laboratory studies.	2003	This study showed that the exposure to low-frequency noise, especially at night, may affect the cortisol response, *i.e.*, lower cortisol levels after awakening were associated with subjective reports of mood and sleep quality.	Ventilation noise and comparison with traffic noise: 12 male subjects with an average age of 24.5 years Ventilation noise: 26 male subjects with an average age of 26 years.
Malgorzata Pawlaczyk-Luszczynska *et al.* (Department of Physical Hazards, Nofer Institute of Occupational Medicine—Lodz, Poland)	Research to investigate the annoyance of LFN at workplaces in control rooms and office-like areas.	2010	There were no differences in the annoyance assessments between the groups (young and old volunteers—females and males). Both groups similarly assessed annoyance from low-frequency noise.	55 young volunteers 70 older volunteers
Ta-Yuan Chang *et al.* (China Medical University—Taiwan, China)	A joint study about occupational noise exposure and incident Hypertension in Men.	2013	High incidence of hypertension in prolonged exposure to noise levels ≥85 dBA.	578 male workers in Taiwan (1998–2008), all subjects were divided into exposure groups (high, intermediate and low).

Source: Adaptation from [16]. * No information.

Exposure to low-frequency noise (LFN) has significant impacts on human health. This impact is absorbed by auditory sensation, which is a function of the perception that encompasses aspects of physiological, pathological and sociological order. There are caveats in relating certain harmful effects to a single source of noise, but human exposure to multiple sources of noise must be used as a criterion [3,6].

The publications on low-frequency noise shown in Table 1 are case-control studies, using large samples and developed by interdisciplinary researchers. In general, they focus on occupational exposure. These studies concluded that the clinical picture is similar in what concerns noise exposure: swelling tenosynovitis, hand pain and hypertension.

Despite all the research on the effects of low-frequency noise and its impact on human health conducted for more than a century, there are still no references to LFN from the electromagnetic field of power poles and power lines.

Electric, magnetic and electromagnetic fields are physical agents associated with the use of electricity for the transmission and transport of energy (low frequency, 60 Hz). These fields interact with living beings in general and the human body in particular, causing harmful effects by inducing electric currents that exceed the skin shield, damaging sensitive cells and organs [3,32]. For this research, we used the criterion for low-frequency noise levels below 200 Hz.

Some studies show that low-frequency noise differs from other environmental noises at comparable levels. Much of the urban noise pollution we are exposed to in our daily environment contains significant energy within this range. Low-frequency noise is very common as background noise in urban environments. The effects of LFN are of particular concern because of their pervasiveness due to numerous sources, efficient propagation and the reduced efficacy of many structures (e.g., walls, houses and hearing protection). For instance, low-frequency noise is a common cause of sleep disturbances, psychological distress, cognitive impairment, increased social conflict, anxiety, emotional instability and nervousness.

3. Experimental Section

In this section, we briefly describe environmental noise measurements and procedures, as well as meteorological data collection.

3.1. Study Area

Guimarães is located in the Ave sub-region of the Braga district in northwestern Portugal. According to the 2011 Census of the National Institute of Statistics, the municipality's population totaled 158,124 inhabitants. Guimarães has a population of 54,097 inhabitants distributed among 20 villages and the population density is 2224 inhabitants/km^2 [33].

The Guimarães municipality is crossed by four lines of 400 kV and nine lines of 150 kV. The village of Serzedelo, located southwest of Guimarães city, has a population of 3680 inhabitants, covers an area of 5.14 km^2 and has a high density of power poles and power lines in its territory. In addition to these facts, there is an electrical substation in Serzedelo with a transformer capacity of 2 $\times$ 360 MVA—Riba de Ave Substation [33] (Figure 1).

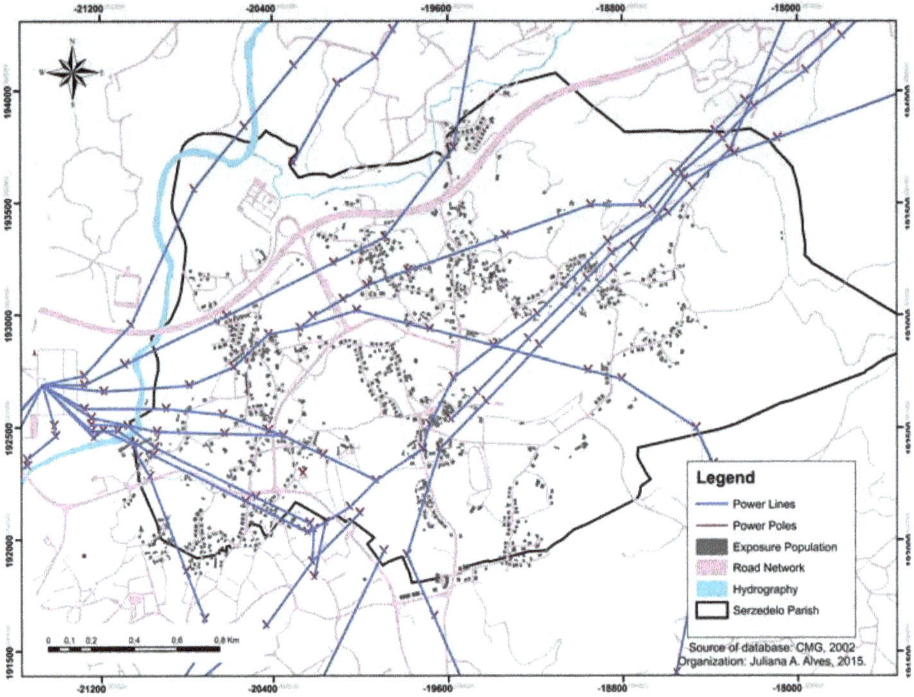

Figure 1. Study area—village of Serzedelo.

The village of Serzedelo has power poles less than five meters away from houses, which does not meet the recommendations of related legislation in force in Portugal (Figure 2). Ninety power poles and 12 power lines cover an area of little more than five km^2, which means 80% of the village is exposed to high and very high voltage.

Figure 2. Power poles and power lines in the village of Serzedelo.

3.2. *Assessment Framework*

The methodology used was based on a procedure developed by the Department of Environment, Food and Rural Affairs (DEFRA), Acoustics Research Center, University of Salford—Procedure for the assessment of low-frequency noise complaints [12].

The noise assessments were made using a class 1 sound level meter with a 1/3-octave filter, the noise indicator recorded was Leq in a range of 10–160 Hz and for an average time of five minutes. The noise levels of the field measurements were compared with the criteria curve (Figure 3). The L10 and L90 were recorded in the same bands to achieve fluctuating characteristics. In the first approach, the noise measurements were taken for 20 min periods, and in the second approach the measurement time was 72 h periods. To evaluate whether an environmental sound could be responsible for the disturbance, the level of recorded sounds was compared with the criterion curve (Figure 3).

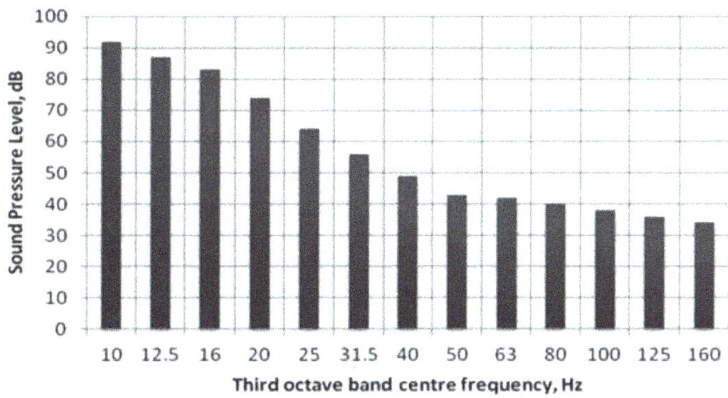

Figure 3. Criterion curve to assess low-frequency noise (Adapted from [12]).

Regarding fluctuations, according to the DEFRA methodology, when the L10–L90 difference exceeds 4 dB, the sound fluctuates and a penalty should be imposed. The DEFRA suggests that 5 dB relaxation may be applied for steady sounds, rather than introducing a penalty for fluctuating sounds. Furthermore, it states that a fluctuating sound with an average level of 5 dB below the threshold would be audible, whereas a steady sound would not. Since the curve values at low frequency are set 5 dB below the threshold, this is again consistent with allowing relaxation for steady sounds.

Using the DEFRA methodology [12], outdoor and indoor measurements were carried out in houses and nearby places. Two approaches were adopted for this research. The first was held in 2014, and the second in 2015. Both assessments were made using a class 1 sound level meter with a 1/3-octave filter. The first approach was completed between June and July 2014 and entailed evaluating nine points, which were divided into two groups, considering their location due to the source under study.

The measuring points were selected based on the methodology used by [11] concerning the exposure to electromagnetic fields in Serzedelo. This included houses near the source (e.g., within 50 m), and houses away from the source (e.g., at a distance equal to or greater than 250 m).

Temperature and relative humidity were measured at an automatic meteorological station located in Merelim, Braga, the nearest meteorological station with available data (about 7 km away). In accordance with DEFRA, we used a sound level meter class 1, with third octave filters from 10 Hz to 160 Hz, a tripod and a field calibrator. The height of the measurements was 1.2 m and was carried out at a distance more than 4 m away from the nearest facade. The following factors were considered to select the measurement points:

(a) Distance to 400 kV power poles of less than 250 m from the "near the source" group.
(b) Distance to 400 kV power poles of more than 250 m from the "away from the source" group.
(c) Avoidance of the influence of other noise sources such as road traffic;
(d) Preference for routes with low road traffic, away from highways and with an absence of noise barriers.

The noise measurements taken in the "near the source" group consisted of six points, outside houses located less than 50 m from the source (Figure 4 and Table 2). Measurements were taken over three days: 26 June, 3 July and 8 July 2014.

Table 2. Characteristics of measurement points in the "near the source" group.

Point	Proximity to the Source	Environmental Area
A	10 m away from a 400 kV voltage power pole	Located near houses and extensive cultivation areas (urban garden).
B	5 m away from a 400 kV voltage power pole	Located on higher ground over houses. During the measurement, the passage of a garbage truck on cobbled ground near the measurement location was recorded. Noise perceived: continuous noise from the power pole. Less intense noise: birds, auto horn and cicada (insect).
C	5 m away from the source	Located near the houses and urban gardens with a high density of 150 kV and 220 kV voltage power poles and power lines. Sounds emitted during the measurement came from a lawnmower, birds, hammering, dogs barking, roosters and sawmill works.
D	3 m away from a high-voltage 400 kV power pole	Located near the houses and comprises an area of the 400 kV power poles corridor. The measurement was carried out near the inter-municipal highway. At this point, there were reports of discomfort from noise on wet, rainy days. In this aspect, residents reported using sleeping pills and experiencing headaches and fatigue.
E	10 m away from the source	Located near the houses and extensive urban gardens. The area had a concentrated, high density of high-voltage power poles and a mobile phone antenna.
F	15 m away from the electrical substation	Located near of the Riba de Ave Substation. The area is located near houses and factories. The closest road, of granite cube, did not record traffic during the measurement period. A microphone shield was used in the sound level meter due to high wind speed.

The "away from the source" group consisted of three measurement points. This group comprised locations between 200 and 300 m away from the influence of power poles and power lines (Figure 4 and Table 3). Measurements were performed over three days: 26 June, 3 July and 8 July 2014.

The second approach comprised three measurement points inside three houses located near the source and in a room where annoyance was detected by the house owner. In point L, the appliances were switched off for 20 min. In points J and M, there were no appliances. The three measurements were divided into three blocks, considering the monitoring dates (February and March 2015) (Figure 5 and Table 4). Point J was measured between 9–12 February, Point L between 11–14 March and Point M between 14–17 March.

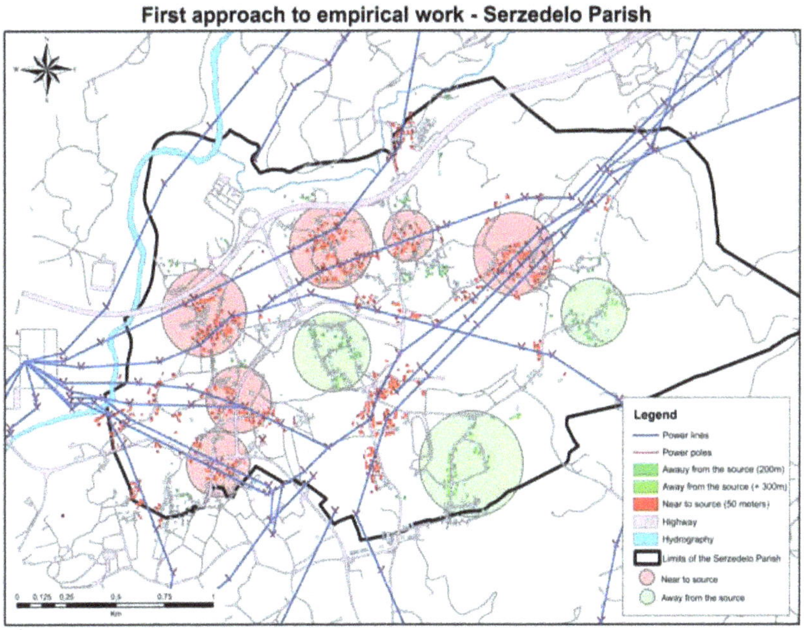

Figure 4. First approach—village of Serzedelo.

Table 3. Characteristics of measurement points in the "away from the source" group.

Point	Proximity to the Source	Environmental Area
G	250 m away from the influence of power poles and power lines	During the measurement, a group of people was talking near the sound level meter and there were also two light vehicles belonging to local inhabitants
H	250 m away from the source	During the measurement, recording sounds came from light vehicles, a child crying and a group of people talking next to the sound level meter.
I	About 400 m away from the source	The measurement was carried out at the side of a granite cube road near some houses. Something to note was the presence of an unidentified background noise.

For the second approach, three analyses were taken and compared with the criterion curve: (1) complete measurement, *i.e.*, comprising the measurement for 72 h (three days); (2) measurement per day, *i.e.*, the values per day, where differentiated, for each frequency band in the criterion curve; and finally (3) measurement between 02:00–04:00, which characterizes night time, according to the DEFRA Guidance.

To complement this approach, an interview was conducted with the person of contact, who also agreed that the measurements could be taken inside the houses. This interview was standardized and had 30 questions following DEFRA guidelines. The questionnaire was structured in three sections. The first one dealt with personal data (seven questions). The second section focused on health and the quality of sleep (eleven questions) and, finally, the third section was concerned with the characteristics of perceived noise (twelve questions).

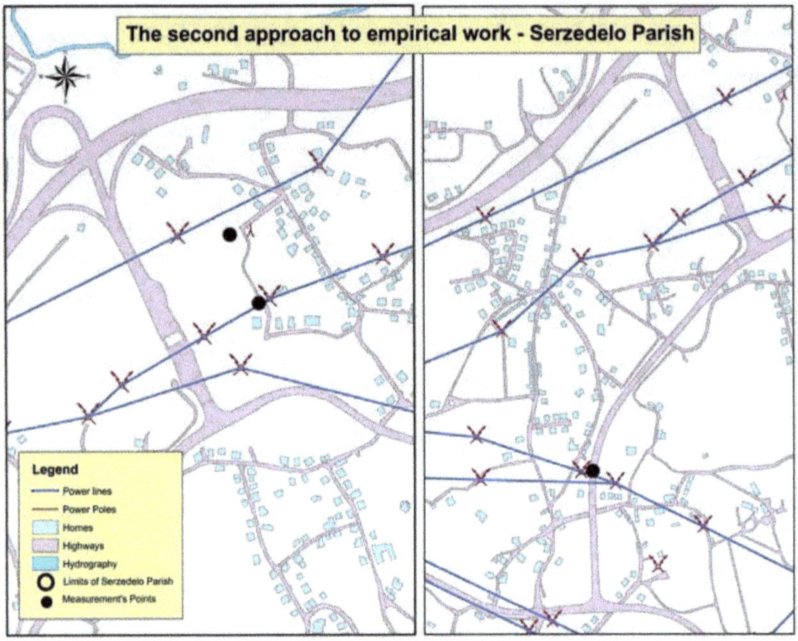

Figure 5. Location of the three measurement points in the second approach.

Table 4. Characteristics of measurement points in the second approach.

Point	Proximity to the Source	Environmental Area	Presence of Obstacles	Reports
J	5 m away from a 400 kV power pole	Located inside a house, in a garage attached to the residence, where noise from the source was detected.	House faced	Noise discomfort from the power pole and some health problems: depression, headaches and insomnia.
L	15 m away from a 400 kV power pole	Located inside the house in a room closest to the source, where the interviewee felt the presence of low-frequency noise the most. There was a mobile phone antenna nearby. The area was located close to a residential area near the A7 highway.	House faced	Noise discomfort from the power pole, especially on humid and rainy days. Health problems included fatigue, headaches and insomnia.
M	10 m away from a 400 kV power pole	Located in an annex close to the source, where the interviewee most felt the presence of low-frequency noise.	House faced	Noise discomfort and stress. In the village, there were cases of cardiovascular disease and stomach cancer.

4. Results and Analysis

4.1. The First Approach

The noise measurements taken in the "near the source" group consisted of six points (points A to F). The "away from the source" group consisted of three measurement points (points G to I). Measurements were performed over three days: 26 June, 3 July and 8 July 2014.

The criterion reference curve was exceeded in all measurement points (dB Leq—Reference), especially between the 50 Hz and 160 Hz frequency bands. According to DEFRA Guidance, the measured levels at these frequencies would be considered audible to most people who are exposed to them. The D, E and F measurement points showed a higher deviation relative to the criterion reference curve (Figure 6a). Particularly in measurement point F, a higher deviation from the reference value was recorded, which can be explained by its location near the Riba de Ave Substation. No association was found between weather conditions and the noise levels measured (Table 5).

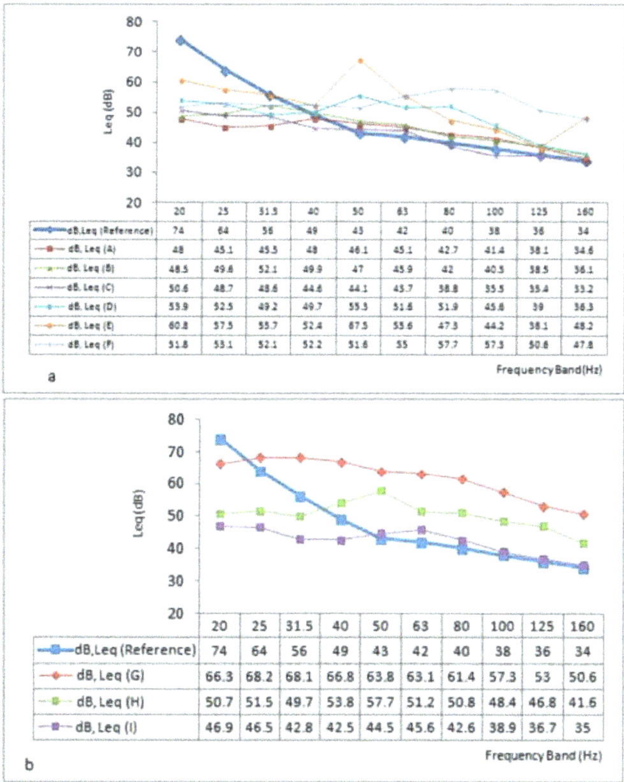

	20	25	31.5	40	50	63	80	100	125	160
dB,Leq (Reference)	74	64	56	49	43	42	40	38	36	34
dB, Leq (A)	48	45.1	45.5	48	46.1	45.1	42.7	41.4	38.1	34.6
dB, Leq (B)	48.5	49.6	52.1	49.9	47	45.9	42	40.5	38.5	36.1
dB, Leq (C)	50.6	48.7	48.6	44.6	44.1	43.7	38.8	35.5	35.4	33.2
dB, Leq (D)	53.9	52.5	49.2	49.7	55.3	51.6	51.9	45.6	39	36.3
dB, Leq (E)	60.8	57.5	55.7	52.4	67.5	55.6	47.3	44.2	38.1	48.2
dB, Leq (F)	51.8	53.1	52.1	52.2	51.6	55	57.7	57.3	50.6	47.8

	20	25	31.5	40	50	63	80	100	125	160
dB,Leq (Reference)	74	64	56	49	43	42	40	38	36	34
dB, Leq (G)	66.3	68.2	68.1	66.8	63.8	63.1	61.4	57.3	53	50.6
dB, Leq (H)	50.7	51.5	49.7	53.8	57.7	51.2	50.8	48.4	46.8	41.6
dB, Leq (I)	46.9	46.5	42.8	42.5	44.5	45.6	42.6	38.9	36.7	35

Figure 6. First approach—(**a**) Measurement point A to F ("near the source" group); (**b**) Measurement point G to I ("away from the source" group).

The criterion reference curve was exceeded in all measurement points (dB Leq—Reference), especially in the G and H measurement points, which presented a higher deviation from the reference value beginning at 25 Hz (for the first) and 40 Hz (for the second). In the I measurement point, the deviation was very low—between 50 Hz and 160 Hz (Figure 6b). We could not establish a relationship between weather conditions and the measured noise levels (Table 5).

Both groups exceeded the values in the criterion curve (dB Leq—Reference). There are two possible explanations for these results: (1) The low-frequency noise detected in the "away from the source" group could result from other sources; (2) The need to redefine the limits of the "near the source" and "away from the source" groups, *i.e.*, what is being regarded as "away from the source" could be classified as "near the source". Given the specific characteristics of low-frequency noise, especially its high capacity for propagation and low absorption by materials and the environment, redefining the limits of the groups could be accounted for.

Table 5. Weather Conditions, Average [34].

		Air Temperature (°C)	Rainfall (mm)	Wind Speeds (km/h)	Relative Humidity (%)	Measurement Points
26 June 2014		17	0.0	1.5	77.5	Point A and F
3 July 2014	**Mean Values**	22.9	0.0	3.5	52.5	Point B, C and H
8 July 2014		28	0.0	13	37	Point D, E, F and G

4.2. The Second Approach

The second approach comprised three measurement points inside three houses (points J–M). For the second approach, three analyses were taken and compared with the criterion curve: (1) complete measurement for 72 h (three days); (2) measurement per day; and finally (3) measurement between 02:00–04:00, which characterizes night time, according to the DEFRA Guidance.

In both analyses of point J, "complete measurement" and "measurement per day", the reference values of the criterion curve were exceeded at the 50 Hz frequency band (Figure 7a,b). Given the lack of home appliances nearby, the high values on the 50 Hz frequency band recorded in the daytime analysis are due to other sources. In the night-time measurement analysis, the values in the criterion curve were not exceeded (Figure 7c).

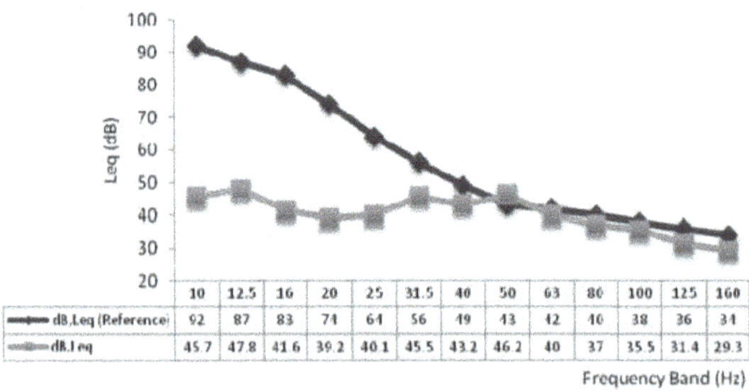

a

Figure 7. *Cont.*

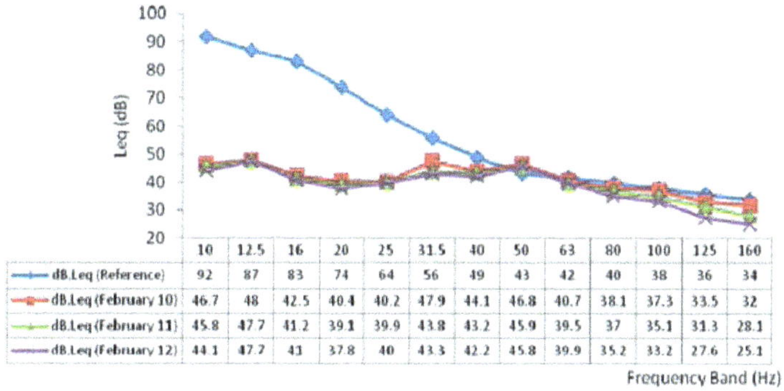

b

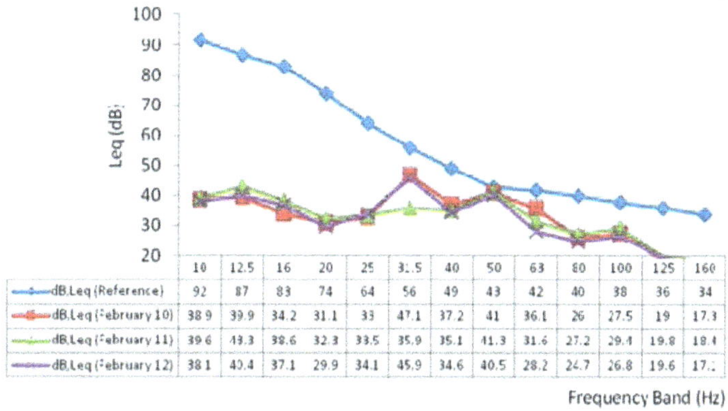

c

Figure 7. Point J (**a**) Complete Measurement (**b**) Measurement per day and (**c**) Measurement at night.

The noise measured over three days showed fluctuating characteristics (Table 6). The evaluation of the fluctuating characteristics of noise was carried out by determining L10–L90 for periods of 15 min, for 72 h and for the 50 Hz frequency band (exceeded bandwidth). The L10–L90 values are greater than 4 dB over 33.90% of the time (Figure 8 and Table 6). During the night period (2:00 to 4:00), for the three days, this value was exceeded for most of this period. In another hand, it was not possible to establish a correlation between weather conditions and the measured noise levels (Figure 9).

Table 6. Point J—Fluctuating Characteristics, percentage of time that L10–L90 ≥ 4 dB.

Frequency (Hz)	10	12.5	16	20	25	31.5	40	50	63	80	100	125	160
L10–L90 ≥ 4	14	41.5	21.1	27.7	27	42.9	38.1	33.9	31.5	32.9	29.8	38.4	34.9

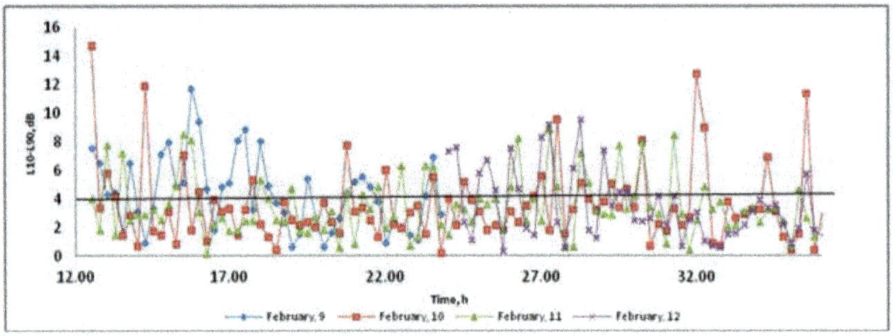

Figure 8. Point J—L10–L90—1/3 Oct 50 Hz.

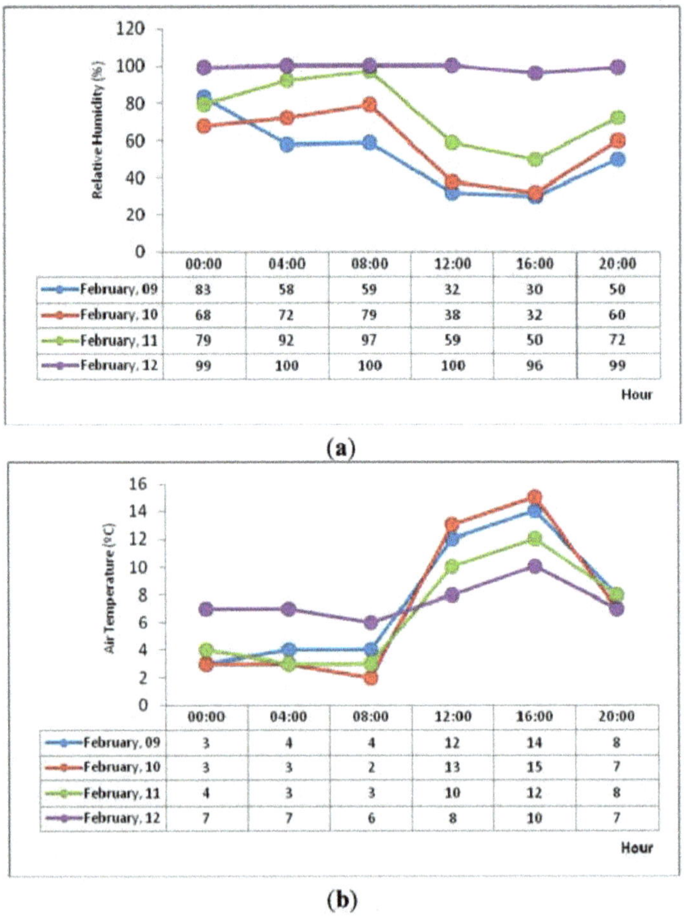

Figure 9. Point J—Weather Conditions [34] (**a**) Relative Humidity and (**b**) Air Temperature.

At point L, the equipment was installed inside the kitchen of the residence, where the interviewee claimed she could hear the noise. Due to the influence of the equipment, an analysis was made for 20 min using shutdown equipment.

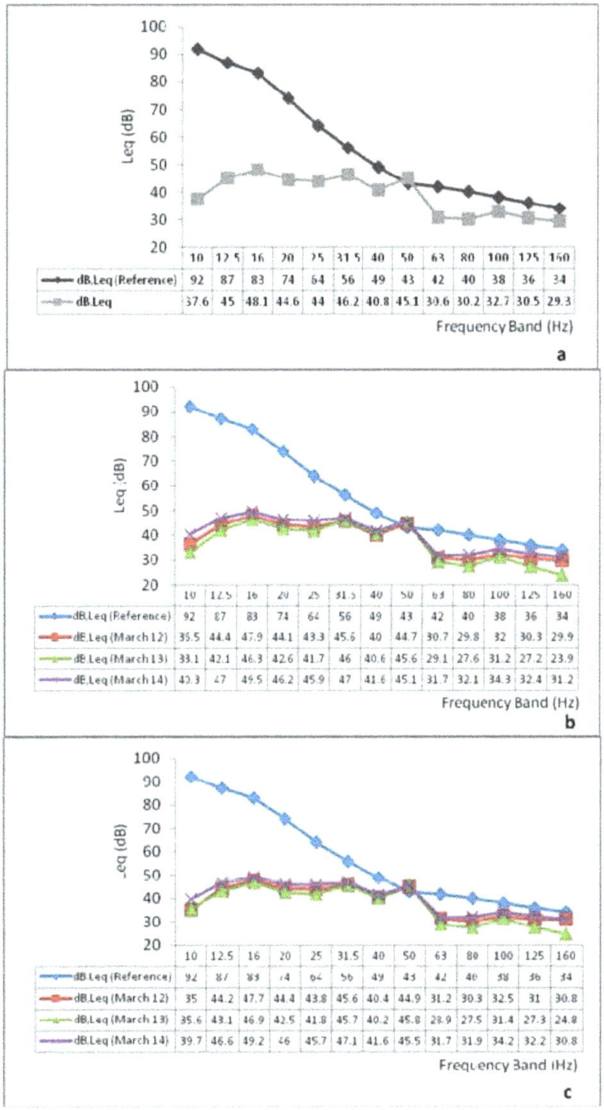

Figure 10. Point L (**a**) Complete Measurement (**b**) Measurement per day and (**c**) Measurement at night.

According to [11], this is the frequency band in which refrigerators operate, which could account for the reference values being exceeded in the 50 Hz frequency band (Figure 11). This deduction can be confirmed with the results obtained after shutdown equipment (Figure 12), where the criterion curve was exceeded in this band (50 Hz). It was not possible to establish a correlation between the air temperature, relative humidity and the measured noise levels (Figure 13).

The fluctuating L10–L90 noise characteristics were determined for periods of 15 min for three days of measurement. The noise during the complete measurement showed fluctuating characteristics during 39.1% of the time for band 50 Hz (Table 7). However, during the night-time measurement (2:00 a.m. to 4:00 a.m.), this value was exceeded for most of this period over the three days (Figure 11).

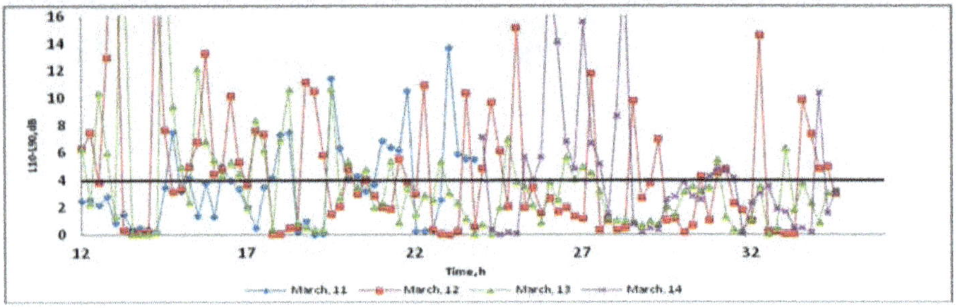

Figure 11. Point L—L10–L90—1/3 Oct 50 Hz.

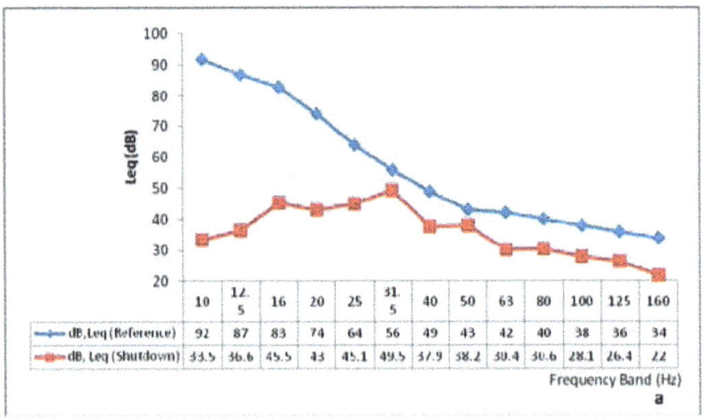

Figure 12. Point L—Shutdown (20 min).

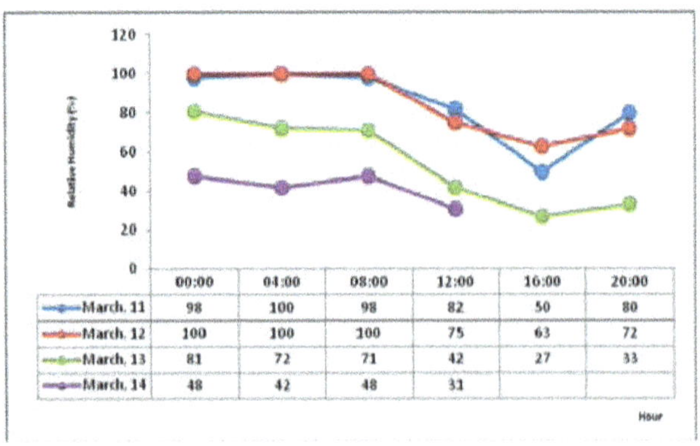

(a)

Figure 13. *Cont.*

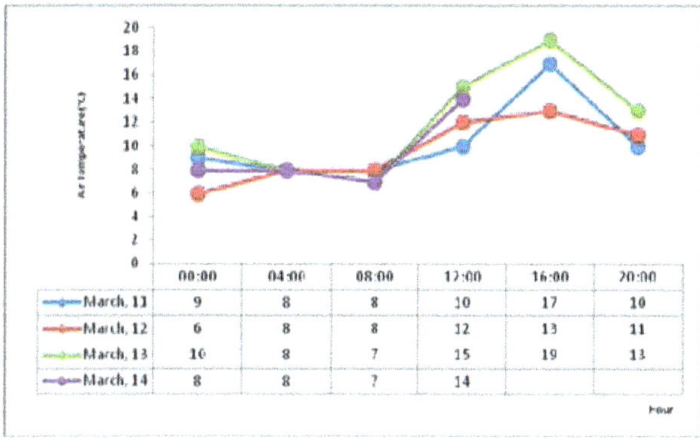

(b)

Figure 13. Point L—Weather Conditions [34] (**a**) Relative Humidity and (**b**) Air Temperature.

Table 7. Point L—Fluctuating Characteristics, percentage of time that L10–L90 $\geq$ 4 dB.

Frequency (Hz)	10	12.5	16	20	25	31.5	40	50	63	80	100	125	160
Complete Measurement L10–L90 $\geq$ 4	22.8	28.0	33.6	25.6	23.4	33.2	33.6	39.1	32.9	42.6	31.8	28.0	22.5
Shutdown (20 min) L10–L90 $\geq$ 4	25.0	50.0	25.0	0.0	25.0	0.0	0.0	25.0	25.0	50.0	50.0	25.0	25.0

At point M, the sound level meter was installed in the outer area of the residence, inside an annex away from electrical appliance interference. In all time scales, the criterion curve was exceeded between the 50 Hz and 160 Hz frequency bands. The measurement shows the highest exceeding values in the second approach, particularly in the 80 Hz frequency band (Figure 14).

We highlight the high noise levels, which exceeded 40 Hz, and records of almost 50% of the time for 50 Hz, 63 Hz and 80 Hz bands (Table 8). Due to the high overdrive bands, an attempt was made to correlate the noise levels measured with the weather conditions. However, this was not possible (Figure 15).

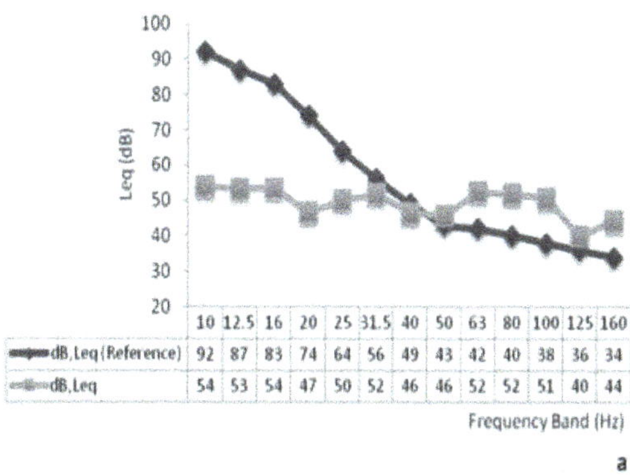

Figure 14. *Cont.*

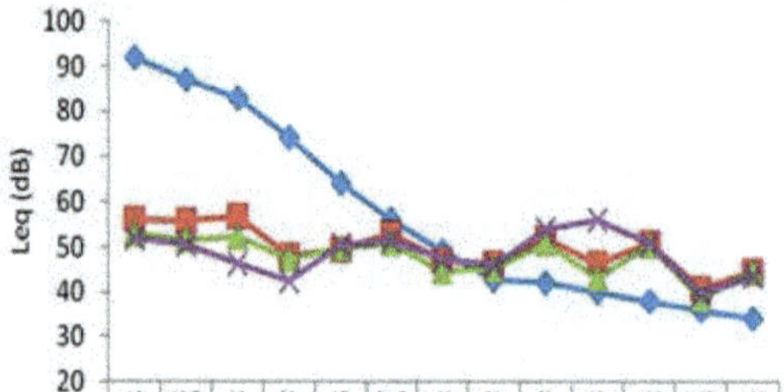

	10	12.5	16	20	25	31.5	40	50	63	80	100	125	160
dB,Leq(Reference)	92	87	83	74	64	56	49	43	42	40	38	36	34
dB,Leq(March 15)	56	55.8	56.7	48.1	49.5	52.8	47	46.2	52	46.3	51	40.9	44.7
dB,Leq(March 16)	53	51.7	52.2	47.5	49.9	50.9	44.4	45.1	50.7	43.2	50.6	38.7	44.5
dB,Leq(March 17)	51.9	50.5	46.2	42.5	50.5	51.7	47.3	45.8	54	55.9	50.7	39.8	43.6

Frequency Band (Hz)

b

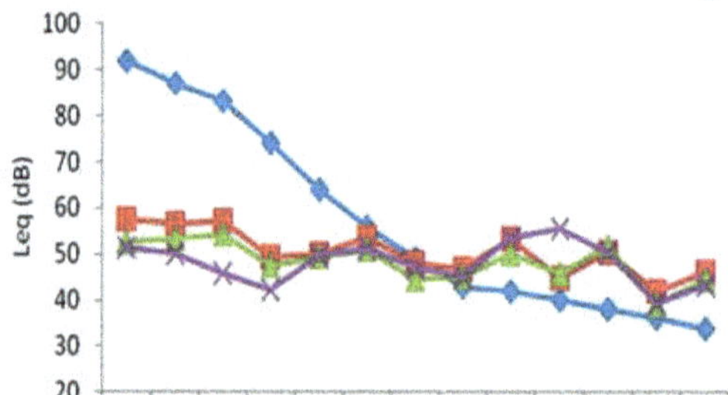

	10	12.5	16	20	25	31.5	40	50	63	80	100	125	160
dB,Leq(Reference)	92	87	83	74	64	56	49	43	42	40	38	36	34
dB,Leq(March 15)	57.5	56.5	57.3	49.3	50	53.5	48	46.9	53.3	45	50.5	42.1	46.2
dB,Leq(March 16)	52.7	53.3	54.3	47.4	49.5	50.9	44.5	45.2	50.2	45.8	51.8	39	44.4
dB,Leq(March 17)	51.4	50.1	45.9	42.4	49.8	51.3	47	45.1	53.5	55.5	50.3	39.4	43.2

Frequency Band (Hz)

c

Figure 14. Point M (**a**) Complete Measurement (**b**) Measurement per day and (**c**) Measurement at night.

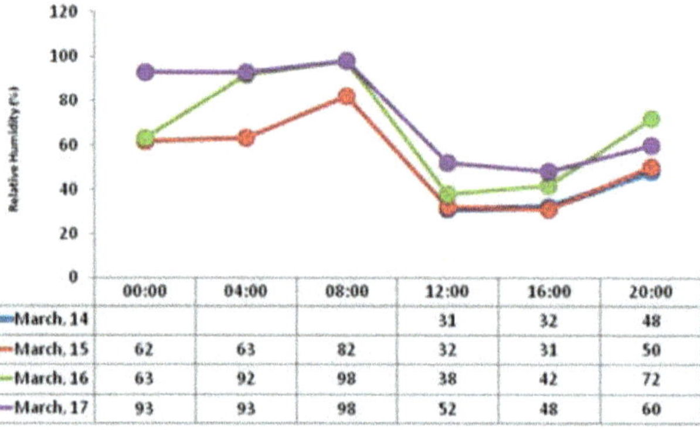

	00:00	04:00	08:00	12:00	16:00	20:00
March, 14				31	32	48
March, 15	62	63	82	32	31	50
March, 16	63	92	98	38	42	72
March, 17	93	93	98	52	48	60

a

Hour

Figure 15. *Cont.*

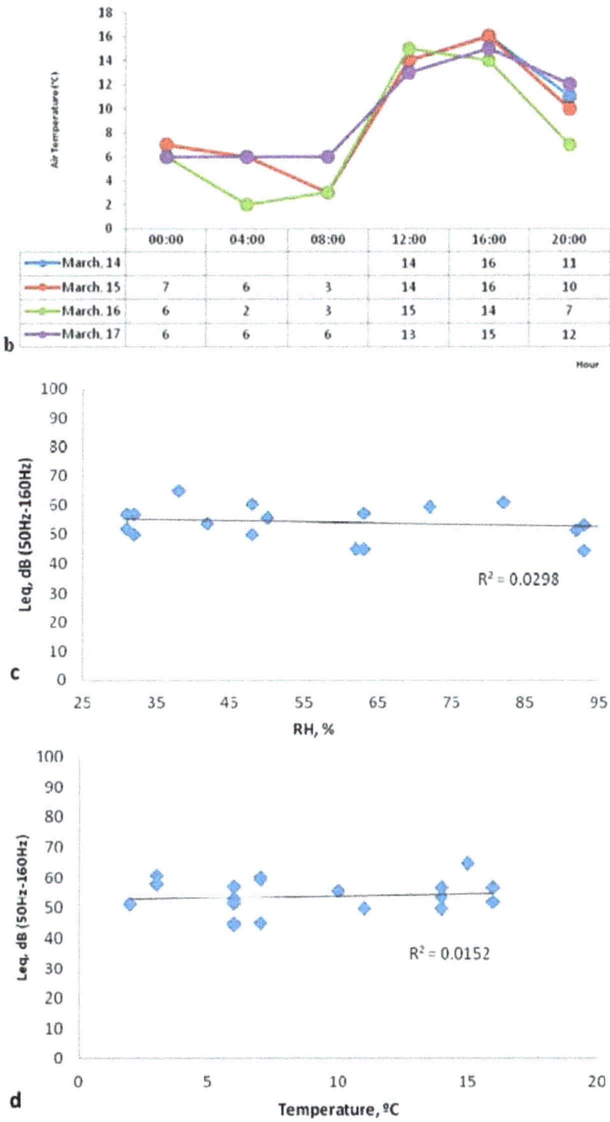

Figure 15. Point M (**a**,**b**) Weather conditions [34]; (**c**) Relative humidity *vs.* Leq (50–160 Hz) and (**d**) Air temperature *vs.* Leq (50–160 Hz).

Table 8. Point M—Fluctuating characteristics, percentage of time that L10–L90 ≥ 4 dB.

Frequency (Hz)	10	12.5	16	20	25	31.5	40	50	63	80	100	125	160
L10–L90 ≥ 4	15.3	27.1	41.7	41.3	43.4	48.6	55.6	40.3	43.8	45.8	33.7	37.2	38.5

For the evaluation of the fluctuating characteristics of the noise, the L10–L90 difference was determined for periods of 15 min for the three days of measurement (Table 8) and exceeded the frequency bands, *i.e.*, the 50–160 Hz range (Figure 16). The exceeded frequency bands presented fluctuating characteristics for 55.6%, 40.3%, 43.8%, 45.8%, 33.7%, 37.2% and 38.5% of the measurement time, respectively, for the frequency bands of 40 Hz, 50 Hz, 63 Hz, 80 Hz, 100 Hz, 125 Hz and 160 Hz.

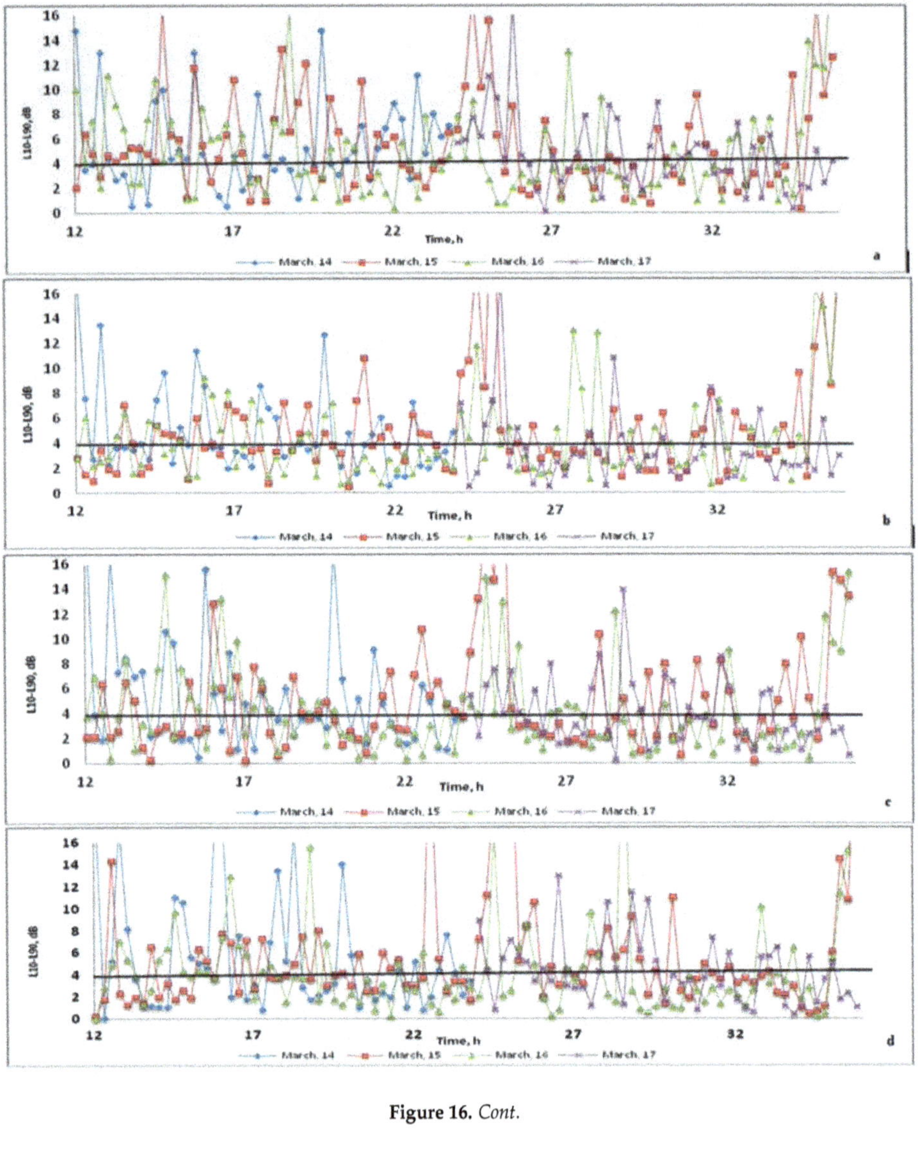

Figure 16. *Cont.*

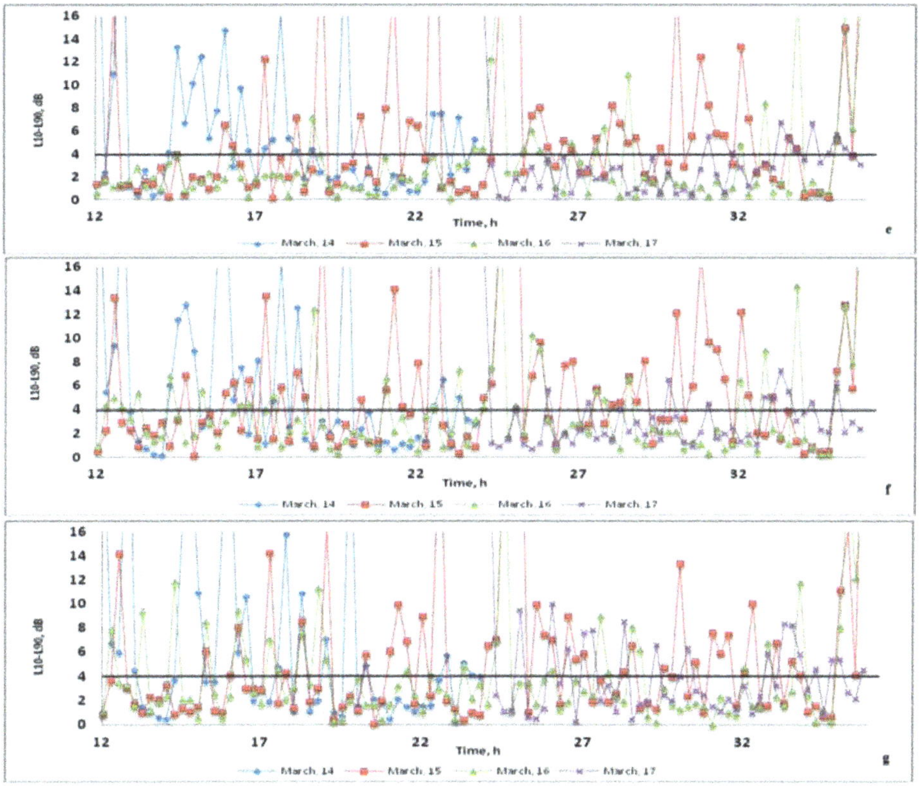

Figure 16. (**a**) Point M—L10–L90—1/3 Oct 40 Hz; (**b**) 1/3 Oct 50 Hz; (**c**) 1/3 Oct 63 Hz; (**d**) 1/3 Oct 80 Hz; (**e**) 1/3 Oct 100 Hz; (**f**) 1/3 Oct 125 Hz and (**g**) 1/3 Oct 160 Hz.

5. Evaluation of the Perception of Noise Discomfort

5.1. The First Approach

The residents of Serzedelo village have expressed their dissatisfaction with power lines in the area through their own initiative, the National Civic Movement Against High-Voltage Power Lines. In 2007, the village's inhabitants filed a petition against the National Energy Company to bury the aerial power lines crossing the village, but to date, nothing has been done. Subsequently, in 2010, a team from the University of Minho (Portugal), supported by the local population and the leader of this movement, conducted an exploratory study in Serzedelo.

The study used two groups: a "near the source" group (118 individuals were interviewed) and an "away from the source" group (55 individuals were interviewed). The most frequent diseases identified were cardiovascular diseases (35.6% in the "near the source" and 42.6% in the other group), and depression (22.9% in the "near group" and 20.4% in the other group). However, there were no significant differences between the two groups [11]. Additionally, self-awareness of general health was not perceived differently by the two groups. Perhaps these kinds of diseases could be related more to the influence of noise than to exposure to electromagnetic fields (power poles and power lines), which was the main focus of the study. Even if at that time the study did not focus on the influence of low-frequency noise pollution, it was concluded that only the "near source" group spontaneously mentioned that the noise was always present in their daily lives (in fact 9.3%), while there was no mention of the noise by residents of the group "away from the source".

5.2. The Second Approach

According to the DEFRA Guidance [12], persons living near the measurement points that mentioned discomfort due to noise should be interviewed to provide other background information to complement the analysis. For this reason, we interviewed three people that were engaged with the three points of measurement of the second approach. Two men and a woman were interviewed, aged between 40 to 70 years old, who had lived in Serzedelo village between 4 and 40 years. Two had a university degree, while the older person had been a postman for 25 years.

For all of them, the noise is continuous in periods of rain, with fog and wet weather. The following narratives express this perception.

> In the evening, I hear a sound and during the day something that seems like sparks (Point L, female, 55 years old).

> I hear a continuous hissing sound (Point J, male, 69 years old).

For all the respondents, noise has been heard continuously and coincides with the time of experience of living in Serzedelo, expressed in the following two narratives.

> I have heard the noise since 1985, when this 400 kV power pole was installed (Point J, male, 69 years old).

> I have heard the noise for several years or more since living in this house (Point M, male, 45 years old).

All the respondents reported that other people who lived with them or nearby hear noise, as did their wives, children, visitors and neighbors. The power poles are seen as the source of the noise by all the respondents.

> The noise comes from electric poles placed to the north and south of the house (Point L, female, 55 years old).

The noise can be heard by respondents inside the house, in the living room, the bedroom and the kitchen. They mentioned that they have adopted strategies to alleviate the noise, such as "sleeping in a different room from the usual", "sleeping in a different position or changing the bed position", "using earplugs" or "going on vacation". The discomfort caused by noise varies between the perceptions from "very uncomfortable" to a "little bother".

6. Discussion

Although there is a study focused on the risk of death in Guimarães [35], using 1997 and 2005 data, which was a response from the regional and local health authorities to the demands of the population of Serzedelo village and to the National Civic Movement Against High-Voltage Power Line demands this was not, in fact, a good solution to the problem. The study did not show significant differences in the risk of death and major causes of death in the population living in Serzedelo village, when compared to the north of the country and with Portugal as a whole. This happened because of the type of approach adopted. Mortality indicators are not sufficient to express the relationship between the low frequency noise, the electromagnetic fields and the different types of cancer as the diseases related to this type of exposure are currently those that involve a lower risk of death. Taking this into account, an approach is needed to address the mortality data which can be obtained by the general official statistics. This inference contradicts the solution that was given by the health authorities, as well as the need to conduct more comprehensive studies.

Compared with the reference values of the DEFRA Guidance [12], noise from power poles exceeds the criterion curve. However, more measurements should be carried out in order to consider other factors that may influence the measured values. A survey investigating noise discomfort experienced by an exposed population, as well as knowledge of the population's health status, and access to morbidity data could complement this analysis.

The concern with the characterization of the surrounding environment is one of the missing aspects in DEFRA methodology. The low-frequency noise levels can vary considerably within a room depending on the measurement location and this can occur when the room dimensions are similar to the wavelength of the sound at these frequencies. The weather seems to be another important aspect that could influence the noise propagation, therefore future research should include measurements both in dry and wet periods. A methodology by season may reveal other noise levels and different perceptions by the population about their discomfort.

7. Conclusions

At the moment we cannot fully answer the fourth question: "Can power poles and power lines affect human health?" This is a complex issue and requires further study. In particular, the issue needs to be developed closely by conducting population surveys. In Portugal, as there are no morbidity data on a village scale for the spectrum of diseases related to this type of problem, the only option is to interview the population. This will follow the same "near the source" and "away from the source" methodology and be based on other variables such as type and duration of exposure to the source, age, self-reported health status and presence of diseases that may influence, among others, depressive disorders and insomnia.

It can be concluded that there is a need to redefine the location range of "away from the source" and "near the source" groups. It makes more sense to opt for people who live far from the source than those who live closer, as they are likely to be less subjected to influences from other sources (e.g., homes with no crossing of highways, people with a fairly healthy diet and a no-stress lifestyle) and residences in rural areas.

No association was found between weather conditions and the noise levels measured. However, we believe that the relationship between low frequency noise levels and weather conditions such as relative humidity, rainfall and wind direction exists, and should be studied to complement this analysis. All environmental characteristics should be taken into account (type of terrain, proximity to other sources of noise, existence of obstacles and factors that facilitate the propagation of noise), especially due to the peculiar characteristics of high propagation of low-frequency noise and its low absorption by materials and the environment.

The limitations of research of this nature focus mainly on assigning disturbing effects on human health to a single factor. Health status is a complex factor that involves multiple aspects, such as lifestyle, as well as genetic and environmental factors. A single indicator may not put into perspective the magnitude of the health problems. Future research along these lines should consider these issues.

Additionally, case-control type studies seem to be the most appropriate for continuing this type of research. It is essential to keep track of the daily lives of the interviewees and monitor their lifestyles (e.g., diet and sleep quality), aspects related to labor dynamics (e.g., the history of activities performed, type of activity currently carried out), genetic predisposition (e.g., history of disease cases in the closest family circle), characterization of the house structure (e.g., position of the bedroom in the residence and amount of electronic equipment). This last item should be thoroughly analyzed. In addition to their bedrooms, where the interviewees spend much of the night, the rooms where they spend most of their time during the day and evening should also be studied in-depth.

This research is part of a more structured investigation that was started in 2009 at the University of Minho, which focused on the influence of low-frequency noise and electromagnetic fields on human health.

Acknowledgments: The authors wish to acknowledge CAPES/Brazil for the doctoral scholarship in Geography for Juliana Araújo Alves (Process: BEX-1684-13/2) and Bruno Ricardo Dias Gonçalves Mendes (Ph.D. Student at the Faculty of Engineering, University of Minho) for the technical support in this research.

Author Contributions: The authors contributed equally to this work.

Conflicts of Interest: The authors declare no conflict of interest.

References

1. Silva, L.T. Environmental Quality Health Index for Cities. *Habitat Int.* **2015**, *45*, 29–35.
2. Clark, C.; Stansfeld, S.; Kephalopoulos, S.; Lekaviciute Gadal, J. (Eds.) *Final Report of the ENNAH Project*; Publications Office of the European Union: Luxembourg, 2013.
3. European Centre for Environment and Health, World Health Organization. *Burden of Disease from Environmental Noise: Quantification of Healthy Life Years Lost in Europe*; Regional Office for Europe: Copenhagen, Denmark, 2011; Available online: http://www.euro.who.int/__data/assets/pdf_file/0008/136466/e94888.pdf (accessed on 29 January 2015).
4. Broner, N. The effects of low frequency noise on people: A review. *J. Sound Vib.* **1978**, *58*, 483–500.
5. Berglund, B.; Hassmén, P.; Job, R.F. Sources and effects of low-frequency noise. *J. Acoust. Soc. Am.* **1996**, *99*, 2985–3002.
6. Berglund, B.; Lindvall, T.; Schwela, D. (Eds.) *Guidelines for Community Noise*; World Health Organization: Geneva, Switzerland, 1999; Available online: http://whqlibdoc.who.int/hq/1999/a68672.pdf (accessed on 28 December 2014).
7. Pawlaczyk-Luszczyńska, M.; Dudarewicz, A.; Szymczak, W.; Sliwińska-Kowalska, M. Evaluation of annoyance from low frequency noise under laboratory conditions. *Noise Health* **2010**, *48*, 166–181.
8. Murphy, E.; King, E.A. An assessment of residential exposure to environmental noise at a shipping port. *Environ. Int.* **2014**, *63*, 207–2015.
9. Castelo Branco, N.A.; Rodriguez, E. The vibroacoustic disease—An emerging pathology. *Aviat. Space Environ. Med.* **1999**, *70*, A1–A6.
10. Castelo Branco, N.A.; Alves-Pereira, M. Vibroacustic disease. *Noise Health* **2004**, *6*, 3–20.
11. Moorhouse, A.; Waddington, D.; Adams, M. Procedure for the Assessment of Low Frequency Noise Complaints. 2011. Available online: http://usir.salford.ac.uk/493/1/NANR45-procedure_rev1_23_12_2011.pdf (accessed on 5 January 2015).
12. Azevedo, B.F.O. O Impacto do Lugar na Saúde da População do Concelho de Guimarães—Estudo de Caso do Electromagnetismo em Serzedelo. 2010. Available online: http://repositorium.sdum.uminho.pt/handle/1822/19766 (accessed on 15 September 2014).
13. Nejadkoorki, E.; Yousefi, E.; Naseri, F. Analysing street traffic noise pollution in the city of Yazd. *Iran J. Environ. Health Sci. Eng.* **2010**, *7*, 53–62.
14. Lee, J.; Gu, J.; Park, H.; Yun, H.; Kim, S.; Lee, W.; Han, J.; Cha, J.-S. Estimation of populations exposed to road traffic noise in districts of seaul metropolitan Area of Korea. *Int. J. Environ. Res. Public Health* **2014**, *11*, 2729–2740.
15. Tenailleau, Q.M.; Bernard, N.; Pujol, S.; Houot, H.; Joly, D.; Mauny, F. Assessing residential exposure to urban noise using environmental models: Dos the size of the local living neighborhood matter? *J. Expo. Sci. Environ. Epidemiol.* **2015**, *25*, 89–96.
16. Leventhall, H.G. Low frequency noise and annoyance. *Noise Health* **2004**, *6*, 59–72.
17. Pawlaczyk-Luszczyńska, M.; Dudarewicz, A.; Waszkowska, M.; Szymczak, W.; Kamedua, M.; Sliwiñska-Kowalska, M. The effect of low frequency noise on human mental performance. *Med. PR* **2004**, *55*, 63–74.
18. Chang, T.-Y.; Hwang, B.-F.; Liu, C.-S.; Chen, R.-Y.; Wang, V.-S.; Bao, B.-Y.; Lai, J.-S. Occupational noise exposure and incident Hypertension in men: A prospective cohort study. *Am. J. Epidemiol.* **2013**, *6*, 818–825.
19. Miedema, H.M.E.; Vos, H. Exposure-response relationships for transportation noise. *J. Acoust. Soc. Am.* **1998**, *104*, 3432–3445.
20. Babisch, W. Traffic noise and cardiovascular disease: Epidemiological review and synthesis. *Noise Health* **2000**, *2*, 9–32.

21. Passchier-Vermeer, W.; Passchier, W.F. Noise exposure and public health. *Environ. Health Perspect.* **2000**, *108*, 123–131.
22. Pedersen, E. Health aspects associated with wind turbine noise—Results from three field studies. *Noise Control Eng. J.* **2011**, *59*, 47–53.
23. Bakker, R.H.; Pedersen, E.; van den Berg, G.P.; Stewart, R.E.; Lok, W. Impact of wind turbine sound on annoyance, self-reported sleep disturbance and psychological distress. *J. Sci. Total Environ.* **2012**, *15*, 42–51.
24. Smith, M.G.; Croy, I.; Ogren, M.; Persson Waye, K. On the influence of freight trains on humans: A laboratory investigation of the impact of nocturnal low frequency vibration and noise on sleep and heart rate. *PLoS ONE* **2013**. [CrossRef]
25. Bluhm, G.; Berglind, N.; Nordling, E.; Rosenlund, M. Road traffic noise and hypertension. *Occup. Environ. Med.* **2007**, *64*, 122–126.
26. Chang, T.-Y.; Beelen, R.; Li, S.-F.; Chen, T.-I.; Lin, Y.-J.; Bad, B.-Y.; Liu, C.-S. Road traffic noise frequency and prevalent hypertension in Taychung, Taiwan: A cross sectional study. *Environ. Health* **2014**. [CrossRef]
27. Kyriakides, K.; Leventhall, H.G. Some effects of infrasound on task performance. *J. Sound Vib.* **1977**, *50*, 369–388.
28. Landström, U.; Kjellberg, A.; Söderberg, L.; Norström, B. The effects of broadband, tonal and masked ventilation noise on performance, wakefulness and annoyance. *Journal Low Freq Noise Vib.* **1991**, *10*, 112–122.
29. Persson-Wayne, K.; Bengtsson, J.; Kjellberg, A.; Benton, S. Low frequency "noise pollution" interferes with performance. *Noise Health* **2001**, *4*, 33–49.
30. Grechkovskaia, N.V.; Parpalei, I.A. The impact of the working conditions on morbidity in workers in jobs hazardous for vibration and noise in aviation enterprises. *Likars'ka Sprava* **1997**, *5*, 20–23.
31. Castelo Branco, N.A. A unique case of vibroacoustic disease. A tribute to an extraordinary patient. *Aviat. Space Environ. Med.* **1999**, *70*, A27–A31.
32. Déoux, S.; Déoux, P. *Ecologia é a Saúde*; Instituto Piaget: Lisboa, Portugal, 1996.
33. Statistics Portugal. Available online: http://www.ine.pt/xportal/xmain?xpid=INE&xpgid=ine_main&xlang=en (accessed on 25 January 2015).
34. Portuguese Institute of Ocean and Atmosphere. Available online: https://www.ipma.pt/en/index.html (accessed on 18 March 2015).
35. Andrade, I.; Sousa, A.; Dias, J.G.; Queirós, L.; Correia, A.M. Risco de Morrer no Concelho de Guimarães: Uma Análise da Mortalidade 1997–2005. Available online: http://portal.arsnorte.min-saude.pt/portal/page/portal/ARSNorte/Conte%C3%BAdos/Sa%C3%BAde%20P%C3%BAblica%20Conteudos/Risco_de_Morrer_Guimaraes_relatorio_1997_2005.pdf (accessed on 29 September 2015).

sustainability

MDPI

Article

Typology of Cities Based on City Biodiversity Index: Exploring Biodiversity Potentials and Possible Collaborations among Japanese Cities

Yuta Uchiyama [1,*], Kengo Hayashi [2] and Ryo Kohsaka [1]

[1] Graduate School of Human and Socio-Environmental Studies, Kanazawa University, Kakuma, Kanazawa 920-1192, Japan; kohsaka@hotmail.com

[2] Research Institute for Humanity and Nature, 457-4 Motoyama, Kamigamo, Kita-ku, Kyoto 603-8047, Japan; kensuke@chikyu.ac.jp

* Author to whom correspondence should be addressed; y-uchiyama@staff.kanazawa-u.ac.jp; Tel.: +81-76-264-5507.

Academic Editor: Tan Yigitcanlar

Received: 31 March 2015; Accepted: 15 October 2015; Published: 23 October 2015

Abstract: A City Biodiversity Index (CBI) has been proposed and applied at the international level to enable local municipalities and cities to manage biodiversity and ecosystem services in a sustainable manner. CBI databases are being constructed as global platforms, though the available dataset is limited. The land-use dataset is one of the datasets that can be utilized to apply the CBI on the national level in countries including Japan. To demonstrate the importance and potential of the CBI under the limitation of the available dataset, we attempted to apply the CBI to the 791 Japanese cities by using available land-use indicators, and categorized the cities based on the indicators. The focus of the CBI is self-assessment, but we propose that grouping of cities with similar profiles is possible and can serve as a basis for potential collaboration. Coordinating policies on various scales is necessary in order to enhance biodiversity on a global scale; one option is to increase collaboration among cities. As a result, we found three groups with similar characteristics amongst cities with forests, paddies, and croplands as major compositions in terms of biodiversity. These findings will contribute to policy formation and efficient information sharing for ecosystem services management.

Keywords: City Biodiversity Index; Satoyama Index; land use mixture; land cover; GIS

---

1. Introduction: Biodiversity and Sustainability

Since 2008, more than half of the world's population has been living in urban areas [1]. Urban population is estimated to continuously increase to 66% of the world population by 2050 [2]. Until the end of the 20th century, urbanization had not gained global attention for its impact on the conservation of natural resources and environment [3–5]. However, since human activities in urban areas influence the global environment by affecting the circulation of substances between urban and non-urban areas [3,6,7], urbanization is a critical issue with regard to environmental sustainability.

At present, reduction of biodiversity is considered one of the most urgent global environmental issues [8]; changes in biodiversity and ecosystem services associated with urbanization is also an important issue [9]. Biodiversity is the basis for ecosystem services and supports human activities in various aspects, including provisioning of resources [10–12]. Ecosystem services for urban populations are frequently obtained from outside city boundaries. These services include the regulation of climate, water, or soil. The primarily benefits are through the trade of resources [3,13–15]. The rapid increase in urban population might mean that more resources need to be obtained from remote areas, leading to changes in the landscape in those remote areas [16]. By this process, urbanization brings about reduction of biodiversity on a global scale.

On the other hand, the Cities and Biodiversity Outlook [9] highlighted that cities could also contribute to the enhancement of biodiversity. Cities include built-up areas, including residential blocks and paved roads, which are frequently developed among agriculture lands, forests, and other natural lands. These developments lead to the formation of complex landscapes involving mosaic of different categories of land use; in areas with such landscapes, biodiversity is relatively high [17–19]. Moreover, in areas surrounding cities, some areas are protected for ecosystem development. Therefore, cities and their surrounding areas can contribute to the enhancement and restoration of biodiversity. However, since recent rapid urbanization leads to landscape changes in areas that have a high potential of biodiversity [10,20–22], adequate managements of urbanization is necessary to preserve city biodiversity.

Global environmental communities such as the Convention on Biological Diversity (CBD) have been increasingly recognizing the loss of biodiversity caused by urbanization as an urgent issue in recent years, particularly since 2008, when the ninth meeting of Conference of the Parties (COP9) officially adopted this issue (Decision XI/28). Therefore, conservation and enhancement of biodiversity and ecosystems in cities are attracting increased attention [5,9,11,23] since they provide environmental as well as socio-economic benefits that can enhance the amenities and the quality of life for urban residents [17]. The ecosystems in a city influence the health and safety of urban populations [24], and can contribute to enhancing resilience against climate changes, natural disasters, or other security issues. In terms of city sustainability, environment, economy, and society need to be considered as the baselines [25]. In this respect, enhancement of city biodiversity is a significant task for urban societies from the viewpoint of sustainability.

Implementation of strategies for enhancement of city biodiversity requires the coordination of local governments [11,26], and the global agenda needs to be connected to local issues [27]. Therefore, accurately monitoring city biodiversity and ecosystem services is necessary [5,28]. For such monitoring, indicators to evaluate city biodiversity and ecosystem services have been developed, such as the city biodiversity index (CBI).

Further, coordinating policies at various scales is necessary in order to enhance biodiversity at a global scale. One such option is increasing collaboration among cities. For example, cities are required to collaborate, share experiences, and establish a network [5]. However, each individual city needs to maintain its own cultural, social, and environmental conditions; this might hinder the formation of such collaborations [29]. The measures and policies used in one city might not be applicable to the others because of the differences in their environmental conditions. Therefore, understanding the unique or similar characteristics of biodiversity and ecosystem services among cities is necessary; this requires comparison of their environmental conditions by using CBI to allow effective exchange of knowledge and policies among cities. However, few studies have compared the global urban biodiversity and ecosystem services by using identical indicators for quantitative measurements.

In this study, we used CBI to quantitatively categorize 791 cities in Japan and identified the common and heterogeneous characters to demonstrate the potential of the CBI by using available land-use indicators. Thus, we identified the groups of Japanese cities based on the land-use indicators. The cities in the same groups can become potential collaborators to share the experiences and knowledge of ecosystem services management, and the collaboration can contribute to the coordination of policies on different spatial scales.

Herein, we provide information regarding CBI and describe the methodologies used, including the datasets and quantifying indicators obtained using the CBI. We then characterized each identified category of cities. We also identified the challenges for urban biodiversity and ecosystem services and possible urban collaboration required for ensuring sustainability.

2. City Biodiversity Index

2.1. Evaluation of City Biodiversity

We reviewed existing literature on the methods used for evaluating biodiversity and ecosystem services for their adequate management.

The importance of biodiversity has been indicated globally; however, loss of biodiversity remains an important issue [30]. Thus, the role of cities in maintaining biodiversity has been emphasized [9]. Some databases allow the evaluation of the impact of human activities on biodiversity and ecosystem services at the macro-regional scale [31]. However, the database for city biodiversity is not yet available. For each regional scale, appropriate indicators (or adjustments) need to be designed urgently [26,29]. Definitions and baseline data, particularly for those related to ecosystem services, are missing for many municipalities.

The research related to the evaluation methods for city biodiversity can contribute to developing global data platforms [32]. The biodiversity in urban areas has been evaluated in order to enhance the quality of life [24]. Ecosystem services, excluding cultural services, are likely not active in urban areas that have a dominant built-up area. However, Larondelle and Haase [33] indicated that services related to the regulation of climate and air quality are implemented in cities. The target cities of their research included Berlin, Helsinki, Salzburg, and Stockholm. Regulating services active in urban forests include reduction of air pollution [34]. Regulation services at forested urban parks might contribute to the enhancement of ecosystem services [35]. Manes *et al.* [36] suggested that tree diversity significantly affected the stability of regulating services, including tropospheric ozone removal, and indicated that analyzing the functions of tree diversity for evaluation for ecosystem services in cities is necessary.

The impacts of urban activities on biodiversity have been investigated in Asian cities [37]. Gómez-Baggethun and Barton [17] indicated the type of ecosystem services that need to be managed during urban planning to enhance the quality of life and resilience. They provided a list of indicators for not only cultural services but also other types of ecosystem services.

Collecting and integrating the existing research results and methods for sharing information and knowledge across cities globally are necessary. Such integration can contribute to efficient and effective management of city biodiversity and ecosystem services. City sustainable index [38] is another major environmental indicator. However, this index does not provide direct evaluation of biodiversity [26]. Therefore, CBI is proposed as an adequate global platform [39].

2.2. Three Key Aspects of City Biodiversity Index

The CBI has three key aspects: (1) native biodiversity, (2) ecosystem service, and (3) governance and management. In all, 23 indicators have been proposed for sustainable management of city biodiversity. These three aspects are necessary to understand biodiversity in urban areas and to ensure its adequate management and conservation. First, the different types of biodiversity existing in urban areas need to be identified (native biodiversity), their importance in terms of ecosystem services needs to be evaluated (ecosystem service), and then methods for monitoring the present biodiversity situation and policies for its management need to be developed (governance and management). Indicators related to area of natural areas and number of native species (vascular plants, birds, and butterflies) are included in the native biodiversity indicators. Carbon storage and the cooling effect of vegetation, and area of parks with natural areas are ecosystem service indicators. The amount spent on biodiversity-related administration, and the status of local biodiversity strategy and action plan are indicators of governance and management. CBI consists of indicators that are organized by the collaboration of different departments in local municipalities to facilitate their communication. In addition, continued monitoring activities by CBI can motivate stakeholders to ensure the conservation of ecosystems. The indicators of CBI can be linked to targets for urban sustainability, and they can play important roles in monitoring and evaluating the strategies used by cities [26].

CBI databases are being constructed as global platforms, although they involve technical and administrative issues [26]. Technical issues include (1) collection of data for indicators, (2) establishment of spatial territories and definitions of indicators, and (3) elucidation of the different ecological background of each city. For the first issue, methods need to be developed to easily collect data by using remote sensing technology and to establish globally shared indicators. Identifying species associated with biodiversity and investigating the relationships between species and ecosystem services are essential. The species indicators are included in the CBI; however, the data of those indicators are not available in the most of the cities, excluding the major cities that have local biodiversity strategies. In the application of the CBI on the national level in Japan, the land-use indicators are limited available indicators. For the second and third issues, considering the spatial- and time-scale dependence of indicators and organizing relevant indicators for each scale are needed. Further, the administrative issues need to be addressed. Governments in cities need to collect and organize relevant data for calculating CBI indicators and facilitate policies and actions for ensuring urban sustainability by linking the indicators to targets. However, capacity shortage of city governments is the main administrative issue. These issues will be discussed in detail in Section 4.

CBI does not intend to compare cities, but attempts to evaluate them individually. Nonetheless, obtaining comparable qualitative data for the indicators of native biodiversity or ecosystem services is possible. For example, the proportion of natural areas, one of the indicators of native biodiversity, can be obtained based on remotely sensed data and can be compared globally. In addition, biodiversity and ecosystem services are strongly related to the proportion of each land use category, including natural areas and their distribution patterns in urban areas. Kadoya and Washitani [40] suggested a correlation among land-use mosaic patterns and biodiversity. Thus, the characteristics of land use in urban areas can be considered the basic information for conserving biodiversity and ecosystem services.

In this study, we attempted to comprehensively identify the characteristics of land use in municipal areas of Japanese cities, based on the proportion of natural areas and other land use categories; further, we determined the degree of land-use mixture. Next, we categorized the cities by using the land-use indicators and identified several types of cities having similar characteristics. Kadoya and Washitani [40] proposed an index that correlates with biodiversity; this index identified the degree of land-use mixture calculated based on the number of land-use categories and proportion of each existing category in a target area. We used this index to evaluate the land-use mixture. This index is mainly used in non-urban areas, but we will extend the application to urban contexts in this study.

3. Categorization of Cities According to Land-Use Indicators

We performed principal component analysis (PCA) of land-use indicators (Table 1) to determine the variables for categorization. The proportion of natural areas, including forests, shrubs, and grasslands, is one of the native biodiversity indicators in CBI. The other native biodiversity indicators are "changes in the number of native species (vascular plants, birds, and butterflies)", "proportion of protected areas", "proportion of invasive alien species", "connectivity measures or ecological networks to counter fragmentation", and "native biodiversity in built-up areas" [39]. The land-use mixture is related to biodiversity in urban regions [40]. We considered other land-use indicators (see Table 1) in addition to those related to CBI, to identify the basis of collaboration among cities with similar ecosystem characteristics. We analyzed 791 cities in Japan to provide a platform for collaboration among cities by categorizing them according to the land-use indicators related to biodiversity potentials. The species indicators are included in the CBI; however, the data of these indicators are not available in the most of the cities. To demonstrate the importance and potential of the CBI, we attempted to apply CBI to the cities by using available land-use indicators. Further consideration is given for the use of CBI as the platform for the collaboration. We used data for the Japanese cities; these cities have an administrative level called "Shi" [41]. Next, we categorized the cities based on cluster analysis by using the results of the PCA.

Table 1. Indicators for categorization of cities.

Indicator	Unit
Average of degree of land use mixture	-
Proportion of forest areas	%
Proportion of natural areas with vegetation excluding forest	%
Proportion of paddy fields	%
Proportion of cropland and other vegetation mosaic	%
Proportion of built-up areas	%

3.1. Data

The global land cover data can be used as land-use distribution data. The high-resolution global land cover data have been developed using recent innovative information technology, and are available freely. We used the data from GLCNMO [42], since these data have relatively high resolution—-15 arc-second; further, these data have been developed based on the collaboration among institutes across several countries. In all, 20 land cover categories are included in these data. Of these, five categories each are for different kinds of forests and natural areas with vegetation excluding forests. Agricultural land covers three categories that include paddy field, cropland, and agricultural land with other vegetation mosaic. The overall classification accuracy of GLCNMO is 77.9% by 904 validation points in the world. For the PCA, we used six indicators that are shown in Table 1. These indicators were calculated for each city by using GLCNMO. In this preliminary study, we used a forest category that included different types of forest. We intend to consider the different types of forests, such as broadleaved deciduous species, broadleaved evergreen species, and conifers in the future to understand the detailed differences among cities characterized by forests. To identify more detailed categories of cities in the further research, we will need to consider shrubs and grasslands separately in the land-use category—natural areas with vegetation excluding forests.

The indicator for land-use mixture was calculated by using the method described by Kadoya and Washitani [40]. Their index is calculated based on the number of land-use categories and proportion of each existing category in a target area. It is calculated by each 6 km square grid. Kadoya and Washitani proposed the grid resolution based on the spatial scale of habitats of the plants and animals. To evaluate the land-use mixture in each city administrative boundaries, we considered the land-cover categories except built-up area as categories that enhance the degree of land-use mixture.

3.2. Results of Principal Component Analysis

We performed PCA on the six indicators. We identified two principal components (Table 2) with a cumulative contribution ratio of 76.7%. The ratio shows that these two principal components can sufficiently explain the differences in cities.

The first component has a strong positive correlation with the degree of land-use mixture, and a negative correlation with its proportion of built-up areas. The second component has relatively strong positive correlation with the proportion of forests and negative correlation with the proportion of paddy and cropland.

The results of the PCA showed that the degree of land-use mixture is one of the important indicators for understanding the characteristics of cities. In addition, the proportions of land-use categories are not alternatives for land-use mixtures, because the former show quantitative characteristics and the latter reflect the qualitative ones.

Table 2. The two principal components.

Eigenvalue and Contribution	PC1	PC2
Eigenvalue	2.6713	1.9294
Contribution	0.4452	0.3216
Cumulative contribution	0.4452	0.7668
Eigenvector		
Landuse mixture	0.5754	−0.1201
Forest	0.3624	0.56
Shrub and Grassland	0.4207	−0.2874
Paddy field	−0.1219	−0.5169
Cropland and Other vegetation mosaic	0.1919	−0.5646
Built-up area	−0.5558	−0.0583
Factor loading		
Landuse mixture	0.9404	−0.1668
Forest	0.5923	0.7778
Shrub and Grassland	0.6876	−0.3993
Paddy field	−0.1993	−0.718
Cropland and Other vegetation mosaic	0.3136	−0.7843
Built-up area	−0.9084	−0.081

3.3. *Result of categorization*

Cluster analysis of the two principal components revealed three categories of cities (see Figure 1). The first principal component values for cities in Category 1 ($N = 93$) are low. Their average degree of land-use mixture is the lowest, and the proportion of built-up area is the highest. Category 2 ($N = 347$) includes cities that have relatively high degree of land-use mixture and high proportion of forest areas. The cities in Category 3 ($N = 351$) have relatively high degree of land-use mixture and high proportion of farmland. Each category has different characteristics in the component of mosaic land use and degree of land-use mixture.

3.4. *Characteristics of Each Category*

To determine the land-use characteristics of each category, we calculated the averages and standard deviations of the six indicators in each category (see Tables 3 and 4). The quartiles, and minima and maxima of the indicators are shown in Figure 2. We used the other land-use indicators along with the ones related to native biodiversity indicators to analyze the land-use characteristics that are related to ecosystem characteristics of the cities. In the discussion of the characteristics of each category, averages across all cities in each category were referred to.

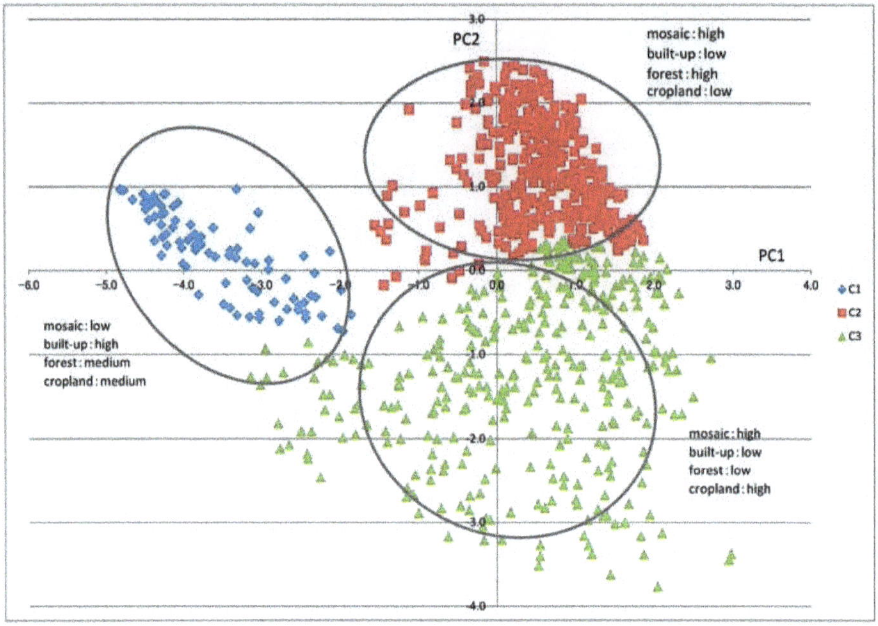

Figure 1. Result of categorization of cities.

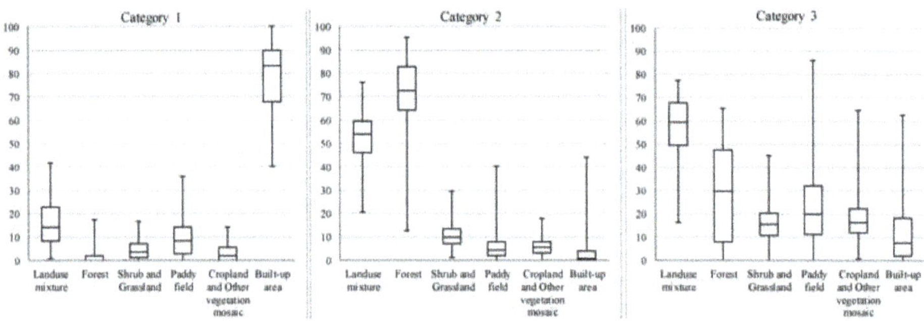

Figure 2. Quartiles and minimums and maximums of the six indicators in each category. Note: Unit: (original values multiplied by 100): Land-use mixture, (%): proportion of forest, shrub and grassland, paddy fields, cropland, and other vegetation mosaic, and built-up area. Horizontal lines in each bar chart show maximum, top 25th percentile, median, bottom 25th percentile, and minimum values.

Table 3. Averages of the six indicators in each category.

Category	No. of City	Landuse Mixture	Forest	Shrub and Grassland	Paddy Field	Cropland and Other Vegetation Mosaic	Built-Up Area
		-	%	%	%	%	%
1	93	0.2	2.1	4.6	9.7	3.5	79.3
2	347	0.5	72.3	10.8	5.7	5.9	4.2
3	351	0.6	27.9	16.0	23.2	19.1	12.3

Table 4. Standard deviations of the six indicators in each category.

Category	No. of City	Landuse Mixture	Forest	Shrub and Grassland	Paddy Field	Cropland and Other Vegetation Mosaic	Built-Up Area
1	93	0.102	3.9	4.0	8.1	3.9	13.4
2	347	0.097	13.5	4.9	5.7	3.5	7.9
3	351	0.132	20.4	7.2	16.3	11.2	13.5

3.4.1. Category 1

The proportion of built-up areas was 79%, and that of forest areas was 2%. Large cities that held central administrative units, such as special wards of Tokyo Prefecture and Osaka City, had relatively high proportion of built-up areas, and they were included in Category 1. Although they had high proportion of built-up areas, the proportion of farmland was not considerably different from that of Category 2. However, cities in Category 1 had low proportion of natural land, and the diversity of land-use category was relatively low. Therefore, the degree of land-use mixture was lower than that of the other categories.

We focused on the municipal areas of the cities; if a city is situated in the center of a large metropolitan area consisting of several municipalities, the proportion of built-up areas of the city might be relatively high. The values of the land-use indicators can be changed depending on the definitions of cities. In the future, we intend to identify the impacts of the definitions on the values of the indicators.

In terms of degree of land-use mixture and proportion of natural land, cities in Category 1 might have less biodiversity, and their ecosystem services might be inactive. The conservation of biodiversity in each land-use category in urban areas is important, as well as the conservation and enhancement of biodiversity in the surrounding areas. Reducing the impact on ecosystems from agglomeration of buildings and paved roads and other anthropogenic objects is necessary; cities in Category 1 had high proportion of built-up areas and might strongly depend on ecosystems in their surrounding areas.

3.4.2. Category 2

The proportion of forest areas was 72%, and that of built-up areas was 4%. The proportion of farmlands was relatively low (12%), and that of natural lands excluding forest areas was 11%. The degree of land-use mixture was relatively high, and the land-use mosaic consisted of natural lands rather than farmlands.

Cities in Category 2 might have high biodiversity and abundant ecosystem services. Cities in Category 1 required management of biodiversity within the group and their nearby areas via the cooperation of its surrounding administrative units. However, the main issue of cities in Category 2 was managing their impact on biodiversity within them.

3.4.3. Category 3

The proportion of forest areas in this category was 28%, paddy fields accounted for 23%, and the built-up areas were 12%. In this category, the proportion of farmlands was relatively high; however, the proportion of a specific land-use category was not extremely higher than that in the other categories. These cities had diverse land-use categories, and the average of the degree of land-use mixture was the highest among all the categories.

Cities in Category 3, which have the most diverse land-use, can have higher biodiversity and more abundant ecosystem services than those in Category 1. However, the built-up areas of cities in Category 3 were surrounded by farmland that could expand easily. Therefore, one of the main issues of these cities was the conservation of ecosystems that depended on farmlands. Thus, if a city could not implement adequate management of farmland, they would risk having negative impacts on their ecosystems.

3.4.4. Regional Characteristics of Japan

The regional land-use characteristics of cities in Japan that have wider mountainous areas often include forest lands with low population density. Even Category 3 cities, which have a high proportion of farmland, have relatively high proportion of forest lands (>25%).

Japan is a part of monsoonal Asia; the land is mostly covered with paddy fields and has high population density like other areas in the region. The proportion of paddy fields in Category 1 that has greater built-up areas was higher than that of cities in Category 2 that have high rates of forest lands. This suggests that paddy fields can exist in regions adjacent to built-up and densely populated areas. By considering these regional characteristics, the policy makers, citizens, and business sectors can implement measures to ensure sustainable management of urban biodiversity and ecosystem services in Japan. These characteristics might not be common among cities having different climatic zones.

3.4.5. Basic Environmental Characteristics

Highly dense paddy fields and high proportions of forest lands are the regional characteristics and basic environmental features of Japan. These might not change easily in a short period. However, the proportion of farmland can change to a great extent, and it can be considered a variable environmental characteristic. Understanding the difference among these environmental characteristics is needed to develop adequate management strategies for city biodiversity. Efficient and effective sharing of knowledge and information can be implemented among cities that have the same basic environmental characteristics and similar variable features.

4. Issues and Prospects

Biodiversity of a city is associated with the sustainability of the city; biodiversity in urban areas can contribute to the enhancement of amenities by increasing cultural services and regulating living environment via regulating services. CBI is the indicator for establishing appropriate managements of biodiversity and ecosystem services in each city. The issues of CBI need to be addressed to ensure city sustainability by implementing adequate managements. There are three main technical issues with CBI, which include (1) collection of data for indicators; (2) establishment of spatial territories and definitions of indicators; and (3) elucidation of the different ecological backgrounds of each city.

In this study, we suggested potential solutions to improve city biodiversity by using land-use indicators that are related to native biodiversity indicators in CBI. The land-use indicators are calculated using global data that can be obtained easily. Thus, the first issue related to data collection can be resolved by using remote sensing data. However, indicators to evaluate the quality of biodiversity are required. We categorized cities based on not only proportions of land-use categories but also degrees of land-use mixture, which revealed the qualitative aspects of land use. These indicators can be used to evaluate the qualitative aspects of biodiversity.

Regarding the second issue, we use administrative units for urban areas. It can be expected that each administrative unit for city government management is different in terms of the amount or nature of human activities. When we identify the relationships among human activities and biodiversity, it is necessary to use urban areas detected by the same definition in terms of manpower.

For addressing the third issue, the characteristics of land use that reflect the characteristics of ecosystems in cities need to be identified. Considering the characteristics of ecosystems that are estimated using land-use characteristics, indicators of CBI can be developed and evaluated.

The second and third issues are related to the spatial- and temporal-scale dependence of indicators. Regarding their spatial-scale dependence, geographic information system (GIS) data that were used in this study cannot be used for analyzing smaller districts in cities; GIS data with greater resolution are needed to monitor and evaluate the more micro-scale ecosystems. The share of each land-use category in an administrative area of a city is changing. If time series land-use data would be applied in this analysis, we might found cities that would move to the other categories from categories to which they belonged in the previous time. We did not use time series data in this preliminary analysis. In further research, time series data will be needed to understand temporal trends in land use of cities.

The third part is of particular importance for reviewing cities nationwide. The uniform application of CBI to cities is likely to result in high scores for cities with green areas, regardless of administrative and citizen efforts. Such categorization of cities might enable collaboration and comparison with existing profiles and conditions.

5. Conclusions

The sustainable use of biodiversity and ecosystem services at the urban level is a global issue. We conducted an empirical study to determine the possibility of CBI application at the national level. To our knowledge, this is the first empirical investigation at the national level that includes both rural and urban areas (except Singapore, which is a state). There are issues in application of CBI, including limitation of available dataset. By using the land-use dataset that can be obtained from remote sensing data, we proposed the solution for the issues related to the limitation of the dataset.

Our results suggest that the Japanese cities can be categorized into three major groups. The major biodiversity components were forest, paddy, and cropland. This categorization might serve as a basis for possible collaboration among Japanese cities that have similar challenges and conditions. The collaborations among cities are required to coordinate policies on various spatial scales to enhance biodiversity on a global scale. The categorization that we attempted can be a preliminary step to establish a method to identify the adequate networks of cities in the world for ecosystem services management.

Many cities have expressed concerns regarding the compilation of data or initiation of their own evaluation of native biodiversity or ecosystem services. Furthermore, for many cities, obtaining funds for activities related to biodiversity conservation is difficult. Given the limitations in budgets and human resources, the simplified and cost-effective measures presented in this study might be useful for the development and application of biodiversity indicators in Japan in the future.

Acknowledgments: This study was conducts as a part of the research project, MEXT Gakushin, KAKEN (26360062 Leader Kohsaka, R), the project of Hokuriku Regional Management Service Association (2015) (Leader Kohsaka, R.) and the project funded by the JFE Steel Corporation (2015) (Leader Kohsaka, R.). Part of project was also conducts as a part of the research project, Environment Research and Technology Development Fund (1-1303; Leader Saito, O.). Thanks are extend to O. Saito and S. Hashimoto for the valuable comments.

Author Contributions: Yuta Uchiyama conceived the paper and analyzed the data; Kengo Hayashi reviewed related studies; Ryo Kohsaka reviewed the discussion and analysis of the CBI application and policies. All authors wrote, reviewed and commented on the manuscript. All authors have read and approved the final manuscript.

Conflicts of Interest: The authors declare no conflict of interest.

References

1. United Nations Population Fund (UNFPA). *State of World Population 2009 Facing a Changing World: Women, Population and Climate*; UNFPA: New York, NY, USA, 2009.
2. United Nations Department of Economic and Social Affairs (UNDESA). *World Urbanization Prospects: The 2014 Revision*; United Nations: New York, NY, USA, 2015.
3. Grimm, N.B.; Faeth, S.H.; Golubiewski, N.E.; Redman, C.L.; Wu, J.G.; Bai, X.M.; Briggs, J.M. Global change and the ecology of cities. *Science* **2008**, *319*, 756–760. [CrossRef] [PubMed]
4. Puppim de Oliveira, J.A.; Balabana, O.; Dolla, C.N.H.; Moreno-Peñarandaa, R.; Gasparatosa, A.; Iossifovaa, D.; Suwaa, A. Cities and biodiversity: Perspectives and governance challenges for implementing the convention on biological diversity (CBD) at the city level. *Biol. Conserv.* **2011**, *144*, 1302–1313. [CrossRef]
5. Wilkinson, C.; Sendstad, M.; Parnell, S.; Schewenius, M. Urban Governance of Biodiversty and Ecosstem Serviecs. In *Urbanization, Biodiversity and Ecosystem Services: Challenges and Opportunities*; Springer: Dordrecht, The Netherlands, 2013; pp. 539–587.
6. Bai, X. Integrating global environmental concerns into urban management: The scale and readiness arguments. *J. Ind. Ecol.* **2007**, *11*, 15–29. [CrossRef]

7. Hardoy, J.; Mitlin, D.; Satterthwaite, D. *Environmental Problems in an Urbanizing World*; Routledge: London, UK, 2001.
8. Rockström, J.; Stffen, W.L.; Noone, K.; Persson, A.; Chapin, F.S., III; Lambin, E.; Lenton, T.M.; Scheffer, M.; Folke, C.; Schellnhuber, H.J.; *et al.* Planetary Boundaries: Exploring the Safe Operating Space for Humanity. Available online: http://www.ecologyandsociety.org/vol14/iss2/art32/ (accessed on 25 March 2015).
9. Secretariat of the Convention on Biological Diversity. Cities and Biodiversity Outlook (CBO), 2012. Available online: https://www.cbd.int/doc/health/cbo-action-policy-en.pdf (accessed on 25 March 2015).
10. McDonald, R.I.; Marcotullio, P.J.; Güneralp, B. Urbanization and Global Trends in Biodiversity and Ecosystem Services. In *Urbanization, Biodiversity and Ecosystem Services: Challenges and Opportunities*; Springer: Dordrecht, The Netherlands, 2013; pp. 31–52.
11. Kohsaka, R. Developing biodiversity indicators for cities: Applying the DPSIR model to Nagoya and integrating social and ecological aspects. *Ecol. Res.* **2010**, *25*, 925–936. [CrossRef]
12. Millennium Ecosystem Assessment (MA). *Ecosystems and Human Wellbeing: Synthesis*; Island Press: Washington, DC, USA, 2005.
13. Alfsen-Norodom, C.; Boehme, S.E.; Clemants, S.; Corry, M.; Imbruce, V.; Lane, B.D.; Miller, R.B.; Padoch, C.; Panero, M.; Peters, C.M.; *et al.* Managing the megacity for global sustainability: The New York Metropolitan Region as an urban biosphere reserve. *Ann. N.Y. Acad. Sci.* **2004**, *1023*, 125–141. [PubMed]
14. Folke, C.; Jansson, A.; Larsson, J.; Costanza, R. Ecosystem appropriation by cities. *Ambio* **1997**, *26*, 167–172.
15. Jansson, Å. Reaching for a sustainable, resilient urban future using the lens of ecosystem services. *Ecol. Econ.* **2013**, *86*, 285–291. [CrossRef]
16. Güneralp, B.; Seto, K.C.; Ramachandran, M. Evidence of urban land teleconnections and impacts on hinterlands. *Curr. Opin. Environ. Sustain.* **2013**, *5*, 445–451. [CrossRef]
17. Gómez-Baggethun, E.; Barton, D.N. Classifying and valuing ecosystem services for urban planning. *Ecol. Econ.* **2013**, *86*, 235–245. [CrossRef]
18. McKinney, M.L. Effects of urbanization on species richness: A review of plants and animals. *Urban Ecosyst.* **2008**, *11*, 161–176. [CrossRef]
19. Muller, N.; Werner, P.; Kelcey, J.G. *Urban Biodiversity and Design*; Wiley-Blackwell: Chichester, UK, 2010.
20. Güneralp, B.; Seto, K.C. Sub-regional Assessment of China: Urbanization in Biodiversity Hotspots. In *Urbanization, Biodiversity and Ecosystem Services: Challenges and Opportunities*; Springer: Dordrecht, The Netherlands, 2013; pp. 57–63.
21. McDonald, R.I.; Formanb, R.T.T.; Kareivac, P.; Neugartena, R.; Salzerd, D.; Fishera, J. Urban effects, distance, and protected areas in an urbanizing world. *Landsc. Urban Plan.* **2009**, *93*, 63–75. [CrossRef]
22. Güneralp, B.; Seto, K.C. Futures of global urban expansion: Uncertainties and implications for biodiversity conservation. *Environ. Res. Lett.* **2013**. [CrossRef]
23. Snep, R.; van Ierland, E.; Opdam, P. Enhancing biodiversity at business sites: What are the options, and which of these do stakeholders prefer? *Landsc. Urban Plan.* **2009**, *91*, 26–35. [CrossRef]
24. Bolund, P.; Hunhammar, S. Ecosystem services in urban areas. *Ecol. Econ.* **1999**, *29*, 293–301. [CrossRef]
25. Elkington, J. *Cannibals with Forks: The Triple Bottom Line of 21st Century Business*; Capstone Publishers: Oxford, UK, 1997.
26. Kohsaka, R.; Pereira, H.; Elmqvist, T.; Chan, L.; Moreno-Peñaranda, R.; Morimoto, Y.; Inoue, T.; Iwata, M.; Nishi, M.; da Luz-Mathias, M.; *et al.* Indicators for Management of Urban Biodiversity and Ecosystem Services: City Biodiversity Index. In *Urbanization, Biodiversity and Ecosystem Services: Challenges and Opportunities: A Global Assessment*; Springer: Dordrecht, The Netherlands, 2013; pp. 699–718.
27. Keating, M. *The Earth Summit's Agenda for Change: A Plain Language Version of Agenda and Three Other Rio Agreements*; Center for Our Common Future: Geneva, Switzerland, 1993.
28. Li, F.; Liu, X.S.; Hu, D.; Wang, R.S.; Yang, W.R.; Li, D.; Zhao, D. Measurement indicators and an evaluation approach for assessing urban sustainable development: A case study for China's Jining City. *Landsc. Urban Plan.* **2009**, *90*, 134–142. [CrossRef]
29. Kohsaka, R.; Okumura, S. Greening the Cities with Biodiversity Indicators: Experience and Challenges from Japanese Cities with CBI. In *Integrative Observations and Assessments, Ecological Research Monographs*; Nakano, S., Yahara, T., Nakashizuka, T., Eds.; Springer: Tokyo, Japan, 2014; pp. 409–424.

30. Butchard, S.H.M.; Walpole, M.; Collen, B.; van Strien, A.; Scharlemann, J.P.W.; Almond, R.E.A.; Baillie, J.E.M.; Bomhard, B.; Brown, C.; Bruno, J.; *et al.* Global biodiversity: Indicators of recent declines. *Science* **2010**, *328*, 1164–1168.
31. Walker, B.; Meyers, J.A. Thresholds in Ecological and Social-Ecological Systems: A Developing Database. Available online: http://www.ecologyandsociety.org/vol9/iss2/art3/ (accessed on 25 March 2015).
32. Niemelä, J. Ecology of urban green spaces: The way forward in answering major research questions. *Landsc. Urban Plan.* **2014**, *125*, 298–303. [CrossRef]
33. Larondelle, N.; Haase, D. Urban ecosystem services assessment along a rural-urban gradient: A cross-analysis of European cities. *Ecol. Indic.* **2013**, *29*, 179–190. [CrossRef]
34. Nowak, D.J.; Crane, D.E.; Stevens, J.C. Air pollution removal by urban trees and shrubs in the United States. *Urban For. Urban Green.* **2006**, *4*, 115–123. [CrossRef]
35. Millward, A.A.; Sabir, S. Benefits of a forested urban park: What is the value of Allan Gardens to the city of Toronto, Canada? *Landsc. Urban Plan.* **2011**, *100*, 177–188. [CrossRef]
36. Manes, F.; Incerti, G.; Salvatori, E.; Vitale, M.; Ricotta, C.; Costanza, R. Urban ecosystem services: tree diversity and stability of tropospheric ozone removal. *Ecol. Appl.* **2012**, *22*, 349–360. [CrossRef] [PubMed]
37. Hou, Y.; Zhou, S.; Burkhard, B.; Müller, F. Socioeconomic influences on biodiversity, ecosystem services and human well-being: A quantitative application of the DPSIR model in Jiangsu, China. *Sci. Total Environ.* **2014**, *490*, 1012–1028. [CrossRef] [PubMed]
38. Mori, K.; Christodoulou, A. Review of sustainability indices and indicators: Towards a new City Sustainability Index (CSI). *Environ. Impact Assess. Rev.* **2012**, *32*, 94–106. [CrossRef]
39. Secretariat of the Convention on Biological Diversity (SCBD). User's Manual for the City Biodiversity Index (CBI), SCBD, 2012. Available online: https://www.cbd.int/subnational/partners-and-initiatives/city-biodiversity-index (accessed on 25 March 2015).
40. Kadoya, T.; Washitani, I. The Satoyama Index: A biodiversity indicator for agricultural landscapes. *Agric. Ecosyst. Environ.* **2011**, *140*, 20–26. [CrossRef]
41. *Population and Number of Household Based on the Basic Resident Registration*; Ministry of Internal Affairs and Communications: Tokyo, Japan, 2014. (In Japanese)
42. Tateishi, R.; Hoan, N.T.; Kobayashi, T.; Alsaaideh, B.; Tana, G.; Phong, D.X. Production of global land cover data—-GLCNMO2008. *J. Geogr. Geol.* **2014**, *6*, 99–122. [CrossRef]

sustainability

MDPI

Article

Incorporating Bio-Physical Sciences into a Decision Support Tool for Sustainable Urban Planning

Zina Mitraka [1,*], Emmanouil Diamantakis [1], Nektarios Chrysoulakis [1], Eduardo Anselmo Castro [2], Roberto San Jose [3], Ainhoa Gonzalez [4] and Ivan Blecic [5]

1. Foundation for Research and Technology—Hellas, Institute of Applied and Computational Mathematics, N. Plastira 100, Vassilika Vouton, P.O. Box 1385, GR-71110 Heraklion, Greece; diamanda@iacm.forth.gr (E.D.); zedd2@iacm.forth.gr (N.C.)
2. Research Centre on Governance, Competitiveness and Public Policies (GOVCOPP) and Department of Social, Political and Territorial Sciences, University of Aveiro, 3810-193 Aveiro, Portugal; ecastro@ua.pt
3. Environmental Software and Modelling Group, Computer Science School, Technical University of Madrid, Campus de Montegancedo, Boadilla del Monte 28660 Madrid, Spain; roberto@fi.upm.es
4. Centre for the Environment, School of Natural Sciences, Trinity College Dublin, Dublin 2, Ireland; agonzal@tcd.ie
5. Department of Architecture, Design and Planning, University of Sassari, Centro Euro-Mediterraneo per i Cambiamenti Climatici, Piazza Duomo 6, 07041 Alghero, Italy; ivan@uniss.it

* Author to whom correspondence should be addressed; mitraka@iacm.forth.gr; Tel.: +30-2810-391-762; Fax: +30-2810-391-761.

External Editor: Tan Yigitcanlar

Received: 28 August 2014; in revised form: 18 October 2014; Accepted: 6 November 2014; Published: 12 November 2014

Abstract: Deciding upon optimum planning actions in terms of sustainable urban planning involves the consideration of multiple environmental and socio-economic criteria. The transformation of natural landscapes to urban areas affects energy and material fluxes. An important aspect of the urban environment is the urban metabolism, and changes in such metabolism need to be considered for sustainable planning decisions. A spatial Decision Support System (DSS) prototyped within the European FP7-funded project BRIDGE (sustainaBle uRban plannIng Decision support accountinG for urban mEtabolism), enables accounting for the urban metabolism of planning actions, by exploiting the current knowledge and technology of biophysical sciences. The main aim of the BRIDGE project was to bridge the knowledge and communication gap between urban planners and environmental scientists and to illustrate the advantages of considering detailed environmental information in urban planning processes. The developed DSS prototype integrates biophysical observations and simulation techniques with socio-economic aspects in five European cities, selected as case studies for the pilot application of the tool. This paper describes the design and implementation of the BRIDGE DSS prototype, illustrates some examples of use, and highlights the need for further research and development in the field.

Keywords: spatial decision support systems; urban planning; sustainability indicators; urban metabolism

1. Introduction

The transformation of landscapes from primarily forest or agricultural uses to urbanised areas modifies energy and material exchanges between the city and its environment. These exchanges define the urban metabolism of a city, which is an important aspect in the functioning of cities [1] and needs to be considered in urban planning. Urban metabolism studies consider a city as a system and distinguish between energy and material flows. Since cities are much more than a mechanism for processing resources and producing wastes, the urban metabolism concept includes livability aspects, referring to the human requirement for social amenity, health and well-being. Sustainability for a city is not only the reduction in metabolic flows, but also the increase of human livability, and economic and social aspects of sustainability need to be integrated along with the environmental ones [2]. The city design (defines the urban form and cover), the urban governance (related to the land use and the anthropogenic emissions) and the local and regional climate dynamics affect the energy, water and carbon exchanges between the urban surface and the atmosphere, modifying the respective urban metabolism components. Therefore, the better understanding and monitoring of energy, water, carbon and pollutants fluxes has the potential to support urban planning and management by: (a) supporting the development of methods for optimization of these fluxes at micro, local and regional scales in the framework of a city-scale climate-change adaptation; and (b) providing the means to support "land-based mitigation", defined as land surface changes producing a modification in urban energy water and carbon exchanges towards slowing the pace of warming in cities. Land-based mitigation complements conventional emissions-based mitigation through addressing the regional to local-scale drivers of climate change that are often the principal driver of ongoing warming trends at these scales [3]. The land-use planning activities of local/regional governments may therefore provide the most direct regulatory means of managing urban energy water and carbon exchanges towards a sustainable urban metabolism, so as to minimize land-based climate, forcing towards the improvement of thermal comfort and air quality, that are both important elements of the quality of life in cities.

Traditional metabolic studies assess the inputs and outputs of food, water, energy, waste, materials, *etc.* from a city, or compare the metabolic processes of several cities [4]. Recent advances in biophysical sciences consider the urban metabolism as the exchange and transformation of energy and matter between a city and its environment, and make possible to provide quantitative estimates of the urban metabolism components at a local scale [5]. Several methods and tools are available for estimating energy, water, carbon and pollutants fluxes, and research efforts in these areas are ongoing. Nevertheless, there is still no standardized method to conduct quantitative urban metabolism studies worldwide [6] and available tools are more scientifically oriented and not user-friendly for planning purposes [6,7]. In addition, comprehensive urban metabolism data at city scale are difficult to obtain, especially for low-income cities. These reasons prevent planners and policy-makers from exploiting existing knowledge and information in urban planning processes.

The rising pressure for urban sustainability confronts planners with the necessity of taking into account the environmental and socio-economic considerations at once. For example, it is not common for urban planners to have background knowledge on urban climatic processes and the interaction with climate experts is difficult because of the different practice backgrounds [5]. Therefore, there is arising need for development of specific evaluation methods and appointed tools to address multiple inter-disciplinary aspects within decision-making regarding urban planning [8]. To support an evaluation, each decision problem needs to be described in a structured way and planning alternatives need to be defined and accompanied by descriptive and quantitative information. Spatial Decision Support Systems (DSS) are powerful decision support tools that comprehensively analyse baseline information and satisfy multiple-period, multiple-objective and multiple-user requirements [9]. Spatial DSS are capable of supporting complex decision-making and of solving semi-structured, or unstructured problems [10]. Such systems help decision-makers in finding concrete solutions for decision issues and facilitate the use of geospatial data and models.

The basis of geospatial decision support is the technology of Geographic Information Systems (GIS). The basic decision supports of GIS include data management to extend human memory, graphic display to enhance visualisation, and spatial analysis functions to extend human computing performance. Beyond these common GIS decision aids, special features include modeling, optimisation and simulation functions required to generate, evaluate, and test the sensitivity of computed solutions. GIS can also support other functions, such as statistical, spatial interaction and location/allocation models. As most territorial and environmental assessments involve several planning alternative options, as well as numerous planners with different views and perceptions, spatial DSS provide effective techniques to assess multiple and/or cumulative impacts, to incorporate stakeholder perceptions and to carry out a vulnerability or suitability analysis in order to evaluate the alternatives under consideration [11]. The use of spatial DSS is lately increasing in impact assessment to support urban planning [12].

The European Seventh Framework Programme (FP7) Project BRIDGE (SustainaBle uRban plannIng Decision support accountinG for urban mEtabolism) introduced the consideration of environmental issues in urban planning support systems, by means of numerical tools. The BRIDGE project did not perform a complete life cycle or a whole system urban metabolism analysis, but rather focused on specific metabolism components, namely energy, water, carbon and pollutants. A corollary of the BRIDGE project was the development of a DSS prototype with the potential to evaluate planning actions that better fit the goal of changing the metabolism of urban systems towards sustainability. State-of-the-art numerical models were employed to simulate the physical flows of energy, water, carbon and pollutants and a Multi-Criteria Evaluation (MCE) approach was adopted to cope with the complexity of interactions between the environmental, social and economic components of the urban metabolism.

The main aim of the BRIDGE DSS is to assist decision-making by providing a structured assessment of alternatives and methods comprising key urban metabolism components for their comparative analysis, ranking and selection among them. An issue to deal with is the dependence of the outcome on the objectives that the user establishes. Different objectives may be set depending on the interests, the needs and the resources of planners and stakeholders, some of which may conflict. Therefore, the optimum solution is a trade-off between objectives, dependent on the end-user (e.g., planner) preferences. The BRIDGE DSS evaluates how Planning Alternatives (PA) modify the physical flows of the urban metabolism components under consideration. To cope with the complexity of urban metabolism issues, objectives are defined in BRIDGE as priorities in relation to interactions between the environmental elements (e.g., fluxes of energy, water, carbon and pollutants) and socio-economic components (e.g., investment costs, housing, employment, *etc.*) of urban sustainability. Indicators, organised in groups, are used to characterise objectives and the evaluation of the performance of each PA is done according to the importance of each objective. The user decides on the relative importance of the objectives and may decide on which indicators to include in the analysis. The combined performance of indicators selected as relevant in each particular case are then used to rank the PA.

The developed BRIDGE DSS prototype and its main components are described in this paper, including the numerical models that were utilised to simulate the physical-flows, the set of indicators developed in the framework of the project based on stakeholder consultation and the MCE approach that was used to evaluate the performance of urban planning actions. Evaluation results for five European cities (Helsinki, Athens, London, Firenze and Gliwice) are also presented and discussed.

2. Materials and Methods

2.1. Study Sites

Five European cities were selected as case studies in the framework of the BRIDGE project in an attempt to capture different environmental and socio-economic behaviours. Helsinki, Finland, was selected as a high latitude city with rapid urbanisation and the requirement of a substantial amount of energy for heating. Athens, Greece, was selected as a mid-latitude Mediterranean city that requires a substantial amount of energy for cooling. London, United Kingdom, was considered as one of the world's mega-cities. Firenze, Italy, was selected as a representative old city with substantial cultural heritage and Gliwice, Poland, a typical eastern European city with dynamic planning process reflecting the economic, social, and political changes held within last two decades.

Stakeholders and potential users were involved in the design of the BRIDGE DSS from the beginning of the project, forming a Community of Practice (CoP) [13]. The CoP consisted of professionals in the field of city planning and researchers participating in the BRIDGE project to establish a learning environment and jointly search for opportunities for improving sustainable urban planning. The participative approach allowed insight into the case studies to be gained in relation to both planning structures and issues, and sustainability considerations. The approach followed in the BRIDGE project helped to provide the stakeholders and users' perspective to the research team, with the likely outcome that the DSS responds to the sustainability objectives in each city, as well as the needs of the users. The identified objectives and the derived indicators were largely shaped by the professional background and perceptions of the case study representatives.

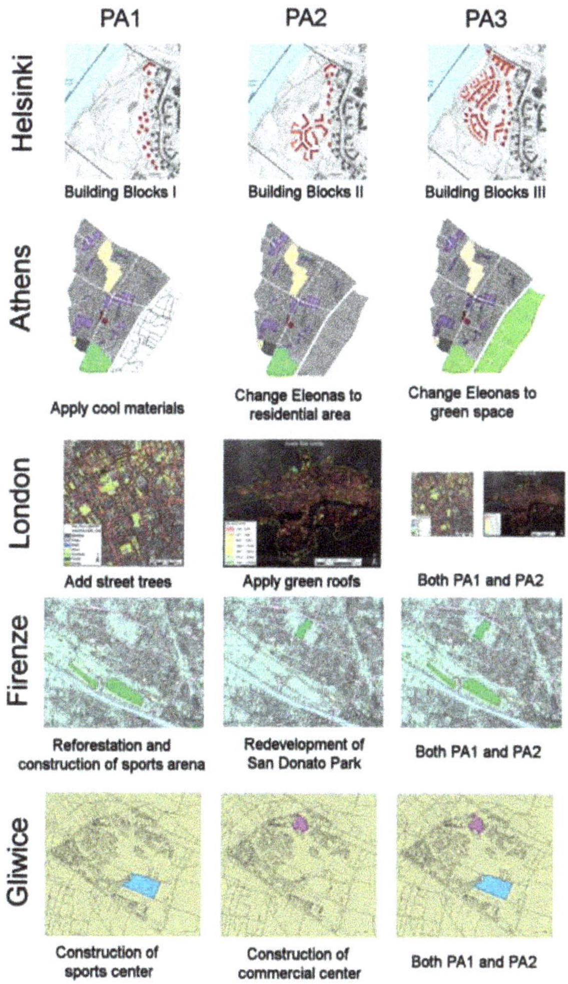

Figure 1. (**a**) Planning alternatives for the five case studies: (**a**) Helsinki, (**b**) Athens, (**c**) London, (**d**) Firenze and (**e**) Gliwice; adapted from [1].

Three PAs for each case study (PA1, PA2 and PA3) were proposed in the framework of the CoP and are shown in Figure 1. The proposed PA for Helsinki referred to replacing a green area (Meri-Rastila) with different kinds of building types and densities (Figure 1a). For Athens, PA1 referred to applying cool materials on all buildings and roads in a central area (the Egaleo municipality), while the other two alternatives concerned the transformation of a brownfield (Eleonas) to either residential area (PA2), or green space (PA3) (Figure 1b). For London, the PA1 was related to the addition of street trees, PA2 to the installation of green roofs of varying slopes, and PA3 to the implementation of both PA1 and PA3 in the central activity zone (Figure 1c). For Firenze, PA1 concerned the complete reforestation of a green area and a sport arena (in the Cascine Park), PA2 the redevelopment of a former industrial area (in San Donato Park) and PA3 the implementation of both PA1 and PA2 (Figure 1d). Finally, for Gliwice, PA1 regarded the construction of a sports centre (Gliwice Central Arena), PA2 the construction of a commercial centre and PA3 again referred to the implementation of both PA1 and PA2 (Figure 1e).

Although all five case studies represent distinct urban structures and planning issues, and the proposed planning alternatives significantly diverge amongst cities, sustainability is at the core of urban planning in all the cities. The planning objectives contemplated in the relevant development plans and the policy priorities identified during the CoP (e.g., addressing greenhouse gas emissions and climate change by enhancing the urban fabric and the energy-efficiency in buildings; reducing traffic-related emissions through improvements in public transport; or/and increasing urban green spaces to address air quality and heat island effect and the associated health implications) illustrate a common line of improvement and promotion of sustainability principles at a European level [12].

2.2. Physical-Flow Numerical Models

Physical-flow numerical models (referred simply as models from now on) were used to simulate the exchange of energy, carbon and pollutants fluxes in urban areas. A large variety of such models exists, from those using simple approximations to very complex ones and the selection and application of specific models depends on the computer power availability, the available data and the objectives of the study. Different types of models from mesoscale air quality models to urban canopy models were implemented, using the cascade modeling technique from large to local scale. Mesoscale meteorological models such as WRF (Weather Research & Forecasting Model) were used to simulate the atmospheric flow in a 3D cube with spatial resolution between 0.2–100 km within domains from 10 km and 50 km to thousands of km [14]. WRF models produced detailed information of all meteorological variables and fluxes involved in the atmospheric flow. Meteorological and atmospheric condition information estimated by WRF provided inputs to chemical transport models such as CAMx (Comprehensive Air Quality Model with Extensions) [15], CHIMERE (Chemistry-transport model) [16] and CMAQ (Community Multiscale Air Quality) [17], which were used to estimate the atmospheric content of carbon and pollutants.

In addition to the application of complex and demanding models, urban canopy models were also used in BRIDGE. These models use less complex approximations, are less demanding in terms of computations and easier to use in decision support tools. Urban canopy models of different turbulence schemes such as LUMPS (Local-scale Urban Meteorological Parameterisation Scheme) [18] and SURFEX (Town Energy Balance—SURFace-atmosphere EXchange Module) [19] were applied for energy, carbon and pollutants fluxes approximation. The ACASA model (Advanced Canopy-Atmosphere Soil Algorithm) [20] was simulated surface-atmosphere interactions and the distribution of trace gases. The second generation Gaussian plume model URBAIR model (Urban Air Quality) [21] was used to evaluate air quality and dispersion patterns, since it is appropriate for distances up to about 10 km from the source. Finally, the SIMGRO model (SIMulation of GROundwater and surface water levels) [22] was used to produce detailed information on all the hydrological processes present in an urban environment. The different model simulations performed for the BRIDGE case studies, generated outputs like the CO_2 flux, CH_4 emissions, PM10, O_3, SO_2 concentrations, sensible and latent heat flux *etc.* A set of aggregation algorithms were developed and applied to estimate absolute indicator values from the model's simulation results, depending on the nature of each indicator.

2.3. Data Collection and Organization

To support both the modeling procedure and the PA evaluation process, data associated to both environmental and socio-economic parameters were collected for the five case studies [23]. Energy, water, carbon and pollutants fluxes were systematically monitored, using both in-situ and remote sensing techniques. Surface fluxes and latent heat, momentum, net urban carbon exchange and aerosols fluxes were measured on a continuous basis by eddy covariance/large aperture scintillometry and by repeated seasonal campaigns of research aircrafts. Both existing and new networks of bio-sensors were used as bio-accumulators assessing air quality. Collected data were used to validate the simulations generated by the models. Satellite, airborne and ground-based remote sensing methods were applied to derive spatial distributions of various geo-physical parameters such as surface albedo and emissivity. These were necessary as input to the models. Additional data were gathered referring to traffic flows, soil and vegetation status and activity, soil moisture, energy exchange and air quality of buildings, surface temperature, indoor pollutants and Particulate Matter (PM) concentrations. The land cover/use dynamics were also estimated and mapped. Socio-economic data concerning parameters like space, mobility, heat and water demand, land-use types, coverage and intensity, building volumes, population density, unemployment rate and education level were also gathered. All the above-mentioned data, along with the model simulation results were organised in databases for the case studies and are part of the BRIDGE DSS.

3. Algorithms Description and Implementation

3.1. Algorithms Description

The overall goal of the BRIDGE DSS was to assist urban planners to better explore the decisions at hand and to analyse the trade-offs between the competing criteria to consider urban metabolism in planning decisions. Therefore, the adapted methodology is based on sustainability objectives defined for specific aspects of urban metabolism. These objectives reflect components of sustainability, namely environmental, social and economic. A set of criteria were associated to the objectives providing a link between the objectives and the indicators. Indicators demonstrate the level of achievement of each criterion in a quantified manner. They intend to reflect the multidimensional nature of the urban metabolism, while making them easily understood by non-experts. The set of indicators developed in the framework of CoPs in BRIDGE project is listed in Table 1 [12]. These indicators are grouped into three main categories that depict the sustainability dimensions of urban metabolism and they are further sub-grouped in a hierarchical mode. This hierarchical grouping allows inclusion of a group or subgroup of indicators at will in the analysis, and accordingly, to adjust their relative importance, providing flexibility concerning the level of detail or aspects considered in the analysis.

The DSS allows the user to decide upon the sustainability objectives according to user needs or preferences and to determine their relative importance. The evaluation of the performance of each PA is done in accordance with the predefined qualitative values (*i.e.*, relative importance) for each criterion to measure the performance of individual alternatives. The developed MCE is then used to measure the intensity of the interactions among the different elements in the system. Different algorithms are used to aggregate the models simulation results, at both geographic (*i.e.*, intervention area and surroundings) and temporal (e.g., annual) levels, resulting in absolute indicator values. Thresholds are used in some cases to establish the status of the indicators' performance, which refer to the maximum values permitted according to legislation. For example the upper limit for PM10 is 50 $\mu g/m^3$, not to be exceeded more than 35 times a calendar year, according to European Directive 2008/50/EC [12]. Model simulations would provide the PM10 concentration per area for a given time interval. The aggregation algorithm counts the times of 50 $\mu g/m^3$ threshold exceedance throughout the calendar year and creates an "indicator map of exceedances". The map indicates the areas where the threshold was exceeded more than 35 times in a year. The computed indicator values are then used in the MCE algorithm to retrieve appraisal scores, as explained below.

Table 1. Set of indicators that are used in the SustainaBle uRban planning Decision support accountinG for urban mEtabolism (BRIDGE) Decision Support System (DSS) Prototype; adapted after [12].

Environmental Indicators	**Social Indicators**
Energy	*Land Use*
Energy consumption by cooling/heating (kWh/m^2); Anthropogenic heat (W/m^2); Bowen ratio (unitless); Percentage of energy from renewable sources (%)	New urbanized areas (m^2); Brownfields re-used (m^2); Density of development (built m^2/total m^2)
Thermal Comfort	*Mobility/Accessibility*
Thermal Comfort Index (Cooling Power); Air Temperature (°C); Number of days above threshold (days/total period)	Quality of pedestrian (qualitative); Length of cycle-ways provided (m); Length of new roads provided (m); Use of public transport (% of total population); Number of inhabitants with access to public transport (inhabitants within 500m of public transport)
Water	*Social Inclusion*
Water consumption (mm^3); Evapotranspiration (mm^3/m^2); Infiltration (mm^3/m^2); Surface run-off (mm^3/m^2); Potential flood risk (peak mm^3/m^2 discharges)	Number of inhabitants with access to services (inhabitants/m^2); Number of inhabitants with access to social housing (inhabitants/m^2)
Air Quality	*Human well-being*
NOx, PM10, PM2.5, O_3, CO, SO_2 concentrations (μg/m^3); NOx, PM10, O_3, SO_2 exceedances (threshold exceeded or not); NOx, PM10, O_3, SO_2 Potential Population Exposure	Number of inhabitants affected by flash flooding (No. of inhabitants); Number of inhabitants affected by heat waves (No. of inhabitants)
Greenhouse Gases	*Economic Indicators*
CO_2, CH_4 Emissions (tonnes)	Cost of proposed development(€ or €/m_2); Effects on local economy (No. of new jobs created); Effects on local economy (€ or €/m_2)

For overcoming problems like simulation uncertainties and the lack of precise socio-economic information regarding the PAs, the developed MCE method compares between alternatives rather estimating absolute appraisal scores. One of the possible PAs is considered as the baseline or reference, which is set by the user—so the other alternatives are compared to that of the reference alternative could be the actual situation (business as usual scenario, where no intervention is done in terms of planning actions), or one of the proposed PAs. Having the indicator values estimated for all PAs, indicator scores are then calculated. Indicator scores (Si) depict their performance compared to the reference situation, and they are calculated taking into account the changes to the indicator value introduced by the PA being analysed (I_{PAi}), compared to the reference situation adopted (I_R). If x stands for the PA in question and R stands for the reference situation, the ith indicator's score is derived using Six = I_{ix}/I_{iR}. The reverse form of the formula ($S_{ix} = I_{iR}/I_{ix}$) is used for indicators having a "positive effect" in sustainability, like for example the "area of green spaces" indicator, which has a positive effect on sustainability if its value is increased).

The user's preferences concerning indicators are defined by weights. Weight is a numeric amount assigned to an indicator, indicating its relative importance in relation to other indicators in the decision situation. The higher the weight, the more important a given indicator is considered. Established weights are normalised, so that their sum in all indicator groups equals one. Weights in BRIDGE DSS are assigned in pairs, defining the relative importance of indicators and indicator groups [24]. Qualitative terms shown in Table 2 are used to assess the relative intensity of preference between two elements. Taking pairs of indicators and asking two questions (e.g., which indicator is more important, C_i or C_j, and how much more), square reciprocal matrices are generated. For a matrix A the element a_{ij} represents the weight given by the user for indicator i compared to j [25]. The diagonal of matrix A is 1 (indicating that each element is the same importance with itself) and $a_{ji} = 1/a_{ij}$. As noted above, the weights w_i are normalised to ensure that the sum of weight equals one, by estimating the eigen vectors v of the maximum eigen value λ of matrix A and then normalising by

$$w_i = \frac{v_i}{\sum_{i=1}^{n} v_j} \tag{1}$$

The scores of indicator groups are calculated using a Cobb-Douglas function combining indicators' scores and their weights, which are defined by the user according to their relative importance. Each group's score (S_n) is given by

$$S_n = S_1^{w_1} \cdot S_2^{w_2} \cdot S_3^{w_3} \cdot \ldots \cdot S_k^{w_k} \tag{2}$$

where S_i is the ith indicator score, w_i is the ith indicator's weight and k is the number of indicators included in the group. One of the advantages of this type of functions is that it enables dealing either with relative or absolute values of indicators, and it is sensitive to the changes in scores (*i.e.*, indicator performance) as well as in weights (*i.e.*, indicator importance/significance). The overall score of each PA is calculated in the same way as the groups' scores, using a function of groups' scores and weights.

In summary, it is based on a value function using scores and weights: the first one translates the relative performance of the PA under evaluation when compared to a reference situation; while the second translates the relative importance of indicators (or indicator groups) ascribed by the user. The BRIDGE DSS results are more than just one appraisal score for each PA. The final appraisal score is just a representation of all the collected information. No confidence measures regarding estimated scores are necessary to compute, because the methodology is based on relative rather than absolute values. The user can assess individual indicator scores, as well as scores of indicator groups, to examine their performance related to the reference situation. Absolute indicators values can also be visualised, in the form of maps to reveal their geographical and temporal variability.

Table 2. Verbal terms shown in the table below are used to assess the intensity of preference between two elements (adapted after [24]).

Value	Verbal Term
0	same importance
1	slightly more important
2	weakly more important
3	weakly to moderate more important
4	moderately more important
5	moderately to strongly more important
6	strongly more important
7	greatly more important
8	absolutely more important

3.2. Implementation

The BRIDGE DSS prototype was developed based on the above described algorithms using Visual Basic programming language and exploiting the available data that were organized in SQL databases. A Graphical User Interface (GUI) was developed as an add-on for ESRI ArcGIS [26] to fully exploit its capabilities on handling geographical data. The role of the GUI is to facilitate the user in selecting the indicators to be used in the analysis as well as to define their relative importance; it is also used for the presentation of the analysis results. An ArcGIS toolbar was created including all menus and options necessary for the implementation of PA evaluations.

Some of the models that were used in the BRIDGE DSS were very demanding in terms of computation power (e.g., WRF model) and this made it impossible to integrate them directly into the developed prototype. Those models can only run in powerful cluster computers; this impeded their integration in the DSS prototype. For the BRIDGE project, model simulation were performed for all the case studies (and their PAs) for a period of one year (hourly time step), and the simulation results were organised in databases which were then integrated into the DSS so they could be accessed for evaluation. BRIDGE databases are structured in order to make them easily accessible by the end-user, by means of the developed GUI, as well as to make it possible to visualise spatio-temporal information in different ways. For the models that were directly integrated in the system (e.g., URBAIR), the user can interfere, parameterise and run at will.

The nature of PA, derived from the CoPs (Figure 1), was different for each case study and thus different implementations were used in the numerical models. The PA for each case study needed to be interpreted so as to fit the specifications of both the models and the system. For Helsinki, the PA were implemented by changing the land uses from green areas to urban areas in the location of new buildings. For Athens, PA1 was implemented by setting the albedo value of buildings to 85% and of roads to 45% and PA2 and PA3 were implemented by changing the land uses of Eleonas from industrial/commercial urban areas to high density urban and grassland respectively. To implement the PA for London, since seven land cover types (buildings, roads, water, grass, conifer, deciduous, shrubs) were considered, a correspondence between the above land cover information and land use types used by the models was established. To take into account the new street trees and green roofs, modifications of the following parameters were made: roof width, road width, and fraction of urban by type. For Firenze, PA1 was implemented by changing land use from sport hall and grassland to deciduous trees, PA2 was implemented by changing land use from industrial to grassland and PA3 was the implementation of both PA1 and PA2 at the same time. For Gliwice, PA1 was implemented by changing the land use from commercial to high density urban area, PA2 was implemented by changing the land use from low density urban to high density urban area and PA3 was the implementation of both PA1 and PA2.

3.2.1. MCE Procedure

In this section, the procedure of how to run an evaluation using the BRIDGE DSS prototype is described in order to present the system's functionalities. The procedure is graphically represented in Figure 2. The first step in the evaluation procedure is the selection of a specific case study to proceed with. Description of the PA is provided by the system along with the respective spatial data that feature each PA.

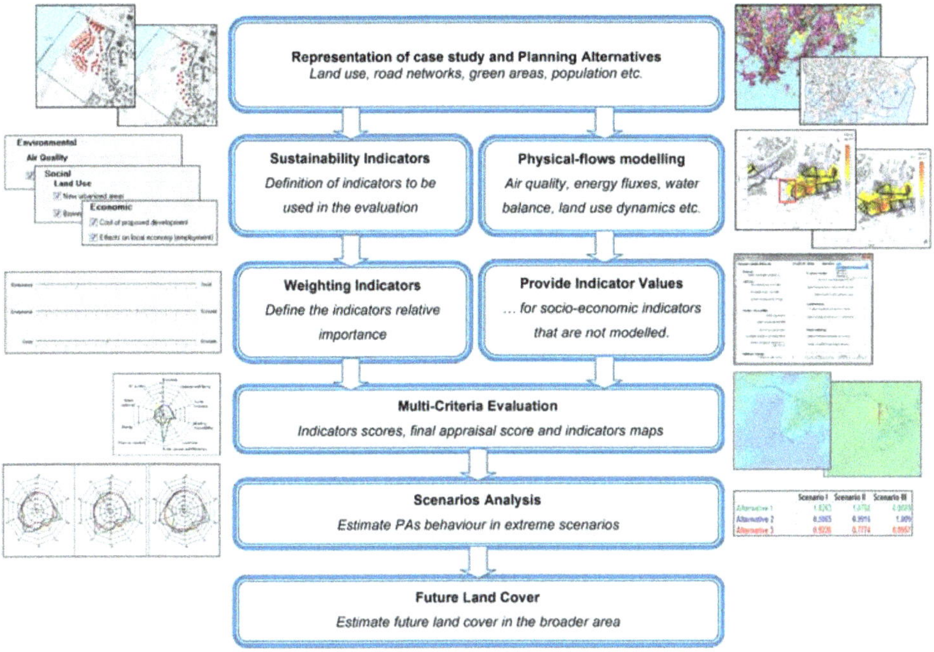

Figure 2. Chart depicting the Planning Alternatives (PA) evaluation procedure used in the BRIDGE DSS prototype.

Subsequently, the user needs to define the indicators to be included in the evaluation. The list of the available hierarchy of indicators is presented in a structured way, as shown in Figure 3, and the user is asked to select among them. The hierarchical organisation of indicators in three main groups related to environmental, social and economic dimensions of sustainability can be observed in this structure. All levels of the hierarchy can be accessed through this window. The relative importance of indicators (and indicator groups) can be defined through this menu using scale bars on the basis of the qualitative terms of Table 2. An example of such a scale bar is shown in Figure 4. If no user preference is set, the indicators are considered of equal importance in the evaluation process. This means, for example, that the indicators in question are all equally important and therefore given the same weight. Tables of normalised weights are computed using the previously described algorithm.

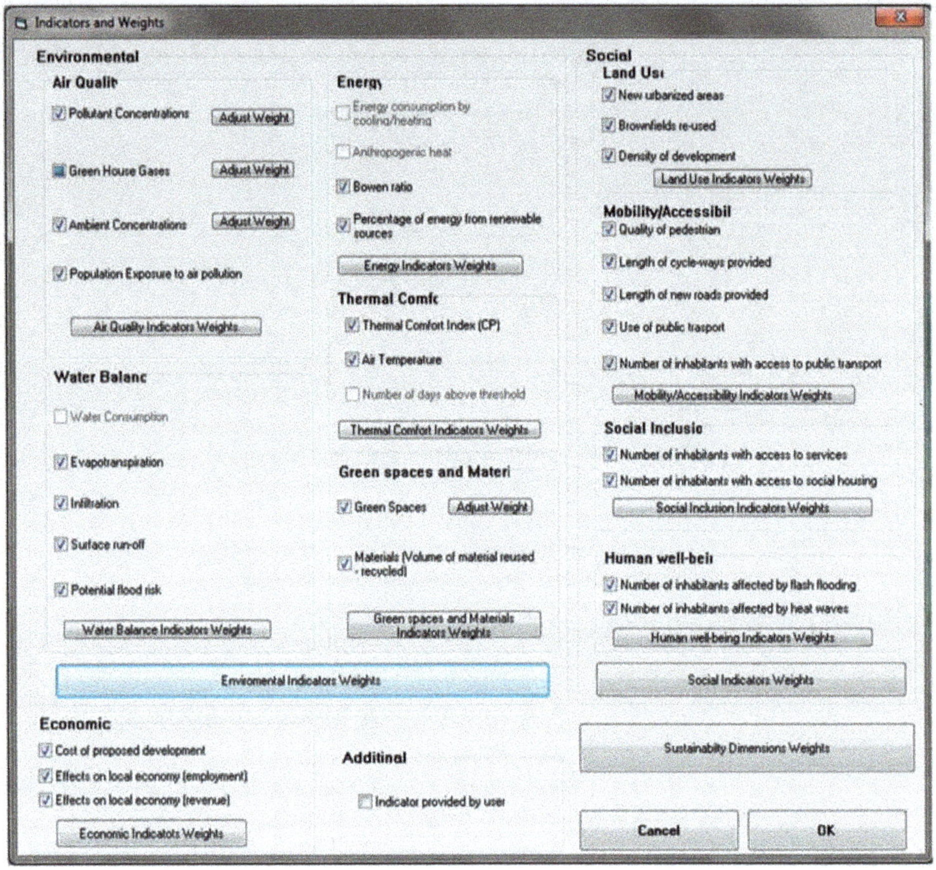

Figure 3. Window for selecting indicators to be used in the evaluation analysis. The indicators hierarchy, shown also in Table 1, is depicted in this selection window.

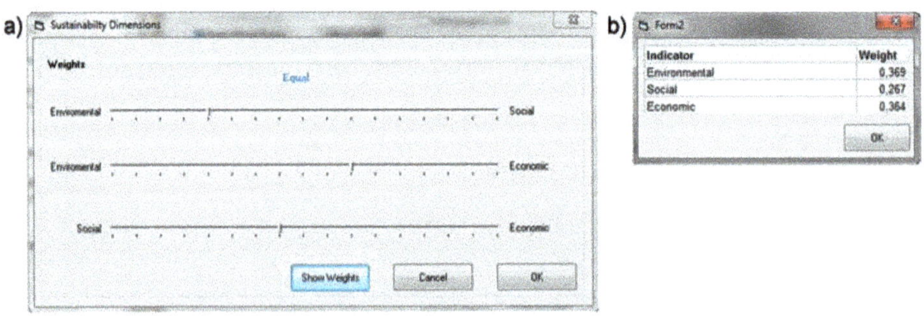

Figure 4. (**a**) Example of scale bars that are used to define the indicators' (or indicator groups') relative importance and (**b**) the normalised weights that are computed based on the user's preferences.

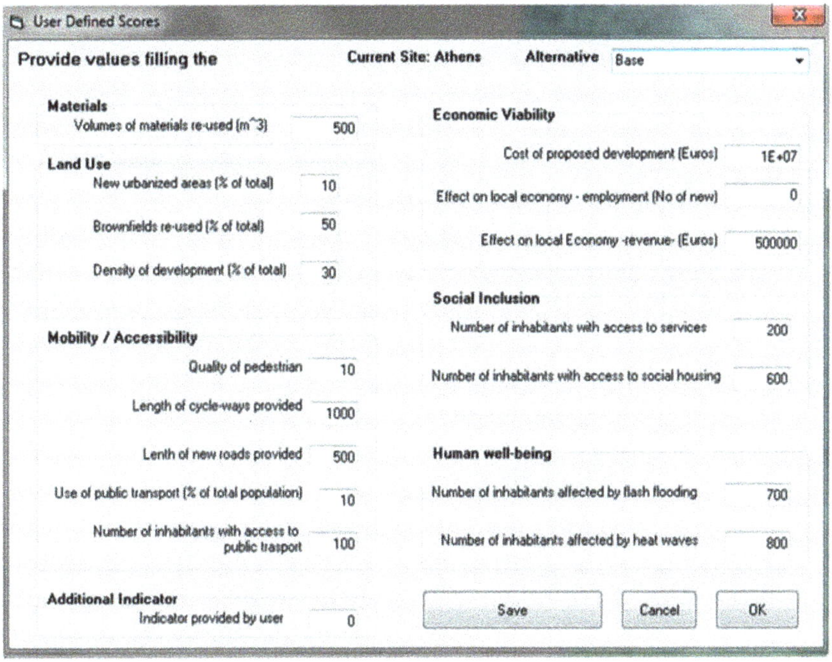

Figure 5. Form that is used for insertion of indicator values that are not modeled in the BRIDGE DSS (*i.e.*, socio-economic indicators).

The next step is the calculation of selected indicator values for each PA. These values are calculated in different ways. Environmental indicators arising from physical flow simulations are calculated by numerical models, while socio-economic indicators reflecting objective values (e.g., number of houses constructed, number of jobs created, *etc.*) are given as data attached to planning alternatives and thus have to be directly inputted by the user. Socio-economic indicator values are not estimated using models in the BRIDGE DSS, as the integration of socio-economic models was out of scope of the BRIDGE project. Therefore, the user is required to provide these values for each PA, using the form shown in Figure 5. There is a choice of providing absolute or relative values. Relative values reflect the relative performance of indicators between the PA in question and the reference situation. Relative values are useful when the absolute values of indicators cannot be defined (e.g., regarding employment indicator, one may not know how many job positions will a PA induce, but might estimate that this be double to the reference situation). Instructions on how to use relative values instead of absolute ones are given in the BRIDGE DSS Prototype User's Guide [27]. Once all the above-described information is made available to the system, the user can run the MCE procedure. A set of different outputs are produced to estimate the PA performance in terms of indicators. The representation of the results is described in the following section.

3.2.2. Evaluation Outputs

The evaluation procedure calculates individual indicator scores, as well as a final appraisal score for each PA. Spider diagrams are used to graphically represent the individual indicator scores of all PA, illustrated in different colors to facilitate comparison. The reference situation is always presented as a circle in the spider diagram, having all appraisal scores equal to one. A final appraisal score for each PA is calculated as a combination of the above scores and weights, as described in the previous section. A final appraisal score higher than one indicates better performance of the PA in question, compared to the reference. The same applies to the underlying indicators scores. Figure 6 shows an example of the evaluation results form. The final appraisal scores for the PA are shown on the top of the window, while the individual appraisal scores are shown on the right. This information is not considered sufficient for taking a decision regarding the best PA, thus the user can also visualise individual indicator maps to have a clear picture of the spatial and temporal distribution of specific indicator values.

Indicator maps are spatio-temporal representations of the indicator values computed in each grid cell using the model simulation results. The desired time period of indicator values to be displayed can be adjusted by the user. The user selects the indicator of interest from a list of available indicators, and the desired time period. Different combinations of time and days can be used to produce different kind of maps (i.e. seasonal maps, daily maps, yearly maps, *etc.*). The type of performed calculations can also be assessed through time and the user can choose between average, minimum, maximum and summary values to be drawn in the relevant indicator map, which is considered necessary due to the differences in the nature of indicators (e.g., EU air quality Directive's reference to annual *vs.* daily values). The option to draw maps of differences between the PA and the reference situation is also provided, which aims at highlighting the differences between PA enabling their comparison.

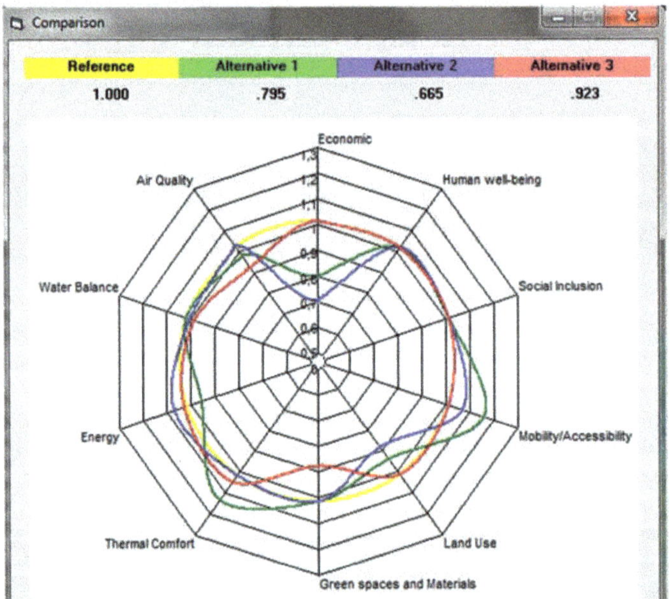

	Base	Alter 1	Alter 2	Alter 3
Enviromental	**1,00**	**1,00**	**1,02**	**0,92**
Air Quality	1,00	0,96	1,01	0,90
Water Balance	1,00	1,00	0,99	0,97
Energy	1,00	0,94	1,05	1,01
Thermal Comfort	1,00	1,12	1,00	1,04
Green spaces and Materials	1,00	1,00	1,00	0,86
Social	**1,00**	**1,02**	**0,97**	**1,00**
Land Use	1,00	0,92	0,87	1,00
Mobility/Accessibility	1,00	1,15	1,06	1,00
Social Inclusion	1,00	1,00	1,00	1,00
Human well-being	1,00	1,00	1,00	1,00
Economic	**1,00**	**0,77**	**0,67**	**1,00**

Figure 6. Example of the Multi-Criteria Evaluation (MCE) result form, depicting the final appraisal score of each PA, the individual computed scores and the spider diagram that graphically depicts the different scores for different PAs.

3.2.3. Further Analysis of MCE Outputs

While the discussion so far has emphasized the capabilities and functionalities of the DSS to evaluate the impact of local-scale PA, these alternatives do not live in a vacuum, or better, they should not be assumed to exist against astatic backdrop. A fair evaluation of projects thus must take into account the evolution, the possible trajectories and future scenarios of the wider urban context, and indeed possibly the whole urban area. That is why BRIDGE DSS features two additional future-scenarios analysis against which (and in the context of which) the PA are evaluated.

The first scenario analysis tool is based on a land-use cellular automata (CA) model, integrated in the BRIDGE DSS prototype. The purpose of the model is to simulate the city-wide land use dynamics, and thus produce scenarios (under different parametric assumptions about the city growth and demand for land uses) of future spatial distribution of city-wide land uses. The model is based on the conventional model of constrained cellular automata, which allocates the demand for land uses to each cell by accounting for the interaction between different land uses and physical, environmental and institutional factors (e.g., zoning, planning regulations) related each cell [28]. Thus, the CA model allows to account for broader effects of planning decisions, in terms of a spatial distribution of land uses over a larger geographical area.

The second wide-scale scenario-analysis tool incorporated in BRIDGE DSS allows the user to evaluate priorities in response to different extreme future scenarios. Three extreme future strategic scenarios were considered in the BRIDGE project, as shown in Table 3 [29]. In Scenario I, climate change is not a threat, there are sufficient energy resources and the cities' success depends on their ability to attract qualified people and firms. Critical issues for sustainability in this case may refer to the efficient use of energy, the transition to renewable resources and cleaner use of fossil energy, as well as attempts for a balanced society and a highly productive economy. In Scenario II, climate change is a burning issue. The cities' attractiveness depends on their capacity to face climate change, while energy is not that big a problem and the economy is growing. To face the climate change threat there is urgent need to absorb greenhouse gases and to reduce the emissions. In Scenario III, the lack of energy is freezing the economy, thus cities' success depends on low costs and energy shortage. Non-renewable sources are reaching the end and the use of renewable sources is insufficient. Reduced mobility leads to urban concentration, resources are diverted for fast increase of renewable energy sources and social inequality is increased.

Table 3. Extreme future scenarios, as these are defined in BRIDGE.

Scenarios	Climate Change	Energy/Technological Development	Economy
Scenario I: BRIDGE in Wonderland	+	+	+
Scenario II: Climate change is a burning issue	-	+	+
Scenario III: Lack of energy in freezing the economy	+	-	-

Model projection for the year 2030 allowed assessing the performance of the environmental indicators for the reference situation and for all PA considered under each of the considered future scenarios. For these projections, assumptions on environmental conditions were made, based on the Intergovernmental Panel on Climate Change (IPCC) scenarios A2, A1F1 and B1 [30], which address the characteristics of the strategic scenarios developed in the BRIDGE project. More specifically, the IPCC scenario A2 (medium emission scenario) was used in the BRIDGE models to simulate the fluxes in case of the Scenario I, the IPCC scenario A1F1 (worst emission scenario) was used in case of Scenario II, whereas the IPCC scenario B1 (best emission scenario) in case of Scenario III [29,31].

For the evaluation of PA against the strategic scenarios, the environmental indicators scores were calculated based on model simulation results for the year 2030. As noted above, model simulations were performed using the IPCC scenarios, whereas the estimations of socio-economic indicators values were defined by the CoPs. In the absence of environmental, energy and economic constraints, the focus of urban policy would be the prevention of other types of problems, and the increase of quality of life in general. The evaluation of the PAs in each case study against strategic scenarios is done by adjusting the indicators relative importance considering the extreme situation outlined by the each scenario. Windows identical to those used for adjusting indicators relative importance for the present conditions (Figure 4) can be used to adjust the indicators' weights. The MCE process, adjusted for the extreme conditions of scenarios and the new user preferences, runs again and results are produced for each strategic scenario as shown in Figure 7.

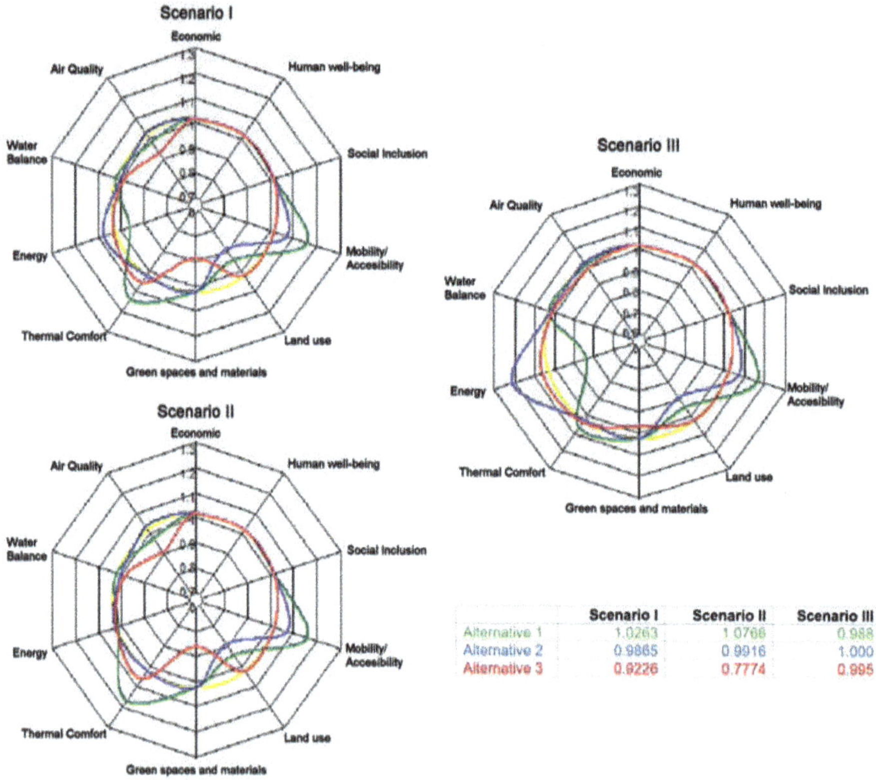

	Scenario I	Scenario II	Scenario III
Alternative 1	1.0263	1.0766	0.988
Alternative 2	0.9865	0.9916	1.000
Alternative 3	0.9226	0.7774	0.995

Figure 7. Example of the scenario analysis result form.

4. Results and Discussion

The BRIDGE DSS prototype was developed with the potential to evaluate PA and determine the alternative that better fits the goal of changing the metabolism of urban systems towards sustainability. Examples of this DSS results are presented and discussed in this section, aiming at highlighting the system's main capabilities and advantages, as well as the limitations and the induced need for further research.

The BRIDGE DSS assesses the performance of a PA in terms of sustainability according to the user preferences. As an example, Figure 8a shows the estimated MCE results for the case study of Athens, with all the indicators included in the analysis having equal importance. In this case PA3 was found to have the highest final appraisal score (0.923), compared to the reference alternative, followed by PA1 (0.795) and PA2 (0.665). However, if the user changes the relative importance of some indicators the results change. For instance, given that Athens' citizens face thermal discomfort issues quite often during summer, a possible user might consider thermal comfort to be absolutely more important than water and energy balance. In such cases, as it can be noted in Figure 8a, PA1 gains a higher appraisal score in terms of thermal comfort (1.12) than PA2 (1.00) and PA3 (1.04). By changing the user's preferences in this way, the individual and final appraisal scores also change (Figure 8b), but PA3 remains with the highest final score (0.949), followed by PA1 (0.838) and PA2 (0.661). This example demonstrates the performance of the MCE algorithm used in the BRIDGE DSS.

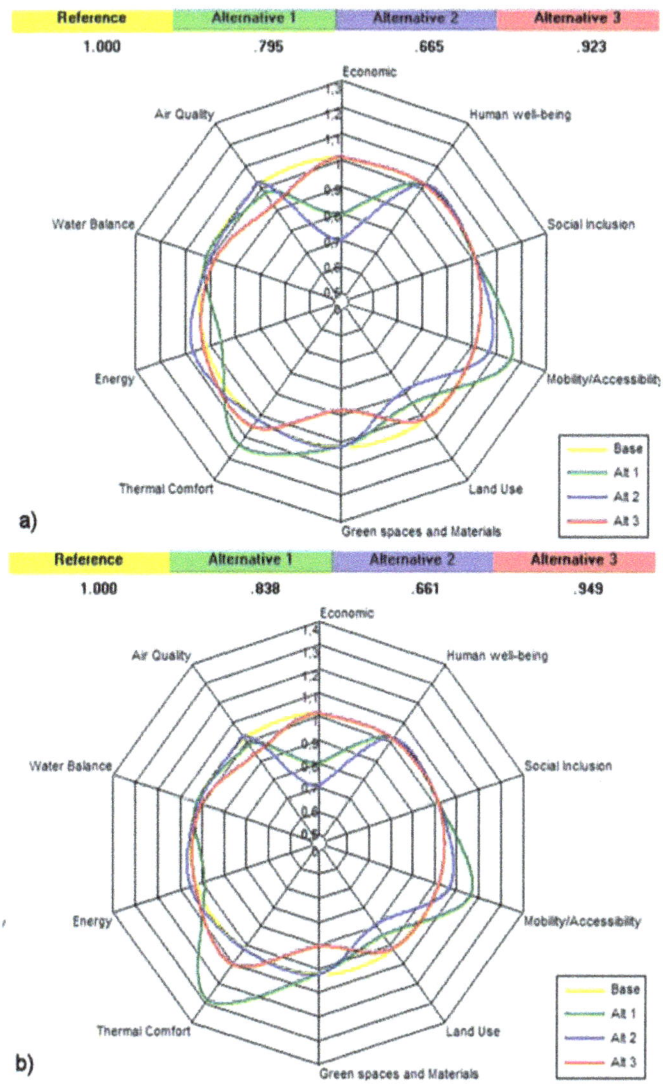

Figure 8. MCE results for the PAs of Athens as estimated by the BRIDGE DSS (**a**) by applying equal importance to indicators and indicator groups, (**b**) by changing the importance of thermal comfort as absolutely more important compared to energy and water balance.

Additional tools are provided by the BRIDGE DSS in order to assess the behavior of the PA with regard to the availability of indicators in space and time. The desired indicator along with the time period can be selected by the user according to a given interest for more detailed examination. Figure 9 shows the spatial distribution of air temperature in the case study of Athens for the reference situation and for each PA (top of the figure) for a summer period between 10:00 and 13:00. The capability of visualising differences between the reference situation and the PAs is also provided by the system. This capability enables automatic identification of spatial differences as well as quantification of those differences. Figure 9 (bottom) shows differences of air temperature between the PAs and the reference situation for the same time period.

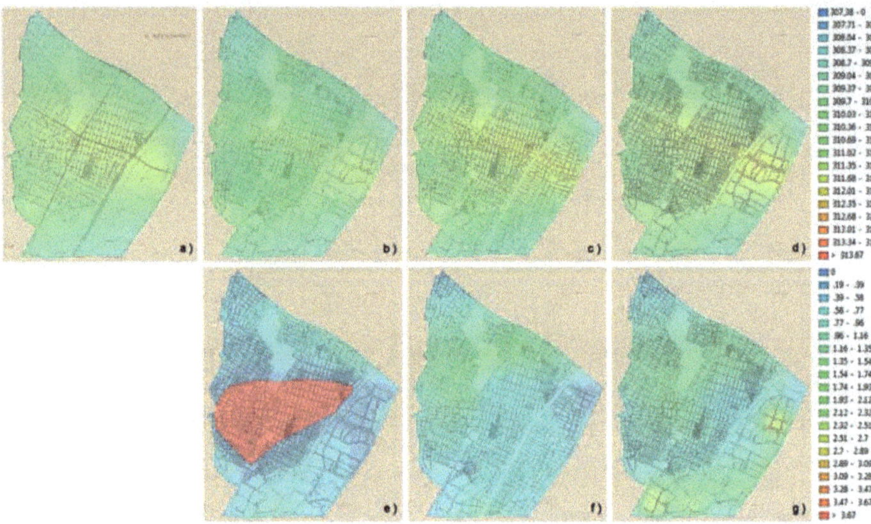

Figure 9. Air Temperature indicator map for the case study of Athens: Spatial distribution of air temperature maximum values for the reference situation (**a**) and for each PA (**b–d**) for summer period between 10:00 and 13:00. The differences between the reference situation and each PA are also illustrated (**e–g**).

Legislation exists to ensure good air quality in urban areas (*i.e.*, the European Directive 2008/50/EC). The established restrictions refer to concentration limits in the area, specifying also the permitted time limit. Spatial and temporal patters of indicators referring to exceedances of established limits are also estimated by the BRIDGE DSS. Such indicator values and their temporal and spatial distribution are highly useful for urban planning actions. Figure 10 shows an example of exceedances in the case study of Helsinki, for the reference situation (a) and the PAs (b–d). Red cells represent areas where the threshold has been exceeded for atmospheric Particulate Matter of diameter of 10 μm or less (PM10).

Figure 10. Example of indicator map of exceedances. Number of exceedances of PM10 threshold in the area of Helsinki for (**a**) the reference situation, (**b**) PA1, (**c**) PA2 and (**d**) PA3. Red cells represent the areas where the established EU threshold is exceeded, at least one time, during a time period of one year.

In addition, the BRIDGE DSS allows the evaluation of PAs in three strategic scenarios (Table 3). The evaluation of PAs in extreme future conditions aims at characterising the PAs as either robust, unstable, or unclear. Robust alternatives are those that present the best score in all situations, unstable results regarding the strategic scenarios analysis are considered those that require deepen knowledge about future evolution before a decision is taken, while an unclear evaluation is characterised as one where the appraisal scores are very similar, indicating the need to use more and better information For example, Table 4 shows the results of strategic scenarios analysis of the case studies of Gliwice and Helsinki. It is clear that for Gliwice, the PA3 was the robust option: whatever the context where the political decision is taken, the best alternative was the construction of both the sports centre and centre of new technologies. However, concerning Helsinki, the PA presented similar performances in the three strategic scenarios. When there were no economic or environmental constraints in the scenarios, the results for all scenarios were almost equal: there were no clear gains of increasing the constructed area. For strategic scenarios II and III, the results pointed to the project with more inhabitants and built up area (PA3), but with a marginal advantage over PA2.

Another tool, embedded in the BRIDGE DSS is the cellular automata module. This module allows the user to estimate future land uses in the broader area, taking into account the foreseen planning action. Figure 11 shows an example of change in the land uses in the next two decades in Athens with and without the implementation of the PAs shown in Figure 1. The cellular automata is useful for assessing the effect of a planning intervention on the land uses of the surrounding areas for a longer time period. In the case of Athens for example, PA1 does not have any effect on land uses since it only refers to changes in materials used for the same structures. On the other hand PA2 and PA3 do result on some land use changes, but only close to the area of interventions.

Table 4. Strategic Scenarios evaluation results for Gliwice and Helsinki.

Case Study	Scenario	PA1	PA2	PA3
Gliwice	Scenario I	1.00	1.01	1.33
	Scenario II	1.00	1.02	1.05
	Scenario III	1.00	1.08	1.37
Helsinki	Scenario I	1.00	0.99	0.98
	Scenario II	1.00	1.10	1.12
	Scenario III	1.00	1.14	1.15

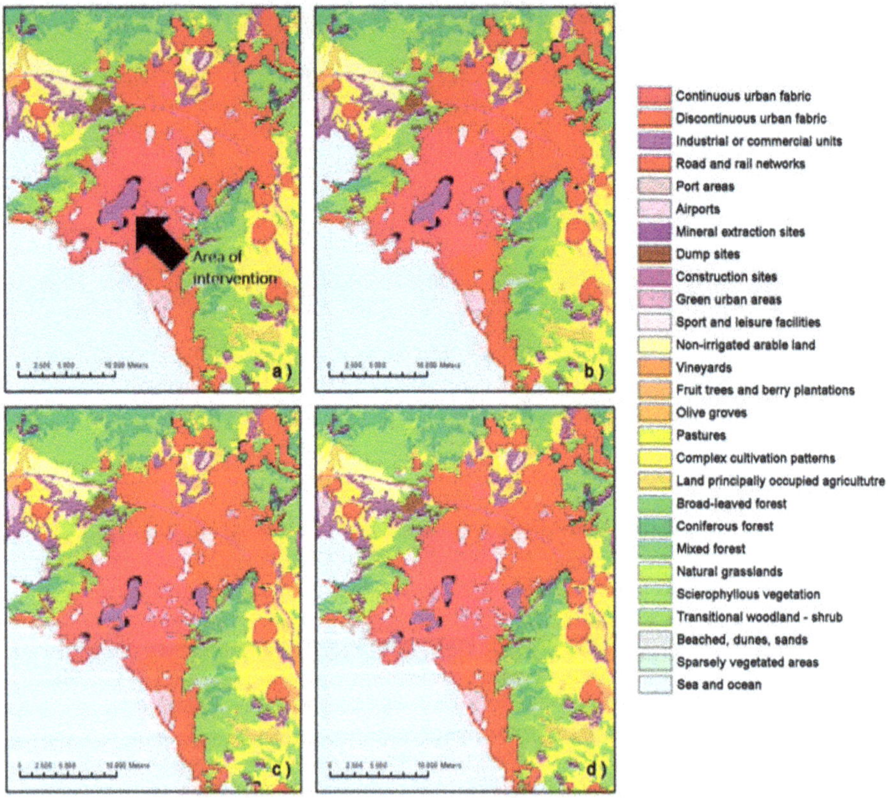

Figure 11. Estimated future land uses for the case study of Athens for (**a**) the reference situation, (**b**) PA1, (**c**) PA2 and (**d**) PA3.

The BRIDGE DSS also allows the user to parameterise and run some models. Those models, as mentioned earlier, are not highly demanding in terms of computation. The attempt of developing such a prototype illustrated the restrictions of implementing numerical physical-flow models in real-time applications. Mesoscale weather models used for simulating energy fluxes and pollutants are yet highly demanding in computer power and user skills, and thus their implementation in the DSS is difficult. A possible solution is to run those models off-line on powerful computer clusters for many possible scenarios and then include the simulation results in a database, as in the case of BRIDGE DSS.

Advances in numerical modeling, as well as in computer power, may allow more simulation models to be embedded into DSS in the future.

5. Conclusions

Several studies have addressed urban metabolism issues but few have integrated the development of methods for the combined analysis of fluxes between a city and its environment and the implementation of numerical tools for the assessment of planning alternatives, based on environmental and socio-economic indicators. The BRIDGE project integrated bio-physical observations with socio-economic data to evaluate PAs and thus addressed and jointly examined the three components of sustainability: environmental, social and economic.

The main outcome of BRIDGE was a spatial DSS. Therefore, the GIS technology was fully exploited to integrate all datasets, to analyse the various spatial entities, to prepare the inputs for the physical flow models and the decision making algorithms, to store the results and to visualise them. A MCE method was implemented in the GIS platform on which the PAs evaluation was based, providing also the option to explore the behavior of PA in extreme future scenarios. In addition to the MCE results, a cellular automata module was also integrated in the system, as a GIS application, allowing the user to estimate the impact of each PA on future land use arrangements in the broader geographical area.

The evaluation of the performance of each PA is based on the relative importance ascribed to each objective by the end-user, facilitating the integration of public concerns into decision-making. The BRIDGE DSS enables end-users to evaluate several urban PA based on previously defined sustainability objectives and indicators, by examining how each PA modifies specific urban metabolism components (energy, water, carbon and pollutants fluxes) towards sustainability. The developed decision methodology allows evaluation of a set of different alternatives, given different planning priorities. In this way, it enables quantified estimates of the effects of different combinations of planning objectives on different alternatives, promoting informed decisions.

The tool has therefore the potential to support sustainable urban planning by informing and enhancing planning processes through the detailed quantitative assessments of environmental aspects of a pair with socio-economic considerations. Although, further development of the tool is necessary until it reaches an operational state, it has great potential in supporting planning decisions accounting for urban metabolism, even in low-income cities where sustainability is critical to the provision of basic services. In this case, it can be used to support qualitative estimations of the impact of different planning alternatives, based on assumptions on the modifications they cause to energy, water, carbon and pollutants fluxes.

Acknowledgments: The research leading to these results has been funded partially by the European Community's Seventh Framework Programme (FP7/2007-2013) under grant agreement n° 211345 (BRIDGE Project) and partially by the General Secretariat for Research and Technology of Greece (Project code 1309: EO4SEB). The authors would like to thank the reviewers for their fruitful comments and insights.

Author Contributions: Zina Mitraka is the main author of this manuscript, she is responsible for the conceptual design and actively involved in all steps of its implementation. Emmanouil Diamantakis is the developer of the BRIDGE DSS prototype. Nektarios Chrysoulakis is the BRIDGE project coordinator and actively involved in the DSS implementation. Eduardo Anselmo Castro is responsible for the decision algorithms development as well as the scenario analysis methodology. Roberto San Jose is responsible for the physical numeric modeling processes carried out. Ainhoa Gonzalez is responsible for developing the indicators list in collaboration with the CoPs, developed the methodological framework and assisted in developing the decision algorithms. Ivan Blecic carried out the work related to the cellular automata.

Conflicts of Interest: The authors declare no conflict of interest.

References

1. Chrysoulakis, N.; Lopez, M.; San Josè, R.; Grimmond, C.S.B.; Jones, M.; Magliulo, E.; Klostermann, J.; Synnefa, A.; Mitraka, Z.; Castro, E.; *et al.* Sustainable Urban Metabolism as a link between bio-physical sciences and urban planning: The BRIDGE project. *Landsc. Urban Plan.* **2013**, *112*, 100–117. [CrossRef]

2. Newman, P.W. Sustainability and cities: Extending the metabolism model. *Landsc. Urban Plan.* **1999**, *44*, 219–226. [CrossRef]
3. Stone, B. Land Use as Climate Change Mitigation. *Environ. Sci. Technol.* **2009**, *43*, 9052–9056. [CrossRef] [PubMed]
4. Kennedy, C.; Cuddihy, J.; Engel-Yan, J. The changing metabolism of cities. *J. Ind. Ecol.* **2007**, *22*, 43–59. [CrossRef]
5. Eliasson, I. The use of climate knowledge in urban planning. *Landsc. Urban Plan.* **2000**, *48*, 31–44. [CrossRef]
6. Stevens, D.; Dragicevic, S.; Rothley, K. *iCity*: A GIS–CA modelling tool for urban planning and decision making. *Environ. Model. Softw.* **2007**, *22*, 761–773.
7. Carsjens, G.J.; Ligtenberg, A. A GIS-based support tool for sustainable spatial planning in metropolitan areas. *Landsc. Urban Plan.* **2007**, *80*, 72–83. [CrossRef]
8. Cowen, D. GIS *versus* CAD *versus* DBMS: What are the differences? *Photogramm. Eng. Rem. S.* **1988**, *54*, 1551–1555.
9. Jankowski, P. Integrating Geographical Information Systems and Multiple Criteria Decision-making Methods. *Int. J. Geogr. Inf. Syst.* **1995**, *9*, 251–273.
10. Batty, M. Planning Support Systems, Progress, Predictions and Speculations on the Shape of Things to Come. In *Planning Support Systems for Cities and Regions*; Brail, R.K., Ed.; Lincoln Institute of Land Policy: Cambridge, MA, USA, 2008.
11. Nyerges, T. *Regional and Urban GIS: A Decision Support Approach*; Guilford Publications, Inc.: New York, NY, USA, 2010.
12. González, A.; Donnelly, A.; Jones, M.; Chrysoulakis, N.; Lopes, M. A decision-support system for sustainable urban metabolism in Europe. *Environ. Impact Asses.* **2013**, *38*, 109–119. [CrossRef]
13. Gonzalez, A.; Donnelly, A.; Jones, M.; Klostermann, J.; Groot, A.; Breil, M. Community of Practice approach to developing urban sustainability indicators. *JEAPM* **2011**, *13*, 591–617.
14. Grell, G.A.; Emeis, E.; Stockwell, W.R.; Schoenemeyer, T.; Forkel, R.; Michalakes, J.; Knoche, R.; Seidl, W. Application of the multiscale, coupled MM5/chemistry model to the complex terrain of the VOTALP valley campaign. *Atmos. Environ.* **2000**, *34*, 1435–1453. [CrossRef]
15. Morris, R.E.; Yarwood, G.; Emery, C.; Koo, B.; Wilson, G.M. Development of the CAMx One-Atmosphere Air Quality Model to Treat Ozone, Particulate Matter, Visibility and Air Toxics and Application for State Implementation Plans. In Proceedings of the Air and Waste Management Association Air Quality Model Guidelines Conference, Mystic, CT, USA, 22–24 October 2003.
16. Vautard, R.; Honore, C.; Beekmann, M.; Rouil, L. Simulation of ozone during the August 2003 heat wave and emission control scenarios. *Atmos. Environ.* **2005**, *39*, 2957–2967. [CrossRef]
17. Byun, D.W.; Ching, J.K.S. (Eds.) *Science Algorithms of the EPA Models-3 Community Multiscale Air Quality (CMAQ) Modeling System*; U.S. Environmental Protection Agency, Office of Research and Development: Washington, DC, USA, 1999.
18. Loridan, T.; Grimmond, C.S.B.; Offerle, B.D.; Young, D.T.; Smith, T.E.L.; Järvi, L.; Lindberg, F. Local-scale urban meteorological parameterization scheme (LUMPS): Longwave radiation parameterization and seasonality related developments. *J. Appl. Meteorol.* **2012**, *51*, 219–241. [CrossRef]
19. Masson, V. Urban surface modelling and the meso-scale impact of cities. *Theor. Appl. Climatol.* **2006**, *84*, 35–45. [CrossRef]
20. Pyles, R.D.; Weare, B.C.; Paw, U.K.T.; Gustafson, W. Coupling between the University of California, Davis, Advanced Canopy-Atmosphere-Soil Algorithm (ACASA) and MM5: Preliminary Results for July 1998 for Western–North America. *J. Appl. Meteorol.* **2003**, *42*, 557–569. [CrossRef]
21. Borrego, C.; Martins, H.; Tchepel, O.; Salmim, L.; Monteiro, A.; Miranda, A.I. How urban structure can affect city sustainability from an air quality perspective. *Environ. Modell. Softw.* **2006**, *21*, 461–467. [CrossRef]
22. Walsum, P.E.V.; Groenendijk, P. Quasi Steady-State Simulation of the Unsaturated Zone in Groundwater Modeling of Lowland Regions. *Vadose Zone J.* **2008**, *7*, 769–781. [CrossRef]
23. Staszewski, T.; Bubak, A.; Nikinmaa, E.; Synnefa, A.; Santamouris, M.; Grimmond, S.; Toscano, P. GIS data and maps on spatial, socio-economic development and impact indicators. Document deliverable D.3.3.3. Sustainable urban planning decision support accounting for urban metabolism, EU Seventh Framework Programme. Available online: http://www.bridge-fp7.eu/images/reports/BRIDGE%20D.3.3.3.pdf (accessed on 4 November 2014).

24. Saaty, T.L. What is the analytic hierarchy process? In *Mathematical Models for Decision Support NATO ASI Series*; Springer: Berlin, Germany, 1988; Volume 48, pp. 109–121.
25. Mitraka, Z. DSS Design Report. Document deliverable D.6.1. Sustainable urban planning decision support accounting for urban metabolism, EU Seventh Framework Programme. Available online: http://www.bridge-fp7.eu/images/reports/BRIDGE%20D.6.1.pdf (accessed on 4 November 2014).
26. ESRI (Environmental Systems Resource Institute). *ArcMap 9.3*, ESRI: Redlands, CA, USA, 2009.
27. Mitraka, Z. Final DSS Prototype—Users Guide. Document deliverable D.6.3. Sustainable urban planning decision support accounting for urban metabolism, EU Seventh Framework Programme. Available online: http://www.bridge-fp7.eu/images/reports/BRIDGE%20D.6.3%20Final%20DSS%20Prototype%20-%20Users%20Guide.pdf (accessed on 4 November 2014).
28. Blecic, I.; Cecchini, A.; Trunfio, G.A. A General-purpose Geosimulation Infrastructure for Spatial Decision Support. In *Transaction on Computational Science VI*; Springer: Berlin, Germany, 2009; Volume 5730, pp. 200–218.
29. Castro, E. Strategic Scenario Analysis. BRIDGE Deliverable D.7.1. Sustainable urban planning decision support accounting for urban metabolism, EU Seventh Framework Programme. Available online: http://www.bridge-fp7.eu/images/reports/BRIDGE%20D.7.1.pdf (accessed on 4 November 2014).
30. IPCC (Intergovernmental Panel on Climate Change). *Emissions Scenarios*; Nakicenovic, N., Swart, R., Eds.; Cambridge University Press: Cambridge, UK, 2000; p. 570.
31. San Jose, R. Model Implementation Report. BRIDGE Deliverable D.4.2. Document Reference: 211345-017-TR-UPM. Available online: http://www.bridge-fp7.eu/images/reports/BRIDGE%20D.4.2.pdf (accessed on 4 November 2014).

sustainability

MDPI

Article

Regional Open Innovation Roadmapping: A New Framework for Innovation-Based Regional Development

Wim Schwerdtner [1,*], Rosemarie Siebert [2], Maria Busse [2] and Ulf B. Freisinger [3]

1 Faculty of Life Sciences, Economics of Horticultural Production, Humboldt Universität zu Berlin, Unter den Linden 6, Berlin 10099, Germany

2 Leibniz Centre for Agricultural Landscape Research (ZALF), Institute of Socio-Economics, Eberswalder Straße 84, Müncheberg 15374, Germany; rsiebert@zalf.de (R.S.); maria.busse@zalf.de (M.B.)

3 Freie Universität Berlin, Kaiserswerther Str. 16-18, Berlin 14195, Germany; ulf.freisinger@fu-berlin.de

* Author to whom correspondence should be addressed; wim.schwerdtner@agrar.hu-berlin.de; Tel.: +49-30-2093-46339; Fax: +49-30-2093-46331.

Academic Editors: Tan Yigitcanlar and Md Kamruzzaman

Received: 24 November 2014; Accepted: 13 February 2015; Published: 25 February 2015

Abstract: To foster sustainable regional development, many regions rely on innovations. To safeguard the generation of innovations and their market introduction, companies have increasingly used technology roadmapping and open innovation. The project INNOrural (Innovations for sustainable rural development) expanded these concepts by applying them to regions. This led to the rise of the "Regional Open Innovation Roadmapping" framework for innovation-based regional development (ROIR). This framework was tested by conducting two innovation roadmapping processes in the model region of Märkisch-Oderland (MOL), Germany: the certification of regional wood fuel and the establishment of a competence center for precision farming technology. Both innovation ideas were selected during the roadmapping process by applying a sustainability assessment. After 12 months, two complete roadmaps were ready for implementation. Key principles of ROIR were identified, including the use of a clear and replicable sustainability assessment method, the involvement of all relevant stakeholder groups in the early process and the cooperation between regional and subject experts. Generally, the broader adaptation of ROIR for additional regions will be useful. Nevertheless, the ROIR processes need to be evaluated in depth to develop a better understanding and to provide evidence of the benefits and limitations of this approach.

Keywords: sustainable regional development; land use governance; regional stakeholder involvement; precision farming (PF); wood fuel; technology roadmapping (TRM); open innovation (OI); Germany

1. Introduction

Competitiveness often depends on the successful development of innovations and their introduction to the market. This applies to companies, business sectors and certain regions, as well. In addition, due to sustainability issues raised by regional developments, innovations are becoming more important [1,2]. This can be described as innovation-based regional development and applies to both industrialized countries and emerging states [3–5]. Currently, innovation is a key issue in regional development programs, such as the European RIS (Regional Innovation Strategies)/RITTS (Regional Innovation and Technology Transfer Strategies and Infrastructure) and LEADER (Liaison Entre Actions de Développement de l'Économie Rurale) and the German federal program, INNOregio (Unternehmen Regionen) [6–9]. Since the launch of RIS and RITTS in the early 1990s, the idea of innovation as a regional necessity and motor for sustainable development has been recognized. In numerous European regions, the European Commission aimed for the regions to develop their own innovation strategies and improve their innovation support infrastructure under the terms of "appropriate innovations". Thus, the link to recent programs, initiatives and projects on regional innovations was provided, which also incorporated a strategic and analytic approach and the inclusion of a broad range of stakeholders. Generally, safeguarding the development and introduction of inventions on the market (innovating as an implementation process) requires systematic approaches, such as project planning, participatory rural appraisal or regional foresight. Nowadays, foresight has become an established instrument of regional policy [10] and RIS [11]. Foresight is defined as "a systematic, participatory, future intelligence gathering and medium-to-long-term vision building process aimed at present-day decisions and mobilizing joint actions" [12] (p. 3). To this end, many foresight projects use a kind of "knowledge triangle", meaning the cooperation of the representatives of the research, education and innovation spheres [13]. In the EU, regional foresight is used frequently for regional development, e.g., for exploring capacities and identifying needs (U.K., North West region) [14], identifying societal needs and patterns of the evolution of emerging technologies (Italy, Lombardy) [15], analyzing social change and its impacts (Germany, Rhineland Palatinate) [16], stimulating regional innovation and strengthening the regional economic system against global competition (Italy, Trento) [17]. Lately, the European Commission has made efforts for a more comprehensive use of foresight approaches within the EU. Under the Research Framework 7 program, the European Foresight Platform (EFP) aims at supporting future decision making [18].

From a foresight point of view, technology roadmapping (TRM) can be seen as a means to transact regional innovation or foresight projects, as depicted by Kindras *et al.* [10] in their example of the Samara region (Russia). TRM derives from a business- and technology-driven perspective, being a tool that supports the planning and coordination of development processes and the introduction of innovations to the market [19–22]. TRM is a flexible technique that can be used for a wide range of situations. It provides a structured and often graphical means for exploring and communicating the relationship between evolving and developing markets, products and technologies over time [23–25]. In a broader sense, TRM covers the whole range of roadmapping that addresses technology, products, processes, market drivers, technical skills, projects, *etc.* [23]. Even though it is frequently used by a broad spectrum of firms and business sectors, the research in this field is relatively sparse [26]. Furthermore, it has not been widely applied to geographical issues and regional sustainability. Only a few studies have dealt with the elements of TRM in urban or regional contexts. Van den Bosch *et al.* [27] applied a bottom-up approach to generate a roadmap for the Rotterdam case study of the transition to a fuel cell transport system. Another study by Lee *et al.* [28] introduced an integrated service-devices-technology roadmapping process for smart city development in Korea. In the Samara case study, scenarios were integrated into a roadmap to identify innovation potentials for strategic decisions [10]. The adoption of TRM for open regional development strategies with a strong consideration of the region's potentials does not yet exist. This gap has not been identified in the TRM literature nor TRM review papers (*cf.* [29]). In addition, Caetano and Amaral [30] found that there are only a few proposals for partner selection and participation methods in roadmapping.

The openness of innovation processes for external expertise, which includes the broad participation of stakeholders, is part of the open innovation (OI) approach. Opening up innovation processes purposely for additional knowledge and ideas from outside has become an important strategy for leading industries to cope with changing environmental conditions and to compete effectively in the market [31,32]. The OI approach can be defined as "... the use of purposive inflows and outflows of knowledge to accelerate internal innovation, and expand the markets for external use of innovation" [33]. It includes not only the opening of external knowledge at the initial stages of the innovation process, but also the continual participation of internal and external stakeholders at all stages. OI is closely connected with establishing and using organizational networks and benefits from cooperation with customers, suppliers, research institutes and teaching institutions to enhance the innovation capability of an organization [32].

In addition to the main focus of OI on a single firm level, some studies have focused on OI in a regional setting. Schaffers *et al.* [34] explored business models to launch living labs for rural and regional development. Due to the close cooperation of users and technology providers, the living lab concept is strongly linked to the OI approach. In EURIS (European Collaborative and Regional Open Innovation Strategies) [35], a recent European project, OI strategies are used to open regional innovation systems and to enhance the innovation ecosystems of various European regions. Belussi *et al.* [36] investigated the existence of an Open Regional Innovation System in Emilia Romagna (Italy) to show how firms adopt the OI strategy in order to overcome firm and regional boundaries.

One proposal to merge TRM and OI has been recently elaborated by Caetano and Amaral [30]. The authors adapt roadmapping to open innovation environments by building a method that is applicable for organizations (primarily SMEs) that pursue technology to push innovation strategies. However, this paper assumed that the integration of the open innovation strategy and the TRM approach could be powerful for seeking sustainable open regional development. From the benefits of integrating the two approaches, the INNOrural project created the methodological framework, Regional Open Innovation Roadmapping (ROIR). ROIR is designed to utilize the innovation potentials of a region for its sustainable development and to maximize the prospect of innovation success.

This paper demonstrates how TRM and elements of OI were combined to form the approach ROIR. It examines the hypothesis that ROIR is able to provide a suitable framework for safeguarding innovation-based regional development projects. The application of ROIR and the development of its processes are illustrated by two case studies in the German region of Märkisch-Oderland (MOL), which tested the approach in practice and examined its advantages and difficulties. Both roadmapping processes concern non-technical or so-called social innovations in order to promote regional sustainability, in which the first addresses the development and certification of a supply chain for wood fuel and the second addresses the implementation of a competence center for precision farming technology (PF) for the state of Brandenburg in the same region. Both innovation examples are often seen as promoters for the sustainable use of natural resources, such as wood fuel, e.g., saving non-renewable energy sources and reducing greenhouse gas emissions [37,38], and PF, e.g., reducing the application amount of chemical fertilizer by adapting fertilization to local field conditions [39–42]. Furthermore, this paper shows the added value offered by ROIR for sustainable regional development.

2. Methodological Framework

The integration of open innovation elements in technology roadmapping (TRM) in the context of sustainable regional development is called Regional Open Innovation Roadmapping (ROIR). This framework was designed and tested in the INNOrural project. Its emphasis shifts from the roadmap as a graphical product for the developing process of the roadmap [29]. In terms of the organization, communication work during the roadmapping process is usually a more important output than the final roadmap [26,27].

Following Chesbrough *et al.* [33] on OI and Phaal *et al.* [23] on TRM, ROIR can be defined as a strategic innovation planning process ("strategic roadmapping") (*cf.* [21]) in which a roadmap for future innovation opportunities or a specific innovation is developed. Thus, the planning process describes in advance all phases of the entire innovation development chain, including R&D, prototype, implementation/mass production and market introduction in detail and usually in diagrammatic form. ROIR uses purposive in- and out-flows of knowledge to increase the internal innovation capability of the organization and to expand markets for the external use of innovation. The roadmap chart (the outcome of the process) is a time-based chart, covering a number of layers and including both commercial and technological perspectives ("multi-layer roadmap") [23]. Both the planning process, as well as the roadmap incorporate the following three groups of contributors (Figure 1):

(1) Internal contributors, which are in-house departments, like R&D, marketing, sales, controlling;

(2) External contributors, which are partner organizations, like trading partners, public authorities;
(3) Relevant stakeholders, such as single citizens and interest groups, which could be affected directly or indirectly to some extent by the development and/or market introduction of innovation.

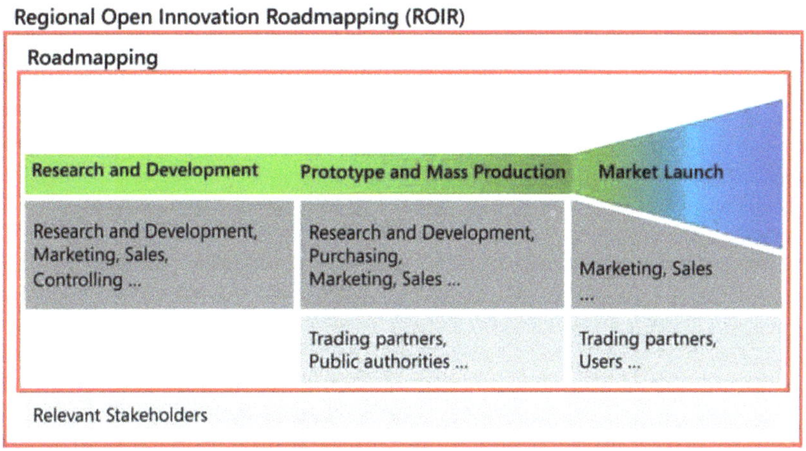

Figure 1. Contributors to the innovation process.

ROIR uses a high level of expert knowledge. Experts are defined as individuals possessing an advanced knowledge in limited areas and having the capability to think strategically on a meta-level [43]. One defining characteristic of the ROIR process is the cooperation between regional experts (representatives of stakeholder groups that provide deep knowledge on all matters concerning the region) and subject experts (individuals qualified to contribute additional expertise to the development and launch of the innovation, who are not necessarily from the region).

OI processes can generally be initiated in two ways: by self-recruitment and by recruitment-by-invitation. For INNOrural, recruitment-by-invitation was chosen, because the aim was to create and test an easy-to-use approach that could be conducted anywhere with minimal resources (time, money and staff). In recruitment-by-invitation, only representatives of stakeholder groups that are relevant to the process are invited to take part. The main advantage is the greater efficiency of gathering a few dozen people instead of a hundred or thousand and building and maintaining a communication process with the participants. The main obstacle is the identification of relevant stakeholder groups, which has led to the opposition of the groups not invited. Failing to identify relevant stakeholders may result in a lack of legitimation. Therefore, particular attention should be made to carefully select representatives of stakeholder groups.

Additionally, the project launched a web-based discussion platform for open innovation processes. This interactive platform was open for all stakeholders and citizens at any time, where people could inform themselves about the current project status and contribute their own ideas.

In the literature, the suggested roadmapping processes differ regarding the number of included phases, which range from three [25,44] to eight [28]. However, at a minimum, most roadmapping processes are composed of the following three main steps: preliminary activity, development of TRM and follow-up activity [28]. The INNOrural project adopted this proceeding to its specific conditions and requirements by structuring ROIR in four phases: (1) identification of regional experts; (2) selection of regional innovations (The "subject experts" are missing in Figure 2, because at this stage of the process (Phase 1), the "regional experts" had to decide upon possible innovations. The "subject experts" were invited in Phase 3 to contribute to the specific, now selected innovations); (3) the regional innovation concept; and (4) the roadmap as a product of the process (Figure 2).

Figure 2. The ROIR process.

3. Model Region, Märkisch-Oderland District

The district of Märkisch-Oderland (MOL) in Brandenburg State was selected to serve as the study region. This state is located in the eastern part of Germany, surrounding the capital, Berlin. Before reunification in 1990, it was part of the German Democratic Republic. Studies have shown that Brandenburg is stagnating at a low level with respect to R&D activities, patent applications, competitiveness and economic growth [45–47]. The district of Märkisch-Oderland (MOL) within this state is located east of Berlin and borders Poland to the east. The peri-urban part of the district next to Berlin is home to business and industry, whereas the eastern rural area is mainly agricultural land. In 2008 (the decision to select MOL as the research area was made in 2009 on the basis of these up-to-date data of that time), the GDP per person was 15,000 EUR, which is approximately half of the average GDP for Germany (28,200 EUR) and below the average level of Brandenburg (19,700 EUR) [48]; this indicates the rather poor economic situation of the district. MOL has the highest unemployment rate in Brandenburg (in 2008, its rate was 13.5% *versus* the national rate of 12.0% [48]). In the rural part of the district, the unemployment rate in 2008 was nearly three-times higher than the rate in peri-urban Berlin [49]. The economic potential and future perspectives of the district were estimated to be rather weak. In a 2008 ranking of 439 regions in Germany, MOL held the 379th position, which is a slight increase of 21 ranks compared with a 2004 study [50]. Using the typology created by Muller *et al.* [51] to describe regional innovation capacities in Europe's new member states, Märkisch-Oderland can be categorized as an E-region, a lagging agricultural region with a relatively underdeveloped economy and structural problems linked to the loss of systemic integration.

Similar to the rest of Germany, precision farming technology (PF) (Precision farming technology is an information-guided management concept in plant production, which allows a precise and site-specific cultivation. It is based on satellite-supported positioning systems, GPS and sensor technologies. PF includes automatic data collection and processing, track guiding systems, site-specific techniques, fleet management, field robots, *etc.*) in MOL failed to penetrate the market, stagnating at approximately 9% of the national market (*cf.* [42,52]). Compared to the U.S. and other European countries, such as Denmark, the U.K. and Sweden, the adoption rate of precision farming technologies in Germany has been relatively slow [53–55]. Experts have stressed the advantages of this technology, but farmers have been difficult to convince.

Another example is the bioenergy region initiative. MOL joined the federal funding program to establish bioenergy within the region based on wood fuel. Despite strong efforts, the region had difficulty convincing consumers of the benefits of wood fuel.

4. The Regional Open Innovation Roadmapping (ROIR) Process in the INNOrural Project

The overall project INNOrural lasted two years. The ROIR process described below took twelve months.

4.1. Phase 1: Identification of Regional Experts (Three Months)

The first phase started with extensive literature research about the study region to understand its socio-economic structure, eminent branches, authorities, organizations/unions, networks, citizens' initiatives, topics of main discussions in the region and its parliament and existing and failed initiatives of regional development. This preliminary research laid the foundation for over 20 interviews that were conducted to obtain more in-depth insights and understanding. After the first phase, 18 people were selected as regional experts. The number of people selected depended mainly on three factors: (1) size of the region; (2) complexity/heterogeneity of the region, *i.e.*, the number of different stakeholder groups; and (3) the structure and character of the region. Additionally, the decision concentrated on the most legitimate regional experts, either by election or qua office. Due to the rural and agricultural character of MOL, the majority of the regional experts had an agricultural or land use background. The selected regional experts were from the county's office for agriculture, the department of the state's ministry for agriculture, the farmers' association, the horticultural association, the local natural reserve, environmental associations, the chamber of industry and commerce and a federal agricultural research institute located in the region.

4.2. Phase 2: Selection of Regional Innovations (Three Months)

The second phase marked the beginning of roadmapping and consisted of two day-long workshops. The first workshop served as a kick-off to explain and discuss the ROIR approach with the regional experts selected in Phase 1. Following the approach of Phaal *et al.* [19] and Phaal and Muller [26], INNOrural aimed at answering the following three questions: (1) where are we now; (2) where do we want to go; and (3) how can we get there? The answer to the first question is mostly based on an intensive situation analysis, for instance involving "strengths, weaknesses, opportunities, and threats" (SWOT) [56]. Because all of our experts were very familiar with the situation of the model region, the situation analysis would have provided only a minor effect in making implicit knowledge explicit. Moreover, it would probably have limited the process to solutions close to the framing conditions. To be more effective and to find more creative innovations, the situation analysis step was skipped. The first workshop included a brainstorming session in which 32 innovation ideas were generated that seemed to be most suitable for MOL (Appendix 1). The innovation ideas covered a wide spectrum, including air boats, theme-oriented tourism, new greenhouse technologies, regional food, green energy, municipal solar collectors, precision farming technology and a wood fuel value chain. Some had been in discussion in the region, but never gained momentum.

Following the first workshop, a so-called inter-phase before the next workshop was used to assess the collected 32 innovation ideas in terms of their contribution to sustainable regional development.

Thus, the regional experts were asked to perform an impact analysis of the suggested 32 innovation ideas in written form, using the EU SENSOR (Sustainability Impact Assessment: Tools for Environmental, Social and Economic Effects of Multifunctional Land Use in European Regions framework and the ex ante Sustainability Impact Assessment Tools (SIAT) [57]. Table 1 shows this procedure for "wood fuel".

Table 1. Sustainability impact factors for the innovation "wood fuel" (n = 18) [58].

Main Goals	Aims	Weighting Factor	Rating of the Inno-Vation	Impact Factor
Economic goals	Positive employment effects Strengthening the attractiveness of the region as tourist destination Strengthening the attractiveness of the region as business location Economic cooperation/networking Creation of a unique selling proposition (USP) Advantage in competition Net value added Technical feasibility	4.55	0.87	3.96
Ecologic goals	Further ecologic development of landscape and agriculture Cooperation between landowners/land users and nature conservancy Resource conservation	2.18	0.67	1.46
Social goals	Strengthening of identification Potential for settlement of conflicts Acceptance Strengthening the attractiveness as lebensraum Range of the innovation Education and training measures Promotion and maintain the cultural heritage	2.27	0.43	0.98

First, the 18 regional experts disposed each of the nine points on the three main goals regarding sustainability aspects (economic, ecologic and social), then calculated the arithmetic mean of expert ratings. Table 1 shows that the economic goals gained had a mean of 4.55, a much higher rating than mean social goals of 2.27 and mean ecologic goals of 2.18.

Secondly, each expert selected three favorite innovation ideas out of the list of 32 and estimated how likely each idea would meet every single aim (*i.e.*, minor goals, like "positive employment effects") using a rating scale from +2 ("very likely to meet aim") to −2 ("very likely to not meet aim"). The estimations were summed and averaged to determine the result ratings for every assessed innovation idea relating to the three main goals. As the second column of Table 1 notes, the innovation idea "wood fuel" rates higher economically (0.87) than it does ecologically (0.67) or socially (0.43).

For this work, the impact factor is the product of multiplying the ratings with the weighting factors (*cf.* [26]). Thus, "wood fuel" reached 3.96 in terms of economic goals, 1.46 ecologically and 0.98 socio-economically. Summing up the three impact factors yields the total impact factor for every proposed innovation. This approach gave "wood fuel" an impact factor of 6.4.

All of the results of the impact analysis were re-distributed to the regional experts within the inter-phase to be used to determine their favorite.

In the second workshop, the stakeholders selected two ideas to be developed in the further roadmapping process. Initially, they decided to concentrate on the four highest ranked ideas (Figure 3). In an intensive discussion, the pros and cons of every idea were deliberated. After a feasibility check by the regional experts, two ideas were selected for the further roadmapping: (1) "certification of wood fuels"; and (2) the "establishment of a competence center for precision farming technology". Tourism networks were not considered, because their main representatives were unable to attend the roadmapping process continuously. This exemplifies the importance of stakeholders and their selection for the outcome of the overall process. The fourth innovation idea, participative water management, was dropped due to changing general political conditions regarding the subject.

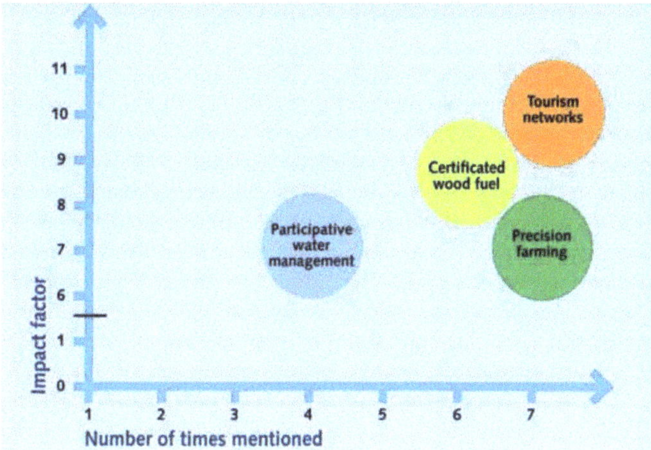

Figure 3. Selection of innovation ideas by impact factor and mentions (n = 24) [58].

4.3. Phase 3: Regional Innovation Concept (Six Months)

The concept phase marks the central unit of the roadmapping process. Since the regional experts chose two innovations, both roadmapping processes were conducted simultaneously during this phase, providing the opportunity to draw comparative conclusions for both. Due to space limitations, only the roadmapping for certification using wood fuel as an example will be described in detail for Phase 3. Additionally, the main differences from roadmapping for "establishing a competence center for precision farming technology" will be discussed in the following section.

Since a roadmapping process aims at responding to the main triggers that obstruct or promote innovations, the concept phase laid out first focuses on the identification of those main triggers for the introduction of certified wood fuel in MOL. To do so, additional knowledge about marketing and market research was needed to complement the expertise of regional experts. This was provided by so-called "subject experts", who were scientists and consultants on regional development and marketing attending the third workshop. Together with the regional experts, they identified three main obstacles for potential consumers to respond to in a regional market for wood fuel: (1) lack of awareness of wood fuel as a heating alternative; (2) insecurity of the wood supply; and (3) insecurity of prices.

To overcome these problems, the regional stakeholders decided to establish a coordinated regional supply chain and an individual certificate for regional wood fuel from the region of MOL. This was initiated to address the identified triggers in order to provide more reliable information for consumers on all aspects of wood fuel heating, to guarantee a continuous supply of wood and to ensure predictable and stable prices.

For the certification, stakeholders from all stages of the supply chain convened at the fourth workshop, including forest rangers, wood processors and energy consultants. An additional pool of subject experts was provided by the national network for wood fuel. The subject experts contributed their subject-based expertise to the development and launch of the innovation. Together with the regional experts, they provided the core set of expertise for the roadmapping process.

In the beginning, the regional experts decided on the institutional frame for implementing certification at the end of the roadmapping process. This is a crucial turning point during any roadmapping process, because the regional experts need to commit and take responsibility for the outcome of the process. Roadmapping processes will fail inevitably if a formal structure is not created for implementing the roadmap. At this point, key people will develop an eminent role, often serving as the "crystallization point" that the necessary structure will evolve around. Luckily, a key person emerged during this phase. The regional experts of wood fuel agreed to the commission of the existing bioenergy agency of MOL as the organization for the implementation phase.

Now, the main question was whether to join an existing certification program for biomass fuel, such as the "Forest Stewardship Council" (FSC) [59] or the "Programme for the Endorsement of Forest Certification Schemes" (PEFC) [60], and to adapt these programs to the region's purposes or to create an independent label with its own criteria. The situation and the possibilities were discussed, assisted by the input of two certification consultants. The ultimate decision of the experts and supply chain stakeholders was to establish an independent label. Certifying authorities should be the biofuel network, since the energy agency was basically a product of the network and dependent on state funding (the bioenergy region program). The possible objectives of such a label were specified, discussed and ranked by the regional experts using a rating scale from +2 ("very relevant objective") to zero ("irrelevant objective") (Table 2). Thus, the region of origin, the physical quality and the security of supplies were regarded as most essential.

Table 2. Ranking of objectives for the certification of wood fuels (n = 18) [58].

Rank	Mean	Aims
1	2.0	Region of origin
2	1.8	Physical quality
3	1.7	Security of supplies
4	1.7	Sustainability (economical)
5	1.4	Sustainability (ecological)
6	1.3	Control and warranty
7	1.2	Sustainability (social)
8	1.1	Publicity of regional products
9	1.1	Transparency
10	0.9	Costs of certification

The fifth workshop was designed to formulate a more precise concept to implement the objectives. In order to assess the achievements of the objectives, both expert groups (regional and subject experts) set up concrete indicators. The objective "region of origin" was therefore underpinned by the obligation to offer a minimum of 50% region-based raw material. "Physical quality" had to be assured by using legal classification, and the "security of supplies" was provided by long-term contracts and security supply storage. In particular, physical quality, the security of supply and social sustainability evoked controversial discussions. However, all of the issues were successfully resolved by the end of the workshop.

In the following interphase, different expert subgroups developed parts of the certification statutes, applying the set up indicators to different parts of the supply chain, like wood processors or energy consultants, going in-depth to solve the last questions, *i.e.*, how to maintain the security supply storage or how to secure supplies when streets are blocked due to hard winters. Solutions to all questions were then merged in the sixth workshop. Certification consultants accompanied this process with a consistency check to ensure that all parts would work flawlessly together during the implementation. The last step focused on marketing the certificate.

All regional experts then agreed on the final product, a concise roadmap to introduce and communicate the certification of wood fuels in MOL (Figure 4). The focus was on three products: wood chips, split logs and heating installation. Since a pellet-producing facility does not exist in MOL, this market could not be addressed by individually certified products. The remaining steps concerned the registration at the German Patent and Trademark Office in Munich and certification of the first supply chain members.

4.4. Phase 4: Roadmap (One Week)

The final assembly was under the patronage of the political district administrator and all of the participants of the two INNOrural processes gathered together. Regional experts, subject experts and additional consultants presented both roadmaps for the region of MOL. Hereafter, the organizations commissioned to carry out the implementation of both roadmaps introduced the implementation plan to the public and the press to obtain broader impact, e.g., to convince both consumers and companies of the benefits of certified wood fuel. In the afternoon, the experiences and outcomes of ROIR were discussed. All of the participants agreed that ROIR had helped them select and plan innovations they considered to be highly useful for their region. Without ROIR, the same result would not have been achieved due to the lack of know-how and manpower to organize the process.

4.5. Comparison to Precision Farming Roadmapping

The second roadmapping process concerned the "establishment of a competence center for precision farming technology" (Table 3). The objective was to promote the new farming technologies in Brandenburg State. The installation of a competence center was considered crucial to overcome the following implementation obstacles for farmers: (1) lack of information regarding the range of products, profitability and constraints; (2) lack of practical skills; and (3) lack of support. Due to high investment costs, most farmers had not been introduced to the new technology.

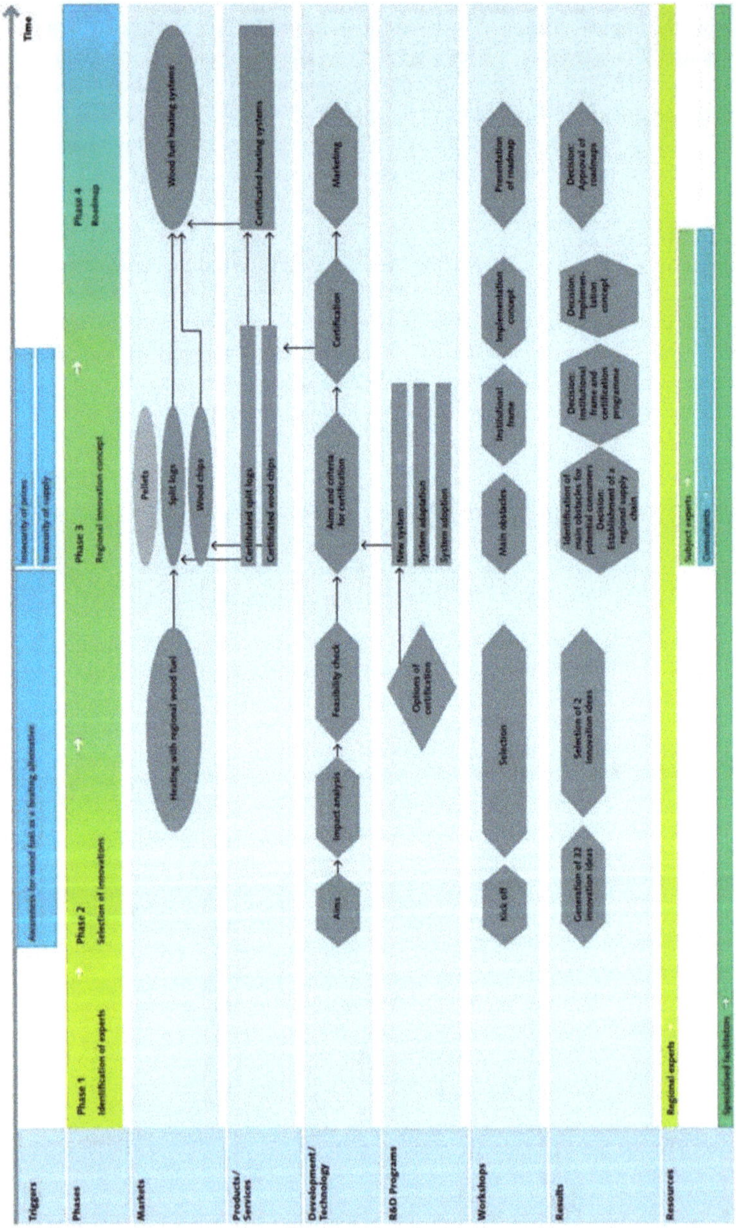

Figure 4. The complete ROIR process.

In contrast to wood fuel roadmapping, an efficient stakeholder network for precision farming did not exist in MOL. Single stakeholders had to be identified and convinced to help build a network of farmers, technology developers, industry representatives, farm consultants and a national agricultural research institute to be located in the region. They agreed on designing a roadmap for a three-year pilot phase. This roadmap laid the basis for a project proposal to request financial support from the European LEADER program for installing the competence center. Its purpose will be to disseminate knowledge about precision farming technologies throughout Brandenburg State. In contrast to the wood fuel process, the implementation of the precision farming roadmap relied on outside funding.

Table 3. Comparison of both roadmapping processes ("wood fuels" and "precision farming").

Criteria	Wood Fuels	Precision Farming
Objective	Promotion of wood fuel as a heating alternative by establishing a regional supply chain certificate for regional wood	Promotion of new farm technologies by establishment of a competence center for precision farming technology
Target group	Consumers	Farmers
Identified obstacles	Lack of awareness of wood fuel as a heating alternative Insecurity of prices Insecurity of supply	Lack of information Lack of practical skills Lack of support
Initial situation	Stakeholder network existed	No stakeholder network existed
Key stakeholders	Forest rangers Wood processors Energy consultants	Farmers Technology developers Industry representatives Farm Consultants National agricultural research institute
Workshops	Six (kick-off, selection, main obstacles, institutional frame, implementation concept, presentation of roadmap)	Six (kick-off, selection, main obstacles, institutional frame, implementation concept, presentation of roadmap)
Implementation obstacles	None	Dependence on outside funding
Implementation	Successful	Successful

In both roadmapping processes, specific key regional experts were essential for their success. These individuals strongly believed in the innovations and were economically connected to the outcome. Their attitude also stimulated others to join the process. A careful identification of stakeholders therefore proved to be one of the major success factors.

Both roadmapping processes were supported by external moderators and scientists with process competence. This helped the stakeholders concentrate on their issues without being side-tracked. The moderators added an outside perspective and addressed neglected topics.

5. Discussion

Generally, the ROIR approach is based on an innovation-oriented regional development perspective, which has a long tradition in the European policy. In the 1990s, the broadly aligned EU programs for innovation-oriented regional development RIS and RITTS were implemented in numerous European regions. The objective of RIS was to build partnerships among key innovation actors and to draw up regional innovation strategies. The early contribution of RIS/RITTS to innovation has been recently extended to EU strategic and funding instruments, such as the Structural Funds under RTDI (Research, Technological Development and Innovation) priorities [61], as well as in further regional innovation activities and externally founded projects, such as EURIS [35]. The mentioned objectives of the RIS/RITTS initiative represent a crucial link to the INNOrural project.

Comparing ROIR with regional foresight, many similarities can be found, e.g., regarding objectives, issues and methods. Kindras *et al.* [10] stated that innovation in regional foresight interlinks all types of factors and actors, aiming at creating a network among key actors of regional innovation systems (mainly firms, research organizations, public institutions, financial companies and technology intermediates). This fully applies to ROIR, as well. However, there is also a significant difference between the two approaches. Regional foresight aims at identifying future innovation potentials and development priorities for strategic decisions in regions [10,15,17], whereas the ROIR process is more implementation-oriented by intending to start and realize innovation activities.

Regional foresight studies use different methods, such as expert panels, group discussions, scenarios or, in rare cases, roadmapping [10,62]. In the Samara case study, regional experts were involved to build scenarios and integrate these scenarios into a roadmap (*cf.* [10]). In ROIR, the experts not only served to identify regional targets for socio-economic development, but also assessed the sustainability aspects of innovative ideas. The *ex ante* sustainability assessment in ROIR determines which innovative idea will be pursued.

An interesting framework that considers open innovation and tries to find solutions to cope with regional problems and to foster regional development is the living labs concept, which was examined in the C@R (Collaboration and Rural) project [34]. Further similarities between ROIR and living labs are the openness for a broad range of stakeholders, the partnership and network creation and operation across different development stages, the detection of potential future business or innovations and the feasibility analysis of inventions. However, living labs is seen as a more user-centric approach. Furthermore, the living labs approach provides less guidance, such as a business plan for implementation and a standardized methodology. Thus, instruments and forms of participation are not clearly defined. Through the application of the TRM approach, ROIR provides a strategic and structured methodology that is relevant and helpful for the transferability of the approach to other regions

Regardless of its final implementation, the ROIR process and the final roadmap was highly appreciated by the participating stakeholders. It allowed stakeholders to reflect on their respective situation and options regarding sustainable regional development and in the face of new challenges. Van den Bosch *et al.* [27] (p. 1033) stated that their roadmapping approach "facilitated interaction between stakeholders, which not only made it possible to formulate a common vision, but also helped in developing new relationships between different stakeholders." With ROIR, new cooperation, networks and even a wood supply chain were established. Lee *et al.* [28] also confirmed that roadmapping contributes to the formation of stakeholder networks. Currently, the open innovation strategy is being considered for building and reinforcing strong networks [35]. What made the INNOrural project special was that it fell back on existing networks. Thus, existing innovation ideas and inventions were continued, refined and, to a certain extent, realized.

Generally, roadmapping offers various opportunities regarding the methodology [28,29], but workshops and interviews with experts or stakeholders are very common [10,19,26,27]. The development of innovation in regional contexts can only be successful with collective action from different stakeholders. Van den Bosch *et al.* [27] confirmed this in their case study and underlined that it is crucial to obtain the commitment of stakeholders.

ROIR provides a framework for communication within the region and across regional boundaries by allowing actors and stakeholders of a region who had rarely met to now get together to recognize and find ways to solve their problems. External experts bring specific subject expertise and experiences from other regions into the discussion. Regarding smart city development, Lee *et al.* [28] stated that with the integration of participation methods (e.g., workshops, interviews and surveys), the roadmapping process becomes a communication platform for knowledge exchange on the city and inter-city levels.

Often, roadmapping processes are regarded as time-consuming and expensive, especially when using participatory methods. Nonetheless, strategic planning, like TRM, aims at saving the time and resources of companies and organizations in the long run. Within the wide range of TRM approaches, some use more basic and easy-to-adopt methods, like the fast start method [23]. ROIR is also designed as a lean, easy-to-use and manageable process for conducting expert workshops as a key method within the roadmapping process. The use of less complex methods, such as expert panels and group discussions, is also seen as beneficial in regional foresight studies (*cf.* [10]).

The ROIR framework may appear to be rather simple, but this simplicity is one of its advantages: it helps to increase transparency about the process and the results among all participants; and it can be used by non-experts, which is a precondition for a wider application in other regions. It is theoretically an interdisciplinary effort placed into the context of its primary concept, TRM. Nonetheless, most applications still remain confined to a specific sector [20] and/or companies [29]. ROIR, however, overcomes these limitations, because it is interdisciplinary, trans-sectoral and, to a certain extent, trans-regional. On the one hand, this offers different "cultures" to approach challenges, which can complement each other. On the other hand, cultural differences can cause difficulties, especially in the starting phase. All participants need to find their role within the process and to agree on norms and rules [63]. Stakeholders of a larger region are by far more heterogeneous than those of any particular business sector. Therefore, these first phases are crucial for overall roadmapping and require much more attention. TRM is mainly about technological innovations for technological problems. In contrast, ROIR for sustainable regional development also addresses non-technical, so-called "social innovations", *i.e.*, the build-up of networks or supply chains under consideration of sustainability aspects. These mainly focus on transactions [64,65]. Such transactions require special explanatory efforts, because they are intangible (as opposed, for instance, to the construction of an engine or a building). This intangibility is one reason for the widespread underestimation of social innovation potentials and for the difficulties in funding those approaches.

To gain legitimacy within the region, ROIR must provide transparency, comprehensibility and rationality about (1) the procedure and methodology within the roadmapping process and (2) the selection of stakeholders and experts:

(1) Regarding the procedure, INNOrural used a clear and reasonable method to assess the sustainability of innovation ideas, which consider ecological, economic and social aspects. This method was applied in alignment with the *ex ante* Sustainability Impact Assessment Tools (SIAT) for land use in European regions [57].
(2) Regarding the selection of stakeholders, the process must also be open to supplemental stakeholders and recommendations from outside during the process. In INNOrural, this was achieved by integrating external experts (so-called subject experts) and providing a web-based communication platform. Supporting staff (e.g., external moderators and scientists) should be on hand for at least some time during the implementation phase to provide continuity.

In regards to the role of the researcher in this study, the developed project methodology can be seen as a form of action research, in which the researcher facilitates the roadmapping processes. Van den Bosch *et al.* [27] also conducted action research and stated that the researcher therefore has influence on the process. From this finding, the authors conclude "that there might be a role for universities and other independent knowledge institutes in mobilizing stakeholders in early stages of innovation processes." [27] (p. 1035).

6. Conclusions

TRM and OI are broadly used to safeguard the development and market introduction of company innovations. ROIR merges the benefits of both approaches and applies them to the regional scale as a tool to provide innovation-based regional development that includes an *ex ante* sustainability assessment of innovation ideas. This approach was used to identify promising innovation ideas with regard to sustainable regional development and to test these in two parallel roadmapping processes in the economically stagnating region of Märkisch-Oderland (MOL) in Germany. The two roadmapping processes are the certification of wood fuel and the establishment of a competence center for precision farming technology. Both processes lasted twelve months each.

Methodologically, ROIR provided a systematic approach to overcome obstacles in the innovation process for the implementation and diffusion of innovations. It helped to address high complexity in an easy-to-use procedure, even by stakeholders not familiar with this process. Some of the main aspects of ROIR to support success are the inclusion of all relevant stakeholder groups early on in the process to portray the heterogeneity of the region and the cooperation between regional and subject experts to provide the core set of expertise that is indispensable for an open roadmapping process to succeed. Furthermore, the identification and inclusion of key individuals, the use of existing networks and structures and the use of external facilitators to support the process were helpful in the RIOR process.

From the case study perspective, both ROIR processes seem to be successful. The certification of wood fuel value chain and the installation of a precision farming competence center for Brandenburg State both made the step from plan (roadmap) to reality within three years. The wood fuel value chain has been attracting a rising number of participants. Now, a majority of companies concerned with wood fuel within the study area hold certificates, gaining a market coverage from 25% (wood sticks) to more than 50% (energy consultants and installation). Following the ROIR, the establishment of a competence center for precision farming technology was provided with an EU LEADER program fund, supporting the competence center and twelve demonstration farms for five years.

Generally, a broader adaptation of ROIR for additional regions will be useful, e.g., the ROIR framework was applied by the ZFARM project (Zero Acreage Farming—Städtische Landwirtschaft der Zukunft) [66] in an urban agriculture context.

Nevertheless, the ROIR processes need to be evaluated in depth to develop a better understanding and to provide evidence of the benefits and limitations of the ROIR approach. Carvalho *et al.* [29] (p. 13) already identified a gap in TRM research, in which the benefits of TRM were described "primarily on the basis of the perceptions of the stakeholders who were involved" and were not measured quantitatively. With the outcome of this evaluation, further research is recommended in regards to larger and more complex regions or for more controversial innovations. The proposed evaluation, the transfer to broader contexts and the generalizability demonstrate that ROIR may be a powerful tool for sustainable regional development. On a case study level, the positive impact of ROIR was visible and perceptible.

Acknowledgments: The study received funding from the Innovation and Technology Analysis (ITA) Program of the German Federal Ministry of Education and Research. All figures and tables were designed by Raimund Spierling.

Author Contributions: All authors were involved in the foundation items. All authors read and approved the final manuscript.

Conflicts of Interest: The authors declare no conflicts of interest. The founding sponsors had no role in the design of the study; in the collection, analyses or interpretation of data; in the writing of the manuscript; nor in the decision to publish the results.

References

1. Von Malmborg, F. Stimulating learning and innovation in networks for regional sustainable development: The role of local authorities. *J. Clean. Prod.* **2007**, *15*, 1730–1741.
2. Gerstlberger, W. Regional innovation systems and sustainability—Selected examples of international discussion. *Technovation* **2004**, *24*, 749–758. [CrossRef]
3. Schwerdtner, W. *Erfolgsfaktoren im Regionalmarketing läNdlicher Räume: Eine Empirische Studie der LEADER ± und REGIONENaktiv-Regionen in Deutschland und Österreich*; Books on Demand: Norderstedt, Germany, 2008.
4. Etzkowitz, H.; Klofsten, M. The innovating region: Toward a theory of knowledge-based regional development. *RD Manag.* **2005**, *35*, 243–255.
5. Porter, M.E.; Stern, S. Innovation: Location Matters. *MIT Sloan Manag. Rev.* **2001**, *42*, 28–36.

6. Boekholt, P. From planned economy to global markets: The RITTS project in south Brandenburg. In *Regional Innovation Strategies: The Challenge for Less-Favoured Regions*; Morgan, K., Nauwelaers, C., Eds.; Routledge: London, UK, 2004; pp. 138–159.
7. Koschatzky, K. The Role of Clusters and Regional Networks in Economic Transformation—Empirical Evidence and Conclusions from the East German Innovation System. In *Innovation Networks & Knowledge Clusters. Findings and Insights from the US, EU and Japan*; Carayannis, E.G., Assimakopoulos, D., Kondo, M., Eds.; Palgrave McMillan: Basingstok, England, 2008; pp. 210–229.
8. European Commission; Directorate-General for Regional Policy (Eds.) European Comission Innovative Strategies and Actions. Results from 15 Years of Regional Experimentation. European Comission Working Document. Available online: http://ec.europa.eu/regional_policy/archive/funds/2007/innovation/guide_innovation_en.pdf (accessed on 31 January 2015).
9. Charles, D.R.; Nauwelaers, C.; Mouton, B.; Bradley, D. *Assessment of the Regional Innovation and Technology Transfer Strategies and Infrastructures (RITTS) Scheme—Final Evaluation Report*; ECSC-EC-EAEC: Brussels, Belgium; Luxembourg, 2000.
10. Kindras, A.; Meissner, D.; Vishnevskiy, K.; Cervantes, M. *Regional Foresight for Bridging National Science, Technology and Innovation with Company Innovation: Expriences from Russia*; Basic Research Program Working Papers. Series: Science, Technology and Innovation, WP BRP 29/STI/2014; National Research University Higher School of Economics: Moscow, Russia, 2014.
11. European Commission; Directorate-General for Research (Eds.) FOR-RIS: Experiences and Ideas for Developing Foresight in a Regional Innovation Strategy Context (RIS/RITTS). Blueprints for Foresight Actions in the Regions. Available online: Ftp://ftp.cordis.europa.eu/pub/foresight/docs/blueprint-for-ris.pdf (accessed on 31 January 2015).
12. Gavigan, J.; Scapalo, F.; Keenan, M.; Miles, I.; Farhi, F.; Lecoq, F.; Capriati, M.; Bartolome, T. (Eds.) *A Practical Guide to Regional Foresight, FOREN Final Report*; IPTS: Sevilla, Spain, 2001.
13. Georghiou, L.; Keenan, M. Towards a typology of evaluating foresight exercises. In Proceedings of the EU-US Seminar: New Technology Foresight, Forecasting and Assessment Methods, Seville, Spain, 13–14 May 2004.
14. Puglisi, M.; Marvin, S. Developing urban and regional foresight: Exploring capacities and identifying needs in the North West. *Futures* **2002**, *34*, 761–777. [CrossRef]
15. Vecchiato, R.; Roveda, C. Foresight for public procurement and regional innovation policy: The case of Lombardy. *Res. Policy* **2014**, *43*, 438–450. [CrossRef]
16. Cuhls, K.; Kolz, H.; Hadnagy, C.M. A regional foresight process to cope with demographic change: Future radar 2030 (Zukunftsradar 2030). *Int. J. Foresight Innov. Policy* **2012**, *8*, 311–334. [CrossRef]
17. Koschatzky, K. Foresight as a governance concept at the interface between global challenges and regional innovation potentials. *Eur. Plan. Stud.* **2005**, *13*, 619–639. [CrossRef]
18. European Foresight Platform (EFP). Available online: https://ec.europa.eu/jrc/en/scientific-tool/european-foresight-platform (accessed on 30 January 2015).
19. Phaal, R.; Farrukh, C.J.P.; Probert, D.R. Strategic roadmapping: A workshop-based approach for identifying and exploring strategic issues and opportunities. *Eng. Manag. J.* **2007**, *19*, 3–12.
20. Phaal, R. Public-Domain Roadmaps. Available online: Http://www.ifm.eng.cam.ac.uk/ctm/trm/documents/published_roadmaps7_6_09.pdf (accessed on 20 April 2014).
21. Weissenberger-Eibl, M.; Speith, S. Flexibles Roadmapping—Eine Methode für die Vorausschau und Technologieplanung im Umfeld technologischer Durchbrüche. In *Vorausschau und Technologieplanung*; Gausemeier, J., Ed.; Heinz Nixdorf Institut: Paderborn, Germany, 2006; pp. 396–424.
22. Nathasit Gerdsri, R.S.V. Dealing with the dynamics of technology roadmapping implementation: A case study. *Technol. Forecast. Soc. Chang.* **2009**, *76*, 50–60. [CrossRef]
23. Phaal, R.; Farrukh, C.J.P.; Probert, D.R. Technology roadmapping—A planning framework for evolution and revolution. *Technol. Forecast. Soc. Chang.* **2004**, *71*, 5–26. [CrossRef]
24. Petrick, I.J.; Echols, A.E. Technology roadmapping in review: A tool for making sustainable new product development decisions. *Technol. Forecast. Soc. Chang.* **2004**, *71*, 81–100. [CrossRef]
25. Garcia, M.L.; Bray, O.H. *Fundamentals of Technology Roadmapping*; SBDD: Albuquerque, NM, USA, 1997.
26. Phaal, R.; Muller, G. An architectural framework for roadmapping: Towards visual strategy. *Technol. Forecast. Soc. Chang.* **2009**, *76*, 39–49. [CrossRef]

27. Van den Bosch, S.J.M.; Brezet, J.C.; Vergragt, P.J. How to kick off system innovation: A Rotterdam case study of the transition to a fuel cell transport system. *J. Clean. Prod.* **2005**, *13*, 1027–1035. [CrossRef]
28. Lee, J.H.; Phaal, R.; Lee, S.-H. An integrated service-device-technology roadmap for smart city development. *Technol. Forecast. Soc. Chang.* **2013**, *80*, 286–306. [CrossRef]
29. Carvalho, M.M.; Fleury, A.; Lopes, A.P. An overview of the literature on technology roadmapping (TRM): Contributions and trends. *Technol. Forecast. Soc. Chang.* **2013**, *80*, 1418–1437. [CrossRef]
30. Caetano, M.; Amaral, D.C. Roadmapping for technology push and partnership: A contribution for open innovation environments. *Technovation* **2011**, *31*, 320–335. [CrossRef]
31. Gassmann, O.; Enkel, E.; Chesbrough, H.O.W. The future of open innovation. *R&D Manag.* **2010**, *40*, 213–221. [CrossRef]
32. Chesbrough, H.O.W. The era of open innovation. *MIT Manag. Rev.* **2003**, *44*, 35–41.
33. Chesbrough, H.O.W.; Vanhaverbeke, W.; West, J. (Eds.) *Open Innovation: Researching a New Paradigm*; Oxford University Press: Oxford, UK, 2006.
34. Schaffers, H.; Guerrero Cordoba, M.; Hongisto, P.; Kallai, T.; Merz, C.; van Rensburg, J. Exploring Business Models for Open Innovation in Rural Living Labs. Sophia-Antipolis, France, 2007. Available online: http://researchspace.csir.co.za/dspace/bitstream/10204/1583/1/Schaffers_2007.pdf (accessed on 20 April 2014).
35. EURIS Project Embracing OPEN INNOVATION IN EUROPE—A Best Practices Guide on Open Innovation Policies. Available online: http://www.euris-programme.eu/themed/euris/files/docs/108/193/euris_guide_1.pdf (accessed on 20 April 2013).
36. Belussi, F.; Sammarra, A.; Sedita, S.R. Learning at the boundaries in an "Open Regional Innovation System": A focus on firms' innovation strategies in the Emilia Romagna life science industry. *Res. Policy* **2010**, *39*, 710–721. [CrossRef]
37. Laborde, D. *Assessing the Land Use Change Consequences of European Biofuel Policies*; ATLASS Consortium; IFPRI: Washington, DC, USA, 2011.
38. Gomiero, T.; Paoletti, M.G.; Pimentel, D. Biofuels: Efficiency, Ethics, and Limits to Human Appropriation of Ecosystem Services. *J. Agric. Environ. Ethics* **2010**, *23*, 403–434. [CrossRef]
39. Pedersen, S.M.; Kirketerp Scavenius, I.M. Environmental Impact with Environmental Indicators—With Precision Farming and Controlled Traffic Systems. FutureFarm Report. Available online: http://www.futurefarm.eu/system/files/FFD5.6_Environmental_impact_final.pdf (accessed on 31 January 2015).
40. Ancev, T.; Whelan, B.; McBratney, A. Evaluating the benefits from precision agriculture: The economcis of meeting traceability requirements and envrionmental targets. In Agriculture '05, Proceedings of the 5th European Conference on Precision Agriculture, Uppsala, Sweden, 9–12 June 2005; Staffort, J., Ed.; Wageningen Academica Publishers: Wageningen, The Netherlands, 2005; pp. 985–994.
41. Yu, M.; Segarra, E.; Lascano, R.; Booker, J. Economic Impacts of Precision Farming in Irrigated Cotton Production. *Tex. J. Agric. Nat. Resour.* **2003**, *16*, 1–14.
42. Werner, A.; Schwarz, J.; Dreger, F. IT-Farming: Linking agriculture production with the value chain in food production to enhance sustainability in rural land use. In *Enhancing the Capacities of Agricultural Systems and Producers*; Margraf: Weikersheim, Germany, 2008; pp. 58–74.
43. Bogner, A.; Littig, B.; Menz, W. *Interviews mit Experten: Eine Praxisorientierte Einführung*; Qualitative Sozialforschung; Springer VS: Wiesbaden, Germany, 2014.
44. Strauss, J.; Radnor, M.; Peterson, J.J. Plotting and navigating a non-linear roadmap: Knowledge-based roadmapping for emerging and dynamic environments. In Proceedings of the East Asian Conference on Knowledge Creation Management, Singapore, 6–7 March 2008.
45. Berthold, N.; Kögel, D.; Kullas, M. *Die Bundesländer im Innovationswettbewerb*; Bertelsmann-Stiftung: Berlin, Germany, 2009.
46. Deutsches Patent- und Markenamt. In *Jahresbericht 2009*; DPMA: München, Germany, 2009.
47. Koschatzky, K.; Lo, V.; Stahlecker, T. *Innovationsbedingungen und Innovationspotentiale in Ostdeutschland. Exemplarische Analyse von Drei Grenzregionen*; Fraunhofer-Institut für System- und Innovationsforschung ISI: Karlsruhe, Germany, 2006.
48. Statistisches Bundesamt. In *VGR des Bundes—Bruttowertschöpfung*; Bruttoinlandsprodukt: Bruttowertschöpfung in den Ländern der BRD 1991 bis 2013; Statistisches Landesamt Baden-Würtemberg: Stuttgart, Germany, 2014.

49. Landesamt für Bauen und Verkehr. In *Kreisprofil Märkisch-Oderland 2009*; Landesamt für Bauen und Wohnen: Hoppegarten, Germany, 2009.
50. Prognos, A.G. *Zukunftsatlas 2007: Ergebnisübersicht Gesamtranking*; Prognos: Berlin, Germany, 2007.
51. Muller, E.; Doloreux, D.; Heraud, J.; Jappe, A.; Zenker, A. Regional Innovation Capacities in New Member States: A Typology. *J. Eur. Integr.* **2008**, *30*, 653–669. [CrossRef]
52. Rogers, E.M. *Diffusion of Innovations*, 5th ed.; Free Press: New York, NY, USA, 2003.
53. Reichardt, M.; Jürgens, C. Adoption and future perspective of precision farming in Germany: Results of several surveys among different agricultural target groups. *Precis. Agric.* **2009**, *10*, 73–94. [CrossRef]
54. Fountas, S.; Blackmore, S.; Ess, D.; Hawkins, S.; Blumhoff, G.; Lowenberg-Deboer, J.; Sorensen, C.G. Farmer Experience with Precision Agriculture in Denmark and the US Eastern Corn Belt. *Precis. Agric.* **2005**, *6*, 121–141. [CrossRef]
55. Srinivasan, A.; Srinivasan, A.E. *Handbook of Precision Agriculture: Principles and Applications*; Food Products Press: Bringhamton, NY, USA, 2006.
56. Kotler, P.; Armstrong, G. *Principles of Marketing*, 14th ed.; Prentice Hall: Boston, MA, USA, 2011.
57. Helming, K.; Tscherning, K.; König, B.; Sieber, S.; Wiggering, H.; Kuhlman, T.; Wascher, D.; Perez-Soba, M.; Smeets, P.; Tabbush, P.; *et al.* Ex ante impact assessment of land use changes in European regions—The SENSOR approach. In *Sustainability Impact Assessment of Land Use Changes*; Helming, K., Pérez-Soba, M., Tabbush, P., Eds.; Springer Berlin Heidelberg: Berlin, Germany; Heidelberg, Germany, 2008; pp. 77–105.
58. Schwerdtner, W.; Freisinger, U.B.; Siebert, R.; Werner, A. *Regional Open Innovation Roadmapping for Sustainable Regional Development—A Practical Guide*; Leibniz-Zentrum für Agrarlandschaftsforschung (ZALF): Müncheberg, Germany, 2010.
59. FSC Council—Forest Stewardship Council. Available online: Http://www.fsc-deutschland.de/ (accessed on 20 April 2014).
60. PEFC—Programme for the Endorsement of Forest Certification. Available online: Http://www.pefc.org (accessed on 20 April 2014).
61. European Commission Regional Innovation Scoreboard 2014. Available online: Http://ec.europa.eu/enterprise/policies/innovation/files/ris/ris-2014_en.pdf (accessed on 20 April 2014).
62. Roveda, C.; Vecchiato, R.; Vercesi, P. Relationships between national and regional foresight: Lessons from experience. *Int. J. Foresight Innov. Policy* **2004**, *1*, 318–324. [CrossRef]
63. Tuckman, B.W. Developmental sequence in small groups. *Psychol. Bull.* **1965**, *63*, 384–399. [CrossRef] [PubMed]
64. Williamson, O.E. Technology and transaction cost economics: A reply. *J. Econ. Behav. Organ.* **1988**, *10*, 355–363. [CrossRef]
65. Williamson, O.E. *The Economic Institutions of Capitalism*; Free Press: New York, NY, USA; London, UK, 1998.
66. Specht, K.; Siebert, R.; Hartmann, I.; Freisinger, U.B.; Sawicka, M.; Werner, A.; Thomaier, S.; Henckel, D.; Walk, H.; Dierich, A. Urban agriculture of the future: An overview of sustainability aspects of food production in and on buildings. *Agric. Hum. Values* **2014**, *31*, 33–51. [CrossRef]

sustainability

MDPI

Article

Scaling-up Strategy as an Appropriate Approach for Sustainable New Town Development? Lessons from Wujin, Changzhou, China

Hao Chen [1], Qiyan Wu [2,3,*], Jianquan Cheng [2,4], Zhifei Ma [5] and Weixuan Song [6]

1 School of Architecture and Urban Planning, Nanjing University, Nanjing 210093, China; chenhao_us@smail.nju.edu.cn
2 School of Geographical Sciences, Nanjing Normal University, Nanjing 210023, China
3 Key Laboratory of Environment Change and Resources Use in Beibu Gulf (Ministry of Education) Nanning 530001, China
4 School of Science and the Environment, Manchester Metropolitan University, Manchester M1 5GD, UK; j.cheng@mmu.ac.uk
5 Department of Geography, Nanjing Normal University, Nanjing 210023, China; fennudelaoma@yeah.net
6 Nanjing Institute of Geography and Limnology, Chinese Academy of Sciences, Nanjing 210008, China; wxsong@niglas.ac.cn

* Author to whom correspondence should be addressed; 09349@njnu.edu.cn or chiyanwu@gmail.com; Tel./Fax: +86-25-8589-4173.

Academic Editors: Tan Yigitcanlar and Md. Kamruzzaman

Received: 19 December 2014; Accepted: 5 May 2015; Published: 8 May 2015

Abstract: China has achieved rapid urbanization and unprecedented economic booming over the past three decades. Numerous cities and towns dreamed of cloning the miracles of Shenzhen and Pudong, Shanghai, in terms of their international development. However, inappropriate development strategies have meant that the majority of fast expanding urban suburbs or newly developed towns suffer a high ratio of vacant dwellings in real estate markets and a massive loss of farmland. The frequent exposure of these empty cities to mass media or the public has urged urban governments to impose fiscal austerity. These unexpected and negative consequences of urban development have explicit conflicts with sustainability. This paper aims to provide a political economy view of these unsustainable outcomes of new development. To achieve this, the processes and agendas of new city or town planning in Wujin District, Changzhou City, are analyzed and evaluated from the perspective of scale theory. Extensive interviews conducted with local politicians at different levels, planners, real estate agents and local residents facilitate the interpretation of these processes and agendas. It is argued that the legends of Shenzhen and Pudong, Shanghai originate from a modified neoliberal capitalism intervention at the right time and place, with which other peer cities are not comparable. It is concluded that the scaling-up strategy is not appropriate for the local new town development of Wujin, which has led to unsustainable outcomes—empty cities and towns—and created important lessons for the sustainable development of Chinese cities.

Keywords: scaling-up strategy; sustainable development; local new town; neoliberal capitalism; Wujin; China

1. Introduction

Debates over scale theory have been actively discussed since the 1980s, including the extension of its meaning from traditional mapping solutions to socio-spatial production in 1990s [1–3], and the disputes between the horizontal *vs.* vertical scales of spaces and social entities [4,5]. Nevertheless, these debates do not agree that scaling-up strategies are extensively applied as panaceas in the restless tide of urban development, urbanization movements and scale reformation in many newly industrialized countries (NICs) in the context of globalization or internationalization [6,7]. It is certain that China, as a protagonist of this arena, has undoubtedly been proficient in applying these strategies to pursue fast urbanization and fascinating economic growth in the new millennium of the reform era [8]. Numerous cities and towns envisaged repeating the miracles of Shenzhen and Pudong, Shanghai in their internationalization paths with sustainable development rationales, green technology and aggressive scale reformation strategies [9].

However, inappropriate development strategies have led to the emergence of many socio-spatial problems either in economic growth or in urban and regional development [10]. Within an urban society, social differentiation and residential segregation have been deteriorating the system [11–13] and separate housing and job spaces have exacerbated traffic congestion and produced serious air pollution in these fast growing cities in China [14]. On the urban and regional scales, the majority of fast growing suburbs or newly developed towns have suffered a high ratio of vacant dwellings in real estate markets and a massive quantity of loss of farmland [15]. The frequent exposure of these empty cities or towns to mass Media and the public [15] have urged their urban governments to impose fiscal austerity. Undoubtedly, these unexpected and negative consequences of urban development have explicit conflicts with sustainability, which is defined by PNAS as: " ... dealing with the interactions between natural and social systems, and with how those interactions affect the challenge of sustainability ... meeting the needs of present and future generations while substantially reducing poverty and conserving the planet's life support systems" [16].

Therefore, it might be questioned what has contributed to the sharp contrast between the beautiful boulevards, amazing daytime views and darkness of empty cities at night? What has contributed to the contrast between urban fiscal austerity and fast urbanization? Why have these internationalized cities or towns planned with sustainability rationales and equipped with green technology become empty cities?

In sum, what have made the urban development strategies unsustainable in planning practice? This paper unfolds the causes behind these questions from the perspective of scale theory. After this introduction, Section 2 aims to give a concise review of the conception of "scale", debates on the scale theory and scaling-up strategies in the context of Chinese urban development. Section 3 focuses on a study area and case study methodology. In Section 4, the practices of scaling-up planning of Wujin City are analyzed at all phases and the unrealistic dream of scaling-up strategy is criticized, followed by addressing a paradox regarding the unsustainability of green technology in this case.

2. Scale and Scaling-up in the Context of Chinese Urban Development

During the evolution of the conception and theories of Scale, there have been several well-known theories in the published international literature, such as scale jumping [3], scale penetrating [17] and scale overlapping [9], which provides a theoretical framework for analyzing the scale strategies in neoliberal urbanization policy in China. Howitt (2003) [1] contends that "scale" is far beyond a single definition, and has become a central theoretical process since 1980s. Other than the traditional definition in cartography as map resolution, scale has been deeply involved into social production [4,18], where scale is not understood as a hierarchy of environmental (or physical), social-economic and cultural (e.g., body and home) activities but can be interpreted as the outcomes of these activities and processes [18–20].

This definition of scale links to the ideology of social-spatial dialectic in which the social, economic and environmental processes and spaces and their scales interweave mutually and interact with each other in a causal direction [2,20]. Thus, Swyngedouw (2004) further pointed out in a historical materialism view that the social, economic and culture phenomena or processes are driven or operated "at a variety of interlocked and nested geographical scales" [21] (p. 129). Meanwhile, Brenner (2001) [22] applied the Structuration Theory of Giddens (1984) [23] into the conception of scale and argued that the social production of scale is the "scalar structuration", which is endogenously embedded in and predicted on the relationships between hierarchical scales.

Unquestionably, there have been ongoing debates on scale in the process of theorizing scale. Marston (2000) criticized that the historical materialism view has neglected the social reproduction in the study of uneven development of capitalism on the production of scale [4]. Furthermore, Marston *et al.* (2005) analyzed the dominant research framework of scale and hierarchical conception of scale, and suggested replacing it with an ontology [5]. However, Harvey (2012) suspected the productiveness of this transformation on some particular milieu [24]; for example, Ostrom (1990) argued that the nested hierarchy of administration system on the design of commons governance should be avoided [25].

Nevertheless, what is the most important in urban governance practice? Smith (1992) [4] expanded the three-scale structure proposed by Taylor (1982) [26], with elaboration of scale's relationship, to the discontinuous and contradictory feature of capital reproduction process, and creatively contributed to the conception "politics of scale" by defining the geographical scale as "the boundaries and bounds the identities around which control is exerted and contested" [4]. Smith (1993) further contended that the processes of scale formation are cut through by all manners of fragmenting, divisive and differentiating processes (e.g., nationalism, localism, class differentiation and competition). Scale mediates between cooperation and competition, between homogenization and differentiation, and between empowerment and disempowerment [27].

Within the entrepreneurialism practice of neoliberal agenda [28], competing for scarce and mobile capital with their peers at same political scale for urban transformation encourages or motivates urban authorities to choose different strategies to attract foreign direct investment (FDI) and enhance their competitive advantages at larger geographical scales. This has been very prevailing in western society as well as in contemporary China [29]. Those urban political reforms have originated from a new rationality of governance practice on system of city governing county or "shi-guan-xian" policy since early 1980s, based on both localization and promotion of state authority activating local economy.

Chung (2007) [30] unscrambled these endeavors as impacts on the existing urban scale and the relationships between cities and other institutions but Ma [29,31] degraded the phenomena as a result of spatial pattern reconstruction. In fact, those endeavors could be extended or further epitomized as a pack of following scale strategies [9]:

First, Jumping of Scales, was described by Smith (2000) as a conception of a process that signals how political power, affixed at a particular geographical scale, is expanded to another scale [3].

That is to say, scalar political strategies are actively mobilized as parts of strategies of empowerment and disempowerment. The social power within and between scales changes with the scalar structure. Shen (2007) [9] articulated that scaling-up from county level to prefecture level, prefecture level to sub-provincial level, and sub-provincial level to provincial level [12] does not lead to spatial expansion, but empowers a city to control over land and economic interests in the newly developed areas instead. In contrast, when scaling-up the territorialization process, some cities, e.g., Kaifeng City in Henan province, have experienced a downsized re-territorialization from a sub-provincial city to a prefecture, and then suffered a decline continuously. Thereby, it is not surprised that scaling-up strategy has been restlessly applied in urban development practice under the orientation of urban growth machine and urban regime since the Chinese urban decentralization occurring in mid-1990s [32–34].

Second, as Brenner (1999) [17] argued that "world city" formation is the ultimate aim of urban planning and regional development in most cities, the strategy of penetrating scale is broadly embodied in majority of urban plans and growth agendas during the city internationalization movement, ranging from developed coastal areas to undeveloped inland area across China. This strategy refers to the aspiration to exert local influence to other scale, particularly, penetrating the influence or force of global or international scale into regional or urban scale.

Third, the strategy of overlapping of scale is also extensively utilized by lower-scale states to share a part of power and resources with higher-scale states through a top-down channel. For example, Shen (2007) [9] stated that a sub-provincial city in China is under administration of a province but it possesses much of provincial power. As a result, the sub-provincial city occupies a massive quantity of development zones, industry parks and technology zones, which are generally established on lower geographical scales, but are empowered by higher-level scale and control over a larger scale of resource in a nested bureaucratic system.

However, these ambitious strategies are not always successfully implemented. Recently, more and more mass media reports reveal that the internationalization movement of Chinese urban development is going far away from their baselines [15]. This is because nearly all urban areas in either coastal areas or inland regions on the reports have large-scale vacant housing, contributing to the formation of empty cities or towns everywhere [35]. This will be explained in detail in the following sections.

3. Study Area and a Case Study Method

3.1. Study Area

Changzhou City is located in the center of northern sub-region of Yangtze Delta in eastern China and at the lower reaches of Yangtze River (Figure 1). Wujin has become a district of Changzhou city since 2002 after a very flexuous administrative boundary evolution [36]. During the period from 1949 to 1983, Wujin was a county with a total area of 1677 km^2, separated from Changzhou (Figure 2A), a county-level city, financially and administratively.

Figure 1. Location of the study area—Wujin.

However, after the administration reform in 1980s [37], Wujin County was administered by Changzhou city and the latter was promoted as a prefecture-level city in 1983 under the policy of "system of city governing county" [9,30]. As administered by Changzhou city, Wujin County first relocated its government site from a city to a county town - Hutang in 1993, then was transformed as a county-level city in 1995 (Figure 2B), finally, was merged into Changzhou city as a urban district in 2002 (Figure 2C).

Wujin was famous by its "Sunan" model of bottom-up urbanization [38,39] and ranked the second among the One Hundred Strongest Counties of China in 1992 and the eighth during the period from 2003 to 2005. It has a total population of 1.6 million and a total GDP of 153.67 billion yuan RMB, which amounts to 46.2% and 50.9% of whole Changzhou City respectively [40].

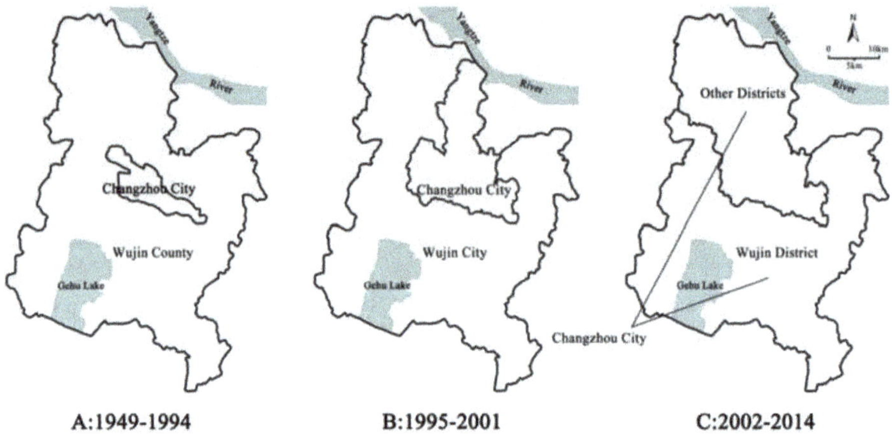

Figure 2. Changes of administrative structure of Changzhou and Wujin.

3.2. A Case Study Method

The research presented in this paper is based on Yin's (2003) case-study methodology [41]. Case studies are used to provide information for decision-making or to discover causal links in settings where cause-and-effect relationships are complicated and not readily known [41]. This method enables us to understand how local governments have selected the different scale strategies for new town development and why they failed to clone the trajectory of Shenzhen as well as Pudong and fall into the trap of unsustainability, with multiple sources of evidence. Such understanding allows us to illustrate and confirm the theoretical propositions previously enunciated in this paper.

The sampling of this case study, particularly the selection of interviewees, was based on previous research and consultancy projects with local governments, from which effective collaboration and efficient communication with relevant organizations have been successfully developed since the early 2000s. The case study data were collected from 49 face-to-face interviews at conferences, in offices or tea bars, or interviews by telephone, since the study involves different ideas, perspectives, opinions and interests from local business communities, developers and individuals. In this context, the interviewees include a set of opposite actors, namely, representatives of prefecture government *vs.* city government, the parishes officers of district government, some independent professionals *vs.* planners involved in local scaling-up strategies. The mixture of so diverse actors aims to investigate the interrelation of different actors in the mega-event. Therefore, it should be ensured that the criteria of actors' selection are relevant, neutral and objective. The interviews lasted for 20 to 90 minutes and then were transcribed into subsequent memo or research reports (see Table A1).

Additional data sets were acquired from: (1) reports of master planning of Wujin; (2) documents and bulletins from Changzhou Municipal Bureau of Planning and Development and Wujin District government; (3) the Statistical Yearbooks (China, Jiangsu province, Changzhou municipality and Wujin District, *etc.*) from 1993 to 2014 respectively. At last, 23 semi-structured surveys were conducted to unfold the failure of scaling-up strategy in social-spatial dimension upon individual cognitions, involving cognitive maps of new Wujin and public attitudes of urbanization process of Wujin from local communities.

4. The Practices of Scaling-up Planning in Three Phases

4.1. Is it Another "Great Leap Forward"?

Wujin's scale formation from a local to an international scale experiences three phases and four periods in the Chinese intuitional context of urban economic reform initiated in mid-1980s and of tax reform in mid-1990s (Figure 3):

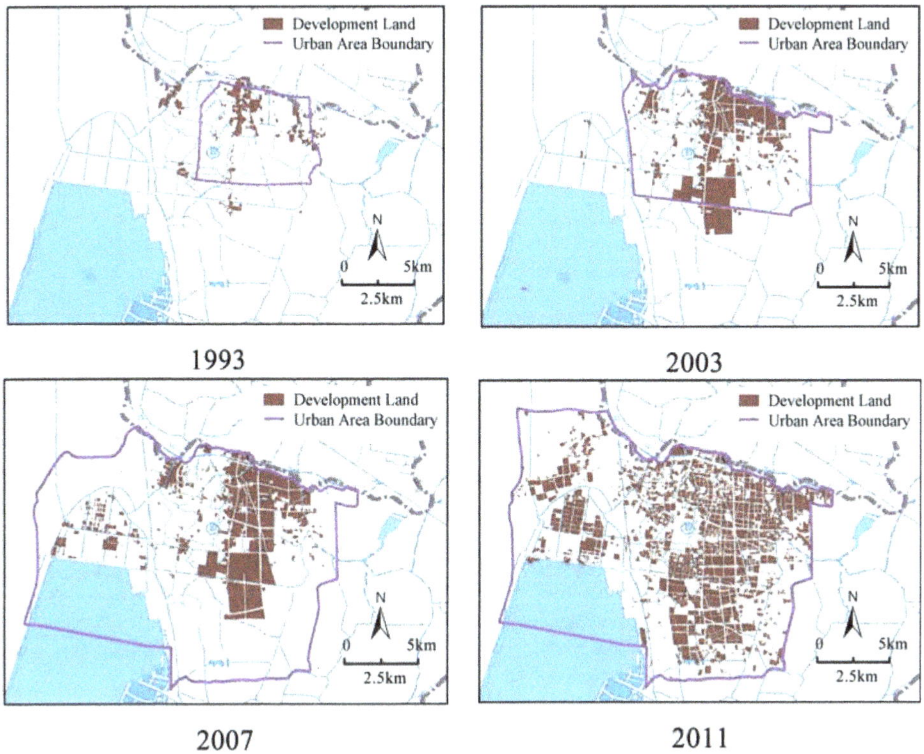

Figure 3. The rescaling of expansion of Wujin during 1993–2011.

4.1.1. Bottom-up Urbanization Before 2003

Although Wujin County was ranked the 2nd of the one hundred strongest counties across China in 1990s and has been famous for its "Sunan" model of bottom-up urbanization [38,39], it had to share a county town with the prefectural-level city—Changzhou due to complicated changes of administration (Figure 2). In the mid-1990s, Wujin has gained its financial independence motivated by the decentralization of central administration. The selection of building Hutang town as its capital was the protruded planning agenda in the master plan and the follow-up construction plan during the period from 1992 to 2003.

In this phase, Hutang Town was planned at a local scale and constructed as a common county capital of Wujin County. As an urban planner (Interviewee E1), who was involved with Wujin Master Plan in that phase, reviewed: *"although it was based on a massive scale of economic activities . . . ., [Wujin] was compiled to the standard of middle or small city [construction regulation of Ministry of Construction of the*

People's Republic of China]. It [the city proper around Hutang capital] was planned to have a total area of 25 km² with a total population of 220,000 . . . ..a typical small city at local scale . . . " (also see, Wujin Master Planning Report 1993–2020) [42].

4.1.2. Scaling-up Formation During 2003 to 2013

Encouraged by the successful scaling-up strategy of Shenzhen City expanding rapidly from a local village to a regional mega city as well as the marvelous transformation of Pudong from a desert area into an international financial and trade center, it is reasonable to foster its own scaling-up dream due to the strong economic power of Wujin. As such, Wujin has become one of many cities participating in the campaign of scaling-up "Great Leap Forward" since 2000s. This campaign can be divided into two periods of planning agenda: scale-jumping (2003–2009) and scale-penetrating (2009–2013).

4.1.3. Scale-Jumping Strategy Based Planning Agenda (2003–2009)

The administrative changes of Changzhou have made Wujin jump from a county to a county-level city, and then become a central district of prefecture-level city of Changzhou. Against this background, Wujin was transformed from a center of local-scale county-level city to a center of a semi-regional-scale prefecture-level city. A new scaling-up strategy was explicitly demonstrated in the Strategic Plan for Wujin District in 2003. The plan aimed to reposition the regional roles of Wujin from a comprehensive local central place to an important node of the Yangtze Delta, by twinning with Changzhou city at a final stage. Further, in the plan, Wujin, taking the regional role, was supposed to be more modernized, larger and more attractive. To meet these objectives, it is imperative to expand the city with a higher standard of urban development and construction. As a result, the population scale planned for the year 2020 was changed from 220,000 to 680, 000 population relative to those in the city planning of 1993 and correspondingly the total built-up area planned for 2020 increased from 25 km² to 82 km² [43]. It indicates that Wujin would be up-scaled from a middle-size city to a large-size city in China's city system, accordingly gaining a high level of political-economic mobilization in the Chinese institutional context [9,28,29].

Furthermore, Wujin was also proposed to be an eco-environmental friendly and greener city to attract more population and capital investment [43]. As such, ecological development was considered in the local planning or urban construction. As one planner who was interviewed explained (Interviewee E2):"*To be a functional center in the region, Wujin should be greener and more livable in the first instance*". The pursuit of these aims make it necessary to increase land consumption, develop more green space and enhance the quality of urban construction.

4.1.4. Scale-Penetration Based Planning and Overlapping Scale Redemption Agenda (2009–2013)

During the new town movement, the land consumption of Wujin in 2008 has reached the peak of land development set for six years (2014–2020), which has invalidated the strategic planning formulated in 2003. As a tool for fighting back against the depression caused by the global financial crisis in 2008, Beijing strived to stimulate an export-inclined economy with a big budget throughout the whole country [44]. The authority of Wujin district government then decided to take advantage of this national policy and made a great leap into the scale of world city through new strategic planning, which began in May and was completed by December 2009 [45]. In the planning scheme, Wujin adopted a penetrating strategy to locate itself into a global world system [46] as a global node:

> *"In a flat world reconstructed by the process of globalization, the cities on the bottom of national hierarchical city system (political system or size system) could have an opportunity to fulfill important functions and play nodal roles in the global city system. Wujin, undoubtedly as such a city, is capable of penetrating into the global city system . . . because of its advantageous location in the Yangtze Delta and its economic strength."*
>
> —Strategic Planning Report for Wujin New City (2009–2030) [45] (also see, Interviewee E3)

Wujin was planned to be one of the cities with global influences in the world but at a lower-scale in national administrative or city-size system, such as example cities of Davos in Switzerland, Seattle in United States, and Kunshan in China [47]. Based on this penetrating strategy, it is projected that Wujin would be a driver of regional economic development and one of the most attractive destinations for foreign direct investments and migrations in the Yangtze Delta as well as Changzhou municipal area. Thereby, the plan has updated the targets of population growth and land development in 2020 up to 950,000 populations and 135 km^2, respectively.

This changed objective also stimulated the restructure of city form, formulating a polycentric spatial structure including a recreational business district (RBD) in front of Gehu Lake that is located in the west, and a tech-industrial town (TIT) in the south, as well as CBD in Hutang central area.

In addition, an "eco-friendly" and "most livable" proposal incorporated into the penetrating strategy into global circuits [45], became a substitute for the "simple greening" strategy formulated in 2003. As an official of planning bureau assessed the plan (Interviewee B5), "*Wujin city will be not simply greener, but also have a better ecological system and more harmonious relationship between growth and ecological environment. Thus, it is imperative to adopt not only greening technology but also eco-technology, such as water resources preservation technology, circular economic technology, and low-carbon technology. These endeavors in ecology construction would enable Wujin to ultimately become one of the most livable cities in the Yangtze Delta as well as one of most well-known livable cities in the world*".

Unfortunately, both the tough economic situation in real estate market and strict land use policy by central government made the new town movement of Wujin face a challenging budget deficiency around 2012. As an officer in Wujin (Interviewee B3) released: " ... *if our finance is no longer dependent on the land leasing incomes, the annual revenue of our functional zone would be only 0.8 billion yuan RMB; it is not sufficient to pay the debt which was produced by the excessive public infrastructure investments in previous years ... only pay its interest; if we still some leftover, we may refund our stockholders a little bit* ... ". This has induced the empowerment of scale overlapping strategy in Wujin. That means, Changzhou city paid for Wujin's debt and changed its urban development strategy from north back to south, then fully integrated Wujin into Changzhou's city proper geographically. However, due to unnegotiable political interest conflicts, the scale overlapping strategy was not properly implemented (IntervieweeB2). " ... *son's mergers father's? which is a metaphor of relation between Changzhou and Wujin in history. It [Changzhou] was a part of us [Wujin]! ... humph, [Changzhou's]economy, I do not think, is better than ours ... if so(merges Wujin into Changzhou), they should pay the bill[means huge debt and other cost]* ... " while the officer disagreed with this argument with laugh during the interview. The official in Changzhou Planning Bureau (Interviewee B7) and the planner in a state-run planning institute of Changzhou (Interviewee E4) also expressed similar viewpoints over the possibility of overlapping each other between Changzhou and Wujin for both historic and realistic conflicts of interest.

4.2. Evaluating Outcomes of Scaling-up Strategies

4.2.1. Consequence (or Outcome) of Great-Leap Land Consumption

The scaling-up strategy has stimulated the acceleration of land development in Wujin since 2003, and has ultimately led to an un-balanced relationship between land consumption and population growth (Figure 4). The built-up area of Wujin was only 47 km^2 in 2003 but 102 km^2 in 2008 already. Such rapid growth has surpassed the growth boundary and invalidated the land development limit set in the strategic plan (2003). Comparatively, its population in 2008 only reached 400,000, 58.8% of the planned population for 2020. As the consequence of low-density urban sprawl, population density in urban areas has dropped rapidly from 6600 in 2003 to 1980 population/km^2 in 2008, which is seriously threatening the sustainability of rural development.

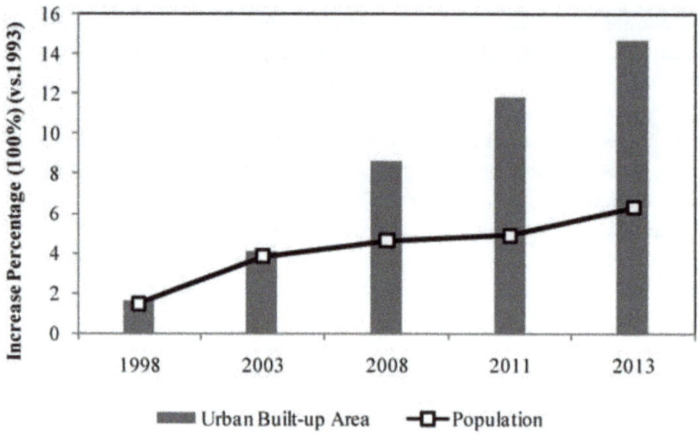

Figure 4. Increase percentages of land consumption and population growth from 1998 to 2013 compared to the data in 1993. Note: data from Wujin County Master Planning Report (1993–2020) [42]; Wujin City Master Planning Report (1998–2020) [48]; Urban Construction Development Report of Wujin Urban Area 2013 [49].

However, the land demand for urban development in Wujin has never ceased but instead kept an average growth rate of 6.8 km^2 per year from 2009 to 2013. These lands were mainly provided for developing the RBD area in the West and TIT area in the South. As a result, the built-up area has reached 136 km^2 in 2013, while the population density in Wujin's urban area continually increased to 1677 population/km^2 in 2013.

4.2.2. Social-Spatial Consequence of Continuous Urban Sprawl

The great-leap land consumption has caused large-scale urban sprawl since 2003, which inevitably led to a series of unsustainable land use, economic and social development.

First, the excessive land consumption has led to a lower efficiency of land resource utilization as well as a waste of public investment. Over the past ten years, a large amount of land resources have been used for housing, shopping malls, open spaces, and other public facilities. Among all the types of land resource uses in Wujin during 2003 to 2013 (Table 1), housing use ranked the second, green open space the third, and transport the fourth. However, all these land resource utilization suffered a very low rate of growth. For example, there is a huge gap between the actual *versus* designed traffic loadings at many avenues (Figure 5). An official survey [50,51] reported that the average rate of residential vacancy in Wujin was 21.7% in 2009, which increased to 22.5% in 2010. The actual vacancy rate should exceed the official rate based on our interview with the managers of real estate agencies or managers of property management companies in the 16 representative zones (Table 2; also see, Figure 5). One real estate sale agent commented (Interviewee A6) " … *eh, it is difficult to make a judgment of how many people live in their residential areas … We sold out around half [of property] … Some [empty dwellings] are soled but nearly half of their owners never shown up [so, they need to pay the strata fee].*"Similarly, a female store owner confirmed the high rate of housing vacancy (Interviewee D5):"*I have lived here for eight years … till now maybe only two third [housing are occupied by residents] … the price was around 5000 Yuan RMB per square meter … now around 6000 [the average housing prices in Nanjing and Shanghai had 3 to 4-fold increases during this period] [50] … just for living … nobody invests on it …* ".

Table 1. Changes of land use structure.

Urban Land Use	1993 (km^2)	2003 (km^2)	2013 (km^2)	Growth (1993–2003) (km^2)	Growth (2003–2013) (km^2)
Housing	4.33	15.61	33.76	+11.28	+18.15
Manufacturing	2.87	17.46	43.08	+14.59	+25.62
Transport	0.62	11.57	18.97	+10.95	+7.4
Green land	0	0.19	13.58	+0.19	+13.39
Commercial and Public facilities	1.64	8.53	13.32	+7.49	+4.79

Note: data from Wujin County Master Planning Report (1993-2020) [42]; Wujin District Strategic Planning Report 2003 [43]; Urban Construction Development Report of Wujin Urban Area 2013 [49].

Furthermore, the rapid urban growth has resulted in a crisis of social identity, undermining the sustainability of social development. The sharp contrast between increasing urban sprawl and declining population growth not only enlarged the geographical scale of Wujin, but also distorted the cognitive scale of residents as well as their community sense in Wujin (Figure 6).

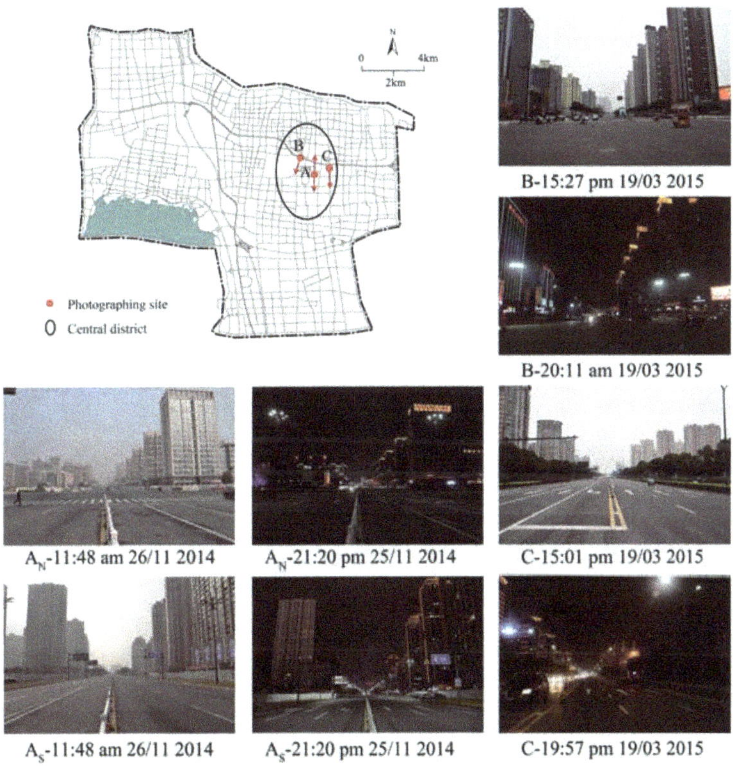

Figure 5. The city view both in daytime and nighttime at a site of downtown area.

Table 2. The surveyed housing vacancy rates in the 16 residential zones.

	Zones	Construction year	Residential vacancy rate
1	Chengzhonghuayuan zone	1990	<10%
2	Sijixincheng zone	2002	10%
3	Tian' anbieshu zone	2004	19.5%
4	Xinchengnandu zone	2005	13%
5	Changhonghuayuan zone	2006	43%
6	Daxuexincun zone	2006	48%
7	Cailingjiayuan zone	2006	30%
8	Nandianyuan zone	2007	30%
9	Yucheng zone	2007	30%
10	Hupanchunqiu zone	2008	35%
11	Tianjuanfeng zone	2010	45%
12	Hongjianyipin zone	2010	50%
13	Moershangpin zone	2010	66%
14	Xinchenggongguan zone	2011	35%
15	Laimengcheng zone	2011	50%
16	Xingheguoji zone	2013	90%

In a mental map on the spatial extent of Wujin new town created from a survey of 23 inhabitants (both locals and migrations) (Interviewees C1–C16; D1–D7), more than 75% interviewees defined the new town into a region no more than 50 km^2, and located in the built-up area prior to 2003. In addition, more than 60% these interviewees stated that they seldom travelled to the western RBD area and the southern TIT area. Nearly 50% these interviewees argued that they had little sense of community belonging in Wujin because of inappropriate spatial scale for living and a high rate of residential vacancy. The poor community sense has made Wujin not an ideal destination for migrants and this has adversely aggravated the degree of unsustainability. An official survey (2011) on the origins of buyers of commercial housing during 2006–2010 revealed that more than 75% to 85% buyers were locals, 10% to 15% buyers from other districts of Changzhou, and only 5% buyers from outside of Changzhou city [52]. These figures also revealed that Wujin has not become a real regional or global node attracting migrants from outside, and accordingly evidenced the failures of scaling-up strategies initiated in 2003.

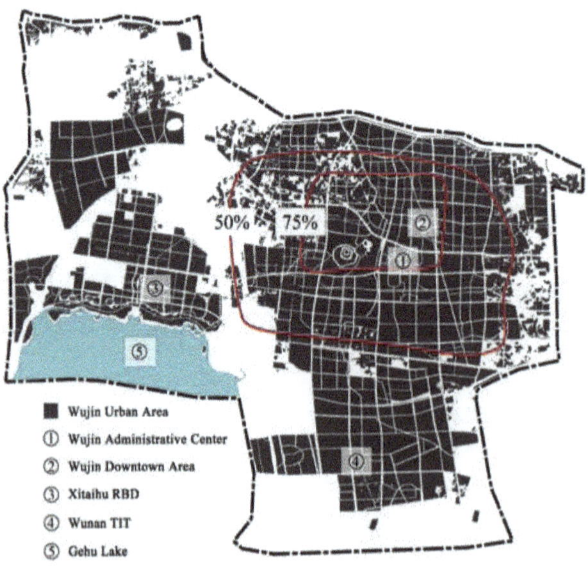

Figure 6. The cognitive map of local community.

4.3. Evaluating Land Revenue Collection

Undoubtedly, a massive gap exists between local revenue generation capacities and the requirements of scaling-up strategy, which allure and push local governments to raise funds in numerous forms away from formal state channels [53]. Since the urban economic reform in the mid-1980s and the follow-up tax sharing reform in the mid-1990s [37], local governments such as Wujin, were anxious to pursue a pro-growth development under the circumstance of decentralization. As the star of counties with the fastest economic growth in China, Wujin's GDP and local revenue in 2003 was 31.6 and 4.0 billion *yuan* RMB [54], much higher than the national average at county level [55]: 3.19 and 0.13 billion *yuan* RMB as well as the provincial average [56]: 12.2 and 0.7 billion *yuan* RMB respectively.

As such, there was not a big financial issue in 2003 for Wujin to confidently practice its scaling-up strategy. In fact, the logic behind the scaling-up strategy adopted by many local governments including Wujin, is transforming massive non-urban land into towns more than that required for urban development and gaining mortgage from banks before the excess land is sold out to real estate developers. Consequently, local authorities transform the loans from banks into infrastructure, public facilities, open spaces, and other fixed capitals to increase the value of urban land as well as its leasing price. Local authorities then lease the expensive land to developers and reclaim the proliferating land premium to repay the debt from banks. When the real estate market is growing, local governments are able to achieve the institutional led excessive "rent-gap" between farmland compensation fee and the proliferating land leasing fees [57–59].

However, the implementation of the strategy is very much dependent on the real estate market situation. Wujin has seemed to suffer a looming financial problem since 2012, caused by the declining real estate market. In the unexpected situation, both the new towns driven by scaling-up strategy and the game of institutional led excessive "rent-gap" are defunct. Local authorities have to face the unbearable financial crisis, confirmed by a local government officer (Interviewee B1) when commenting upon their endeavors of underplaying the looming crisis: " ... *We [the Wujin government] do have some financial issues these years ... eh ... but, anyway, Wujin still has a huge number of population and an abundant amount of [farm] land for development ...* ". However, an mid-ranked official agreed (Interviewee B3):" ... *The debt of our zone [only a developing zone of new town of Wujin] has reached around 15 billion yuan RMB right now [Nov 2014] ... With the declining real estate market and land market, the annual revenue of new town is as little as only 0.8 billion yuan RMB ... maybe it depends on if the central government wishes to pay the debt at last or helps local government keep the housing price up ...* ". A real estate developer confirmed (Interviewee A2):" ... *After 2012, the [real estate market] business has declined steadily ... [housing price] too high? I don't think so, at least our profit is lower than before ... eh ... maybe the compensation fee and tax is much higher than before ... .it is not our [real estate business] nowadays ...* ". A Regional manager of a nation-wide real estate company also expressed his disappointment over real estate market of Wujin (Interviewee A1), "*if this situation [real estate market declined] goes on, we may have to withdraw from Wujin*".

5. Discussion

5.1. Dream or Nightmare? An Unbroken Bubble of Scaling-up Strategy or Planning Competitions

The case study of Wujin may be a special case of many urbanization practices in China broadly utilizing scaling-up strategy. The National Development and Reform Commission (NDRC) Report (2013) [60] revealed that numerous new towns have been built in the surveyed 12 provinces during 2000 to 2012: 200 new towns in 156 prefecture-level cities, at least 67 new towns in 161 county-level cities and 55 in 12 sub-province-level cities with an average size of 63.6 km^2, which equals to 55.3% of average built-up area of original towns. In summary, the factors contributing to the failure of Wujin are listed as follows.

Although Wujin was merged into Changzhou city as a central district, it was unable to acquire substantial financial support from Changzhou city government as a subprime center. For historic reasons, the development strategy made by Changzhou city was to develop the northern city proper and build a sub center in Xinbei District (Figure 1C). As such, most public resources such as Changzhou's administration center, large-scale gymnasium, hospitals, cultural centers, transport hubs, and other public facilities have been developed into the new northern center rather than into the area of Wujin in south (IntervieweeB4, B6, B7 and E4). Undoubtedly, Wujin's development was marginalized to a peripheral position in Changzhou city. The sharp contrast between the top-down strategy of Changzhou and the bottom-up strategy of Wujin determined the fate of Wujin's scale-jumping strategy.

Wujin was not the only city adopting the penetrating strategy across the Yangtze Delta. In fact, the new towns in Shanghai, Suzhou and Nanjing cities have also implemented the scaling-up strategies with support from their higher-level (or mother) cities, and were all expected to be a functional node within the global system. Comparatively, Wujin had two inherent disadvantages over others. For example, Wujin had a disadvantageous administrative status comparing to other new cities and towns in Shanghai and Nanjing. Changzhou was just a normal prefecture-level city in Jiangsu Province, compared with Shanghai, provincial-level city and Nanjing, a sub provincial-level city, who both could have more political and economic resources and more development opportunities. Wujin also had a geographical disadvantage. Contrasting to its above-mentioned cities, Wujin has a very poor spatial accessibility to the global city—Shanghai and is therefore less able to integrate into internationalized economy and society (Figure 1B). Thus, the economic structure of Changzhou (and Wujin) is more likely localized rather than globalized, same as other cities, which can be inferred from the comparisons of two statistics related to economic openness level (Figure 7). Consequently, Wujin was deemed a loser in the scaling-up competitions.

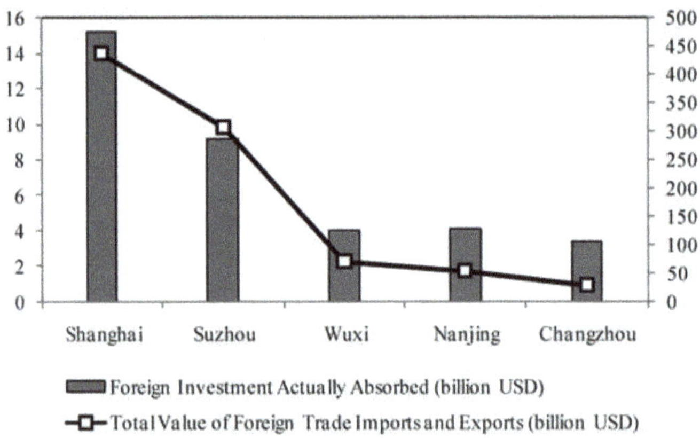

Figure 7. Comparison of economic openness level between Changzhou and its competitors in 2012. Note: data from Shanghai Statistical Yearbook 2013 [61] and Jiangsu Statistical Yearbook 2013 [55].

These lessons from the Wujin case reveal that globalization may present opportunities for local development but may be traps for local strategic decision-making. A successful practice of jumping to a higher scale or penetrating into the global system is not solely dependent on local initiatives. The winners should have global economic restructure and function as nodes of international capital production [24,62]. The failure of Wujin demonstrates that the grand scaling-up strategies are just illusory stories for most local cities, made up by global powers. It is infeasible for a local city to mobilize all of its resources for the illusory dreams. The losers in the gamble would inevitably face a cumbersome social-spatial unsustainability.

5.2. The development of Green Technology

It is undeniable that Wujin has deployed many greening and ecological strategies during the scaling-up practices since 2003. Therefore, why have these strategies not facilitated the production of sustainable outcomes?

Following these "greening", "eco-friendly" and "most livable" strategies, Wujin has made a number of greening endeavors such as developing more green space, reserving waterfront space, and building wide green belts on roads or streets, which has resulted in a substantial increase of green land during 2003 to 2013 (Table 2). Moreover, Wujin also deployed several ecological technologies including the developments of its public transport system, green architecture technology, and low-carbon technology. For example, Wujin was a pioneering city in running bus rapid transit (BRT) system and a pilot city in developing a green architecture industry in China. Wujin also invested 0.2 billion *yuan* RMB in developing a low-carbon small town, which was the first low-carbon project in Jiangsu Province (Interviewee B5).

Through these greening and ecological practices, Wujin was awarded a series of prizes for its green or livable development. For example, Wujin gained its first national award of "national ecological demonstration zone" by State Environmental Protection Administration of China (SEPA) for its endeavors in tree planting in 2006. Continuously, Wujin won an award of "international garden city" in the 11th International Garden City Final sponsored by the international Federation of Parks and Recreation Administration (IFPRA) in 2007. In 2010, Wujin achieved "Dubai International Award for Best Practices to Improve the Living Environment", which was issued by UN-HABITAT.

Unfortunately, these excellent efforts in greening and eco-friendly developments have not automatically led to successful practice for sustainable urban development. This is because all of these efforts and actions only rested on the ambitions of local government for marketing pro-growth urban development strategy, and failed to mobilize the whole society to autonomously participate in these, and they are never rooted in the daily practice of local residents. Thereby, the strategies of "let it be greener", "let it be eco-friendly", and "most livable" are not realistic manifestos to make a city sustainable but masks of entrepreneurialism policies to make the city greener and more attractive to capitals and high-skilled labors, to start the strategies of jumping scale or penetrating scale in a globalization context. Thereby, the greening strategy neither endorsed the scaling-up strategy, nor cut down the deficit budget of local government; instead it is more likely to provide a justification to overdraw local resources.

Furthermore, the investment in green technologies for implementing sustainable urban development strategies was far beyond the durable financial power of local governments. As a result, the local governments had to sell out more land plots for fundraising and then remarkably reduce the property price in real estate market. These actions jeopardized its revenue and threatened the development of agricultural sector and local environment. Consequently, these kinds of sustainable urban development strategy have inevitably led to unsustainable outcomes in planning practice.

6. Conclusions

Motivated by the successful scaling-up strategies of Shenzhen and Pudong, Shanghai in China, numerous cities and towns dreamed of repeating their paths of internationalization through similar economic policy, advanced green technology and eco-environment friendly rationale. Unfortunately, nearly all of fast booming suburbs or newly developed towns have, to some degree, faced a series of sustainable development challenges, such as a high ratio of vacant dwellings, a massive loss of farmland, and nearly bankrupted local governments. In the case study of Wujin, the unsustainable consequences of scaling-up strategies and sustainable technologies in urban development are twofold:

Given that the legends of Shenzhen and Pudong, Shanghai originate from a modified neoliberal capitalism [62] intervention at right time and in right place, their successful practices are definitely unrepeatable. Mr. Deng Xiaoping has demonstrated his outstanding political-economic wisdom on the design and development of Shenzhen in 1980s, and then of Pudong in 1990s. That is, both cases were embedded with incomparable state scale and economic scale, which are not available for any other peers in China. Although it was one of the most economically wealthiest county in China, Wujin was restrained in its embarrassing administration level, less motivated economic capacity, and less attractable regional housing market, making it impossible to practice scale jumping, scale penetrating and even the scale overlapping to achieve its scale formation in the context of internationalization.

Furthermore, the politician, bureaucrats and planners of Wujin have realized its weaknesses in scaling-up strategy; as complementary alternatives, they have practiced an eco-environment friendly strategy with extensive uses of advanced green technology into urban development. However, it is unreasonable to mistake "the light green" as "deep green", either sustainability [63]. Consequently, the slogans "think globally and act locally" proposed by the founder of "Friends of the Earth", David R. Brower would be the best philosophy for these local cities and towns than its reverse one "act globally and think locally".

The analyzed case of Wujin may have important implications for the sustainable development of Chinese cities, which are undertaking rapid urbanization under the dual forces of market and government intervention. It is clear that local governments should learn the lessons from Wujin that economic unsustainability jeopardized sustainable urban development though it is still early to conclude that the Wujin's case is unsustainable when the city has been developing very fast. Continuous monitoring of its development process and even quantitative models of evaluation would contribute to the appropriate design of local development strategies. Some potential efforts would be further developed in the future to reveal the sustainable and unsustainable practice of Wujin and its peers under the roofs of *policy boosterism* effort [64] or framework of *sustainability fix* strategy [65].

Acknowledgments: The research is supported jointly by National Natural Science Foundation of China (No. 41271176, 41361022 and 41201161), Chinese Minister of Education Project of Humanities and Social Sciences (No. 12YJAZH159), National Philosophy and Social Science Foundation of China (No. 12BSH027), a Project Funded by the Priority Academic Program Development of Jiangsu Higher Education Institutions (PAPD), and Jiangsu Center for Collaborative Innovation in Geographical Information Resource Development and Application.

Author Contributions: Mr. Hao Chen organized the field survey, wrote some sections and commented on the manuscript. Prof. Qiyan Wu as the correspondence author conceptualized and structured the paper, and wrote some sections. Dr. Jianquan Cheng revised and constructively commented on the paper. Mr. Zhifei Ma conducted the field survey and processed the maps and diagrams. Dr. Weixuan Song gave some constructive comments.

Conflicts of Interest: The authors declare no conflict of interest.

Appendix A

Table A1. List of interviews from 2009–2014.

	Venues	Data	Interviewees	Contents
A	**Local Business Community (8 interviewees, A1–A8)**			
1	Office	23 November 2014	A regional manager of a nation-wide real estate company	Sales performance; business environment and future investment consideration
2	Restaurant	23 October 2014	A local real estate developer	The development of real estate market in Wujin
3	Marketing hall	23 November 2014	A sale manager of a local real estate enterprise	Sales performance recent years
4	Store	23 November 2014	A local small business owner	Business environment; interest of investment in real estate
5	Community	23 November 2014	An employee of a trans-region company	Business and living environment in Wujin

Table A1. *Cont.*

	Venues	Data	Interviewees	Contents
6	Scale office	23 November 2014	A branch manager of a second-hand property agency	Business performance and appraisal over the second real estate market of Wujin
7	Park	23 November 2014	A shopkeeper in a community store	Business performance recent years
8	Restaurant	23 November 2014	A manager of a restaurant in the central business district (CBD)	Business performance; business environment in Wujin
B	**Officials (7 interviewees, B1–B7)**			
1	Restaurant	23 October 2014	An official of tourism bureau, Wujin	Developing issue of Wujin; problems in restructuring economy
2	Restaurant	23 October 2014	An alcalde	Developing issue of Wujin, opinions on scale overlapping strategy from a Wujin perspective
3	Hotel	22 November 2014	An official in a functional zone	Finance issue of Wujin
4	Office	24 November 2014	Official A in a department of district government	History of planning-making
5	Office	24 November 2014	Official B in a department of district government	"Green" strategy in planning
6	Office	24 November 2014	An vice-director of a planning institute in Wujin	Problems of excessive land consuming
7	Office	24 November 2014	An official of planning bureau, Changzhou	Conflicts of spatial development strategies between Changzhou and Wujin; opinions on scale overlapping strategy from a Changzhou perspective
C	**Local residents and managers of community property management companies (23 interviewees) (C1–C23)**			
1–4	Sijixincheng zone; Chengzhonghuayuan zone;	22 November 2014	Two residents and two managers of property management companies	Vacancy rate; mental map; community sense; future plan of property investment
5–10	Nandianyuan zone; tianjuanfeng zone; xinchengnandu zone; xinchengnandu zone	22 November 2014	Three residents and three managers of property management companies	Vacancy rate; mental map; community sense; future plan of property investment
11–16	Laimengcheng zone; Xingheguoji zone; Moershangpin zone	22 November 2014	Three residents and three managers of property management companies	Vacancy rate; mental map; community sense; future plan of property investment
17–23	Other 8 zones	22 November 2014	One resident or manager of property management company in each zone	Vacancy rate

Table A1. *Cont.*

	Venues	Data	Interviewees	Contents
D	**Migrant workers (7 interviewees) (D1–D7)**			
1–3	Taxi; hotel; restaurant	23 November 2014	Three migrant workers in service sector	Attraction of Wujin's living environment and their settlement plans; mental map
4–5	Street; store	23 November 2014	Two migrant workers in manufacturing sector	Attraction of Wujin's living environment and their settlement plans; mental map
6–7	open spaces	23 November 2014	two migrant worker in high-tech industries	Attraction of Wujin's living environment and their settlement plans; mental map
E	**planners (4 interviewees) (E1–E4)**			
1	In a meeting	15 October 2009	A planner joined in the master planning 1993	Details of the planning process
2	By telephone	26 November 2014	A planner joined in the strategic planning 2003	Details of the planning process
3	In university	4 November 2014	A planner joined in the strategic planning 2009	Details of the planning process
4	By telephone	26 November 2014	A planner in a state-run planning institute of Changzhou	Opinions on scale overlapping strategy from a Changzhou perspective

References

1. Howitt, R. Scale. In *A Companion to Political Geography*; Agnew, J., Mitchell, K., O'Tuathail, G., Eds.; Blackwell: Oxford, UK, 2003; pp. 138–157.
2. Lefebvre, H. *The Production of Space*; Blackwell: Malden, MA, USA, 1991.
3. Smith, N. Scale. In *The Dictionary of Human Geography*, 4th ed.; Johnston, D., Gregory, G., Pratt, G., Watts, M., Eds.; Blackwell: Oxford, UK, 2000; pp. 724–727.
4. Marston, S.A. The social construction of scale. *Prog. Human Geogr.* **2000**, *2*, 219–242. [CrossRef]
5. Marston, S.A.; Jones, J.P.; Woodward, K. Human geography without scale. *Trans. Inst. Br. Geogr.* **2005**, *4*, 416–432. [CrossRef]
6. Armstrong, W.; McGee, T.G. *Theatres of Accumulation: Studies in Asian and Latin American Urbanization*; Routledge: New York, NY, USA, 2007.
7. McCann, P.; Acs, Z.J. Globalization: countries, cities and multinationals. *Reg. Stud.* **2011**, *1*, 17–32. [CrossRef]
8. Luo, X.; Cheng, Y.; Yin, J.; Wang, Y. Province-leading-county as a scaling-up strategy in China: The case of Jiangsu. *China Rev.* **2014**, *1*, 125–146.
9. Shen, J. Scale, state and the city: Urban transformation in post-reform China. *Habitat Int.* **2007**, *3*, 303–316. [CrossRef]
10. Wu, Q.; Cheng, J.; Liu, D.; Han, L.; Yang, Y. *"Kunming—A Regional International Mega city in Southwest China" in Urban Development Challenges, Risks and Resilience in Asia Mega Cities*; Singh, R.B., Ed.; Springer: Tokyo, Japan, 2015; pp. 323–351.
11. Huang, Y.; Li, S.M. *Housing Inequality in Chinese Cities*; Routledge: New York, NY, USA, 2007.
12. Li, S.M. Housing inequalities under market deepening: The case of Guangzhou, China. *Environ. Plan. A* **2012**, *12*, 2852–2866. [CrossRef]

13. Wu, Q.; Cheng, J.; Chen, G.; Hammel, D.; Wu, X. Socio-spatial differentiation and residential segregation in the Chinese city: Based on the 2000 community-level census data: A case study of the inner city of Nanjing. *Cities* **2014**, *39*, 109–119. [CrossRef]
14. Huang, R.J.; Zhang, Y.; Bozzetti, C.; Ho, K.F.; Cao, J.J.; Han, Y.; Daellenbach, K.R.; Slowik, J.G.; Platt, S.M.; Canonaco, F.; *et al.* High secondary aerosol contribution to particulate pollution during haze events in China. *Nature* **2014**, *514*, 218–222. [PubMed]
15. Yu, H. Size and Characteristic of Housing Bubbles in China's Major Cities: 1999–2010. *China World Econ.* **2011**, *6*, 56–75. [CrossRef]
16. Kates, R.W. What kind of a science is sustainability science? *Proc. Natl Acad. Sci. USA* **2011**, *49*, 19449–19450. [CrossRef]
17. Smith, N. Geography, difference and the politics of scale. In *Postmodernism and the Social Sciences*; Doherty, J., Graham, E., Malek, M., Eds.; Macmillan: London, UK, 1992; pp. 57–59.
18. Smith, N. *Uneven Development: Nature, Capital and the Production of Space*; Blackwell: Oxford, UK, 1984; pp. 34–65.
19. Swyngedouw, E. Neither global nor local: "Globalization" and the politics of scale. In *Spaces of Globalization*; Cox, K.R., Ed.; The Guilford Press: New York, NY, USA, 1997; pp. 137–166.
20. Swyngedouw, E. Scaled and geographies: Nature, place, and the politics of scale. In *Scale and Geographic Inquiry: Nature, Society, and Method*; Sheppard, E., McMaster, R.B., Eds.; Blackwell: Malden, MA, USA, 2004; pp. 129–153.
21. Brenner, N. The limits to scale? Methodological reflections on scalar structuration. *Prog. Human Geogr.* **2001**, *4*, 591–614. [CrossRef]
22. Giddens, A. *The Constitution of Society: Outline of the Theory of Structuration*; Polity: Cambridge, UK, 1984.
23. Harvey, D. *Rebel Cities: From the Right to the City to the Urban Revolution*; Verso: New York, NY, USA, 2012.
24. Ostrom, E. *Governing the Commons: the Evolution of Institutions for Collective Action*; Cambridge University Press: Cambridge, UK, 1990.
25. Taylor, P. A materialist framework for political geography. *Trans. Inst. Br. Geogr.* **1982**, *7*, 715–734. [CrossRef]
26. Smith, N. Homeless/global: Scaling places. In *Mapping the Futures: Local Cultures Global Change*; Robertson, G., Tickner, L., Eds.; Routledge: London, UK, 1993; pp. 87–120.
27. Harvey, D. From managerialism to entrepreneurialism: the transformation in urban governance in late capitalism. *Geogr. Ann.* **1989**, *1*, 3–18. [CrossRef]
28. Ma, L.J.C. Urban administrative restructuring, changing scale relations and local economic development in China. *Polit. Geogr.* **2005**, *24*, 477–497. [CrossRef]
29. Chung, H. The change in China's state governance and its effects upon urban scale. *Environ. Plan. A* **2007**, *39*, 789–809. [CrossRef]
30. Ma, L.J.C. Urban transformation in China, 1949–2000: A review and research agenda. *Environ. Plan. A* **2002**, *34*, 1545–1569. [CrossRef]
31. Zhu, J. Local growth coalition: The context and implications of China's gradualist urban land reforms. *Int. J. Urban Reg. Res.* **1999**, *25*, 534–548. [CrossRef]
32. Zhang, T. Urban development and a socialist pro-growth coalition in Shanghai. *Urban Aff. Rev.* **2002**, *4*, 475–499. [CrossRef]
33. He, S.; Wu, F. Property-led redevelopment in post-reform China: A case study of Xintiandi redevelopment project in Shanghai. *J. Urban Aff.* **2005**. [CrossRef]
34. Brenner, N. Globalisation as reterritorialisation: The re-scaling of urban governance in the European Union. *Urban Stud.* **1999**, *3*, 431–451. [CrossRef]
35. Business Insider. Available online: http://www.businessinsider.com/pictures-chinese-ghost-cities-2010-12?slop=1#slideshow-start (accessed on 20 November 2014).
36. Zhang, J.; Wu, F. China's changing economic governance: Administrative annexation and the reorganization of local governments in the Yangtze River Delta. *Reg. Stud.* **2007**, *1*, 3–21.
37. Qian, Y. How Reform Worked in China? In *In Search of Prosperity: Analytic Narratives on Economic Growth*; Rodrik, D., Ed.; Princeton University Press: Princeton, NJ, USA, 2003; pp. 297–333.
38. Ma, L.J.C.; Fan, M. Urbanization from below: The growth of towns in Jiangsu, China. *Urban Stud.* **1994**, *10*, 1625–1645. [CrossRef]

39. Cui, G.; Ma, L.J.C. Urbanization from below in China: Its development and mechanisms. *Acta Geogr. Sinica* **1999**, *2*, 106–115. (In Chinese)
40. Changzhou Municipal Bureau of Statistics, National Bureau of Investigation Corps Changzhou. *Changzhou Statistical Yearbook (2013)*; China Statistics Press: Beijing, China, 2013.
41. Yin, R. *Case Study Research: Design and Methods*, 3rd ed.; Sage: London, UK, 2003.
42. Wujin Urban Planning Bureau (WUPB). *Wujin County Master Planning Report (1993–2020)*; WUPB: Changzhou, China, 1995.
43. Wujin Urban Planning Bureau (WUPB). *Wujin District Strategic Planning Report (2003)*; WUPB: Changzhou, China, 2003.
44. Lardy, N.R. *Sustaining China's Economic Growth after the Global Financial Crisis*; Peterson Institute for International Economics: Washington, DC, USA, 2012.
45. Wujin Urban Planning Bureau (WUPB). *Wujin District Strategic Planning Report (2009)*; WUPB: Changzhou, China, 2009.
46. Friedmann, J. The world city hypothesis. *Dev. Change* **1986**, *1*, 69–83. [CrossRef]
47. Chien, S. Institutional innovations, asymmetric decentralization, and local economic development: A case study of Kunshan, in post-Mao China. *Environ. Plan. C* **2007**, *25*, 269–290. [CrossRef]
48. Wujin Urban Planning Bureau (WUPB). *Wujin City Strategic Planning Report (1998–2020)*; WUPB: Changzhou, China, 1999.
49. Wujin Urban Planning Bureau (WUPB). *Urban Construction Development Report of Wujin Urban Area 2013*; WUPB: Changzhou, China, 2014.
50. Jinling Evening News. Available online: http://jlwb.njnews.cn/html/2014-07/18/content_1659081.htm (accessed on 28 November 2014). (In Chinese)
51. NetEase Finance. Available online: http://money.163.com/13/0226/08/8OKKF2EJ00253B0H.html (accessed on 28 November 2014). (In Chinese)
52. Wujin Urban Planning Bureau (WUPB). *12th Five-Year Plan for Housing Development in Wujin Urban Area*; WUPB: Changzhou, China, 2011.
53. Xu, J.; Yeh, A.G.O. City repositioning and competitiveness building in regional development: new development strategies in Guangzhou, China. *Int. J. Urban Reg. Res.* **2005**, *29*, 283–308. [CrossRef]
54. Wujin District Bureau of Statistics, National Bureau of Investigation Corps Wujin. *Wujin Statistical Yearbook (2004)*; Wujin District Bureau of Statistics: Changzhou, China, 2004.
55. Comprehensive Department, National Bureau of Statistics (CD-NBS). *China Statistical Yearbook for Regional Economy (2004)*; China Financial & Publishing House: Beijing, China, 2004.
56. Jiangsu Province Bureau of Statistics, National Bureau of Investigation Corps Jiangsu. *Jiangsu Statistical Yearbook (2004–2013)*; China Statistics Press: Beijing, China, 2004–2013.
57. Lin, G.C.; Yi, F. Urbanization of capital or capitalization on urban land? Land development and local public finance in urbanizing China. *Urban Geogr.* **2011**, *1*, 50–79. [CrossRef]
58. Ding, C.; Lichtenberg, E. Land and urban economic growth in China. *J. Reg. Sci.* **2011**, *2*, 299–317. [CrossRef]
59. Wu, Q.; Waley, P. Coalescing for local growth in Kunming, China. *Int. J. Urban Reg. Res.* **2015**, unpublished work.
60. China Center for Urban Development (CCUD). Available online: http://www.ccud.org.cn/2013-09-26/113349759.html (accessed on 20 November 2014).
61. Shanghai Municipal Bureau of Statistics, National Bureau of Investigation Corps Shanghai. *Shanghai Statistical Yearbook (2004–2013)*; China Statistics Press: Beijing, China, 2004–2013.
62. Harvey, D. *A Brief History of Neoliberalism*; Oxford University Press: New York, NY, USA, 2005.
63. Davidson, K.; Gleeson, B. The sustainability of an entrepreneurial city? *Int. Plan. Stud.* **2014**, *2*, 173–191. [CrossRef]
64. McCann, E. Policy boosterism, policy mobilities, and the extrospective city. *Urban Geogr.* **2013**, *1*, 5–29. [CrossRef]
65. While, A.; Jonas, A.E.; Gibbs, D. The environment and the entrepreneurial city: Searching for the urban 'sustainability fix' in Manchester and Leeds. *Int. J. Urban Reg. Res.* **2004**, *3*, 549–569. [CrossRef]

sustainability

MDPI

Article

A Framework for Sustainable Urban Water Management through Demand and Supply Forecasting: The Case of Istanbul

Murat Yalçıntaş [1,†], Melih Bulu [2,†,*], Murat Küçükvar [3,†] and Hamidreza Samadi [3,†]

1 Business Administration, Istanbul Ticaret University, Istanbul 34445, Turkey; myalcintas@ticaret.edu.tr
2 School of Management and Administrative Sciences, Istanbul Sehir University, Istanbul 34662, Turkey
3 Department of Industrial Engineering, Istanbul Sehir University, Istanbul 34662, Turkey; muratkucukvar@sehir.edu.tr (M.K.); hamidrezasamadi@std.sehir.edu.tr (H.S.)

* Author to whom correspondence should be addressed; melihbulu@sehir.edu.tr; Tel.: +90-216-559-9085; Fax: +90-216-474-5353.

† These authors contributed equally to this work.

Academic Editors: Tan Yigitcanlar and Md. Kamruzzaman

Received: 30 March 2015; Accepted: 3 August 2015; Published: 13 August 2015

Abstract: The metropolitan city of Istanbul is becoming overcrowded and the demand for clean water is steeply rising in the city. The use of analytical approaches has become more and more critical for forecasting the water supply and demand balance in the long run. In this research, Istanbul's water supply and demand data is collected for the period during 2006 and 2014. Then, using an autoregressive integrated moving average (ARIMA) model, the time series water supply and demand forecasting model is constructed for the period between 2015 and 2018. Three important sustainability metrics such as water loss to supply ratio, water loss to demand ratio, and water loss to residential demand ratio are also presented. The findings show that residential water demand is responsible for nearly 80% of total water use and the consumption categories including commercial, industrial, agriculture, outdoor, and others have a lower share in total water demand. The results also show that there is a considerable water loss in the water distribution system which requires significant investments on the water supply networks. Furthermore, the forecasting results indicated that pipeline projects will be critical in the near future due to expected increases in the total water demand of Istanbul. The authors suggest that sustainable management of water can be achieved by reducing the residential water use through the use of water efficient technologies in households and reduction in water supply loss through investments on distribution infrastructure.

Keywords: water supply; demand; time-series forecasting; ARIMA; urban water sustainability; Istanbul

1. Introduction

In 1995, the former World Bank Vice-President Ismail Serageldin claimed that "*the wars of the next century will be about water*" [1]. Whether this hypothesis comes true or not, we are living in a world becoming constrained by water and the human beings are facing with serious social and economic problems related to accessibility of clean water resources. Despite occupying as little continental surface as 2%, more than 50% of the world population today lives in cities and this number is expected to reach 70% in 2050 [2]. In the future, cities will continue to play a major economic role, as well as contribute to further environmental degradation through water resource consumption, climate change and pollution [3,4]. Sufficient water supplies with a reasonable cost will continue to become one of the top agenda items for city decision-makers due to overcrowded cities and the climate change threats the water supply for most of the cities in Turkey. Opposite to common beliefs, Turkey is not a water rich country and the World Water Foundation (WWF)'s recent report on the water footprint of Turkey shows that country might be faced with serious water shortages by 2030 due to rising population, consumption, industrialization, and agricultural production [5]. In this regard, the use of forecasting models for understanding the long-term sustainability of water demand and supply becomes necessary.

In the literature, forecasting models are widely used in environmental studies to understand the level of pollution and identify the potential risks for depletion of the limited resources supplied by the ecological systems. As commonly used forecasting techniques, traditional methods such as time series, regression and an autoregressive integrated moving average (ARIMA) as well as soft computing techniques such as fuzzy logic, genetic algorithm, and artificial neural networks are being extensively used for a time-series demand forecasting [6–8]. Especially for urban water demand modeling, the ARIMA model has performed more accurately than time-series and multiple regression methods when forecasting demand based on climate variables [9]. ARIMA models are extensively used in time-series forecasting, especially in water and energy forecasting [10,11]. For instance, Praskievicz and Chang [12] used daily and monthly data from 2002 to 2007 to conduct a statistical analysis of seasonal water consumption in Seoul, South Korea. Significant improvement of the modeling of seasonal water use was achieved by developing the ARIMA models, which account for autocorrelation in the time series and explain up to 66% of the variance in water use. Ediger *et al.* [13] developed forecasting model for production of fossil fuel sources and compared the performance of two forecasting techniques: regression analysis and ARIMA. In other study, Ediger and Akar [14] used the ARIMA and seasonal ARIMA (SARIMA) methods to estimate the primary energy demand in Turkey. The authors concluded that the ARIMA forecasting of the total primary energy demand appears to be more reliable than the summation of the individual forecasts. Unakıtan and Türkekul [15] also assessed annual time series data for energy consumption using the ARIMA model. Erdoğdu [16] studied on the future growth of gas demand in Turkey and developed a forecasting model using an ARIMA framework. However, time-series ARIMA forecasting considers only the past values without showing determinants explicitly. In addition, this method is found to be suitable when climatic and socio-economic data are not available. Hence, the accuracy of ARIMA forecast may be limited due to the presence of high correlations among determinants and nonlinear relationships between water demand and its determinants [17].

Due to the aforementioned limitations of linear time series forecasting techniques, non-linear models are also used in demand forecasting in environmental studies. For instance, in predictive models for forecasting hourly urban water demand, artificial neural networks, projection pursuit regression, multivariate adaptive regression splines, compare a series of predictive models for forecasting water demand. Consequently, the results which obtained are nearly reliable in comparison by common forecasting models [18]. In other work, 39 multiple linear regression models, 9 time series models, and 39 artificial neural network models were developed to forecast the water demand and their relative performance was also compared. The neural network showed a better performance for prediction of daily summer water demand compared to multiple linear regression and time series analysis [19]. In other research, the performance of regression, time series, and neural network models are analyzed for short-term peak water demand forecasting. The significance of climatic variables such as rainfall and maximum temperature on water demand management is also investigated. The artificial neural network models consistently outperformed the regression and time-series models developed in this study [20]. To forecast urban annual water demand, Huang *et al.* [17] proposed a combination of models including wavelet transform (WT) and kernel partial least squares-autoregressive moving average (KPLS-ARMA). The combined models are applied to understanding the nonstationarity and forecasting the annual water demand of Dalian City, China. The results showed that the performance of combined WT, KPLS and ARMA outperforms other forecasting models. In other work, artificial neural network and ARIMA models are used for energy consumption forecasting. Four main factors were considered, including population, gross domestic product, exports, and total visitor arrivals. The results indicates neural network models release acceptable forecast accuracy when single predictor is considered but the forecast accuracy of artificial neural network does not improve extremely as the number of predictors increase [21].

In this research, the water supply and demand forecasting model is built based on the ARIMA method. Although several techniques including neural networks and multivariate regression analysis are used in the literature for time-series forecasting, with a given limited water demand and supply data for the period between 2006 and 2014, the authors used ARIMA in their analysis. As discussed initially, ARIMA forecasting is extensively used in urban water demand forecasting. However, there is still an important knowledge gap on the use of forecasting models for urban water supply and demand, simultaneously. This research used ARIMA for both demand and supply forecast covering the period between 2015 and 2018 and tried to estimate the long-term viability of urban water management practices in Istanbul looking at the gap between demand and supply. To address these research objectives, first, data including water demand categories and supply sources are collected. Then, the ARIMA forecasting model is developed based on a time-series real data obtained from the Istanbul Metropolitan Municipality. In this regard, this paper has the following research objectives:

(1) analyze the water demand of Istanbul based on five consumption categories such as residential, commercial, industrial, agriculture, park and gardens and others,
(2) analyze the water supplied to Istanbul based on different supply sources such as pipeline, dams and underground,
(3) present the share of the water demand categories, supply sources, and water losses between period of 2006 and 2014,
(4) evaluate the sustainability of water supply polices for until 2018 by looking at the gap between water supply and demand,
(5) highlight policy areas need urgent attention in order to sustain the current urban water management practices and provide a vital guidance and analytical framework for city water planners for future.

The rest of the paper is organized as follows. First, the data collection process is described and the ARIMA forecasting model is briefly explained in a step-by-step manner. Next, the findings are presented for water demand categories and supply sources as well as forecasting results for the period covering 2015 and 2018. Finally, the findings are summarized and the future work is pointed out.

2. Data Collection

This work is based on a real dataset obtained from the Istanbul Water Supply and Sewage Administration (ISKI) which is governed by the Municipality of Metropolitan Istanbul [22]. Since the data for demand and supply was available starting from the 2006, the initial dataset covered the period between 2006 and 2014. The dataset is first divided into two main categories: water demand and water supply. Each category is then divided into several sub-categories based on water demand type and supply source. First, the water demand data account for the total annual amount of water used by the following consumption categories (see Table 1):

Table 1. Water demand categories and their descriptions.

Demand Categories (*)	Description of Each Category
Residential	The amount of annual water used by households (invoiced water expense).
Commercial	The amount of annual water used by commercial facilities such as hospitals, schools, banks, hotels, offices, *etc.*
Industrial	The amount of annual water used by industrial facilities such as power plants, textile factories, cement plants, *etc.*
Outdoor	The amount of annual water used outdoor activities such as garden irrigation and/or parks.
Agricultural	The amount of annual water used for agricultural irrigation in rural areas.
All other unspecified	All other unspecified annual water use categories.

(*) The amount of demand categories is presented in terms of meter cubes (m^3).

Second, the water supply represents the total annual amount of water supplied to city which includes the following supply sources (see Table 2):

Table 2. Water supply sources and their descriptions.

Supply Sources (*)	Description of Each Supply Source
Pipeline and Regulator	The amount of annual water supplied by the Yesilçay and Melen regulators.
Dams	The total annual water supplied by dams such as Alibeyköy, Büyükçekmece, Darlık, Istrancalar, Kazandere, Ömerli, Pabuçdere, Elmalı, Sazlıdere and Terkos. This is the largest water supply source for the city.
Underground	The amount of annual water withdrawal from the underground reservoirs.

(*) The amount of water supply sources is presented in terms of meter cubes (m^3).

3. Methods

In this research, the ARIMA is utilized as reliable technique to forecast the demand and supply of water in Istanbul and consequently estimate the sustainability of demand and supply in the long run. The ARIMA is a commonly used time-series forecasting methodology in environmental studies [23,24]. The general model introduced by Box and Jenkins includes autoregressive as well as moving average parameters, and implicitly contains differencing in the formulation of the model [25]. In general, the three sorts of parameters in the model are known as: the autoregressive parameters (p), the number of differencing passes (d), and moving average parameters (q). In general, the ARIMA method consists of five main iterative procedures which are listed as follows:

(1) Stationarity checking and differencing,
(2) Model identification,
(3) Parameter estimation,
(4) Diagnostic checking, and
(5) Forecasting

ARIMA has been emanated from the autoregressive model (AR) and the moving average model (MA) and the integration of the AR and MA result in the ARIMA model, which was introduced in 1926, 1937, and 1938, respectively [13]. Although this technique has been extensively used for a time series forecasting, ARIMA model requires instruction in statistical analysis as well as appropriate knowledge of the field of application. Furthermore, the accessibility of an easy to use but adjustable specialized computer program is essential [26]. Therefore, one of the software which is used for developing ARIMA forecasting model is the R statistical software package.

In the ARIMA forecasting, the first step starts with creating statistical control charts to illustrate the trend of data before running the ARIMA. Hence, an initial step in analyzing time series data is running I-MR charts which show the control limits on the chart based on the mean of the moving range (the absolute difference between each consecutive pair of points). If there is no independence between the points, we have the so-called condition autocorrelation which there is little difference in each consecutive pair of points. Moreover, we used first autocorrelation (ACF) and partial autocorrelation (PACF) to understand the correlations between data at different points in time lagged by one or more periods. In the next step of constructing the ARIMA model, the data needs to be stationary. In other words, there should be no trend in the process either upwards or downwards [13,14]. To achieve this condition, certain stability in the process mean needs to be achieved which the differencing data. Next, we run first autocorrelation (ACF) and partial autocorrelation (PACF) to understand the correlations of differencing. The autoregressive part of the ARIMA model predicts the value at time *t* by considering previous values in the series at time *t-1*, *t-2*, *etc.* The moving average (MA) uses past residual values which show the differences between the actual value and the predicted value based on the model at time *t*, so that the ARIMA model allocated (p ,d, q), which are named as the number of different passes, moving average parameters, and autoregressive parameters, respectively [21,23]. After checking the data and identifying the aforementioned parameters, we run the R-software to forecast a set of factors: (1) water demand of city; (2) water supply; and (3) water supply without the pipeline projects for the period between 2014 and 2018. The visual check of the accuracy of forecasts is used for determining whether or not the current exponential smoothing model fits the data. Hence, the precision of the forecasting are evaluated by different methods. In this paper, we simply used the Mean Absolute Percentage Error (MAPE) for testing the ARIMA model performance. MAPE represents the accuracy as a percentage and value is simply calculated using the following formulation [27]:

$$\frac{1}{n}\sum_{t=1}^{n}\left|\frac{At-Ft}{At}\right| \quad (1)$$

where A_t is the actual value, F_t is the forecast value and n is the number of data points.

4. Results and Discussion

In this paper, the results are presented in two sub-sections. In the Section 4.1, water demand and supply characteristics of Istanbul are described in details. In the Section 4.2, a time-series ARIMA forecasting results are discussed for both demand and supply categories.

4.1. Water Demand and Supply Analysis

4.1.1. Total Water Demand and Percentage Contribution of Demand Categories

In this section, the total water demand between 2006 and 2014 is analyzed. The results show that residential water consumption is higher than all other consumption categories. This consumption category is followed by commercial activities whereas industrial, parks and gardens, agriculture categories and others are found have a lower contribution to total water demand. Among these categories, only residential consumption has shown a steady increase between 2006 and 2014, and the remaining categories have an increasing and decreasing trend. Between 2006 and 2014, the net water demand shows the highest amount in 2013 and 2014 while the minimum amount of water demand is observed in 2008. The total residential water use has a little fluctuation between 2006 and 2010 ranging between 3.86×10^8 m^3 and 3.96×10^8 m^3. Furthermore, the total water demand related to commercial activities is found to be the largest in 2008 and 2010, and after 2010, the total commercial water use has showed a declining trend. Overall, the total amount of water consumed in Istanbul reached to peak in 2013 and 2014 at 5.84×10^8 m^3 and 5.82×10^8 m^3, respectively.

In addition, the percentage contribution of all water demand categories for the period between 2006 and 2014 is analyzed. The results indicate that residential water consumption is responsible for approximately 80% of total water demand with an exception of 2008. For this year, the share of residential water use is found to be responsible for nearly 70% of total water use, and 15% of total water demand is attributed to the commercial activities. On the other hand, the percentage share of industrial and agriculture activities are less than 1% of total water demand in 2013 and 2014. This finding clearly showed that the water demand of Istanbul is largely dominated by household use and industrial and agriculture activities have a lower impact on the net water demand of the city. Although Marmara region has the largest industrial facilities in Turkey, the industrial sectors are not water intensive as compared to residential consumption. When compared to other cities, Istanbul is found have a high unit price of water which makes Istanbul unaffordable city for water intensive industrial production. Therefore, it is important to note that reducing the residential water use can be the most critical policy strategy to minimize the net water footprint of the city in the future.

4.1.2. Total Water Supply and Percentage Contribution of Supply Sources

Sakarya and Melen regulators (pipeline projects), dams, and underground represent the main water supply sources. The results showed that the largest amount of water supplied to city is obtained from the dams. Starting from 2011, the Sakarya and Melen regulators have become the second important water supply source after the dams. The amounts of water obtained from these two regulators have steadily increased between 2011 and 2014. In addition, starting from 2010, there is an increasing trend for the water supplied to city for both dams and regulators. After dams and Sakarya and Melen regulators, Pabuçdere, Kazandere and Istrancadere are the most important water supply sources for Istanbul. Based on the results, in 2014, the highest amount of water is supplied to city with a value of 9.24×10^8 m^3 while the least amount of water is supplied in 2006 with a value of 7.32×10^8 m^3. This result indicates that there is approximately 25% increase in the net amount of water supplied to city between 2006 and 2014.

In addition, the percentage share of each water supply source for the period during 2006 and 2014 is presented. The results clearly show that dams are the main water supply source of the city. For instance, in 2006, more than 95% of the water is supplied by the dams. The share of water obtained from the underground reservoirs is always less than 5% of total water supply. Starting from 2011, the share of water supplied by pipeline projects has started to increase and almost 25% of total water is supplied by the Sakarya and Melen regulators. Between 2010 and 2014, the percentage share of water supplied by dams has steadily decreased and there has been an increasing share of water obtained from the pipeline systems. The findings of city water supply indicate that water supply of Istanbul is largely dominated by dams and regulators. These facts might bring several important risks in the long run. For instance, the water level in dams in Istanbul is able to show significant fluctuations over the last decade that can create a risky condition for city water supply. In addition, the water is supplied by regulators are obtained from the neighbor city Sakarya and a conflict related to water trading between these cities may put a risk for meeting the total water demand in the future.

4.1.3. Total Water Coming and Supplied to City *versus* Water Demand

Figure 1a presents the net amount of water coming to city, supplied to city, and water demand between 2006 and 2014. This analysis will be important to understand the gap between incoming water *versus* supplied water and water demand. The results show that in 2006 and 2010, the difference between incoming water and supplied water was minimum at 6.45×10^7 m^3 and 7.35×10^7 m^3, respectively. In other words, for 2006 and 2010, the amount of water coming to city just met with the supply requirement. Starting from 2010, the gap between incoming and supplied water has steadily increased until 2014. For this year, the difference between incoming water to city and water supplied to city is found to the largest compared to other years. When compared to 2014, the net water surplus (difference between incoming water and water supplied to city) is found to be 9 times more than the surplus in 2006. The construction of Sakarya and Melen regulators has played a significant role in this result. Especially, starting from 2011, there has been a growing water supply to city from these regulators. Overall, the amounts of incoming water, supplied water, and demand have shown a steadily increasing trend after 2009.

In addition, the trend for total water supply and demand and the gap between them are also investigated. In this research, the difference between the net water supplied to city and water demand is called a water loss. The findings show that the difference between demand and supply was lower in 2008 compared to other years. Starting from 2009, the supply and demand gap, in order words water loss, has started to increase and reached to its maximum value in 2014. Compared to 2006 levels, the total water loss is found to be around 1.5 times more in 2014. In 2013 and 2014, the net water loss is calculated as 3.26×10^8 m^3 and 3.42×10^8 m^3, respectively. On the other hand, the difference between supply and demand is found as 2.34×10^8 m^3 in 2006.

Figure 1b also presents the water demand, supply and incoming water to city without Sakarya and Melen regulators. This analysis is important to understand the trend without these pipeline projects and see whether city-planners are able to meet supply and demand requirements without these investments. The findings showed that the difference between incoming water and water supplied to city are lower in 2009 and 2010. After 2009, the amount of water supplied to city has grown continuously until 2014. Especially, the incoming water has reached its peak in 2014 and the gap between incoming water and supplied water is found to be the largest in 2014. When compared to Figure 1a, the total incoming water to city is lower; however incoming water amount is still able to meet the supply requirements. As presented in Figure 1a, in 2014, the total incoming water is 1.6 times more than the water supplied to city whereas this ratio is found as 1.2 times when there is no Sakarya and Melen pipeline projects for the city (see Figure 1b). However, to see the long-term sustainability of the incoming water, the ARIMA forecasting results are presented in the Section 4.2.

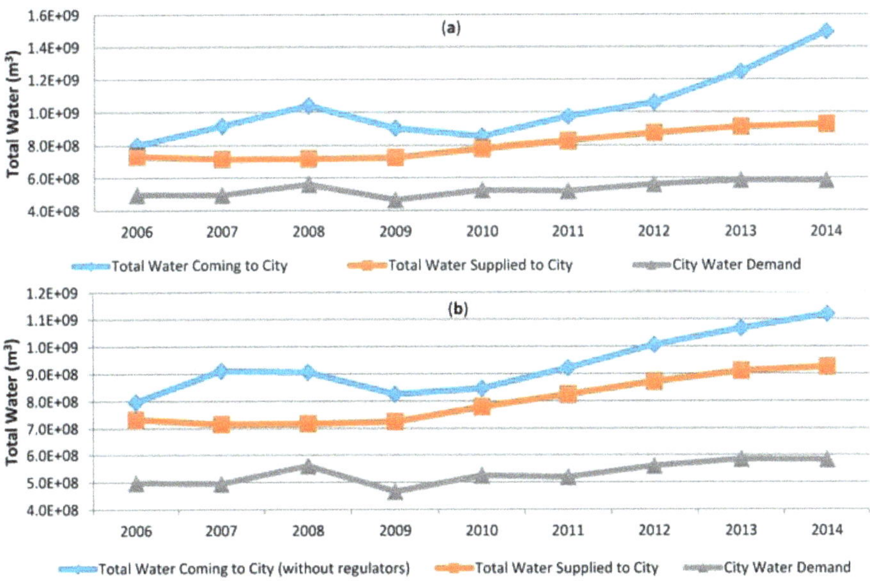

Figure 1. Water supply and demand between 2006 and 2014 (m^3). (**a**) With Sakarya and Melen regulators; (**b**) Without Sakarya and Melen regulators.

4.1.4. The Net Water Loss Analysis

In this section, the net water loss of water supply is analyzed considering the gap between the amount of demand and water supplied to city between 2006 and 2014. In addition, three important metrics such as water loss to demand ratio, water loss to supply ratio, and water loss to residential demand ratio are presented.

First, Figure 2a presents the trend of water supply, demand and loss between 2006 and 2014. This analysis revealed important insights regarding the time-series change of water supply and demand as well as water loss. The results indicate that the total water loss has a minimum value in 2008 at 1.56×10^8 m^3. After 2008, the total water loss has shown a growing trend until 2014. Similar to water loss, the net amount of water supplied to city also has an increasing trend after 2010. Interestingly, in 2014, the gap between demand and supply is found to be the largest with a comparison with other years. Figure 2b also shows the two important ratios such as *water loss to supply ratio* and *water loss to demand ratio*. First, the *water loss to supply ratio* is calculated by dividing the net water loss to net water supplied to city. The results showed that in 2013 and 2014, almost 35% of total supplied water is lost in the system. The water loss to supply ratio is found to be minimum in 2006, 2007 and 2008; however starting from 2010, there has been an increasing trend in the loss ratios.

In addition to water loss to supply ratio, the *water loss to demand ratio* is analyzed between 2006 and 2014. This ratio accounts for the net water loss against total water demand for the city. This ratio is found to have a minimum value in 2006, 2007, and 2008. On the other hand, after 2010, the water loss to supply ratio is found to be nearly 60% of total water demand of city. Similar to water loss to supply ratio, the highest value is observed in 2014 (see Figure 2b). Finally, Figure 2c depicts the *water loss to residential demand* ratio which is calculated by dividing the net water loss into the net residential water demand in city during 2006 and 2014. The results showed that this ratio was too high for some years. For instance, in 2014, the total water loss accounts for approximately 75% of total residential consumption in Istanbul. In other words, the city wasted 75% of household water use in the supply network. The value of this indicator is found to be the lowest in 2008 and stable between 2011 and 2014 accounting for over 70% of household water demand (see Figure 2c).

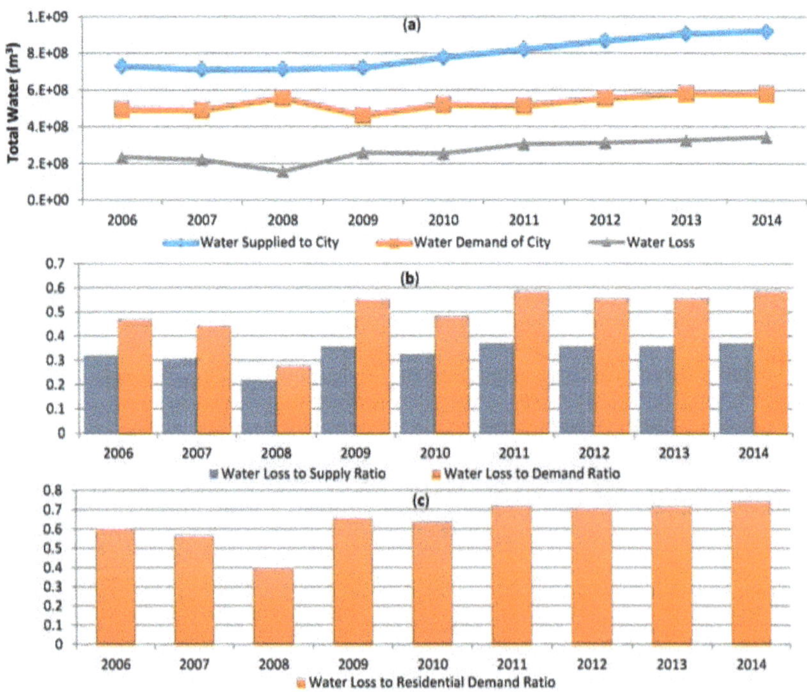

Figure 2. Water supply, demand and loss characteristics between 2006 and 2014. (**a**) Water supply, demand and loss (m^3); (**b**) Water loss to supply ratio and water loss to demand ratio; (**c**) Water loss to residential demand ratio.

4.2. ARIMA Demand and Supply Forecasting Results

After analyzing the water demand categories, supply sources and water loss amounts, it is important analyze the future of water demand and supply and the gap between these parameters in the long run. Therefore, the authors developed the ARIMA based time series forecasting model covering the period between 2015 and 2018. First, Figure 3a presents the testing data for ARIMA fit *versus* actual data between 2007 and 2014. The results showed that ARIMA model outputs and actual data have a good fit for the testing period. To understand the error between ARIMA outputs and actual data, the mean absolute percentage error (MAPE) is used as a statistical error check parameter. The MAPE results are presented in Table 3 for three estimates: demand, supply, and supply without pipeline projects. Based on the study findings, MAPE is calculated as 6.30 for demand forecast.

Table 3. MAPE values of demand and supply forecasts.

	Demand	Supply	Supply Without Pipeline Projects
MAPE	6.30	2.57	5.82

After testing the ARIMA model, the forecasting results are presented in Figure 3b. The forecasting results are illustrated for three estimates: upper limit, normal forecast and upper limit. This illustration helped us to reflect inherent uncertainties for forecasting outcomes. According to results, in 2018, the total water demand of city reached 7.72×10^8 m^3 for the upper limit forecast, 5.82×10^8 m^3 for the normal forecast, and 3.91×10^8 m^3 for the lower limit forecast. For the upper limit demand forecast, the net water consumption is expected to increase 1.32 times when compared to 2014 levels. For normal forecast value, this value is found to be similar to the water demand in 2014. Between 2015 and 2018, the upper limit forecast of water demand is ranged between 6.77×10^8 m^3 and 7.72×10^8 m^3.

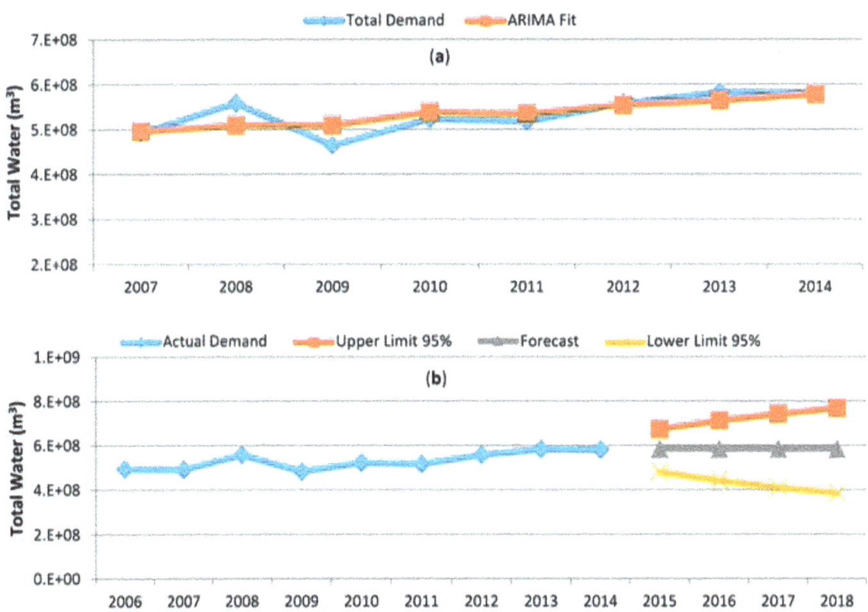

Figure 3. ARIMA forecasting results for water demand (m^3). (**a**) Actual demand *versus* ARIMA fit; (**b**) Demand forecasting data for upper, normal and lower limits.

In addition to water demand forecasting, the water supply forecasting results are also presented in Figure 4. Similar to demand forecasting, first, the ARIMA fit *versus* actual supply data between 2007 and 2014 is illustrated. The findings showed that the ARIMA forecast and real data have a good fit and little fluctuations are observed between actual data and ARIMA outputs. To estimate the error between ARIMA fits and actual data, MAPE is utilized. According to the findings, MAPE is calculated as 2.57. Hence, it is observed that water supply forecast has a better performance than the water demand forecast for testing data (see Table 3).

After testing the ARIMA model for supply data, the forecasting results are also illustrated in Figure 4b. Similar to water demand forecast, the forecasting results are visualized for three estimates such as upper limit, normal forecast, and upper limit. Based on the results, for the year 2018, the total water supplied to Istanbul is expected to reach to 1.11×10^9 m^3 for upper limit forecast, 1.02×10^9 m^3 for normal forecast, and 9.25×10^8 m^3 for lower limit forecast. For the upper limit supply forecast, the net water supply is expected to increase 1.21 times in comparison with the net supply in 2014. When we look at the normal forecast value, it is observed that this value is found to be 1.1 times more than the water supply in 2014. Between 2015 and 2018, the normal forecast of supply is ranged between 9.48×10^8 m^3 and 1.02×10^9 m^3.

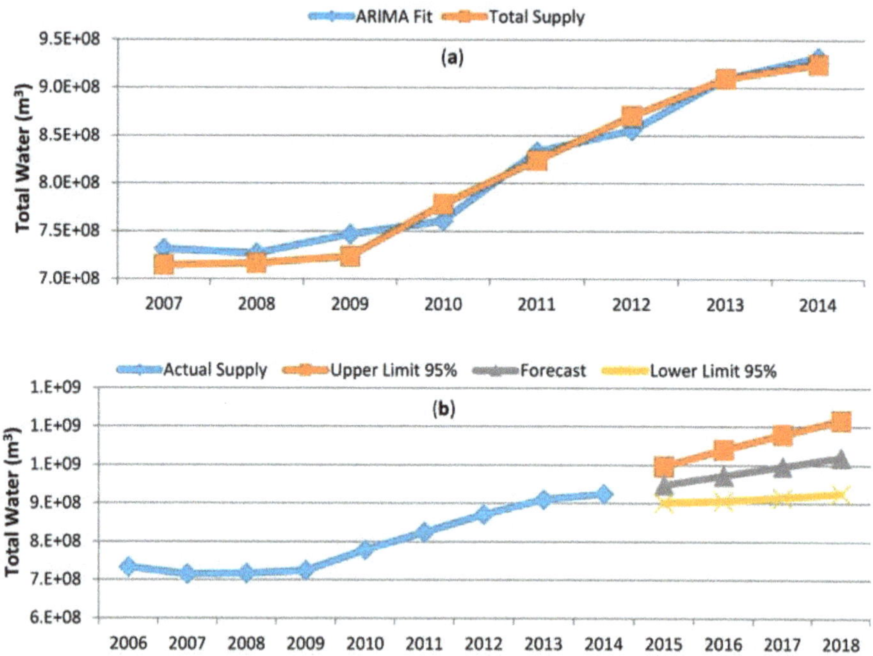

Figure 4. ARIMA forecasting results for water supply (m^3). (**a**) Actual supply *versus* ARIMA fit; (**b**) Supply forecasting data for upper, normal and lower limits.

Figure 5a shows the water demand and supply forecasts for the period between 2015 and 2018. The forecasting results are presented for three estimates: upper limit, normal forecast, and lower limit. The results presented in Figure 5a are based on water supplied to city with the Sakarya and Melen regulator projects. The forecasting results showed that supplied water in 2018 is able meet the water demand in this year. This result is found to be valid for both upper and lower limit forecasts and in both conditions, water supplied to city is significantly greater than the water demand in Istanbul. In 2018, the lower limit forecast of supplied water is found to be 1.19 times higher than the upper limit forecast of water demand. Furthermore, the lower limit forecast of water supply is 1.58 times greater than the normal forecast of water demand. Therefore, it is likely to conclude that the city is not expected to have a water shortage to meet the demand if the water supply is not affected by unforeseen environmental and climatic conditions.

Although Figure 5a presents the estimates of water demand and supply amounts, it is still important to see the future of city water supply without pipeline projects. The findings revealed interesting insights and show that for upper limit forecasts, the water supply is found to be quite larger than water demand until 2018. On the other hand, upper limit demand forecast values are getting closer to lower limit supply forecast values. In 2018, the value of lower limit supply forecast is found to be 8.43×10^8 m^3 and the upper limit demand forecasting is calculated as 7.72×10^8 m^3 (see Figure 5b). Therefore, it is important to note that pipeline projects might be highly critical for meeting the increasing city water demand. Without these projects, the city may suffer from water shortage in the near future. This case might also bring unforeseen risks for sustainable supply of city water due to fact that any conflict between neighboring cities can result in serious disputes related to water trade between Istanbul and water exporter cities.

Figure 5. ARIMA forecasting results for water demand and supply. (**a**) Demand *versus* supply (m^3); (**b**) demand *versus* supply without regulators (m^3).

5. Conclusions and Recommendations

This research is a first and critical attempt towards understanding the present and future of water demand and supply capacity of Istanbul and developing a time-series demand and supply forecasting model for sustainability of urban water management policies. Although water continues to become a highly critical and scarce resource for cities, there is unfortunately a little research done for analyzing the long-term sustainability of urban water management in Istanbul. Especially, by investigating the main demand categories and supply sources, the findings will guide the city decision makers to identify main areas that need an urgent attention in the long run. In addition, forecasting the future water demand and supply amounts, the decision makers will be able to assess the effectiveness of their water conversation and supply policies which can lead to management of the city water in a sustainable way. Hence, the methodology applied in this paper can be a practical framework for other metropolitan cities towards achieving sustainable urban water management. Overall, the findings of the current research highlighted the following points:

- The residential water consumption is found to be dominant accounting for approximately 80% of city water use. Hence, giving a high priority for reducing household water use and incentives for water efficient household equipment can be listed among the sound water reduction strategies. Specific to Istanbul, the water use related to industrial, commercial and agricultural activities are comparatively lower than residential use, and the importance of water reduction polices addressing these activities are not likely to diminish the net water demand of the city as much as policies addressing reductions in residential use. The water efficiency related targets can be achieved by using a range of water efficient components in toilets, showers, kitchen taps, basin taps, dishwashers, washing machines, and baths and the importance of using water efficient equipment and residential water conservation strategies and their impacts are widely discussed in the literature [28–30].
- The total water supplied to Istanbul is largely supplied by dams located in the city. This indicates that city's water supply is highly sensitive to changing climatic conditions and rainfall patters. As we are facing with dangerous climate change worldwide, the impacts of global climate change might affect the long-term sustainability of water supplied from the dams. This is because high temperatures and drought are able to bring water reserves to low levels in city dams. Although the municipality built pipeline projects in order to supply additional water to city from neighboring cities, this case can also create conflicts between Istanbul and water exporters. Political conflicts or unexpected water shortages in neighboring cities can be risk for Istanbul when supplying water from Sakarya and Melen regulators. The forecasting results also showed that the city might not meet with the future demand without these pipeline projects. Hence, it is likely to conclude that the water supply of Istanbul might be subject to important risks in the upcoming decades.
- The water sustainability metrics that are used in this research also revealed important insights regarding water loss characteristics of Istanbul. First, water loss to demand ratio showed that in 2011 and 2014, the total water loss accounts for around 60% of total water used in Istanbul. In other words, the city wasted more than 50% of the residential water due to deficiencies in the water delivery infrastructure. Although city planners seek to minimize the water use in households, the distribution loss remained another critical area needs an urgent attention to reduce the demand on clean water as well as additional water supply. Hence, reducing the water loss should be priority area for decision makers for sustainable future of the urban water and the significance of several water loss reduction strategies and investments are discussed in previous studies [31–33]. However, high upfront investment needs to improve the current water supply network and having historical ruins preventing new underground infrastructure projects are listed among the main obstacles for city decision makers.
- The authors urge that the water supply of water in Istanbul is primarily dependent on dams and therefore the amount of water stored in dams will be critical to supply enough water to city. Especially, the fluctuations in stored water in city dams are enormously high which can lead to problematic cases for certain periods. At this point, it is not sustainable for Istanbul to continue with its current water loss ratio ranging between 35% and 37%. In 2014, the Turkish government representatives proposed the first official regulation for water loss control. Starting from 2014, decreasing the water loss has become mandatory for the water utilities in Turkey. In addition, according to the 10th Development Plan of Turkey covering the period between 2014 and 2018, preventing water losses will be a main priority of Turkish government [34,35]. The findings of this paper revealed that water loss in distribution network is still high and continues to result in huge economic losses for the city. Although policies addressing residential water reduction are among the top agenda items in a sustainable urban water management, the investments for the

renovation of the water distribution network infrastructure should also have a similar priority in order to minimize the net water consumption effectively.

It should be kept in mind that current research is based on the limited dataset covering the period from 2006 to 2014. Although this work is a first attempt toward developing water supply and demand projection model for Istanbul, there is obviously further research needs on working with the long-term data series and addressing uncertainties in water supply and demand variables. For instance, the supply of water is highly dependent on the rainfall patters and temperature increases that are expected to have considerable fluctuations over the next decade. In addition, rising investments through mega-projects such as construction of The Yavuz Sultan Selim Bridge, The Third Airport with 150 million passenger capacities and The New City Project have a strong potential for increasing the city population as well as demand for clean water in the near future. Hence, much more attention should be given to probabilistic forecasting methods in order to reflect the role of uncertainty in future supply and demand forecasts of water in Istanbul [6,36]. In addition, real time water demand and supply tracking and management of city water through information technology (IT) are likely to be inevitable in the future and the importance of the algorithm-based IT applications and knowledge-based urban development is highlighted in the previous studies [37,38].

Last but not least, the traditional forecasting models including multivariate regression analysis, ARIMA, as well as more advanced models including soft computing, expert systems and artificial neural networks are mostly utilized for both short and long-term water supply and demand forecasts. However, these forecasting models are usually looking at cause and effect relationship without considering the dynamic complex interactions between the parameters of water demand and supply such as income level, water demand rate, increasing temperature, *etc.* Therefore, for future research, the authors suggest to model the water supply and demand projections of Istanbul using the system dynamics approach. In this way, the model will be able to take into account the interactions among economic and social dimensions, offering a realistic platform for practical use for sustainable urban water management by city-planners [39–42].

Acknowledgments: This material is based upon work supported in part by the Istanbul Water and Sewage Administration (ISKI). The authors are grateful to ISKI Strategic Management and Planning Department Head, Mehmet Palan for his support in data collection process.

Author Contributions: Murat Yalçıntaş carried out the analysis including work related to data collection and contributed to writing of the manuscript. Melih Bulu supervised the research and contributed to the literature review and conclusion sections. Murat Küçükvar worked on the selection of the water supply and demand parameters and contributed the model development part of the manuscript. Finally, Hamidreza Samadi run the ARIMA forecasting model using the R-software and obtained the forecasting results. All of the authors contributed equally to this research. All authors have read and approved the final manuscript.

Conflicts of Interest: The authors declare no conflict of interest.

References

1. Chakraborty, R.; Serageldin, I. Sharing of River Waters among India and its Neighbors in the 21st century: War or Peace? "The wars of the next century will be about water." . *Water Int.* **2004**, *29*, 201–208. [CrossRef]
2. UNEP. *Cities and Green Buildings: In the Transition to a Green Economy, a UNEP Brief by Moustapha Kamal Gueye*; UNEP-Economics and Trade Branch: Geneva, Switzerland, 2009; p. 3.
3. Wagner, I.; Breil, P. The role of ecohydrology in creating more resilient cities. *Ecohydrol. Hydrobiol.* **2013**, *13*, 113–134. [CrossRef]
4. Goonetilleke, A.; Yigitcanlar, T.; Ayoko, G.A.; Egodawatta, P. *Sustainable Urban Water Environment: Climate, Pollution and Adaptation*; Edward Elgar Publishing: Cheltenham, UK, 2014.
5. WWF. Su ayak izi raporu: Su,üretim ve uluslararası ticaret ilişkisi. World Water Foundation. Available online: http://awsassets.wwftr.panda.org/downloads/su_ayak_izi_raporweb.pdf (accessed on 10 February 2015).
6. Donkor, E.A.; Mazzuchi, T.A.; Soyer, R.; Alan Roberson, J. Urban water demand forecasting: Review of methods and models. *J. Water Resour. Plan. Manag.* **2012**, *140*, 146–159. [CrossRef]

7. Suganthi, L.; Samuel, A.A. Energy models for demand forecasting. *Renew. Sustain. Energy Rev.* **2012**, *16*, 1223–1124. [CrossRef]
8. Zhou, S.L.; McMahon, T.A.; Walton, A.; Lewis, J. Forecasting operational demand for an urban water supply zone. *J. Hydrol.* **2002**, *259*, 189–202. [CrossRef]
9. House-Peters, L.A.; Chang, H. Urban water demand modeling: Review of concepts, methods, and organizing principles. *Water Resour. Res.* **2011**, *47*. [CrossRef]
10. Billings, R.B.; Jones, C.V. *Forecasting Urban Water Demand*; American Water Works Association: Denver, CO, USA, 2008.
11. Wang, Y.; Wang, J.; Zhao, G.; Dong, Y. Application of residual modification approach in seasonal ARIMA for electricity demand forecasting: A case study of China. *Energy Policy* **2012**, *48*, 284–294. [CrossRef]
12. Praskievicz, S.; Chang, H. Identifying the relationships between urban water consumption and weather variables in Seoul, Korea. *Phys. Geogr.* **2009**, *30*, 324–337. [CrossRef]
13. Ediger, V.Ş.; Akar, S.; Uğurlu, B. Forecasting production of fossil fuel sources in Turkey using a comparative regression and ARIMA model. *Energy Policy* **2006**, *34*, 3836–3846. [CrossRef]
14. Ediger, V.Ş.; Akar, S. ARIMA forecasting of primary energy demand by fuel in Turkey. *Energy Policy* **2007**, *35*, 1701–1708. [CrossRef]
15. Unakitan, G.; Türkekul, B. Univariate Modelling of Energy Consumption in Turkish Agriculture. *Energy Sources Part B* **2014**, *9*, 284–290. [CrossRef]
16. Erdogdu, E. Natural gas demand in Turkey. *Appl. Energy* **2010**, *87*, 211–219. [CrossRef]
17. Huang, L.; Zhang, C.; Peng, Y.; Zhou, H. Application of a Combination Model Based on Wavelet Transform and KPLS-ARMA for Urban Annual Water Demand Forecasting. *J. Water Resour. Plan. Manag.* **2014**, *140*, 04014013. [CrossRef]
18. Herrera, M.; Torgo, L.; Izquierdo, J.; Pérez-García, R. Predictive models for forecasting hourly urban water demand. *J. Hydrol.* **2010**, *387*, 141–150. [CrossRef]
19. Adamowski, J.F. Peak daily water demand forecast modeling using artificial neural networks. *J. Water Resour. Plan. Manag.* **2008**, *134*, 119–128. [CrossRef]
20. Bougadis, J.; Adamowski, K.; Diduch, R. Short-term municipal water demand forecasting. *Hydrol. Process.* **2005**, *19*, 137–148. [CrossRef]
21. Lai, S.L.; Liu, M.; Kuo, K.C.; Chang, R. Energy Consumption Forecasting in Hong Kong Using ARIMA and Artificial Neural Networks Models. *Appl. Mech. Mater.* **2014**, *672*, 2085–2097. [CrossRef]
22. ISKI. Guncel su verileri. The Istanbul Metropolitan Municipality, Turkey. Available online: http://www.iski.gov.tr/web/barajdoluluk.aspx (accessed on 5 January 2015).
23. Diaz-Robles, L.A.; Ortega, J.C.; Fu, J.S.; Reed, G.D.; Chow, J.C.; Watson, J.G.; Moncada-Herrera, J.A. A hybrid ARIMA and artificial neural networks model to forecast particulate matter in urban areas: The case of Temuco, Chile. *Atmos. Environ.* **2008**, *42*, 8331–8340. [CrossRef]
24. Pao, H.T.; Tsai, C.M. Modeling and forecasting the CO_2 emissions, energy consumption, and economic growth in Brazil. *Energy* **2011**, *36*, 2450–2458. [CrossRef]
25. Pankratz, A. *Forecasting with Univariate Box-Jenkins Models: Concepts and Cases*; John Wiley & Sons: Hoboken, NJ, USA, 2009; Volume 224.
26. Melard, G.; Pasteels, J.M. Automatic ARIMA modeling including interventions, using time series expert software. *Int. J. Forecast.* **2000**, *16*, 497–508. [CrossRef]
27. Makridakis, S.; Wheelwright, S.C.; Hyndman, R.J. *Forecasting Methods and Applications*; John Wiley & Sons: Hoboken, NJ, USA, 2008.
28. Fidar, A.; Memon, F.A.; Butler, D. Environmental implications of water efficient microcomponents in residential buildings. *Sci. Total Environ.* **2010**, *408*, 5828–5835. [CrossRef] [PubMed]
29. Lee, M.; Tansel, B.; Balbin, M. Influence of residential water use efficiency measures on household water demand: A four year longitudinal study. *Resour. Conserv. Recycl.* **2011**, *56*, 1–6. [CrossRef]
30. Racoviceanu, A.I.; Karney, B.W. Life-cycle perspective on residential water conservation strategies. *J. Infrastruct. Syst.* **2010**, *16*, 40–49. [CrossRef]
31. Farley, M.; Trow, S. (Eds.) *Losses in Water Distribution Networks: A Practitioner's Guide to Assessment, Monitoring and Control*; IWA Publishing: London, UK, 2003.
32. Fontana, N.; Giugni, M.; Portolano, D. Losses reduction and energy production in water-distribution networks. *J. Water Resour. Plan. Manag.* **2011**, *138*, 237–244. [CrossRef]

33. Mutikanga, H.E.; Sharma, S.K.; Vairavamoorthy, K. Methods and tools for managing losses in water distribution systems. *J. Water Resour. Plan. Manag.* **2012**, *139*, 166–174. [CrossRef]
34. Kalkınma Bakanlığı, T.C. Onuncu Kalkınma Planı 2014–2018. Available online: http://www.kalkinma.gov.tr/Lists/Yaynlar/Attachments/518/Onuncu%20Kalk%C4%B1nma%20Plan%C4%B1.pdf (accessed on 15 January 2015).
35. Water Loss Forum. Post Show Report. Available online: http://www.waterlossforum.org/files/1st_waterlossturkeyforum_postshowreport.pdf (accessed on 15 January 2015).
36. Kurunç, A.; Yürekli, K.; Çevik, O. Performance of two stochastic approaches for forecasting water quality and streamflow data from Yeşilırmak River, Turkey. *Environ. Model. Softw.* **2005**, *20*, 1195–1200. [CrossRef]
37. Bulu, M.; Önder, M.A.; Aksakalli, V. Algorithm-embedded IT applications for an emerging knowledge city: Istanbul, Turkey. *Expert Syst. Appl.* **2014**, *41*, 5625–5635. [CrossRef]
38. Yigitcanlar, T.; Bulu, M. Dubaization of Istanbul: Insights from the knowledge-based urban development journey of an emerging local economy. *Environ. Plan. A* **2014**, *47*, 89–107. [CrossRef]
39. Ahmad, S.; Simonovic, S.P. Spatial system dynamics: New approach for simulation of water resources systems. *J. Comput. Civ. Eng.* **2004**, *18*, 331–340. [CrossRef]
40. Qi, C.; Chang, N.B. System dynamics modeling for municipal water demand estimation in an urban region under uncertain economic impacts. *J. Environ. Manag.* **2011**, *92*, 1628–1641. [CrossRef] [PubMed]
41. Mirchi, A.; Madani, K.; Watkins, D., Jr.; Ahmad, S. Synthesis of system dynamics tools for holistic conceptualization of water resources problems. *Water Resour. Manag.* **2012**, *26*, 2421–2442. [CrossRef]
42. Zarghami, M.; Akbariyeh, S. System dynamics modeling for complex urban water systems: Application to the city of Tabriz, Iran. *Resour. Conserv. Recycl.* **2012**, *60*, 99–106. [CrossRef]

sustainability

MDPI

Article

Neighborhood Sustainability Assessment: Evaluating Residential Development Sustainability in a Developing Country Context

Tan Yigitcanlar [1,*], Md. Kamruzzaman [1] and Suharto Teriman [2]

1 School of Civil Engineering and Built Environment, Queensland University of Technology, 2 George Street Brisbane, QLD 4001, Australia; md.kamruzzaman@qut.edu.au

2 School of Planning and Surveying, University of Technology MARA, Seri Iskandar, Bandar Baru Seri Iskandar, Perak 32610, Malaysia; suhar429@perak.uitm.edu.my

* Author to whom correspondence should be addressed; tan.yigitcanlar@qut.edu.au; Tel.: +61-7-3138-2418; Fax: +61-7-3138-1170.

Academic Editor: Marc A. Rosen

Received: 12 December 2014; Accepted: 16 February 2015; Published: 3 March 2015

Abstract: Rapid urbanization, improved quality of life, and diversified lifestyle options have collectively led to an escalation in housing demand in our cities, where residential areas, as the largest portion of urban land use type, play a critical role in the formation of sustainable cities. To date there has been limited research to ascertain residential development layouts that provide a more sustainable urban outcome. This paper aims to evaluate and compare sustainability levels of residential types by focusing on their layouts. The paper scrutinizes three different development types in a developing country context—*i.e.*, subdivision, piecemeal, and master-planned developments. This study develops a "Neighborhood Sustainability Assessment" tool and applies it to compare their sustainability levels in Ipoh, Malaysia. The analysis finds that the master-planned development, amongst the investigated case studies, possesses the potential to produce higher levels of sustainability outcomes. The results reveal insights and evidence for policymakers, planners, development agencies and researchers; advocate further studies on neighborhood-level sustainability analysis, and; emphasize the need for collective efforts and an effective process in achieving neighborhood sustainability and sustainable city formation.

Keywords: sustainability assessment; sustainable urban development; neighborhood sustainability; neighborhood sustainability assessment index; sustainable city; Ipoh; Malaysia

1. Introduction and Background

Sustainability has been a contested concept with many definitions since Brundtland report and hardly any consensus over a single term that can facilitate an easy measurement of the concept [1,2]. Consequently, the concept has been expanded with various disciplinary scopes [3,4]. In this paper, neighborhood sustainability is defined as the process of developing a neighborhood level urban form or built environment that meets the needs of its residents whilst avoiding unacceptable social and environmental impacts both locally and in a broader context [5]. By urban form, we refer to the spatial distributions of different land uses connected together with physical infrastructures and associated transport networks [6]. The way these features are distributed within a neighborhood has profound impact on sustainability both locally and globally. For example, research has shown that the availability of goods and services (e.g., diverse land uses) within local areas enables residents to participate fully in society (*i.e.*, meets the local needs for jobs, recreation, social, health activities), and in turn, contributes to economic and social sustainability locally [7]. In contrast, a lack of local opportunities encourages motorized travel and thereby affects the environmental sustainability both locally (e.g., noise, habitat fragmentation, increased impervious surface and consequent damages in water quality and the formation of urban heat island) and globally (e.g., air pollution and climate change) [8,9]. Therefore, different urban forms contribute differently to sustainability and research studies around the globe have indicated that the built environment is the most promising sector for a rapid transition to sustainability [10].

The need for a sustainable urban form at the local level has long been advocated by the United Nations [11] through its "Local Agenda 21" programs. Neighborhoods are considered as the building blocks of cities where most development (e.g., new buildings) takes place, and therefore, the overall sustainability of a city depends on the sustainability of its neighborhood [12]. However, past studies on sustainability assessment have focused on either the city level e.g., [13,14] or building level e.g., [15]; whereas the assessment of neighborhood sustainability, an intermediate level, has received very little attention in general and in the context of developing countries in particular [10,16].

Limited research to-date suggests that sustainable neighborhoods have a significant positive impact on property prices [17], and that people living in sustainable neighborhoods are happier [18] and enjoy a better quality of life and place [13]. Consequently, neighborhoods are increasingly gaining attention as planning units of great potential for contribution to sustainable urban development [19]. At the same time, an increasing urge for tools to assess their sustainability is recorded worldwide [10]. Neighborhood sustainability assessment (NSA) tools are defined as a set of criteria and themes; and are used to: (a) Evaluate and rate the performance of a given neighborhood; (b) Assess the neighborhoods' position on the way towards sustainability, and; (c) Specify the extent of neighborhoods' success in approaching sustainability goals [16].

NSA tools have conveniently been used to benchmark the sustainable efficiency of neighborhood developments [20]. For example, Han *et al.* [21] estimated sustainability level of an eco-community (*i.e.*, Xihe in China), and found that it achieved only a moderate sustainability level despite the community was built to become a sustainable neighborhood. NSA tools have also been used to provide greenness certificates of neighborhoods by respective authorities [17,22]. Using a NSA tool, Li *et al.* [23] compared the sustainability levels of 52 mining communities and found that four of them have reached a strong level of sustainability, 11 have achieved a satisfactory level, and the remaining 37 are still weak in their sustainability endeavors. More importantly, the availability of a NSA tool helps authorities to focus development towards sustainable outcomes. For example, after analyzing 19 housing developments throughout England, Smith *et al.* [22] found that in the absence of appropriate NSA tools, even where there is a desire to create a more sustainable solution, many schemes are falling short of their potential.

A number of NSA tools are currently operational around the world. The well-known ones include but not limited to the followings: LEED ND, UK; BREEAM for Communities, UK; CASBEE-UD, Japan; ECC, USA; HQE2R, European Union; Ecocity, European Union; SCR, Australia; QSAS, Qatar; Green Mark for Districts, Singapore; NSF, New Zealand; HK-BEAM, Hong Kong; EcoEffect, Sweden; EcoProfile, Norway, and; Escale, France (see, [10,16,20,22] for a review). These tools have broadly been categorized into: (a) Third-party assessment tools, which are spin-offs of building assessment tools and assess the sustainability beyond a single building (see, [22]), and; (b) Tools, which are embedded into neighborhood-scale plans and sustainability initiatives to assess their sustainability performance [16]. After critical reviews of these tools, researchers have raised several concerns about their methodology, applicability and transferability to another context. Sharifi and Murayama [16] found that most of them are weak in taking into account the different dimensions of sustainability (e.g., economic, social, environmental, and institutional). They have also noted that most of these tools possess ambiguities in terms of criteria weighting, scoring, and rating system with no mechanism for local adaptability and participation.

Furthermore, the transferability of NSA tools has been questioned particularly with respect to the selection of sustainability assessment criteria [24]. This is particularly true in case of new residential development. For example, Säynäjoki *et al.* [20] assessed the applicability of the LEED-ND, BREEAM for Communities, and CASBEE-UD tools in the context of new residential development in Finland and found that the consideration of some of the suggested mandatory criteria of the tools are not feasible and relevant in the local context. In addition, many internationally available NSA tools do not sufficiently explain how and why the criteria were chosen, and the methodology used to determine the requirements is also not clear [20]. For example, Smith *et al.* [22] have identified that the inclusion of landscape related criteria are often ignored in these tools. The issue of transferability exacerbates due to the complexity associated with defining a neighborhood in different contexts [10]. These findings imply that any realistic and reliable assessment should take account of the specificities of local context and varying needs of different stakeholders [19].

The quest for sustainability of residential neighborhoods is more than a century old [25], and mostly relates to integrating land use, transport systems and the environment [26–28]. The Garden City movement led by Sir Ebenezer Howard is considered as an early initiative and emerged as a response to unsustainable condition of the then residential neighborhoods; and consequently, the concept of the three magnets was developed to combine the nature and environment with economic and social life [29]. Since then various neighborhood development models have emerged and practiced in different contexts and branded as, for example, cohousing, the common interest development, the gated community, the smart community, traditional neighborhoods, neo-traditional neighborhoods, conventional suburban neighborhood, eco-community, ranchette development, subdivision development, piecemeal development, and master-planned development [21,25,30–32]. Although the main purpose of all these models is to provide housing, their urban forms differ significantly, particularly in terms of layout design (e.g., density, street network, pedestrian access to transit and commercial stores, land use mix, gardens, parks and other attributes that characterizes spaces between homes) [30,33]. Relatively recent research has shown that these features significantly contribute to sustainable urban development [34,35]. Although a residential neighborhood is an outcome of the synergy and combination of these individual features, scant evidence was found in the literature investigating the overall impacts of these residential models on sustainability. Rather research studies to date have focused on analyzing the sustainability of two broad classes of urban forms—*i.e.*, compact and sprawling developments. As a result, a growing interest is evident in the literature on the increased importance of identifying various urban form typologies and their inter- and intra-urban scale interactions [36].

Against the backdrop of above urgencies, Frame and Vale [35] (p. 287) have stated that "there is a dearth of design and assessment tools for the residential built environment and of indicators to monitor progress towards sustainable development". House building industries have already been criticized for their "build and walk away" trading ethos where the emphasis is predominantly on manufacturing rather than design and planning, and thereby, very little response to the sustainability agenda [25]. The problem is even more severe in the context of developing countries where most of the residential development models are borrowed from the developed nation and are being implemented and marketed as sustainable model without being assessed their sustainability outcome in a local setting [37, 38]. A World Bank report shows that some 90% of global urban growth now takes place in developing countries—and between the years 2000 and 2030, developing countries are projected to triple their entire built-up urban areas [39]. This unprecedented urban growth possesses great concerns for policymakers on how to steer growth in a sustainable way in future, because urban growth is attractive as it leads economic growth of cities [14,40]. Despite some similarities in sustainability principles of neighborhoods between developed and developing countries, the differences are even larger and the resources to deal with them are considerably scarce in developing countries. Nevertheless, the urbanization can provide an opportunity for developing countries by practicing sustainability principles in their residential developments and thereby avoiding problems that experienced by the developed nations [41].

The research reported in this paper aims to contribute to the efforts in bridging the sustainability assessment knowledge gap by investigating the sustainability outcomes of three popular residential development models (*i.e.*, subdivision, piecemeal, and master-planned developments) from an exemplar developing country context—*i.e.*, Malaysia. This way the paper contributes to the sustainability assessment literature in the mostly neglected geographic lacuna of developing countries. Malaysia is a representative case study from the developing country context as it has been suffering from high population increase, rural to urban migration, and deforestation with major causes from large-scale land development, mining and dam construction and logging. Much like the rest of the developing countries, these have caused loss of biodiversity, erosion, wildlife being threatened, siltation of rivers and water pollution. As stated by Sumiani *et al.* [42], "Malaysia, being one of the Asian countries that is rapidly developing, increasingly facing the tension between the economic incentives and the claim for ethical consciousness with regard to accounting for the environment" (p. 897).

The study develops a NSA tool to assess and/or compare sustainability levels of abovementioned residential development models. The main rationale behind developing a new assessment tool is to factor in local characteristics most appropriately—by involving a mixture of local and international experts in the formation of the tool—in sustainability evaluation, and thus provide a more reliable output to inform decision makers for effective and efficient actions and solutions. The tool is not only helpful in assessing the sustainability of current practices, but also potentially can act as an integrated residential design and development guide and expedites a fundamental shift in where and how people live in developing countries.

2. Literature Review

2.1. Neighborhood Sustainability Assessment Frameworks and Tools

Few studies have indicated that a good NSA tool should possess the following characteristics: (a) Sustainability coverage—consideration of the major themes of sustainability of neighborhoods based on which their performance to be measured in a comprehensive and integrated way; (b) Inclusion of pre-requisites—benchmark strategies to assure the achievement of a certain level of performance; (c) Adaptation to locality—consideration of the context-specific needs and priorities in the assessments; (d) Scoring and weighting—rigorous methods to be used to score and weigh different criteria; (e) Participation—mechanisms to involve different stakeholders during the development and operational stages; (f) Presentation of results—reporting of assessment results in a way meaningful to decision makers, and; (g) Applicability—practicability of the NSA tools and strategies to increase their applicability (see [16,23]). Gibson *et al.* [43] provides a similar criteria and processes for sustainability assessment. Furthermore, Reith and Orova [44] provide an extensive comparison of the existing five assessment systems, CASBEE-UD, the 2009 and 2012 versions of the BREEAM Communities, LEED-ND, and DGNB-UD. They criticize these tools by stating, certain areas of sustainable urban development are not covered or do not get enough attention by the NSA systems, thus, further studies can discuss the possibilities and methods for including new indicators that broaden their coverage area.

2.1.1. Themes and Coverages

Themes are considered as the high-level issues or concerns of sustainability. Common themes of neighborhood-wide sustainability assessment includes building energy and water efficiency, energy production and supply, water and waste management systems, transportation solutions and footpaths that discourage personal car-use, promote walking and cycling, connectivity, urban density, site ecology, mixed use, health and well-being (e.g., quality of life of residents), and involvement of the public [20,22]. Again, each theme can have one or more criteria to evaluate. Each criterion including context-specific criteria has, in turn, one or more indicators, which are variables that provide specific measurements [16].

2.1.2. Indicators and Indices

Three levels of indicators are used in NSA tools, which correspond to the level of themes—*i.e.*, individual indicator; thematic indicators; and composite indicators [23]. Individual indicators form the first step in aggregating quantitative information. They include large lists of indicators covering a wide range of issues to improve the integration of environmental concerns into policies. Thematic indicators are individual indicators grouped around a specific theme. Composite indicators are formed when thematic indicators are compiled into a synthetic index, and presented as a single composite measure.

Five important characteristics of the different indicators used in the NSA tool include: (a) Policy relevance (monitor key outcomes, policy or legislation and measure progress towards goals); (b) Analytical validity (accessible and measurable, clearly defined and reproducible, representative of the system being assessed); (c) Systematic (capture systems information, including system variables, system levels and component systems); (d) Simplicity and operability (unambiguous, understandable, practical, clearly display the extent of the sustainability, appeal to the public and reflect the interests of different stakeholders, contain as few indicators as possible, but no fewer than necessary), and; (e) Cost effectiveness (require a limited number of parameters to be established, use existing data and information wherever possible) [23].

The process used to develop sustainability indicators has been debated in the literature—from the top, initiated primarily by governments and based on expert input (expert-led), or from the bottom (citizen-led) drawing on local networks and involving the public voice. These tensions between expert-led *versus* citizen-led processes of sustainability assessment seemed to be solved through the integration of the two approaches—so called joined-up approach. Finally, previous research has also shown that the assessor, his/her point of view and time of assessment often play a prime role in the assessment results, because they influence the criteria and benchmarks that are considered. Consequently, a transparent, objective and plural (or promoted in a multi-agent contest) assessment has recently been found necessary [10]. In addition to the indicator development process, citizens can also involve in the development of NSA tools in other three stages. Firstly, at the time of defining the sustainability targets and identifying the core criteria and indicators are going to be assessed. Secondly is during weighing different criteria. Having a consensus based weighting for different categories of indicators, can improve the assessment process. Finally, citizens can participate by providing feedbacks that help planners update the system [16].

2.1.3. Criteria Scoring, Weighting, Normalization, and Aggregation

Criteria scoring and weighting are often a controversial issue in the NSA process [22]. Criteria weighting implies the significance of a criterion amongst all the criteria used within a theme despite this has been identified to be an extremely difficult task and involves subjectivity. This subjectivity also frequently holds during the scoring process of a criterion. The subjectivity associated with scoring and weighting of different criteria has made this practice vulnerable to ambiguity. When subjectivity exists, research studies often used an expert-led approach such the Analytic Hierarchy Process [21,23]; and Delphi [21]. Recently, studies have highlighted that a consensus-based approach is helpful in such a situation in order enhance the transparency, which is pointed out to be an essential characteristic of scoring and weighting systems [16]. Standardization or normalization of criteria score is also a common practice in the NSA process, which helps to make the criteria comparable. Different normalization techniques have been used in the literature including standard deviation, min-max, categorical scale, and above and below mean [21,45]. In the NSA system, the weighted sum method is usually used for the derivation (aggregation) of composite or thematic scores based on normalized criteria scores and criteria weights [21,22]. Sometimes, the composite score is again classified (e.g., equal interval classification) to denote the level of sustainability of a neighborhood in a more understandable Likert-scale format (e.g., excellent, good, average, poor, bad) [21,23].

2.2. Characteristics of Residential Development Models

This section reviews literature on the sustainability issues of residential development types or models. However, the review is limited to only the three types of models that were adopted as case studies in this research—*i.e.*, piecemeal, subdivision and master-planned developments. Piecemeal development refers to houses that are developed in a piecemeal way and adds to the existing building clusters of a neighborhood. These are small-scale residential construction on vacant lot or a series of lots adjacent to existing residential development [46]. Such development takes the form of duplex, triplex or quadruplex on a single lot or single-family houses or townhouses on a number of lots. These provide potential buyers with a variety of options, vitality, viability and access to existing facilities such as schools, parks and emergency services. A major difference between piecemeal developments and infill developments is that the former bears no formal objectives of infill development [47]. Such an objective is important to create a complete, well-functioning neighborhood, and with attention to the essential design elements that fits the existing context in order to gain neighborhood acceptance. Piecemeal development is often not considered as a desirable feature for a neighborhood, because it lacks the coherence of a neighborhood. However, many argue that such limitation can be overcome with proper planning; and thereby, piecemeal development provides opportunities for residents to live close to existing amenities and workplace and consequently support local commercial establishments.

Residential subdivision refers to the division of a land into two or more residential lots, permitting the construction of buildings as stipulated in the building codes. Residential subdivisions take a number of different forms, ranging from large lots (over 0.4 ha), standard lots (0.27 ha), and small lots (less than 450 m^2) [48]. Developers of a subdivided lot usually provide infrastructure to the lot including streets, sewers, and water mains [49]. Standard subdivisions involve sub-dividing a site with the primary goal of maximizing the number of lots conditional on local regulations. However, such arrangements disregard site-specific features and thereby, detrimental to natural landscape. An alternative is to subdivide a certain portion of land for residential development and keeping aside the remaining lands to protect natural areas and green spaces [50]. However, the appeal of subdivision development lies to its low-density arrangements that provide rural style living, flexible building-design with increased privacy.

Master-planned developments are defined as large scale integrated housing developments on large tracts of undeveloped, suburban green field land, with mixed housing types, landscape, recreational, commercial, and service facilities [51]. They are developed based on a mechanism of planning control over an entire project site, underpinned by a particular vision for the completed development [52]. Located on the growth frontier of a city's fringe, they sometimes occur on renewal or infill sites, whose essential features include a definable boundary and fairly uniform character [53]. A master-planned development, also referred as master-planned estate or community, requires a larger land for development—in Malaysia usually larger than 800 ha—and includes a balanced mix of land uses for residents to live, work, shop, play, and learn [53,54].

Although a master-planned development provides better amenities that support sustainability compared to piecemeal and subdivision developments, there are buyers who do not opt to buy houses under the master-planned concept for variety of reasons. For instance, although the increased density is compensated for by high quality physical infrastructures and amenities in a master-planned development [52], it has invited criticism relating to loss of privacy and private space. Even though living in an enclosed community can create strong bonding between residents and increase support for each other, it can also create social exclusion with people outside of their boundaries [55]. In terms of socioeconomic characteristics, Ross *et al.* [56] point out that residential segregation by income could promote distrust between groups and decline in overall social connection within communities. Such segregation, no matter how subtle, has the tendency to undermine social cohesion as well as increase social exclusion and is, therefore, detrimental to achieving a socially sustainable society [57]. These issues have been reported to be limited in subdivision developments where the distribution of dwellings is more dispersed and less compact compared to master-planned, which leads to increased privacy.

The appeal of subdivision developments belongs to its low-density arrangements that offer attractive, countryside or rural-style living with increased privacy. However, this has huge implications on the infrastructure and servicing costs, which are increased due to the extensive infrastructure network and municipal amenities serving residential areas with lower densities. The infrastructure and associated public facilities that need to coincide with the entire neighborhood pattern cause inefficiency in the provision. For example, subdivisions that are built further into the countryside not only diminish the rural character of the entire neighborhood, but also increase automobile related travel activities, and its associated monetary costs and environmental externalities. It seems that master-planned developments do not face the critical sustainability issues in a physical context as much as subdivision developments, but rather in respect of socioeconomic issues.

3. Empirical Investigation

3.1. *Overview of Residential Development in the Case Study Context*

This research operationalizes a NSA tool using three residential development models selected from Malaysia as a representative of developing countries. Like most of the other developing countries, urban population in Malaysia has increased tremendously in the last four decades, from slightly over five million (38.8% of total population) in 1980 to nearly 20 million (72.2% of total population) in 2010 [58]. During this period, population growth in urban areas had taken place at a much faster rate than that of rural population. This was largely due to the availability of vast employment opportunities, which fuelled migration of people from rural areas in searching for better quality of life [59]. Population migration has become one of the contributing factors to the speedy progress of urbanization, in the form of rapid development of residential neighborhoods to accommodate the increasing number of urban dwellers. In addition, the expansions of city-regions, increases in the standard of living, and changing lifestyles have collectively led to an increase in housing demand. New residential areas are encroaching onto city fringes towards suburban and green field areas. Both large and small-scale developers have been actively building dwellings in these areas ranging from a few blocks to large master-planned style projects. These residential developments, particularly in major urban areas, represent a large portion of urban land use in Malaysia, and, thus, have become a major contributor to overall urban (un)sustainability. Amongst the various types of residential development, three types have been found to be dominant in prior studies including subdivision, piecemeal, and master-planned developments [60,61]. Table 1 lists the salient characteristics of these developments.

In Malaysia, both piecemeal and subdivision residential developments occur in an ad-hoc manner in the absence of an overall blueprint plan for the residential zone with a minimum development size of 0.4 ha. Master-planned developments on the other hand are based on pre-drawn overall master plan or blueprint plans, typically with a minimum development size of 100 ha. The small-scale piecemeal and subdivision residential developments have created disadvantages to residents because developers can get away from providing basic amenities (such as open spaces and community center), should the number of dwellings fall under 30 units [62]. In contrast, master-planned developments (relatively large in scale) have to provide the necessary amenities as required by the planning standards. Sustainable urban development practice in our case developing country context of Malaysia is extensively reported in the literature [63–68]. Rather than repeating what have been already said, we focus on residential sustainability assessment in a case study location in Malaysia.

Table 1. Salient characteristics of residential development types in Malaysia.

	Subdivision Development	Piecemeal Development	Master-Planned Development
Location	Suburban area	City fringes	Greenfields
Development size	Minimum 0.4 ha	Minimum 0.4 ha	Between 100 and 500 ha
Layout plans prepared by	Local planning authorities and private developers	Small scale private developers	Large scale private developers
Sale type	Vacant lot for single dwelling	Lot and building as completed house units	Lot and building as completed house units
Type of houses	Detached dwelling	Detached, semi-detached, terrace dwellings	Detached, semi-detached, terrace dwellings
Provision of amenities	Not required if less than 30 dwellings	Not required if less than 30 dwellings	Provided by developers as per planning guidelines
House design and construction	Buyers	Developers	Developers
Planning control	General development guidelines	General development guidelines	General and master-planned estate specific guidelines

3.2. Selection of Case Studies

The research develops a NSA tool to evaluate the sustainability of three most common residential development models from Malaysia. To operationalize the NSA tool, this study requires three representative residential developments, one from each development model type—*i.e.*, subdivision, piecemeal, and master-planned. The following criteria were used for the selection of case studies: (a) Located in the same local government area—to make sure they are subjected to the same planning and development regulations, and also have access to the same municipal services and amenities; (b) An appropriate case of the residential development type—to make sure the representativeness of each cases; (c) Have a minimum of 80% completion and take up rate—to make sure the maturity of developments—and; (d) Have data and information availability, local council support and body corporation collaboration with the research team—to make sure access to adequate data for a sound analysis. After a thorough examination of the potential cases all across Malaysia, we selected the following three residential developments from Ipoh City, Perak, Malaysia (Figure 1)—*i.e.*, Kampung Tersusun Batu 5 (subdivision development), Taman Canning or Canning Garden (piecemeal development), and Bandar Seri Botani (master-planned development).

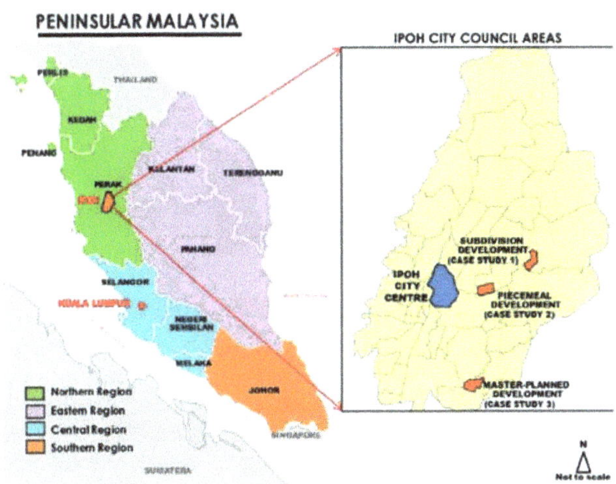

Figure 1. Location of the case study areas in Malaysia.

3.2.1. Subdivision Development

The first case study is a subdivision development, named "Kampung Tersusun Batu 5", located about 5 km to the Northeast of Ipoh (Figure 1). This is a 96.5 ha standard subdivision layout development that sits on a flat area of land bounded by a local highway and pockets of other residential development. The case study comprises 1181 parcels of single story detached houses and associated amenities including pockets of neighborhood parks, open spaces, shop lots and places of worship, and a primary school. The residential parcels were drawn up by the local planning authority in 1998 and were sold to individuals who then built their own houses, subject to local planning standards and guidelines. The typical parcel size is a 500 m^2 rectangular lot shape while corner parcels have an additional 10%–20% extra space. Owing to the type of dwelling, it has an average density of 14.6 dwellings per ha. In this development site, members of the Malay community own most of the houses. Figure 2 illustrates the layout and land use of the development.

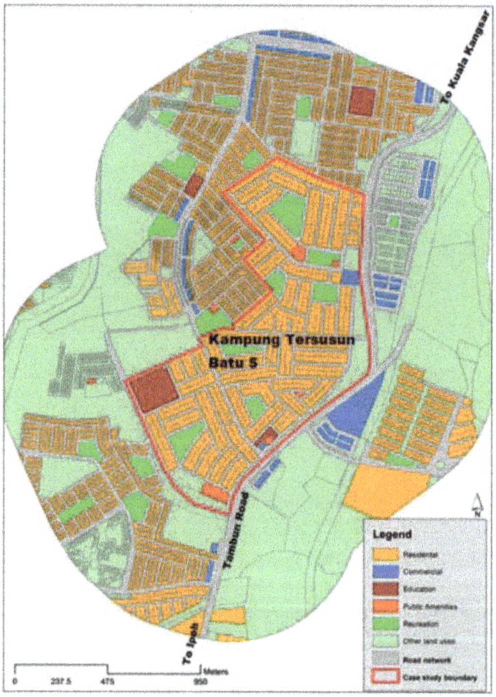

Figure 2. Land use classification of the subdivision development.

3.2.2. Piecemeal Development

This second case study is a piecemeal development called "Taman Canning or Canning Garden", located 3 km to the East of Ipoh (Figure 1). Developed during the mid-1980s, this mixed dwelling type residential area comprises 1555 residential parcels spread on 100.2 ha of relatively flat land. Single and double story terrace houses occupy a total of 44% of the residential parcels. Semi-detached houses occupy 16% of the residential stock, and single story detached houses inhabit 40%. Other land uses include two centralized neighborhood shop blocks, a farmers market, two primary schools, a large neighborhood playfield and pockets of neighborhood parks. The site is surrounded by piecemeal residential developments to the North, military land use to the East and a cemetery to the South. A federal highway separates the site from a large commercial land use to the East of the site.

Development of the site took place in a number of stages by three different developers and spanning over six years. Providing mixed housing options, the site is occupied by the mixed ethnic and cultural groups (*i.e.*, Malay, Chinese, and Indian) and socioeconomic backgrounds. The typical parcel size is 500 m^2 for a detached house, 240 m^2 for a semi-detached house, and 185 m^2 for a terrace house. The high number of terrace houses contributes to its higher average density of 28.3 dwellings per ha. Figure 3 displays the layout and land use of the development.

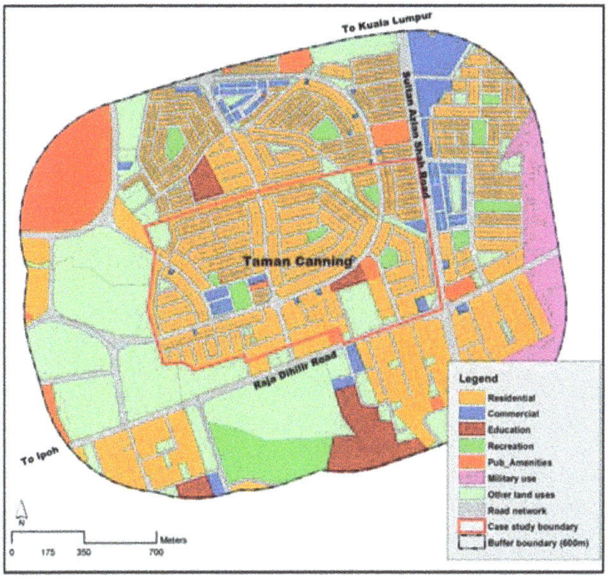

Figure 3. Land use classification of the piecemeal development.

3.2.3. Master-Planned Development

The final case study area sits on a 108 ha former oil palm plantation located 7 km to the South of Ipoh (Figure 1). This is a typical example of large-scale integrated green field development that exists all across Malaysia. This case occupies the first of a three-phase, large 312 ha, self-sustained residential, and light industrial master-planned development project. A total of 74.6 ha (69.2%) of the case study site is allocated to residential and supporting uses including neighborhood parks, roads and public amenities. A commercial precinct, a large local park and an education precinct present the next significant land uses. With an estimated population of 9048 residing in 2262 residential dwellings (1928 terrace houses and 334 semi-detached houses), it is the largest of the three cases in terms of physical size, population and number of residential dwellings. Parcel sizes for terraces house range between 100 and 145 m^2, while for semi-detached houses, the parcel size is 300 m^2. Being developed on a green field site, the master-planned development is still surrounded by agricultural land use and forest areas. Even though the original topography was undulating, the majority of the residential, commercial and education precincts have been flattened. This is typical of any housing developments in Malaysia. The purpose of flattening the land is to optimize time and construction cost, especially the terrace houses dominating the case study landscape. This case study recorded the highest dwelling density among all cases with an average density of 30.3 dwellings per ha. This is not surprising given that terrace houses dominate nearly 90% of the development. Figure 4 shows the layout and land use of the development site.

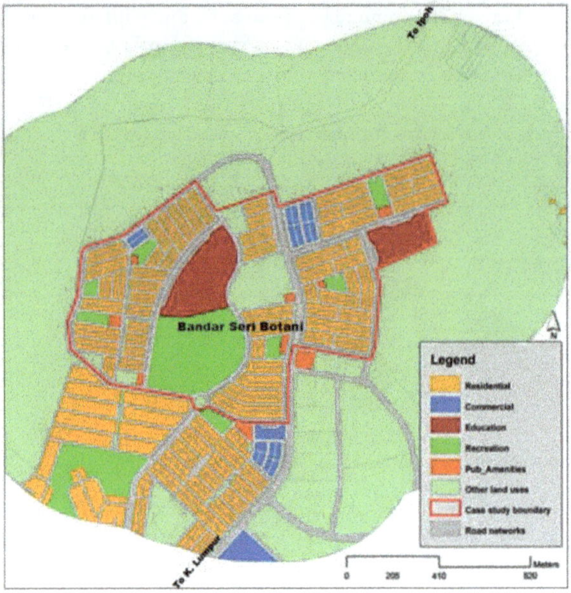

Figure 4. Land use classification of the master-planned development.

3.3. *Development of a Neighborhood Sustainability Assessment Tool*

The research develops a NSA tool to investigate sustainability levels of the selected three residential development models. A four-step process was followed for the development of the NSA tool in this research as outlined below.

3.3.1. Formation of a Set of Sustainability Indicators

A thorough review of the literature was conducted in order to identify a pool of relevant indicators as used in prior studies for the measurement of neighborhood level sustainability. A similar method was used in a number of previous research studies (e.g., [21,69–74]). The initial search identified a total of 128 sustainability indicators in three major sustainability categories of environmental, social and economic (see Table A1). The use of such a vast array of indicators is not uncommon in the literature. However, Frame and Vale [33] have suggested that the use of such a big number of indicators is difficult to interpret and integrate. Consequently, the list was reduced to 38 indicators (see Table A2). In this reduction process, we evaluated each of the 128 indicators based on the criteria of soundness, measurability, robustness, relevance, resilience, availability, and cost-effectiveness in consideration to our case study local context [23].

3.3.2. Delphi Study to Select the Most Relevant Indicators and Their Weights

Delphi method is a critical part of the development of indicator base of the NSA tool in order to make it a local context sensitive tool—in this application local context is Malaysia as a representative example of developing countries. A three round Delphi study was conducted to select the most relevant indicators from the originally selected 38 indicators. A total of 60 experts were involved in the Delphi study—*i.e.*, 29 from Malaysia and 31 from abroad. This balanced distribution of local and international experts—*i.e.*, almost 50% each—assures both local and universal characteristics to be factored in the analysis. The representation of such a large number of experts in the Delphi process was found to be representative of previous studies see [21]. This composition both local and international experts also meets the contextual criterion as discussed previously. Given that sustainability is a complex issue comprising of multiple dimensions, consideration was given to select the experts from diverse background so that the dimensional issues are properly represented in the measurement process. The experts were selected from both private and academic sectors with expertise in urban/environment/social/community planning/science, project management, architecture/design, housing/neighborhood/transport/infrastructure development, civil engineering, sustainability assessment, and policymaking. Upon consensus, the three round Delphi study enabled to reduce the number of indicators from 38 to 18. The indicator reduction process was undertaken as explained below.

In Round I, the indicator number was brought down from 38 to 24 based on a minimum of 75% expert agreement on the relevance and suitability of indicators. In Round II, the number was brought down from 24 to 18 based on a minimum of 75% expert agreement on relevance and suitability. In Round III, experts were given a final chance to reevaluate the shortlisted 24 indicators, and provide the level of importance of each indicator on a 7-point Likert scale (from "1 = very low" to "7 = very high") in terms of their contribution to sustainability in the Malaysian context (see Table A3). The importance scores are used as weighting of the indicators. The weight of indicators ranges between 4.19 and 6.22, when 24 indicators are considered, and 5.08 and 6.02, when 18 indicators are considered. This is to say, if a weighting assignment was requested from the experts for the entire indicator pool (128 indicators) or Round I indicators (38 indicators), the weighting scheme would surely show a distribution with wider in range. In other words, the current flat weighting scheme has no negative impact on the reliability of the results. Table 2 lists the categories, indicators, calculation methods, measurement units of indicators and their weights.

Table 2. Categories, indicators, measures, units and weights of neighborhood sustainability assessment (NSA) index.

Categories	Indicators	Calculations	Units	Weights
Environmental	Land use mix	Total land use mix (LUM) value/Total parcel area Where total LUM = Σk(pk ln pk)/ln N, k = Category of land use; p = proportion of land area devoted to specific land use; N = # of land categories	Index value	5.83
	Dwelling density	Dwelling units/Residential area Where: Residential area include internal street + half width adjoining access roads)	Dwelling unit Per ha.	5.27
	Impervious surfaces	[Total impervious area (TIA)/Total neighborhood area] × 100 Where, TIA = roads, buildings, driveways, sidewalks, drainage, car parks	Percentage	5.21
	Internal connectivity	Total Intersections/(Total Intersections + Cul-de-sac)	Index value	5.86
	External connectivity	Total perimeter length/# entry and exit points	Meter	5.43
	Open space provision	Total open space/total residents	Square meter per person	6.02
	Non-motorized transport	[Total walkway + cycle length]/total street length	Percentage	5.77
	Access to public transport	(ΣDna/ΣDa) × 100 Where Dna = # of dwellings located within a 600 m of a bus stop; Da = Total dwellings	Percentage	5.86
Social	Access to education	(ΣDna/ΣDa) × 100 Where Dna = # of dwellings located within a 600m of a educational facility; Da = Total dwellings	Percentage	5.77
	Access to local services	(ΣDna/ΣDa) × 100 Where: Dna = # of dwellings located within a 600 m of a local service center; Da = Total dwellings	Percentage	5.46
	Access to recreational space	(ΣDna/ΣDa) × 100 Where Dna = # of dwellings located within a 400 m of a park; Da = Total dwellings	Percentage	5.64
	Access to community centers	(ΣDna/ΣDa) × 100, Where Dna = # of dwellings located within a 600 m of a community center; Da = Total	Percentage	5.24
	Access to emergency services	Average response distance from 3 types of emergency services (*i.e.*, police, ambulance, fire department)	Kilometers	5.08
	Crime prevention and safety	Total length of blind frontage/total frontage length	Percentage	5.8
	Traffic calming	Streets segments with traffic safety measures/total street segments	Percentage	5.14
Economic	Commercial establishment types	Number of diverse types of business activities	Number of types	5.51
	Affordable housing	Total affordable houses/Total residential in study area	Percentage	5.69
	Housing option diversity	$1 - \Sigma(n/N)^2$, where n = total dwelling is a category, N = total dwellings in all categories	Index value	5.42

In contrast with the rating system, the budget allocation method was applied to generate weight for the three categories—*i.e.*, environmental, social and economic. The experts in Round II of the Delphi exercise were given 100 points to distribute across the three categories. The exercise constituted the following aggregate category scores: 39.27 for environmental category, 33.01 for social category, and 27.72 for economic category.

3.3.3. Indicator Scoring

"Land use mix" (LUM) score was derived using an entropy equation developed by Frank *et al.* [75] based on five land use classes—*i.e.*, residential, commercial, recreation, education, and public amenities. The criteria score ranges from 0 to 1 in which a higher score represents a better sustainability. "Dwelling density" score was calculated based on number of dwelling density located within a unit of residential land [76]. Like LUM, a higher density represents better sustainability of neighborhoods. A neighborhood with higher LUM and density reduces car-dependency (thereby less emissions) and enhances walking and cycling (thereby better health and wellbeing of residents) [77]. "Impervious surface" area was calculated based on proportion of neighborhood lands covered by impermeable materials (e.g., roads, buildings, car park, and driveways). A lower value of impervious surface represents a higher sustainability level. This is due to the fact that an increase of impervious surfaces result in flash flood due to increased storm-water runoff peaks [35]. "Internal connectivity" refers to the connectedness between two points within a neighborhood. A higher intersection density represents higher connectedness and supports walking and thereby more favorable for a sustainable development whereas a higher cul-de-sac density represents an advance in design efficiency for automobile movement but a retrograde step in design efficiency for pedestrian or transit movement [78]. "External connectivity" of neighborhood eases its connection with surrounding areas refers to the ease of street. In this research, external connectivity was calculated by measuring distance between two entry/exit points around a neighborhood. Therefore, a higher value represents less connectivity in this measure and consequently a lower level of sustainability. The other two environmental indicators used in this research are self-explanatory—*i.e.*, "open space provision" and "non-motorized transport".

In the social dimension, indicators associated with "access to different opportunities and services" (e.g., public transport) were measured by calculating the percentage of dwelling units of a neighborhood that are located within a certain distance from respective services as outlined in Table 2. The distance bands were determined based on the literature. If a higher proportion of dwelling units are located within the specified distance in a neighborhood, that neighborhood possesses a higher sustainability level. In contrast, a shorter average response distance from emergency services indicates a better sustainability level. In this research, the crime prevention through environmental design principle was adopted to assess sustainability level in the "crime prevention and safety" indicator. As suggested by Mackay [79], this research used free from blind frontage as the indicator. The amount of blind frontage was determined by calculating the ratio of blind frontage length to total street frontages. Therefore, a lower percentage of blind frontages indicate better sustainability of a neighborhood. The "traffic-calming" indicator was derived as a result of calculating the ratio of street segments that are equipped with at least a traffic-calming feature [77].

Three criteria were identified to be important by the experts in the economic dimension of sustainability including the "types of commercial establishments" exist, availability of "affordable housing", and the "diversity of housing stock" within a neighborhood. A higher diversity of commercial establishments and housing stocks represents a higher sustainability of neighborhoods. Housing affordability was determined based on the local context and affordable house price was considered between RM 50,000 and RM 60,000—about US$14,000–17,000 [80].

3.3.4. Normalization of the Indicator Scores

The indicator scores were normalized based on the categorical normalization technique [43]. Using the technique, each indicator score was transformed into a numerical scale ranging from 1 to 5. Indicator values of less than 30% received a normalized scale of 1, indicator values between 30% and 50% received a normalized scale of 2, indicator values between 50% and 70% received a normalized scale of 3, indicator values between 70% and 90% received a 4, and values of 90% and higher received a scale of 5.

3.3.5. Calculating Indicator, Category and a Composite Sustainability Score

The weighted sum aggregation method was used to calculate category sustainability level of each case study neighborhood Equation (1). The category scores were subsequently aggregated (weighted) to form a composite sustainability score.

$$Y_j = \sum W_i X_i \tag{1}$$

where Y_j is the aggregated score of category j, X_i is the normalized value of indicator i under Y_j, W_i is the weight of indicator i.

4. Results

The results of our empirical analysis backs up the literature findings of master-planned developments offering a better option for creating sustainable layouts in urban areas [51]. Table 3 displays the raw scores of the indicators, normalized and index scores along with the composite index scores for the three development types, where these findings are further discussed below.

Table 3. Neighborhood Sustainability Assessment Index (NSAI) raw values/scores of the criteria, their normalization, and weighted scores.

Categories	Category Weights	Indicators	Indicator Weights	Raw Indicator Scores of the Cases			Normalized Indicator Scores of the Cases			Weighted Indicator Score of the Cases		
				SDD	PMD	MPD	SDD	PMD	MPD	SDD	PMD	MPD
		Land use mix	5.83	0.47	0.3	0.59	3	1	5	17.49	5.83	29.15
Environmental	39.27	Dwelling density	5.27	14.03	28.3	30.3	1	4	5	5.27	21.08	26.35
		Impervious surfaces	5.21	43.8	54.5	49.4	5	1	3	26.05	5.21	15.63
		Internal connectivity	5.86	0.95	0.89	1	3	1	5	17.58	5.86	29.3
		External connectivity	5.43	349	382	398	5	3	1	27.15	16.29	5.43
		Open space provision	6.02	14.8	5	17.5	4	1	5	24.08	6.02	30.1
		Non-motorized transport	5.77	0	12.3	14.8	1	4	5	5.77	23.08	28.85
		Environmental category total scores of the cases (weighted-sum of the indicators)								123.39	83.37	164.81
Social	33.01	Access to public transport	5.86	59.6	47.7	57.2	5	1	4	29.3	5.86	23.44
		Access to education	5.77	68.6	54.2	96.4	2	1	5	11.54	5.77	28.85
		Access to local services	5.46	91.4	83.6	100	2	1	5	10.92	5.46	27.3
		Access to recreational space	5.64	94.8	67.5	94.3	5	1	5	28.2	5.64	28.2
		Access to community centers	5.24	96.9	66.5	90.2	5	1	4	26.2	5.24	20.96
		Access to emergency services	5.08	3.9	1.7	5.9	3	5	1	15.24	25.4	5.08
		Crime prevention and safety	5.8	3.6	19.8	25.3	5	2	1	29	11.6	5.8
		Traffic calming	5.14	8.9	2.7	19.9	2	1	5	10.28	5.14	25.7
		Social category total scores of the cases (weighted-sum of the indicators)								160.68	70.11	165.33
Economic	27.72	Commercial establishment types	5.51	5	14	14	1	5	5	5.51	27.55	27.55
		Affordable housing	5.69	0	19.6	25.9	1	4	5	5.69	22.76	28.45
		Housing option diversity	5.42	0	0.74	0.73	1	5	5	5.42	27.1	27.1
		Economic category total scores of the cases (weighted-sum of the indicators)								16.62	77.41	83.1
Total	100		100									
		Composite sustainability scores of the cases (weighted-sum of the categories)								10,610.28	7734.076	14,233.16

Note: SDD = subdivision development, PMD = piecemeal development, MPD = master-planned development.

4.1. Subdivision Development

The results indicate that subdivision development is ranked second with an index score of 10,610. Based on the overall normalized indicator scores generated from spatial data analyses, subdivision development records full score of 5 (very good) on six indicators, score of 4 (good) on one indicators, score of 3 (acceptable) on three indicator, score of 2 (low) on three indicators and score of 1 (very low) on five indicators—see the normalized scores in Table 3. The indicator sustainability levels indicate that subdivision development achieves high sustainability on its seven indicators comprising impervious surfaces, external connectivity, access to public transport facilities, access to recreational space, access to community centers, crime prevention and safety, and finally open space provision. On the other hand, the subdivision development achieves low sustainability level due to lacking in access to education facilities, access to local services, traffic calming measures, dwelling density, non-motorized transport, commercial establishment, affordable housing and housing option diversity. The results indicate that in the Malaysian scenario, subdivision development is still regarded as having a fairly acceptable level of sustainability, especially in terms of providing for common neighborhood facilities and access to open space. This is supported by its typically small parcel size configuration of 500 m^2, creating an average density of over 14 dwellings per ha. Such size is much lower than typical subdivision development lots in the North American or Australian examples [62,81].

4.2. Piecemeal Development

The results show that piecemeal development sits on the third place with an index score of 7734 with a much poorer performance compare to the other two development types. Piecemeal development records a full score of 5 (very good) on three indicators, score of 4 (good) on three indicators, score of 3 (acceptable) on one indicator, score of 2 (low) on one indicator and score of 1 (very low) on ten indicators—see Table 3. Looking at the indicator sustainability levels, the piecemeal development achieves high sustainability on access to emergency services, commercial establishment, housing option diversity, dwelling density, non-motorized transport and affordable housing. However, the piecemeal development achieves low sustainability levels on a majority of its indicators (11 indicators) namely, crime prevention and safety, land use mix, impervious surfaces, internal connectivity, open space provision, access to public transport facilities, access to education facilities, access to local services, access to recreational space, access to community centers, and traffic calming measures. Within the Malaysian context, the development of residential neighborhoods in a piecemeal approach is not seen as desirable, because it is considered as lacking in overall planning of the neighborhood that supports and influence sustainability. This explains why the outcomes of the sustainability assessment among the three case studies put piecemeal development in third place, after master-planned and subdivision developments. This is in contrast with the literature findings from the Western experience suggest that with a proper planning, piecemeal development can become a well-functioning residential development and provide opportunities for residents to live close to existing amenities and workplace as well as providing better support for local commercial establishments [82].

4.3. Master-Planned Development

This development type receives the highest index score of 14,233 as the best performing development site and type. Based on the overall normalized indicator scores generated from spatial data analyses, master-planned development records a full score of 5 (very good) on 12 indicators, score of 4 (good) on two indicators, score of 3 (acceptable) on one indicator and score of 1 (very low) on three indicators—see Table 3. Looking at the indicator sustainability levels, a good sustainability achieved by the master-planned development is due to its high scores on 14 indicators, which involves large scale integrated housing developments with mixed of land uses, dwelling density, internal connectivity, open space provision, non-motorized transport, access to education facilities, access to local services, access to recreational space, traffic calming measures, commercial establishment, affordable housing,

housing option diversity, access to public transport facilities and access to community centers. On the other hand, the master-planned development achieves low sustainability level at three indicators namely, external connectivity, access to emergency services and crime prevention and safety. Consistent with the literature [83], the master-planned development concept should be consistently promoted throughout the country not only because of its good sustainability but also because it serves as a mechanism of planning control over an entire project site, underpinned by a particular vision for the completed development. Moreover, sustainable residential design helps to shape strong characters, identity and perception of a place, and create a distinctive master-planned development community, which is equally important for market appeal. The results from this study indicate that master-planned development is the most sustainable neighborhood in Malaysia compared to subdivision and piecemeal developments. However, the result does not indicate in any way the degree to which master-planned development layouts is better than the others. This is because the research only seeks to identify which one of the three types of neighborhood layouts typically found in low-rise residential developments in Malaysia is the most sustainable. Having said that however, the finding provides justification to the policy makers and built environment agencies to encourage more future residential neighborhoods to be developed based on the master-planned concept. This finding also justifies the claims by planners that such comprehensive development of master-planned development by a single agent has the advantages of providing greater design flexibility, better neighborhood environments, exclusive open spaces, and community facilities for the residents [60]. Another reason explaining the higher score of master-planned development lays in the stringent development control mechanism for large-scale developments, including residential master-planned development must adhere to, in the form of an environmental impact assessment (EIA) and social impact assessment (SIA) requirements. EIA and SIA reports are required for residential development of more than 50 ha. Due to its sheer size, master-planned development in Malaysia generally fall within this category and are, therefore, subject to EIA and SIA approval from the relevant Ministries [60]. The reports need to justify that the proposed master-planned development fulfills the criteria required by the relevant Ministries, which helps to explain why master-planned development is generally well-developed compared to the smaller size piecemeal and subdivision developments.

5. Discussion and Concluding Remarks

The literature findings revealed that rapid urbanization has brought environmentally, socially, and economically great challenges to cities and societies. To build a sustainable neighborhood, these challenges need to be faced efficiently and successfully. In this regard the first step of action is to determine the sustainability levels of neighborhoods [84]. From this perspective the literature points to a number of NSA tools. However, as the critique of these tools suggests they have limitations in their indicator systems and adaptation in the developing country context is challenging.

This research contributes to the literature in two ways. A primary contribution of this research is the development of a NSA tool with an intention to be applied in the context of developing countries. Although there are quite a few NSA tools available in practice, these are built focusing on developed countries. As a result, their direct applications were found to be difficult in this research (*i.e.*, developing country context) where the meaning and definition of sustainability vary substantially. For example, an affordable house in a developed country might be extremely unaffordable in this research. Similarly, a 1% reduction in car-based travel might be a significant shift towards sustainability in a developed country whereas this makes no difference in a developing country context where car is not the main mode of transport. In addition, currently available NSA tools often comprise of numerous indicators that requires the availability of extensive database to process and operationalize, which are rarely available to the researchers and/or planning authorities in developing countries. Moreover, research has highlighted several methodological weaknesses of the existing NSA tools as discussed earlier in the paper. These issues necessitate the development of a NSA tool suitable to operationalize in the context of this research.

The NSA tool was developed focusing on the assessment of certain aspect of a neighborhood in this research—namely the urban form of differential residential models/types in developing countries. As a result, the assessment focused only on the design aspects of residential neighborhood types (e.g., layout, road network, buildings, and community facilities). Consequently, some important themes that might be important for other type of assessment were ignored in this research—such as building energy and water efficiency, water and waste management. The NSA tools developed for this research contains only 18 criteria/indicators. They were selected based on a 3 round Delphi study involving both local and international experts. Therefore, although limited in scope, these 18 indicators consist of the most relevant factors associated with sustainability assessment in the context of this research as accepted by both local and international communities. This joined-up process thereby reduces the tensions between expert-led *versus* citizen-led processes of sustainability assessment in this research. In addition, the Delphi method reduces the subjectivity of the criteria weighting in this research by involving both experts and local citizens [21]. The robustness of the applied method was evident in the sensitivity analysis with no changes in the final results when various combinations of weightings were tested (e.g., weighting from local expert only, weighting from international expert only, and a combination of both—not reported in the paper though). Although these findings justify an initial validity, further research should seek to apply the developed NSA tool in another developing country context, or perhaps using a different weighting system (e.g., AHP), to investigate its wider validity.

The second major contribution of this research is to assess the sustainability of three prominent residential development models (*i.e.*, master-planned, subdivision, and piecemeal developments) that are being adopted in an accelerated rate within the urban fabric of developing countries. Although residential sustainability is a century old concept and various residential models have been developed over the years aiming for sustainable outcome, any systematic method to assess an overall residential sustainability level is almost non-existent in the literature [35]. Unlike this research that incorporates an overarching framework of assessment, prior studies focuses only on a (or few) specific element of neighborhood feature (e.g., density) and its influence on certain outcome (e.g., car-ownership). The findings from this research robustly identified that master-planned communities provide option for more sustainable living in the context of this research over sub-division and piecemeal developments. Although these findings are in line with the scant evidence reported in the literature on this topic, which also justifies the validity of the developed tool, a more rigorous validation process by applying the tool against a gold standard (e.g., brown/green field development) is warranted. Note also that despite the results are presented in a quantifiable manner in this research, they represent sustainable utility/rating of a neighborhood, and therefore, cannot be mathematically traded-off (e.g., type A is two times better than type B). For example, although the experts rated the availability of open spaces highly (e.g., 6.02) compared to traffic calming measures (e.g., 5.14), this does not necessarily mean that one hectare of open spaces can be replaced by adding two traffic calming measure.

Despite master-planned communities out-performed in this research, local practitioners and policymakers must pay attention to make this neighborhood type more accessible to the wider communities (e.g., through provisioning of rapid transit system) in order to avoid social exclusion and car-dependency. Although the performance of piecemeal development was found to be poor, this research identified that ample opportunities exist to improve the sustainability performance of this neighborhood type if a focused policy is in place (e.g., in-fill development policy) through, which the development can be regulated or oriented towards important facilities.

This research develops a NSA tool and provides a comparison of sustainability performance of three residential neighborhood types. However, it neither provides an assessment of the neighborhoods' position on the way towards sustainability nor specifies the extent of the neighborhoods' success in approaching and achieving sustainability goals. Such assessment requires to set-up benchmark strategies to assure the achievement of a certain level of performance and the responsibility lies to the local planning authorities. However, the NSA tool developed in this research can be useful to serve as an integrated residential design and development guide and expedites a fundamental shift in where and how people live in developing countries—which was found to be a third policy related contribution of this research.

The findings, within Malaysia as a representative context for developing countries, demonstrated that master-planned development is the most sustainable residential development form followed by subdivision and piecemeal development models. This provides justification for policymakers and built environment (planning and development) agencies to encourage future residential neighborhoods to be developed based on the master-planned concept. The finding substantiates the claims by planners that such comprehensive development of master-planned estates or communities by a single agent has the advantages of providing greater design flexibility, better neighborhood environments, exclusive open spaces, various sustainable development practices, and community facilities for the residents [85]. Unlike many of the developed nations, the concept of master-planned development in Malaysia is still at its infancy, but the continuing national economic growth has encouraged its conception and wider practice. Although in our study master-planned development scores a high overall sustainability ranking in comparison to other two development types, there is surely room for improvement to increase the sustainability levels further. For example, master-planned development practices can learn from subdivision development experiences especially with regard to the provision of external connectivity, crime prevention and safety, and access to emergency services. With regards to the development of residential neighborhoods in a piecemeal approach, a new innovative strategy is needed to improve its sustainability level. The findings indicate that this development type is not seen as a desirable development form in Malaysia and attention needs to be given to the issue of lacking in overall planning of the neighborhood that supports sustainability.

In terms of research limitations, we highlight some of the critical issues as follows: (a) Sustainable urban development surely contains more features than of the physical neighborhood features and layouts that we mainly investigated in this research—especially energy consumption and pollution generated from each buildings; (b) Although the potential correlation between selected indicators may not have a significant impact on the results—due to the nature of investigation being a purely comparative one—it is still important to run appropriate statistical checks; (c) The weighting assignment is mainly based on Delphi expert suggestions, and alternative methods such as Factor Analysis can provide alternates; (d) Malaysia may not be a perfect representation for all of the developing countries—perhaps more suitable case for the developing countries from the Southeast Asia; (e) Based on three case study investigations, it is not possible to reach to a conclusion and claim that master-planned developments provide a more sustainable urban development form, and; (f) Direct replicability of the tool in a different context may be problematic—as the tool requires local experts contribution along international experts in the development of the indicator base. To address some of these research limitations and challenges, we are planning to expand our investigation including more case studies from different cities in Malaysia and other developing countries, incorporating various other aspects of sustainability in the analysis, such as building energy and water use, transport mode preferences of residents, recycling, air pollution and other socioeconomic dimensions of sustainability, and run a number of statistical tests to make sure of the reliability of the results.

Lastly, we underline that sustainability and development are contradicting terms or more correctly an oxymoron. However, this does not diminish the importance of efforts in minimizing the negative effects of urbanization in a rapidly developing world. Therefore, as a concluding remark of the paper we stress the following set of recommendations that are broad, but clearly describe the fundamental steps of an effective process in making a move towards a more sustainable urban neighborhood development also see [86]:

(a) Looking for the big picture;
(b) Understanding the sustainability phenomena clearly;
(c) Understanding the drivers of urban sustainability, and determining key factors and indicators;
(d) Collecting and accessing to the relevant data;
(e) Adopting tools and models and modeling the data;

(f) Defining quality targets for sustainable urban development;
(g) Facilitating the creation of relevant knowledge in the area of sustainable urban development;
(h) Formulating the urbanization policy from a sustainable development perspective;
(i) Changing behaviors and including stakeholder and community views;
(j) Forming collective efforts to develop sustainable urban neighborhoods;
(k) Planning dynamically for sustainable urban development;
(l) Translating the sustainability agenda into a number of strategic initiatives for implementation;
(m) Enhancing the control and monitoring mechanisms, and;
(n) Enabling an iterative policy and plan making process.

Acknowledgments: The authors wish to acknowledge the support of University of Technology MARA and Queensland University of Technology for jointly supporting the research upon which this paper is based. The authors are also grateful to the anonymous reviewers, who provided constructive comments on an earlier version of the paper.

Author Contributions: This paper represents a result of teamwork. Tan Yigitcanlar and Suharto Teriman designed the research; Suharto Teriman collected data and conducted empirical analyses; Tan Yigitcanlar and Md. Kamruzzaman reviewed the literature, and wrote and revised the paper. All three authors read and approved the final manuscript.

Appendix Appendix

Table A1. Indicator pool related to residential development compiled from the literature.

Indicator Categories	Indicators
Environmental indicators related to residential development	Preferred locations
	Population density
	Brownfields redevelopment
	Use mix
	Bicycle network and storage
	Average parcel size
	Steep slope protection
	Developed acres per capita
	Site design for habitat or wetland
	Conforming dwelling density
	Restoration of habitat or wetland
	Non-conforming dwelling density
	Conservation management for habitat or wetland
	Single-family housing share
	Walkable streets
	Mobile home housing share
	Compact development
	Multi-family 2–4 housing share
	Reduce parking footprint
	Multi-family 5+ units housing share
	Street network
	Group quarters housing share
	Tree-lined and shaded streets
	Residential water consumption
	Certified green building
	Residential energy consumption
	Building energy efficiency
	Population density
	Building water efficiency

Table A1. *Cont.*

Indicator Categories	Indicators
Environmental indicators related to residential development	Use mix Water efficient landscaping Average parcel size Resource preservation and adaptive reuse Developed acres per capita Stormwater management Conforming dwelling density Heat island reduction Non-conforming dwelling density Solar orientation Single-family housing share On-site renewable energy sources Mobile home housing share Infrastructure energy efficiency Multi-family 2–4 housing share Recycle content in infrastructure Multi-family 5+ units housing share Light pollution reduction Group quarters housing share Energy efficiency Residential energy consumption Renewable energy Imperviousness Minimum air quality performance Stormwater runoff Day lighting Total suspended solids Site selection Open space Public transport access Park space availability Open spaces, landscaping and heat island effect Residential wastewater production Stormwater management Street centerline distance Avoiding environmentally sensitive areas Sidewalk completeness Access to quality physical activity promoting environment Pedestrian route directness Connectivity through neighborhood design Street network density Sustainability of the physical environment Street connectivity Flexibility of public spaces Bicycle network Mixed use Residential water consumption Connectivity Non-residential wastewater production External connections Brownfields redevelopment Location

Table A1. *Cont.*

Indicator Categories	Indicators
Societal indicators related to residential development	Mixed-use neighborhood centers Connectivity through feeling of safety Mixed-income diverse communities Sustainability of transport Transit facilities Proximity (school/parks/transit) Access to civic and public space Housing proximity to transit Access to recreation facilities Housing proximity to recreation Neighborhood schools Housing proximity to education Existing building reuse Housing proximity to key amenities District heating and cooling Dwellings within 1/8 mi. of 3+ modes Wastewater management Transit stop coverage Solid waste management infrastructure Regional accessibility Sustainable maintenance Home-based vehicle trips Community services and connectivity Non home-based vehicle trips Access to education Home-based vehicle miles travelled Access to childcare/services Non home-based vehicle miles travelled Access to health services Parking demand Access to communication Parking supply Access to quality community facilities Transit service density Connectivity through public transport Rail transit boarding Connectivity through place/social cohesion
Economic indicators related to residential development	Housing jobs proximity Jobs/housed workers balance Local food production Conforming employment density Affordable housing Non-conforming employment density Housing choice Employment proximity to transit Housing proximity to employment center Locations with reduces automobile dependence Employment opportunity

Table A2. Delphi Round I indicators.

Indicators
1. Land use mix diversity
2. Residential dwelling density
3. Impervious surfaces
4. Street connectivity
5. Street route directness
6. Pedestrian accessibilities
7. Pedestrian network coverage
8. Vehicular entry and exit routes
9. Non-motorized transport facilities
10. Open space/active greens per dwelling
11. Open space/active greens per development area
12. Natural topography preservation
13. Sensitive areas/natural environment preservation
14. Vegetation retained to create the development
15. Storm water retention/detention system
16. Tree planting for shades/wind-break
17. Building exposure to natural ventilation
18. Proximity to public transit nodes/system
19. Resident's vehicle kilometer traveled
20. Motor vehicle ownerships
21. Proximity to recreation facilities
22. Proximity to education facilities
23. Proximity to local services
24. Availability of dedicated spaces for public amenities
25. Existence of well-defined boundary
26. Existence of neighborhood central place
27. Availability of existing amenities and services
28. Provision of community centers
29. Provision of religious centers
30. Provision of common recreation facilities for all ages
31. Provision of safety elements for crime prevention
32. Traffic calming measures
33. Separation between pedestrian and motorized traffic
34. Availability of commercial establishments
35. Diversity of housing option
36. Provision of affordable housing
37. Employment opportunities within immediate vicinity
38. Avoidance of high grade land

Table A3. Delphi Round II indicators, and Round III weights and consensus level.

Indicators	Weights	Consensus Levels (%)
1. Land use mix diversity	6.03	87.5
2. Dwelling density	5.47	81.3
3. Impervious surfaces	5.41	84.4
4. Internal connectivity	6.06	90.7
5. External connectivity	5.63	87.6
6. Non-motorized transport facilities	5.97	90.7
7. Environmentally sensitive areas	5.06	59.4
8. Open space provision	6.22	96.9
9. Solar orientation	4.88	62.5
10. Access to public transport facilities	6.06	93.8
11. Access to education facilities	5.97	93.9
12. Access to health facilities	4.78	53.2
13. Access to local services	5.66	93.7
14. Access to recreational space	5.84	97.0
15. Access to community center	5.44	87.6
16. Access to emergency services	5.16	71.9
17. Crime prevention and safety	6.00	96.9
18. Traffic calming	5.34	81.2
19. Commercial establishments	5.50	93.8
20. Skills development centers	4.19	37.5
21. Employment self-containment	4.66	53.2
22. Housing option diversity	5.41	87.6
23. Housing prices diversity	5.28	68.8
24. Affordable housing	5.69	81.3

Conflicts of Interest: The authors declare no conflict of interest.

References

1. World Commission on Environment and Development. *Our Common Future*; Oxford University Press: Oxford, UK, 1987.
2. Masnavi, M. Measuring Urban Sustainability: Developing a Conceptual Framework for Bridging the Gap between Theoretical Levels and the Operational Levels. *Int. J. Environ. Res.* **2007**, *1*, 188–197.
3. Li, X.; Yeh, A. Modelling Sustainable Urban Development by the Integration of Constrained Cellular Automata and GIS. *Int. J. Geogr. Inf. Sci.* **2000**, *14*, 131–152. [CrossRef]
4. Shatu, F.; Kamruzzaman, M.; Deilami, K. Did Brisbane Grow Smartly? Drivers of City Growth 1991–2001 and Lessons for Current Policies. *Sage Open* **2014**, *4*, 1–19. [CrossRef] [PubMed]
5. Hamilton, A.; Mitchell, G.; Yli-Karjanmaa, S. The BEQUEST Toolkit: A Decision Support System for Urban Sustainability. *Build. Res. Inf.* **2002**, *30*, 109–115. [CrossRef]
6. Bertolini, L.; le Clercq, F.; Kapoen, L. Sustainable Accessibility: A Conceptual Framework to Integrate Transport and Land Use Plan-Making: Two Test-Applications in the Netherlands and a Reflection on the Way Forward. *Transport Policy* **2005**, *12*, 207–220. [CrossRef]
7. Hine, J.; Kamruzzaman, M.; Blair, N. Weekly activity-travel behaviour in rural Northern Ireland: Differences by context and socio-demographic. *Transportation* **2012**, *39*, 175–195. [CrossRef]
8. Jabareen, Y. Sustainable Urban Forms: Their Typologies, Models, and Concepts. *J. Plan. Educ. Res.* **2006**, *26*, 38–52. [CrossRef]
9. Newman, P.; Kenworthy, J. The Land Use—Transport Connection: An Overview. *Land Use Policy* **1996**, *13*, 1–22. [CrossRef]
10. Berardi, U. Sustainability assessment of urban communities through rating systems. *Environ. Dev. Sustain.* **2013**, *15*, 1573–1591. [CrossRef]
11. United Nations. Report of the United Nations Conference on Environment and Development, Rio de Janeiro, Brazil, 3–14 June 1992; United Nations: New York, NY, USA, 1992.
12. Choguill, C. Developing Sustainable Neighbourhoods. *Habitat Int.* **2008**, *32*, 41–48. [CrossRef]

13. Alshuwaikhat, H.; Aina, Y. GIS-based urban sustainability assessment: The case of Dammam city, Saudi Arabia. *Local Environ.* **2006**, *11*, 141–162. [CrossRef]
14. Shen, L.-Y.; Ochoa, J.J.; Shah, M.N.; Zhang, X. The application of urban sustainability indicators: A comparison between various practices. *Habitat Int.* **2011**, *35*, 17–29. [CrossRef]
15. Ding, G. Developing a multicriteria approach for the measurement of sustainable performance. *Build. Res. Inf.* **2005**, *33*, 3–16. [CrossRef]
16. Sharifi, A.; Murayama, A. A critical review of seven selected neighborhood sustainability assessment tools. *Environ. Impact Assess. Rev.* **2013**, *38*, 73–87. [CrossRef]
17. Mesthrige-Jayantha, W.; Sze-Man, W. Effect of green labelling on residential property price: A case study in Hong Kong. *J. Facil. Manag.* **2013**, *11*, 31–51. [CrossRef]
18. Cloutier, S.; Larson, L.; Jambeck, J. Are sustainable cities "happy" cities? Associations between sustainable development and human well-being in urban areas of the United States. *Environ. Dev. Sustain.* **2014**, *16*, 633–647. [CrossRef]
19. Sharifi, A.; Murayama, A. Viability of using global standards for neighbourhood sustainability assessment: Insights from a comparative case study. *J. Environ. Plan. Manag.* **2014**, *58*, 1–23. [CrossRef]
20. Säynäjoki, E.; Kyrö, R.; Heinonen, J.; Junnila, S. An Assessment of the Applicability of Three International Neighbourhood Sustainability Rating Systems to Diverse Local Conditions, with a Focus on Nordic Case Areas. *Int. J. Sustain. Build. Technol. Urban Dev.* **2012**, *3*, 96–104. [CrossRef]
21. Han, Y.; Dai, L.; Zhao, X.; Yu, D.; Wu, S. Construction and application of an assessment index system for evaluating the eco-community's sustainability. *J. For. Res.* **2008**, *19*, 154–158. [CrossRef]
22. Smith, C.; Dunnett, N.; Clayden, A. *Residential Landscape Sustainability: A Checklist Tool*; Wiley: Hoboken, NJ, USA, 2008.
23. Li, Z.; Zhao, Y.; Zhao, H. Assessment indicators and methods for developing the sustainability of mining communities. *Int. J. Sustain. Dev. World Ecol.* **2008**, *15*, 35–43. [CrossRef]
24. Haapio, A. Towards sustainable urban communities. *Environ. Impact Assess. Rev.* **2012**, *32*, 165–169. [CrossRef]
25. Clarke, R. *Can The House Building Industry Create a Sense of Community? A Critique of Four Residential Development Models*; University of London: London, UK, 2004.
26. Yigitcanlar, T.; Kamruzzaman, M. Investigating the interplay between transport, land use and the Environment: A Review of the literature. *Int. J. Environ. Sci. Technol.* **2014**, *11*, 2121–2132. [CrossRef]
27. Kamruzzaman, M.; Hine, J.; Yigitcanlar, T. Investigating the link between carbon dioxide emissions and transport related social exclusion in rural Northern Ireland. *Int. J. Environ. Sci. Technol.* **2015**. [CrossRef]
28. Kamruzzaman, M.; Yigitcanlar, T.; Washington, S.; Currie, G. Australian baby boomers switched to more environmentally friendly modes of transport during the global financial crisis. *Int. J. Environ. Sci. Technol.* **2014**, *11*, 2133–2144. [CrossRef]
29. Howard, E. *Garden Cities of To-Morrow*; Swan Sonnenschein & Co: London, UK, 1902.
30. Song, Y.; Quercia, R. How are neighbourhood design features valued across different neighbourhood types? *J. Hous. Built Environ.* **2008**, *23*, 297–316. [CrossRef]
31. Odell, E.; Theobald, D.; Knight, R. Incorporating ecology into land use planning: the songbirds' case for clustered development. *J. Am. Plan. Assoc.* **2003**, *69*, 72–82. [CrossRef]
32. An, L.; Brown, D.; Nassauer, J.I.; Low, B. Variations in development of exurban residential landscapes: Timing, location, and driving forces. *J. Land Use Sci.* **2010**, *6*, 13–32. [CrossRef]
33. Biddulph, M. *Introduction to Residential Layout*; Butterworth-Heinemann: Oxford, UK, 2007.
34. Friedman, A. *Sustainable Residential Development: Planning and Design for Green Neighborhoods*; McGraw-Hill: New York, NY, USA, 2007.
35. Frame, B.; Vale, R. Increasing uptake of low impact urban design and development: The role of sustainability assessment systems. *Local Environ.* **2006**, *11*, 287–306. [CrossRef]
36. Ghosh, S.; Vale, R. Typologies and basic descriptors of New Zealand residential urban forms. *J. Urban Des.* **2009**, *14*, 507–536. [CrossRef]
37. Yigitcanlar, T.; Teriman, S. Rethinking sustainable urban development: Towards an integrated planning and development process. *Int. J. Environ. Sci. Technol.* **2015**, *12*, 341–352. [CrossRef]
38. Yigitcanlar, T.; Lee, S. Korean ubiquitous-eco-city: A smart-sustainable urban form or a branding hoax? *Technol. Forecast. Soc. Chang.* **2014**, *89*, 100–114. [CrossRef]

39. Suzuki, H.; Dastur, A.; Moffatt, S.; Yabuki, N. *Eco² Cities: Ecological Cities as Economic Cities*; World Bank: Washington, DC, USA, 2010.
40. Kotharkar, R.; Bahadure, P.; Sarda, N. Measuring compact urban form: A case of Nagpur City, India. *Sustainability* **2014**, *6*, 4246–4272. [CrossRef]
41. Vehbi, B.; Hoşkara, Ş. A model for measuring the sustainability level of historic urban quarters. *Eur. Plan. Stud.* **2009**, *17*, 715–739. [CrossRef]
42. Sumiani, Y.; Haslinda, Y.; Lehman, G. Environmental reporting in a developing country: A case study on status and implementation in Malaysia. *J. Clean. Prod.* **2007**, *15*, 895–901. [CrossRef]
43. Gibson, B.; Hassan, S.; Tansey, J. *Sustainability Assessment: Criteria and Processes*; Routledge: London, UK, 2013.
44. Reith, A.; Orova, M. Do green neighbourhood ratings cover sustainability? *Ecol. Indic.* **2015**, *48*, 660–672. [CrossRef]
45. Nardo, M.; Saisana, M.; Saltelli, A.; Tarantola, S. *Tools for Composite Indicators Building*; European Commusion: Ispra, Italy, 2005.
46. Farris, T. The barriers to using urban infill development to achieve smart growth. *Hous. Policy Debate* **2001**, *12*, 1–30. [CrossRef]
47. Listokin, D.; Walker, C. *Infill Development Standards and Policy Guide*; Centre for Urban and Policy Research: New Brunswick, NJ, USA, 2007.
48. Austin, M. Resident perspective of the open space conservation subdivision in Hamburg Township, Michigan. *Landsc. Urban Plan.* **2004**, *69*, 245–253. [CrossRef]
49. Thorsnes, P. Internalizing neighbourhood externalities: The effect of subdivision size and zoning on residential lot prices. *J. Urban Econ.* **2000**, *48*, 397–418. [CrossRef]
50. Arendt, R. Linked landscapes: Creating greenway corridor through conservation subdivision design strategies in the northeastern and central United States. *Landsc. Urban Plan.* **2004**, *68*, 241–269. [CrossRef]
51. Urban Land Institute. *Trends and Innovations in Master-Planned Communities*; Urban Land Institute: Washington, DC, USA, 1998.
52. Gwyther, G. Paradise Planned: Community Formation and the Master-Planned Estate. *Urban Policy Res.* **2005**, *23*, 57–72. [CrossRef]
53. Minnery, J.; Bajracharya, B. Visions, planning process and outcomes: Master-planned communities in South East Queensland. *Aust. Plan.* **1999**, *35*, 33–41. [CrossRef]
54. Ewing, R.; Handy, S. Measuring the Unmeasurable: Urban Design Qualities Related to Walkability. *J. Urban Des.* **2009**, *14*, 65–84. [CrossRef]
55. Costley, D. Master-planned communities: Do they offer a solution to urban sprawl or a vehicle for seclusion of the more affluent consumers in Australia? *Hous. Theor. Soc.* **2006**, *23*, 157–175. [CrossRef]
56. Ross, N.; Norbrega, K.; Dunn, J. Income segregation, income inequality and mortality in North American metropolitan areas. *GeoJournal* **2002**, *53*, 117–124. [CrossRef]
57. Ross, N.A.; Houle, C.; Dunn, J.R.; Aye, M. Dimensions and dynamics of residential segregation by income in urban Canada. *Can. Geogr.* **2004**, *48*, 433–445. [CrossRef]
58. Department of Statistics. *Population and Housing Census of Malaysia*; Malaysian Government: Kuala Lumpur, Malaysia, 2010.
59. Jamaliah, J. Emerging trends of urbanisation in Malaysia. *Stat. Malays.* **2004**, *1*, 43–54.
60. Teriman, S. *Measuring Neighbourhood Sustainability: A Comparative Analysis of Residential Types in Malaysia*; Queensland University of Technology: Brisbane, Australia, 2012.
61. Tan, T. Sustainability and housing provision in Malaysia. *J. Strat. Innov. Sustain.* **2011**, *7*, 62–71.
62. Department of Town and Country Planning. *Planning Standards and Guidelines*; Malaysian Ministry of Housing and Local Government: Kuala Lumpur, Malaysia, 1995.
63. Choon, S.; Siwar, C.; Pereira, J.; Jemain, A.; Hashim, H.; Hadi, A. A sustainable city index for Malaysia. *Int. J. Sustain. Dev. World Ecol.* **2011**, *18*, 28–35. [CrossRef]
64. Hezri, A. Sustainability indicator system and policy processes in Malaysia: A framework for utilisation and learning. *J. Environ. Manag.* **2004**, *73*, 357–371. [CrossRef]
65. Hezri, A.; Hasan, M. Management framework for sustainable development indicators in the State of Selangor, Malaysia. *Ecol. Indic.* **2004**, *4*, 287–304. [CrossRef]

66. Joseph, C. Understanding Sustainable Development Concept in Malaysia. *Soc. Responsib. J.* **2013**, *9*, 441–453. [CrossRef]
67. Hezri, A.; Hasan, M. Towards sustainable development? The evolution of environmental policy in Malaysia. *Nat. Resour. Forum* **2006**, *30*, 37–50. [CrossRef]
68. Noor, T.; Vijayaram, M. Need to implement the environmental governance and sustainability mechanisms for sustainable development in Malaysia. *J. US-China Public Adm.* **2011**, *8*, 800–807.
69. Dur, F.; Yigitcanlar, T.; Bunker, J. A spatial-indexing model for measuring neighbourhood-level land-use and transport integration. *Environ. Plan. B* **2014**, *41*, 792–812. [CrossRef]
70. Yigitcanlar, T.; Dur, F. Developing a sustainability assessment model: The sustainable infrastructure land-use environment and transport model. *Sustainability* **2010**, *2*, 321–340. [CrossRef]
71. Dizdaroglu, D.; Yigitcanlar, T.; Dawes, L. A micro-level indexing model for assessing urban ecosystem sustainability. *Smart Sustain. Built Environ.* **2012**, *1*, 291–315. [CrossRef]
72. Dizdaroglu, D.; Yigitcanlar, T. A parcel-scale assessment tool to measure sustainability through urban ecosystem components: The MUSIX model. *Ecol. Indic.* **2014**, *41*, 115–130. [CrossRef]
73. Yigitcanlar, T.; Dur, D.; Dizdaroglu, D. Towards prosperous sustainable cities: A multiscalar urban sustainability assessment approach. *Habitat Int.* **2015**, *45*, 36–46. [CrossRef]
74. Dur, F.; Yigitcanlar, T. Assessing land-use and transport integration via a spatial composite indexing model. *Int. J. Environ. Sci. Technol.* **2015**, *12*, 803–816. [CrossRef]
75. Frank, L.; Andersen, M.; Schmid, T. Obesity relationships with community design, physical activity, and time spent in cars. *Am. J. Prev. Med.* **2004**, *27*, 87–96. [CrossRef] [PubMed]
76. Kamruzzaman, M.; Baker, D.C.; Washington, S.; Turrell, G. Advance transit oriented development typology: Case study in Brisbane, Australia. *J. Transp. Geogr.* **2014**, *34*, 54–70. [CrossRef]
77. Kamruzzaman, M.; Washington, S.; Baker, D.; Brown, W.; Giles-Corti, B.; Turrell, G. Built environment impacts on walking for transport in Brisbane, Australia. *Transportation* **2015**. [CrossRef]
78. Cervero, R.; Gorham, R. Commuting in transit *versus* automobile neighborhoods. *J. Am. Plan. Assoc.* **1995**, *61*, 210–225. [CrossRef]
79. Mackay, M. *Which Suburbs Work: A Comparison between Traditionally Planned Suburbs and Conventional Suburban Development*; Ministry of Planning: Perth, WA, USA, 2001.
80. National Property Information Centre. *Property Market Report*; Malaysian Government: Kuala Lumpur, Malaysia, 2010.
81. Goodman, R.; Douglas, K. Privatised Communities: The use of owners corporations in master-planned estates in Melbourne. *Aust. Geogr.* **2008**, *39*, 521–536. [CrossRef]
82. Baldwin, C.; Osborne, C.; Smith, P. *Infill Development for Older Australians in South. East. Queensland: An Analysis of the Preferences of Older Australians in An Urban Environment*; University of the Sunshine Coast: Brisbane, Australia, 2012.
83. Tilt, J.; Cerveny, L. Master-Planned in Exurbia: Examining the Drivers and Impacts of Master-Planned Communities at the Urban Fringe. *Landsc. Urban Plan.* **2013**, *114*, 102–112. [CrossRef]
84. Yigitcanlar, T.; Dizdaroglu, D. Ecological approaches in planning for sustainable cities: A review of the literature. *Glob. J. Environ. Sci. Manag.* **2015**, *1*, 71–94.
85. Suen, W.; Tang, B. Optimal site area for high-density housing development. *Habitat Int.* **2002**, *26*, 539–552. [CrossRef]
86. Goonetilleke, A.; Yigitcanlar, T.; Ayoko, G.; Egodawatta, P. *Sustainable Urban Water Environment: Climate, Pollution and Adaptation*; Edward Elgar: Cheltenham, UK, 2014.

MDPI
St. Alban-Anlage 66
4052 Basel
Switzerland
Tel. +41 61 683 77 34
Fax +41 61 302 89 18
www.mdpi.com

Sustainability Editorial Office
E-mail: sustainability@mdpi.com
www.mdpi.com/journal/sustainability

www.ingramcontent.com/pod-product-compliance
Lightning Source LLC
Chambersburg PA
CBHW041750220926
51847CB00014B/157
* 9 7 8 3 0 3 8 9 7 9 0 6 7 *